McGRAW-HILL'S HANDBOOK OF ELECTRICAL CONSTRUCTION CALCULATIONS

Other Electrical Construction Books of Interest

Croft & Summers • AMERICAN ELECTRICIAN'S HANDBOOK, 13/e
Johnson & Whitson • ELECTRICAL CONTRACTING FORMS AND PROCEDURES MANUAL
Johnson & Whitson • SUCCESSFUL BUSINESS OPERATIONS FOR ELECTRICAL CONTRACTORS
Johnson • ELECTRICAL CONTRACTING BUSINESS HANDBOOK
Kolstad & Kohnert • RAPID ELECTRICAL ESTIMATING AND PRICING, 5/e
Maybin • LOW VOLTAGE WIRING HANDBOOK
McPartland & McPartland • MCGRAW-HILL'S NATIONAL ELECTRICAL CODE® HANDBOOK
McPartland & McPartland • HANDBOOK OF PRACTICAL ELECTRICAL DESIGN, 3/e
Ray • HOW TO START & OPERATE AN ELECTRICAL CONTRACTING BUSINESS, 2/e
Richter & Schwan • PRACTICAL ELECTRICAL WIRING, 17/e
Traister • HANDBOOK OF ELECTRICAL DESIGN DETAILS
Traister • SECURITY/FIRE ALARM SYSTEMS: DESIGN, INSTALLATION, AND MAINTENANCE, 2/e
Traister • MCGRAW-HILL'S ILLUSTRATED POCKET GUIDE TO THE NEC® TABLES
Traister • MCGRAW-HILL'S ILLUSTRATED INDEX TO THE NEC® CODE
Traister • INDUSTRIAL ELECTRICAL WIRING
Traister • THE ELECTRICIAN'S TROUBLESHOOTING POCKET GUIDE
Whitson • HANDBOOK OF ELECTRICAL CONSTRUCTION TOOLS AND MATERIALS

McGRAW-HILL'S HANDBOOK OF ELECTRICAL CONSTRUCTION CALCULATIONS

Joseph F. McPartland
Electrical Consultant
Tenafly, New Jersey

Brian J. McPartland
Editor of Electrical Contractor's
"Design and Installation Update"
Tappan, New York

Steven P. McPartland
Sales Engineer
GE/Northern Electric Supply Company
South Plainfield, New Jersey

Jack E. Pullizzi
Assistant Facilities Engineer
AT&T Bell Laboratories
Holmdel, New Jersey

Revised Edition

McGraw-Hill
New York San Francisco Washington, D.C. Auckland Bogotá
Caracas Lisbon London Madrid Mexico City Milan
Montreal New Delhi San Juan Singapore
Sydney Tokyo Toronto

Library of Congress Cataloging-in-Publication Data

McGraw-Hill's handbook of electrical construction calculations / Joseph F. McPartland . . . [et al.].—Rev. ed.
p. cm.
Includes index.
ISBN 0-07-046641-6
1. Electric engineering—Estimates—Handbooks, manuals, etc.
I. McPartland, Joseph F. II. McGraw-Hill Book Company.
TK435.M415 1998
621.3—dc21 97-29144
CIP

McGraw-Hill

A Division of The McGraw-Hill Companies

1 2 3 4 5 6 7 8 9 0 DOC/DOC 9 0 2 1 0 9 8 7

ISBN 0-07-046641-6

The sponsoring editor for this book was Zoe G. Foundotos, the editing supervisor was Scott Amerman, and the production supervisor was Pamela A. Pelton. It was set in Times Roman by Renee Lipton of McGraw-Hill's Professional Book Group composition unit.

Printed and bound by R. R. Donnelley & Sons, Inc.

CONTENTS

PREFACE

The revised edition of this calculations book continues to focus on the clear, strong, and inseparable relationship between the proper execution of electrical calculations and profitability, system adequacy, operating continuity, as well as National Electrical Code® (NEC®) compliance, which will limit legal liability. Quick and accurate determination of the minimum sizes and ratings for all aspects of design and installation—which directly impact overall cost—obviously rest in one's knowledge of and ability to apply practical calculations. All quantitative characteristics of an electrical system's components must be carefully calculated to assure the best, most cost-effective combination of materials and labor necessary to construct modern electrical systems.

A common thread from beginning to end in this book is the obligatory consideration of those NEC rules that require, prohibit, limit, or otherwise influence electrical calculations. The instances of controversy are many, and it is possible that more than one conclusion may be drawn. The purpose of the material covered in this book is to expose electrical designers, installers, inspectors, estimators, and other electrical professionals to all aspects of electrical calculations—especially those difficult and confusing decisions related to calculating minimum/maximum ratings and sizes of electrical equipment and conductors. Clearly, calculations are an everyday, yet vitally important, concern for all electrical professionals—engineers, designers, contractors, estimators, inspectors, plant electrical personnel, master/journeyman/apprentice electricians, and others. A certain amount of assistance can be obtained from colleagues, industry friends, and even manufacturers' sales engineers, but the ultimate decision regarding sizes, rating, dimensions, weights, and other technical specifications of selected equipment must be made by the individual(s) professionally—and legally—responsible for the design and/or installation.

The layout of this book lends itself to "classroom" use. That is, this material can be adapted by any group—inspectors, contractors, engineers, apprentices in training, and others—for use in a continuing education program. Any group could, for instance, discuss one, or more, calculation(s) block(s) at a series of meetings using the material given here as a framework for cohesive presentation. Obviously, the amount covered would be dictated by time available.

It should be noted that the calculations given in this books are typical ones that every electrical professional should be able to handle with ease. If you can't, devote some time to becoming more proficient. Review each example calculation carefully, making certain that each step is completely understood. Read the accompanying text, which will provide some insight to the reason behind the rule as well as serve to illuminate those cases that may have more than one interpretation.

No one doubts the value of technical proficiency. However, such expertise can only be gained by a conscious effort on the part of the individual. Only through continuous review and education can one truly keep her- or himself on the leading edge. Failure to do so may be hazardous to one's professional life!

Joseph F. McPartland

Brian J. McPartland

Steven P. McPartland

Jack E. Pullizzi

McGRAW-HILL'S HANDBOOK OF ELECTRICAL CONSTRUCTION CALCULATIONS

CHAPTER 1
BASIC CALCULATIONS

OHM'S LAW

The most basic calculations performed by electrical professionals are associated with the relationship between voltage, resistance, and current as expressed by Ohm's law. Basically stated, Ohm's law shows that the value of current flowing in an electric circuit will be dependent upon the applied voltage and the circuit's resistance. That is,

$$\text{Current} = \text{voltage/resistance} \qquad \text{or} \qquad I = E/R$$

where I is equal to the current in amperes (A), E is equal to the applied voltage in volts (V), and R is equal to the circuit's resistance in ohms (Ω).

Ohm's law can be manipulated to determine any one of those variables when two others are known. For example, if the values of current and circuit resistance are known or given, it is possible to determine the applied voltage as follows:

$$\text{If} \qquad I = E/R \qquad \text{then} \qquad I \times R = E$$

If $I = 20$ A and $R = 6\ \Omega$, $E = 120$ V (20 A × 6 Ω).

Similarly, if the voltage and current are known or given, the circuit's resistance can be determined from the following formula:

$$R = E/I$$

If $E = 120$ V and $I = 20$ A, then $R = 6\ \Omega$ (120 V/20 A).

Ohm's law formulas are summarized in Fig. 1-1. Those formulas can be directly applied to any purely resistive circuit, any purely resistive portion of a circuit, or any direct-current (dc) circuit or portion thereof.

WATT'S LAW

Watts (W) and *kilowatts* (kW; thousands of watts) are the two units used to express electric power. *Power* is the rate at which work is done. And a watt is the measure of the time rate at which electric energy is expended.

Electric lighting and heating devices are rated in watts. Circuits, systems, and electric energy sources are also rated in watts. But motors are correlated to the mechanical loads they must drive and are rated in horsepower. However, horsepower (hp) can be related to watts by the following ratio:

$$1\ \text{hp} = 746\ \text{W or } \tfrac{3}{4}\ \text{kW}$$

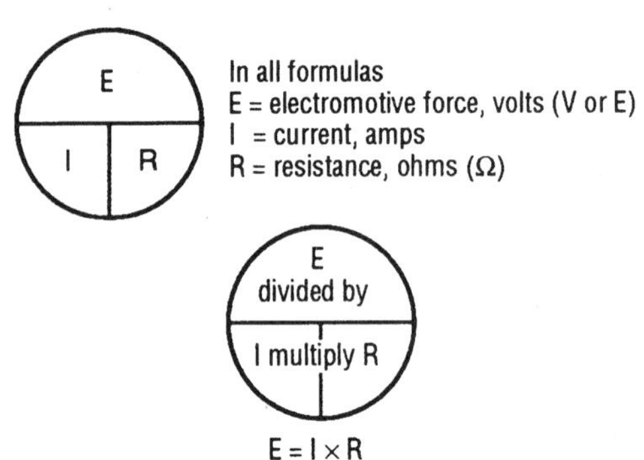

$E = I \times R$

Voltage (volts) = current (amps) × resistance (ohms)

$I = E \div R$

Current (amps) = voltage (volts) ÷ resistance (ohms)

$R = V \div I$

Resistance (ohms) = voltage (volts) ÷ current (amps)

Additional algebraic manipulation of the formulas below will produce others which may be useful in specific applications:

$P = I^2 \times R$	$I = P/E$	$E = P/I$	$R = P/I^2$
$P = E^2/R$	$I = \sqrt{P/R}$	$E = \sqrt{P \times R}$	$R = E^2/P$

These formulas are directly applicable to any resistive circuit, any resistive portion of a circuit, any dc circuit, and any ac circuit or portion of a circuit with unity power factor. Unity power factor is 100% power factor circuits without inductive or capacitive reactance.

Additional formulas can be determined by the use of the more complex Ohm s law circle in Fig. 1-2.

FIG. 1-1 The mathematical relationships between voltage, current, and resistance as described by Ohm's law. In the diagram, the vertical line is to be taken as a multiplication sign and the horizontal line indicates division.

The basic formula (also known as *Watt's law*) used to determine the rate at which electric energy is used—that is, the *power* of an electric circuit—is as follows:

$$P = E \times I$$

where P = power, W
E = applied voltage, V
I = current, A

If the voltage E is equal to 120 V and the current I is equal to 20 A, then the power P is equal to 2400 W (120 V × 20 A).

Inasmuch as the basic equation of Watt's law—$P = E \times I$—involves the voltage and current associated with the circuit in question, the equation can also be algebraically manipulated and written in different forms by substituting equivalent variables from Ohm's law.

For example, Ohm's law shows that $E = I \times R$. By substituting $I \times R$ for E in the equation $P = E \times I$, it can also be said that

$$P = (I \times R) \times I$$

or

$$P = I^2 \times R$$

Therefore, if we have a circuit of 10-Ω resistance with 4 A flowing through it, the power is

$$P = (4\text{ A})^2 \times 10\ \Omega$$
$$= 16\text{ A} \times 10\ \Omega$$
$$= 160\text{ W}$$

There are, of course, a number of other formulas that can be derived from Ohm's and Watt's laws. Those equations are summarized in the *power wheel* shown in Fig. 1-2. Using two known values with one of the equations shown at the outer part of each quadrant will give the value of the particular unknown variable shown at the inner part of that quadrant.

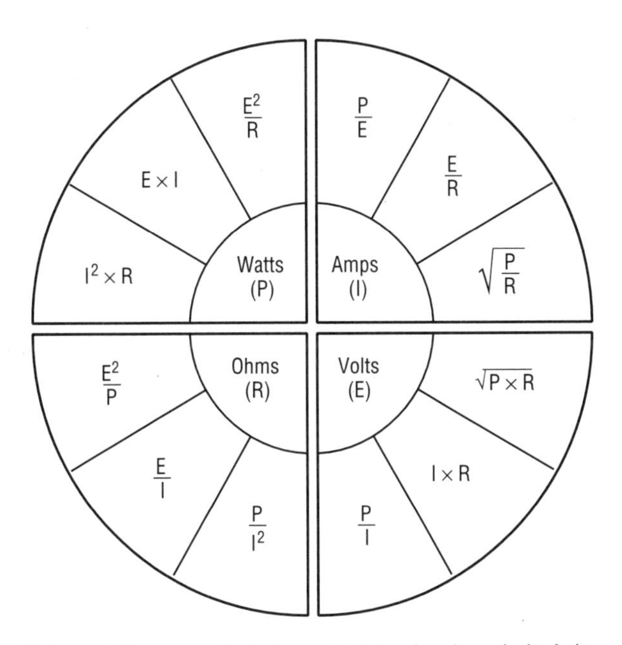

FIG. 1-2 The power wheel shows a variety of mathematical relationships between voltage, current, resistance, and power through algebraic manipulation of Ohm's and Watt's laws.

For example, if we have a load with 20-Ω resistance that is being operated at 120 V, the power consumption can be determined from the equation $P = E^2/R$. Substituting the known values, we can see that

$$P = (120\text{ V})^2/20\ \Omega$$

$$= 14{,}400\text{ V}/20\ \Omega$$

$$= 720\text{ W}$$

The equations shown in Fig. 1-2 can also be directly applied to any purely resistive circuit, any purely resistive portion of a circuit, or any dc circuit, as well as any alternating-current (ac) circuit, or portion thereof, with unity power factor.

APPLICATION IN SERIES CIRCUITS

As previously indicated, the formulas shown in Figs. 1-1 and 1-2 may be applied directly to any dc circuit, while their use on ac circuits is limited. Those limitations are related to additional electrical phenomena that occur when the source of supply is an ac voltage. The following discussion will focus on application of Ohm's and Watt's laws to dc circuits, purely resistive ac circuits, and ac circuits with unity power factor. Other conditions that alter the basic equations when applied to ac circuits will be discussed later.

The fundamental electric circuit consists of a utilization or load device, the conductors that will carry the current to the load device, and a voltage source. Such an arrangement is called a *series* circuit because all resistances are connected in series with one another and the voltage source. As far as the connection of load devices is concerned, series circuits are not generally used for connection of lighting and power loads in a modern electrical distribution system, unless a single load is supplied by an individual, dedicated branch circuit. However, understanding the theory of series circuit operation is essential to developing a full understanding of what takes place in more complicated branch circuits, feeders, and service supplies.

One characteristic of a series circuit is that individual resistances are additive. If two or more resistances are connected in series, the individual resistances are added together to obtain a single, total resistance R_T for the circuit as follows:

$$R_T = R_1 + R_2 + R_3 + R_4 + \cdots$$

Once the effective total resistance is established, the current that flows in any series circuit can be determined from Ohm's law as follows:

$$I = E/R_T$$

As can be seen, evaluation of the circuit shown in Fig. 1-3 considers the resistance of the circuit conductors as well as the resistance of the load device (the incandescent lamp). Therefore, we have three resistances that must be summed before the current I can be determined. As shown, the total resistance R_T is equal to 20.5 Ω, and the resultant current flow is a repeating decimal that we rounded to 0.585 A.

Note that for any series circuit, the current is constant. That is, the same value of current flows from the negative battery terminal through all resistances in the circuit to the positive terminal.

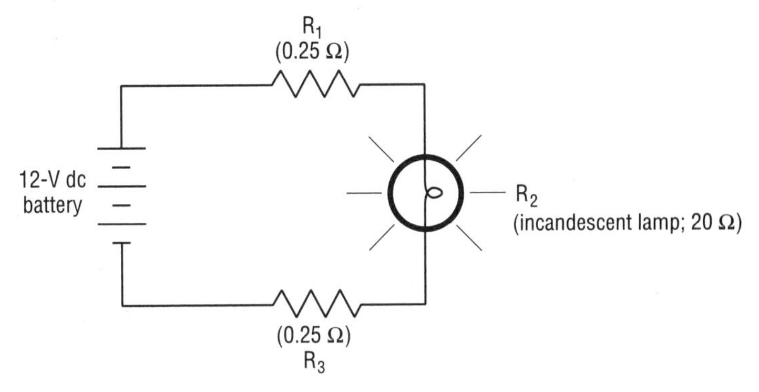

Ignoring the battery's internal resistance, we see there are three resistances in this series circuit. They are:

1. Conductor to the lamp	(R_1)	0.25 ohms
2. The lamp	(R_2)	20.00
3. Conductor from the lamp	(R_3)	0.25
Total resistance	(R_T)	20.50 ohms

Therefore, $I = E/R = \dfrac{12\text{-V dc}}{20.50\text{ ohms}} = 0.585\text{ A}$

FIG. 1-3 DC series circuit analysis using Ohm's law.

It is also important to note that individual *voltage drops* will be equal to the total applied voltage. That is, part of the 12-V direct current is used up forcing the current through each of those individual resistances. And the amount of voltage "used" by each resistance is commonly referred to as the voltage drop.

The voltage drop for each of the individual resistances in the circuit shown in Fig. 1-3 can be determined from Ohm's law as follows:

$$E = I \times R$$

Because we now know the current—and current is constant throughout the circuit—and because the individual resistances were given, we can calculate the voltage drop for the individual resistances of the circuit.

Conductor to lamp:

$$E_1 = I \times R_1 = 0.585\text{ A} \times 0.25\ \Omega = 0.14625\text{ V}$$

Lamp:

$$E_2 = I \times R_2 = 0.585\text{ A} \times 20\ \Omega = 11.7\text{ V}$$

Conductor from lamp:

$$E_3 = I \times R_3 = 0.585\text{ A} \times 0.25\ \Omega = 0.14625\text{ V}$$

Although the sum of E_1, E_2, and E_3 should be equal to the total voltage applied (12 V dc), because the value for current was rounded, there is a slight difference:

$$E_1 = 0.14625$$

$$E_2 = 11.7$$

$$E_3 = 0.14625$$

$$E_T = 11.9925 \text{ V}$$

If the current value had not been rounded off, the sum of those three voltage drops would be equal to the voltage applied. This is an excellent example of how one calculation can affect another. While we can account for this negligible difference, if a mistake had been made when determining the circuit's current—either while adding individual resistances or while doing the division—the value of E_T would not be 12 V dc. When a calculation does not give the expected answer or gives an unreasonable answer, always recheck any calculated values used, as well as the calculation just performed.

As indicated, the equations given in the power wheel may also be applied to any purely resistive circuit. A simple toaster is one example of such a circuit. Consider a toaster that is marked 120 V and 720 W by the manufacturer (see Fig. 1-4). Because the resistance of the heating element within the toaster is fixed by the manufacturer, the current flow and power used depend upon the applied voltage. This is expressed by the equation

$$R = E^2/P$$

$$= (120 \text{ V})^2/720 \text{ W}$$

$$= 14{,}400 \text{ V}/720 \text{ W}$$

$$= 20\ \Omega$$

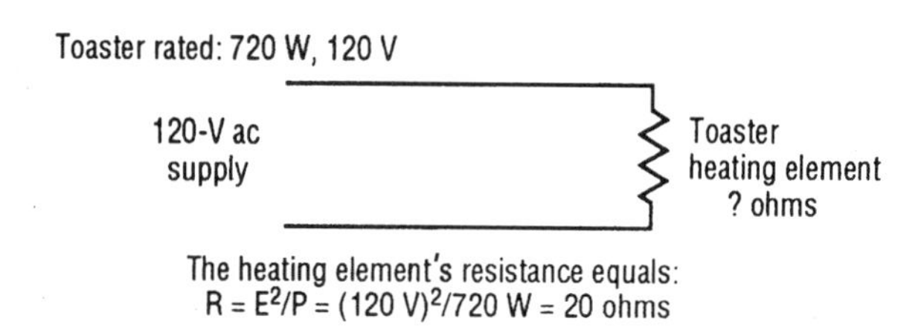

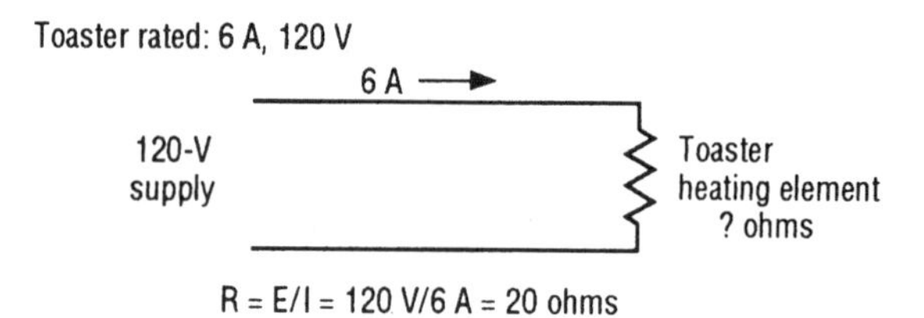

FIG. 1-4 Application of power wheel equations to a purely resistive ac circuit.

If the toaster had been marked with a voltage, say 120 V, and an amperage value, say 6 A, the heating element's resistance could be determined from

$$R = E/I$$
$$= 120 \text{ V}/6 \text{ A}$$
$$= 20\ \Omega$$

Up to this point we've been applying basic calculations to that portion of the circuit attached between the positive and negative terminals of a voltage source. That part is generally referred to as the *external* circuit. It includes the resistances of the load(s) and the supply conductors. However, sources—batteries and generators—also have an *internal* resistance. That internal resistance reacts when current flows to produce a voltage drop within the generator or battery that subtracts from the output voltage. Therefore, if the output voltage is measured without current being supplied to the external circuit, its value will exactly equal that of the voltage generated. If load current is being supplied, then the actual output voltage will be equal to the generated voltage value minus the internal voltage drop.

For example, if a generator has 120-V rated output and an internal resistance of 0.5 Ω and is connected by two conductors, each with a 1-Ω resistance, to supply a load with a 57.5-Ω resistance, the circuit current, output terminal voltage of the generator, voltage drop on the conductors, and the load voltage—that is, the voltage value across the load—can be determined. See Fig. 1-5.

First, establish the current I by adding all resistances and dividing the rated voltage by that sum as follows:

$$I = 120 \text{ V}/(0.5\ \Omega + 1\ \Omega + 1\ \Omega + 57.5\ \Omega)$$
$$= 120 \text{ V}/60\ \Omega$$
$$= 2 \text{ A}$$

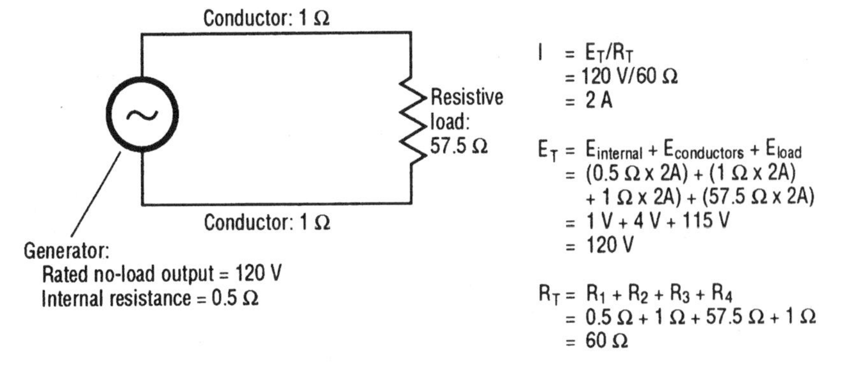

FIG. 1-5 Analysis of both internal and external portions of a series circuit.

Next, the generator output voltage is equal to the rated voltage minus the internal voltage drop:

$$E_{\text{internal}} = 2\text{ A} + 0.5\ \Omega$$
$$= 1\text{ V}$$

The output voltage under load is

$$E_{\text{rated}} - E_{\text{internal}} = E_{\text{output}}$$
$$120\text{ V} - 1\text{ V} = 119\text{ V}$$

Voltage drop in the circuit conductors is equal to their resistance times the current:

$$\text{E}_{\text{conductors}} = (R_1 + R_2) \times I = (1\ \Omega + 1\ \Omega) \times 2\text{ A}$$
$$= 2\ \Omega \times 2\text{ A} = 4\text{ V}$$
$$\text{Voltage across load} = 119\text{ V} - 4\text{ V} = 115\text{ V}$$

By way of review, solve the following basic series circuit problems by using one of the formulas given in the power wheel (Fig. 1-2).

Problem 1. Given $R_T = 20\ \Omega$ and $I = 6$ A, find the circuit voltage E.
Solution. From the power wheel (Fig. 1-2), we see that $E = I \times R$. Substitute the known values for resistance and current as follows:

$$E = I \times R$$
$$= 6\text{ A} \times 20\ \Omega$$
$$= 120\text{ V}$$

Problem 2. Given $E = 120$ V and $R_T = 20\ \Omega$, find the current I.
Solution. When the circuit voltage and resistance are known, the current can be calculated from the equation $I = E/R$. Substitute the known values for voltage and resistance as follows:

$$I = E/R$$
$$= 120\text{ V}/20\ \Omega$$
$$= 6\text{ A}$$

Problem 3. Given $I = 6$ A and $E = 120$ V, find total circuit resistance R.
Solution. The total resistance can be determined from $R = E/I$:

$$R = E/I$$
$$= 120\text{ V}/6\text{ A}$$
$$= 20\ \Omega$$

Problem 4. Using the values for voltage and current given in Prob. 3, find the power P consumed by that circuit.

Solution. The equation that expresses the mathematical relationship between power consumption and a circuit's voltage and current is

$$P = E \times I$$

In this case,

$$P = 120 \text{ V} \times 6 \text{ A}$$

$$= 720 \text{ W}$$

Those other formulas shown in Fig. 1-2 for power P, voltage E, current I, and resistance R are used when a question asks for a specific electrical characteristic and the known or given information can be applied to one of those formulas. For example, consider the information given in Prob. 1. If the question asked for power consumption P instead of the circuit voltage E, it could be calculated by $P = I^2 \times R$ because the value of the circuit's current and its resistance are given. Therefore

$$P = (6 \text{ A})^2 \times 20 \ \Omega$$

$$= 36 \text{ A} \times 20 \ \Omega$$

$$= 720 \text{ W}$$

In a similar manner, if Prob. 2 required solving for power P, the formula $P = E^2/R$ would provide the answer.

$$P = (120 \text{ V})^2/20 \ \Omega$$

$$= 14{,}400 \text{ V}/20 \ \Omega$$

$$= 720 \text{ W}$$

In any case, the equation selected will be dictated by the information that is given or known and the electrical characteristic that must be found. Then, it is only a matter of successfully performing the required mathematical operation(s).

The primary characteristics of a series circuit are shown in Fig. 1-6. Look them over carefully and make certain they are well understood. A sure and confident understanding of these parameters is absolutely essential before you can analyze more complicated circuit arrangements.

Assume that the circuit in Fig. 1-6 is supplied by a 120-V source and that $R_1 = 10 \ \Omega$, $R_2 = 20 \ \Omega$, $R_3 = 10 \ \Omega$, $R_4 = 5 \ \Omega$, and $R_5 = 5 \ \Omega$. Find

- Total resistance
- Circuit current
- Voltage drop at each resistance
- Total power consumption of the circuit

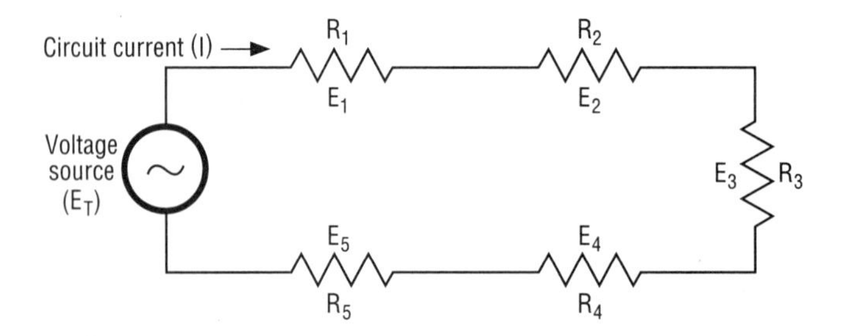

SERIES CIRCUIT CHARACTERISTICS

- One path for current flow from the voltage source through all resistances—conductors and loads—and back to the voltage source.
- The current (I) is constant throughout the entire circuit.
- The total resistance (R_T) is equal to the sum of all resistances ($R_T = R_1 + R_2 + R_3 + \cdots$).
- The sum of the voltage drop at each resistance is equal to the total applied voltage ($E_T = E_1 + E_2 + E_3 + \cdots$).
- The voltage drop at each individual resistance is equal to the circuit current (I) times the individual resistance ($E_1 = I \times R_1$; $E_2 = I \times R_2$; $E_3 = I \times R_3$; etc.).

FIG. 1-6 Characteristics of a series circuit.

Total resistance R_T:

$$\begin{aligned} R_T &= R_1 + R_2 + R_3 + R_4 + R_5 \\ &= 10\ \Omega + 20\ \Omega + 10\ \Omega + 5\ \Omega + 5\ \Omega \\ &= 50\ \Omega \end{aligned}$$

Circuit current I:

$$\begin{aligned} I &= E_T/R_T \\ &= 120\ \text{V}/50\ \Omega \\ &= 2.4\ \text{A} \end{aligned}$$

Individual voltage drops E_1 through E_5

$$\begin{aligned} E_1 &= I \times R_1 \\ &= 2.4\ \text{A} \times 10\ \Omega \\ &= 24\ \text{V} \end{aligned}$$

$$\begin{aligned} E_2 &= I \times R_2 \\ &= 2.4\ \text{A} \times 20\ \Omega \\ &= 48\ \text{V} \end{aligned}$$

Because the resistance of R_3 is the same as that of R_1, we know that the voltage drop at E_3 equals 24 V.

$$E_4 = I \times R_4$$

$$= 2.4 \text{ A} \times 5 \ \Omega$$

$$= 12 \text{ V}$$

Again, we know that E_5 is the same as E and equals 12 V.

The sum of those individual voltage drops would be equal to the applied voltage, or 120 V. This step serves as a check for all previous calculations. That is, if a mistake has been made while determining the total resistance, circuit current, or individual drops, the sum of those values will *not* be equal to the applied voltage. Add the calculated individual drops:

$$E_1 = 24 \text{ V}$$

$$E_2 = 48 \text{ V}$$

$$E_3 = 24 \text{ V}$$

$$E_4 = 12 \text{ V}$$

$$E_5 = 12 \text{ V}$$

$$E_T = 120 \text{ V}$$

Because the total is equal to the applied voltage, it can be said that the previous calculations are accurate.

Total power P:

$$P = E \times I$$

$$= 120 \text{ V} \times 2.4 \text{ A}$$

$$= 288 \text{ W}$$

APPLICATION IN PARALLEL CIRCUITS

Modern utilization devices—lighting fixtures, motors, appliances—are generally connected in parallel on branch circuits. A typical parallel circuit—a small appliance branch circuit supplying a coffee pot, toaster, and electric frying pan—is shown in Fig. 1-7. As can be seen, in a parallel circuit, there is more than one path for current to flow through. That is, current from the source—the 2-wire branch circuit—flows through each of the appliances independently of the other loads. And the total current drawn by each load will be carried by the branch circuit. Therefore, we see that in a parallel circuit the total current will be equal to the sum of the currents drawn by each load. This is one of the differences between series and parallel circuits. This concept is expressed mathematically as

$$I_T = I_1 + I_2 + I_3 + \cdots$$

PARALLEL CIRCUIT *consists of coffee maker, toaster, and frying pan plugged into kitchen appliance circuit. What currents will flow in the portions of the circuit indicated by the arrows in the drawing? Neglect the voltage drop in the conductors.*

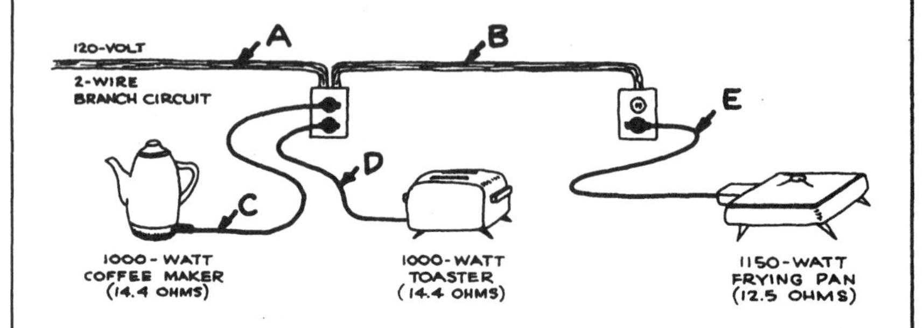

THE FIRST STEP *is to draw a simple schematic diagram to represent the circuit. Since we are neglecting the conductor voltage drop, there is a 120-volt potential across each appliance.*

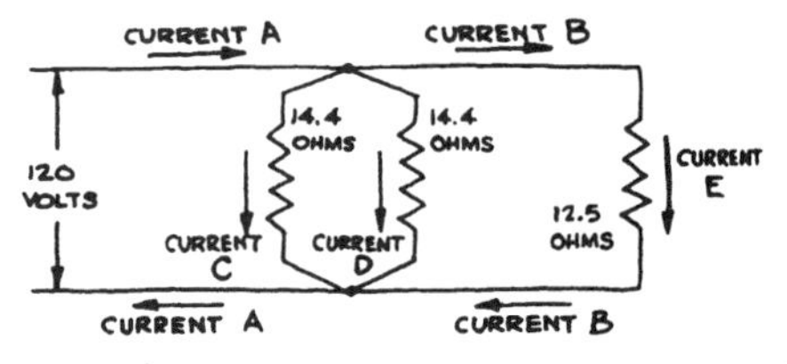

We may apply Ohm's Law in turn to each appliance:

$$\text{CURRENT} = \frac{\text{VOLTS}}{\text{OHMS}}$$

$$C = \frac{120 \text{ VOLTS}}{14.4 \text{ OHMS}} = \boxed{8.33 \text{ AMPERES}}$$

$$D = C = \boxed{8.33 \text{ AMPERES}}$$

$$E = \frac{120 \text{ VOLTS}}{12.5 \text{ OHMS}} = \boxed{9.6 \text{ AMPERES}}$$

$$B = E = \boxed{9.6 \text{ AMPERES}}$$

Current A, the total branch circuit current, is equal to the sum of the currents in each of the parallel branches:

$$A = C + D + E$$

$$A = 8.33 + 8.33 + 9.6$$

$$A = \boxed{26.26 \text{ AMPERES}}$$

We may also find current A by determining R_{eff}, the effective (or equivalent) resistance:

CURRENT A — 120 VOLTS — R_{eff}

$$\frac{1}{R_{eff}} = \frac{1}{R_1} + \frac{1}{R_2} + \frac{1}{R_3}$$

$$= \frac{1}{14.4} + \frac{1}{14.4} + \frac{1}{12.5}$$

$$= .0694 + .0694 + .08$$

$$= .2188$$

$$R_{eff} = \frac{1}{.2188} = 4.57 \text{ OHMS}$$

$$A = \frac{120 \text{ VOLTS}}{4.57 \text{ OHMS}} = \boxed{26.26 \text{ AMPERES}}$$

Obviously, breakers or fuses will open the normal 20-ampere circuit with such a load. This emphasizes the desirability of having two 20-amp kitchen appliance circuits.

FIG. 1-7 Simple parallel circuit application.

Another major difference with parallel circuits is that the voltage is constant if all the resistances are connected to the same point on the supply conductors. Unlike the series circuit, where a proportional value of voltage is dropped across each resistance, the basic theoretical point with parallel circuits is that the value of voltage across all loads is the same. In the example shown in Fig. 1-7, the resistance values of the circuit conductors are not being considered for the sake of simplicity. In fact, that portion of the branch circuit conductor carrying current *A* is in series with the part carrying current *B* and the load carrying current *E*. This circuit is actually a series-parallel circuit. But before it is possible to analyze a series-parallel circuit, it is necessary to understand the nature of a parallel circuit. Therefore, the series resistances will be ignored.

Assuming that the voltage is the same across all parallel-connected loads, the total current can be determined by adding the individual currents drawn. For example, consider the parallel electric heaters shown in Fig. 1-8. The total current can be determined as follows:

First, establish the individual currents drawn by each load. The current drawn by the 1200-W heater equals

$$\begin{aligned} I_1 &= P/E \\ &= 1200 \text{ W}/240 \text{ V} \\ &= 5 \text{ A} \end{aligned}$$

The current drawn by the 2600-W heater equals

$$\begin{aligned} I_2 &= P/E \\ &= 2600 \text{ W}/240 \text{ V} \\ &= 10.833 \text{ A} \end{aligned}$$

PARALLEL ELECTRIC HEATERS

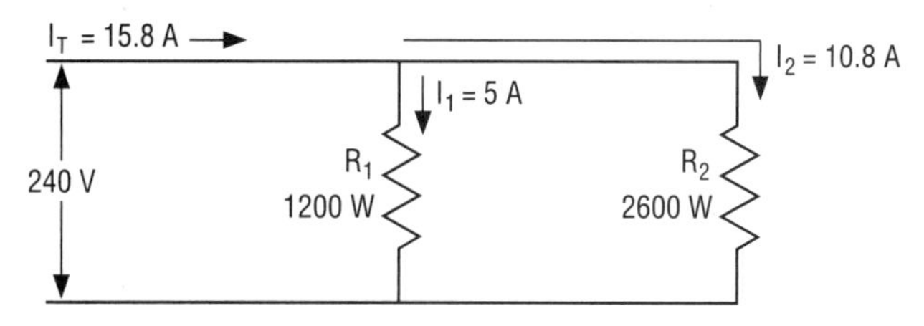

To find the total current drawn (I_T), determine the current drawn by each load using the formula I = P/E and add the individual currents together:

$$I_T = I_1 + I_2 + \cdots$$

FIG. 1-8 Determining total current in a parallel circuit.

which is rounded to 10.8 A. Total current I_T equals

$$I_T = I_1 + I_2$$

$$= 5\text{ A} + 10.8\text{ A}$$

$$= 15.8\text{ A}$$

Where the voltage and power ratings of the loads are known, total current could also be found from the equation $I = P/E$. That is, if it is not necessary to know the individual currents, the total current can be established by dividing the total power by the circuit voltage as follows:

$$I_T = (P_1 + P_2 + P_3 + \cdots)/E$$

$$= (1200\text{ W} + 2600\text{ W})/240\text{ V}$$

$$= 15.833\text{ A}$$

which is rounded to 15.8 A.

Another way of evaluating a parallel circuit involves establishing an equivalent or effective single resistance R_{EFF} that would draw current equal to the sum of the currents flowing through the individual resistances. That is, the parallel circuit is simplified, represented, and analyzed as a single series resistance.

It is important to note that the ohmic value of the effective resistance R_{EFF} will always be less than the smallest resistance of the parallel group of resistances. If the calculated effective resistance is greater than the smallest resistance in the group, a mistake has been made, and the calculation should be checked.

For any group of parallel resistances, the effective resistance R_{EFF} can be found using the following formula:

$$1/R_{EFF} = 1/R_1 + 1/R_2 + 1/R_3 + \cdots$$

To find the effective resistance of the parallel electric heaters shown in Fig. 1-9, first determine the individual resistances. The resistance of the 1200-W heater is

$$R_1 = E^2/P$$

$$= (240\text{ V})^2/1200\text{ W}$$

$$= 57{,}600\text{ V}/1200\text{ W}$$

$$= 48\ \Omega$$

The resistance of the 2600-W heater is

$$R_2 = E^2/P$$

$$= (240\text{ V})^2/2600\text{ W}$$

$$= 57{,}600\text{ V}/2600\text{ W}$$

$$= 22.15\ \Omega$$

which is rounded to 22 Ω.

1200-W AND 2600-W ELECTRIC HEATERS

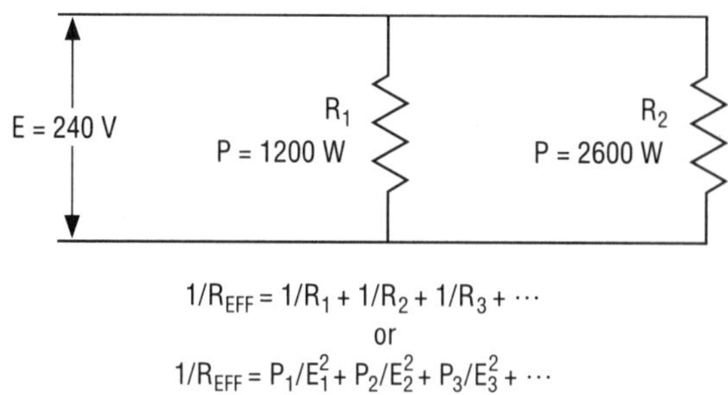

$$1/R_{EFF} = 1/R_1 + 1/R_2 + 1/R_3 + \cdots$$

or

$$1/R_{EFF} = P_1/E_1^2 + P_2/E_2^2 + P_3/E_3^2 + \cdots$$

and where the design voltage of all loads is the same:

$$1/R_{EFF} = \frac{P_1 + P_2 + P_3 + \cdots}{E^2}$$

Application of the above formulas will show that this parallel circuit is the equivalent of a series circuit with a resistance of about 15 ohms.

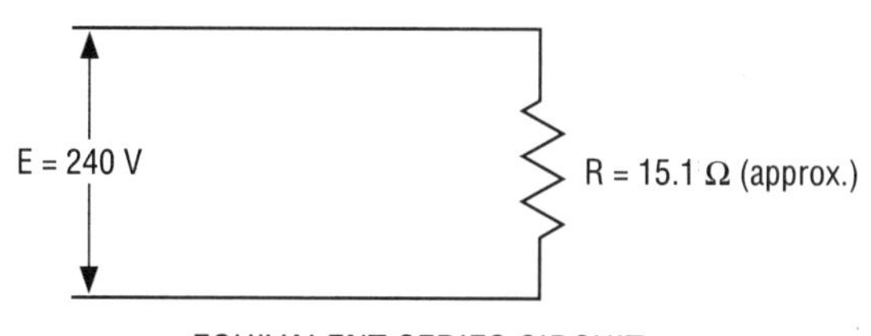

EQUIVALENT SERIES CIRCUIT

FIG. 1-9 Effective resistance R_{EFF} in a parallel circuit.

The effective resistance equals

$$\begin{aligned} 1/R_{EFF} &= 1/R_1 + 1/R_2 \\ &= 1/48 + 1/22 \\ &= 11/528 + 24/528 \\ &= 35/528 \\ R_{EFF} &= 528/35 \\ &= 15.0857\ \Omega \end{aligned}$$

which is rounded to 15.1 Ω.

Thus, 15.1 Ω represents the total effective series resistance of both loads. That is, if the parallel loads were replaced by a single resistance in series with the circuit conductors, the effect on the total circuit current would be the same. This concept will be applied later when series-parallel circuits are analyzed.

The effective resistance R_{EFF} can also be determined by substituting E^2/P for R as follows:

$$1/R_{EFF} = 1/(E_1^2/P_1) + 1/(E_2^2/P_2) + 1/(E_3^2/P_3) + \cdots$$

or

$$= P_1/E_1^2 + P_2/E_2^2 + P_3/E_3^2 + \cdots$$

And if the design voltages of all loads are the same, it can be said that

$$1/R_{EFF} = (P_1 + P_2 + P_3 + \cdots)/E^2 \qquad \text{or} \qquad R_{EFF} = E^2/(P_1 + P_2 + P_3 + \cdots)$$

Apply that formula to the example shown in Fig. 1-9 as follows:

$$1/R_{EFF} = (P_1 + P_2)/E^2$$

$$= (1200 \text{ W} + 2600 \text{ W})/(240 \text{ V})^2$$

$$= 3800 \text{ W}/57{,}600 \text{ V}$$

$$R_{EFF} = 57{,}600 \text{ V}/3800 \text{ W}$$

$$= 15.15\ \Omega$$

Although the effective resistance found by using this method differs slightly from the value established above using the actual resistance values, that difference can be attributed to rounding. If exact values had been used, both answers would have been identical.

Once the effective resistance R_{EFF} is known, the total current of the parallel circuit can be calculated using Ohm's law as follows:

$$I_T = E/R$$

$$= 240 \text{ V}/15.1\ \Omega$$

$$= 15.8940 \text{ A}$$

which is rounded to 15.9 A.

Again, there is a slight difference between the value of total current determined through addition of individual currents (15.8 A) and this answer, but that can be attributed to rounding.

Let's consider a circuit where four 40-Ω loads are connected in parallel. The effective resistance R_{EFF} would be

$$1/R_{EFF} = 1/R_1 + 1/R_2 + 1/R_3 + 1/R_4$$

$$= 1/40 + 1/40 + 1/40 + 1/40$$

$$= 4/40$$

$$R_{EFF} = 40/4$$

$$= 10\ \Omega$$

From this example, it can be seen that when a number of parallel resistances have the same resistance value, the effective resistance of the group is equal to the ohmic value of one resistance divided by the total number of parallel resistances. This aspect of identical resistances in parallel is shown in Fig. 1-10.

IDENTICAL PARALLEL RESISTANCES

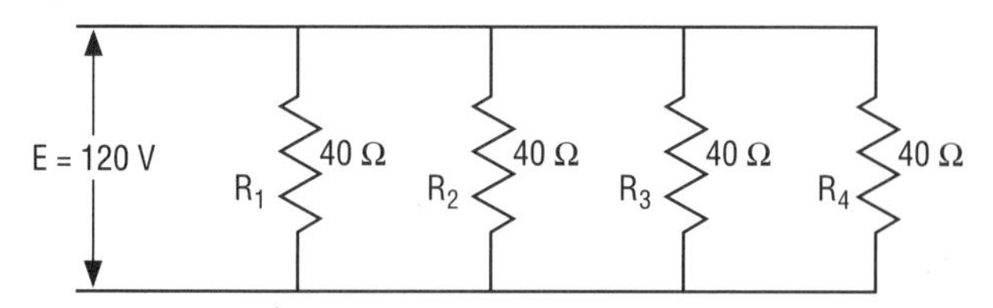

R_{EFF} is equal to the ohmic value of one resistance divided by the total number of resistances:

$$R_{EFF} = 40 \text{ ohms}/4$$
$$= 10 \text{ ohms}$$

FIG. 1-10 Effective resistance R_{EFF} where all parallel resistances are identical can be found by simple division.

Solve the following:

Problem 1. Four loads—two 1200-W and two 2400-W—are connected in parallel on a 240-V circuit. Find the total current.

$$I_T = (P_1 + P_2 + P_3 + P_4)/E$$
$$= (1200 \text{ W} + 1200 \text{ W} + 2400 \text{ W} + 2400 \text{ W})/240 \text{ V}$$
$$= 7200 \text{ W}/240 \text{ V}$$
$$= 30 \text{ A}$$

Problem 2. Three loads of 10, 15, and 30 Ω are connected in parallel on a 120-V circuit. Find the individual currents and the total current.

$$I_1 = E/R_1$$
$$= 120 \text{ V}/10 \ \Omega$$
$$= 12 \text{ A}$$
$$I_2 = E/R_2$$
$$= 120 \text{ V}/15 \ \Omega$$
$$= 8 \text{ A}$$
$$I_3 = E/R_3$$
$$= 120 \text{ V}/30 \ \Omega$$
$$= 4 \text{ A}$$

Total current equals

$$I_T = I_1 + I_2 + I_3$$
$$= 12\text{ A} + 8\text{ A} + 4\text{ A}$$
$$= 24\text{ A}$$

Problem 3. Three resistances of 10, 20, and 40 Ω are connected in parallel. Find the effective resistance of the group.

$$I/R_{EFF} = 1/R_1 + 1/R_2 + 1/R_3$$
$$= 1/10 + 1/20 + 1/40$$
$$= (4 + 2 + 1)/40$$
$$= 7/40$$
$$R_{EFF} = 40/7$$
$$= 5.71\ \Omega$$

Problem 4. Two 2400-W loads are connected in parallel on a 120-V circuit. Find the effective resistance of the group.

$$R_{EFF} = E^2/(P_1 + P_2)$$
$$= (120\text{ V})^2/(2400\text{ W} + 2400\text{ W})$$
$$= 14{,}400\text{ V}/4800\text{ W}$$
$$= 3\ \Omega$$

Problem 5. Five 20-Ω resistances are connected in parallel. Find the effective resistance.

$$\mathrm{R}_{EFF} = \text{value of resistance/no. of identical resistances}$$
$$= 20\ \Omega/5$$
$$= 4\ \Omega$$

As has been shown, there are distinct differences between series and parallel circuits. The primary characteristics of a parallel circuit are summarized in Fig. 1-11. Again, a sure and confident understanding of those properties is imperative.

APPLICATION IN SERIES-PARALLEL CIRCUITS

Although circuits are divided into two basic categories, series and parallel, circuits for power, light, heat, signals, and communications almost always involve both series and parallel connections. Take the simple case of a kitchen appliance circuit supplying, say, a coffee maker, a toaster, and an electric frying pan (see Fig. 1-12). The appliances are

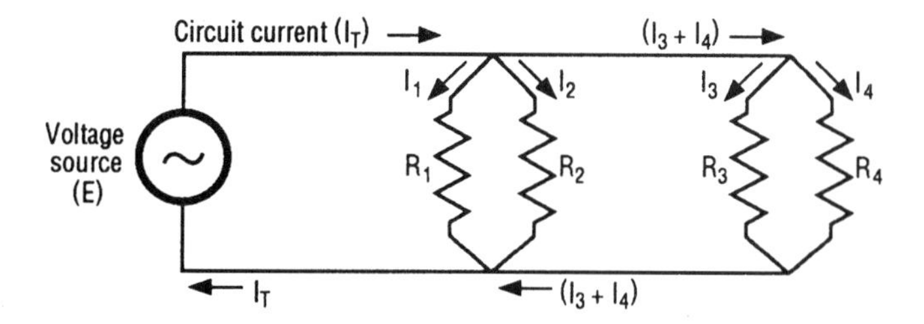

PARALLEL CIRCUIT CHARACTERISTICS

- More than one path for current flow from the voltage source through the loads and back to the source.
  The voltage (E) is constant across all loads.
- The total current (I_T) is equal to the sum of the current drawn by each load.
- The total power (P_T) is equal to the sum of the power consumed by each load ($P_1 + P_2 + P_3 + \cdots$).
- The current drawn by each load is equal to the applied voltage (E) divided by the individual load s resistance.
- The total current (I_T) is also equal to the total power ($P_1 + P_2 + P_3 + \cdots$) divided by the applied voltage (E).
- The reciprocal of the effective resistance ($1/R_{EFF}$) is equal to the sum of the reciprocal of each individual resistance ($1/R_1 + 1/R_2 + 1/R_3 + \cdots$).
- The effective resistance (R_{EFF}) is equal to the voltage squared (E^2) divided by the total power ($P_1 + P_2 + P_3 + \cdots$).

FIG. 1-11 Characteristics of a parallel circuit.

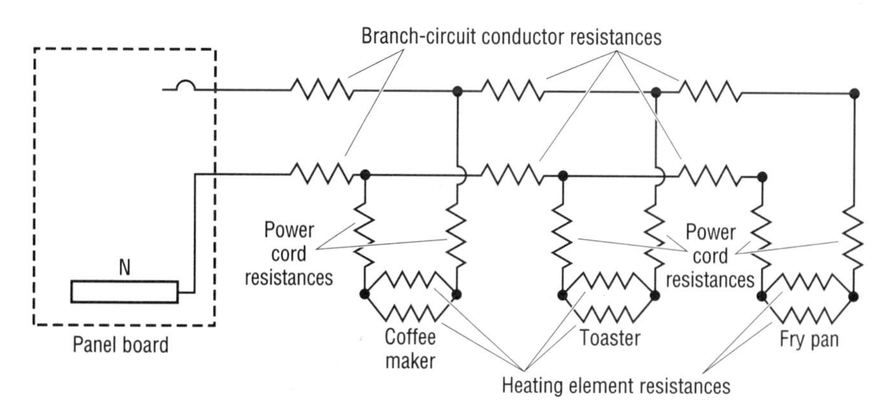

FIG. 1-12 Virtually all modern circuit applications contain both series and parallel resistances.

connected by cord and plug to a receptacle outlet. In each of those appliances, the individual heating elements may be connected in parallel with one another. But those parallel heating elements are actually in series with the power cord.

Then consider the branch circuit conductors from the panelboard to the receptacle outlet. The three appliances—each a series-parallel circuit—are connected in parallel across the branch circuit conductors. But, again, we have the resistance of the branch circuit conductors connected in series with the three parallel-connected appliances. If the electrical system were traced further back toward the generator output or other source of supply, a continuing succession of series and parallel resistances would be found.

A diagram of almost any electrical supply system shows a treelike progression of series and parallel resistances from the generator to the feeders, the branch circuits, and the loads (see Fig. 1-13). In present-day electrical systems, the feeder and branch circuit conductors are considered to be series resistances, as are tap conductors supplying individual load devices, such as lighting fixtures and motors. Load devices (motors, lights, etc.) are generally connected in parallel with one another. At the panelboards, the branch circuit or feeder conductors are connected in parallel with one another on the panelboard busbars. And, at switchboards, the outgoing circuits are also connected in parallel with one another. The generator or utility source provides the voltage which causes current flow through the entire system. This current will be the total amount required by all of the load devices—lamps, motors, etc.—connected at the extremities of the system.

As the total current flows into the system, it is divided among the various parallel circuits and loads. And as any current flows to any part of the circuit, there is a voltage

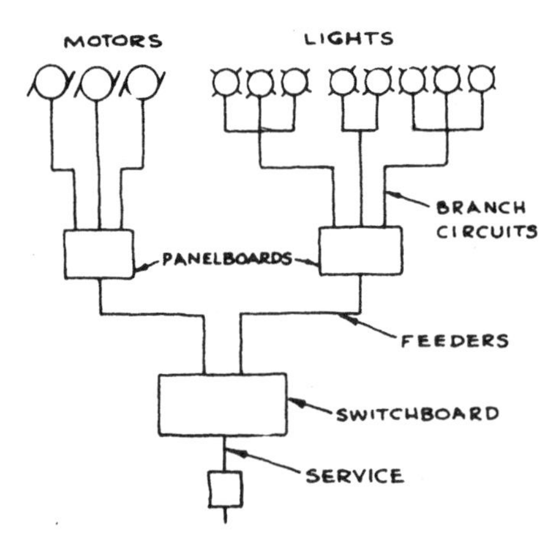

FIG. 1-13 Modern power distribution systems are made up of a number of series and parallel resistances.

drop in accordance with the series formula $E_{VD} = I \times R$. The voltage is, therefore, maximum at the generator or source of supply and is successively reduced from point to point (switchboard to panelboard, panelboard to panelboard, panelboard to load, or load to load) as a result of the various circuit resistances.

Calculation of voltage, current, resistance, and power values in series-parallel circuits is a simple matter of correlating separate series and parallel calculations in accordance with the basic mathematical relationships for each type of circuit. That is, series circuits are analyzed using series formulas, and parallel elements are evaluated using parallel formulas.

Where voltage and resistance values are given or known, it is possible to calculate an equivalent resistance for the entire circuit. Once the equivalent resistance has been established, it is possible to determine total current, individual voltage drops, and power consumption.

In Fig. 1-14*a,* we see a simple series-parallel circuit. Upon closer inspection it can be seen that the conductor resistances are connected in series and the load resistances are connected in parallel. Start with the resistances farthest from the source (R_3, R_4, and R_5). Those three resistances are in series with one another. And because resistances in series are additive, the equivalent or total resistance of that portion of the circuit is $R_3 + \mathrm{R}_4 + \mathrm{R}_5$. Therefore,

$$0.5\ \Omega + 19\ \Omega + 0.5\ \Omega = \text{equivalent resistance}$$

$$20\ \Omega = \text{equivalent resistance } R_{EQ}$$

It can be seen that the effective series resistance R_{EQ} of R_3, R_4, and R_5 is equal to 20 Ω. That single resistance value may now be substituted for R_3, R_4, and R_5 as shown in Fig. 1-14*b*. Farthest from the source, we now have two resistances in parallel with each other—R_2 and R_{EQ}. To determine the equivalent series resistance of two parallel loads, we use the parallel formula:

$$1/R_{EFF} = 1/R_1 + 1/R_2 + 1/R_3 + \cdots$$

In this case,

$$\begin{aligned} 1/R_{EFF} &= 1/R_2 + 1/R_{EQ} \\ &= \tfrac{1}{20}\ \Omega + \tfrac{1}{20}\ \Omega \\ &= \tfrac{2}{20}\ \Omega \\ R_{EFF} &= 20\ \Omega/2 \\ &= 10\ \Omega \end{aligned}$$

The formula used above can be applied to any combination of parallel resistances. But because we had identical resistances in parallel, it would have been possible to establish the effective resistance by dividing the resistance of one load by the number of resistances in parallel, as follows:

For identical resistances,

$$R_{EFF} = \frac{\text{resistance value}}{\text{no. of resistances}}$$

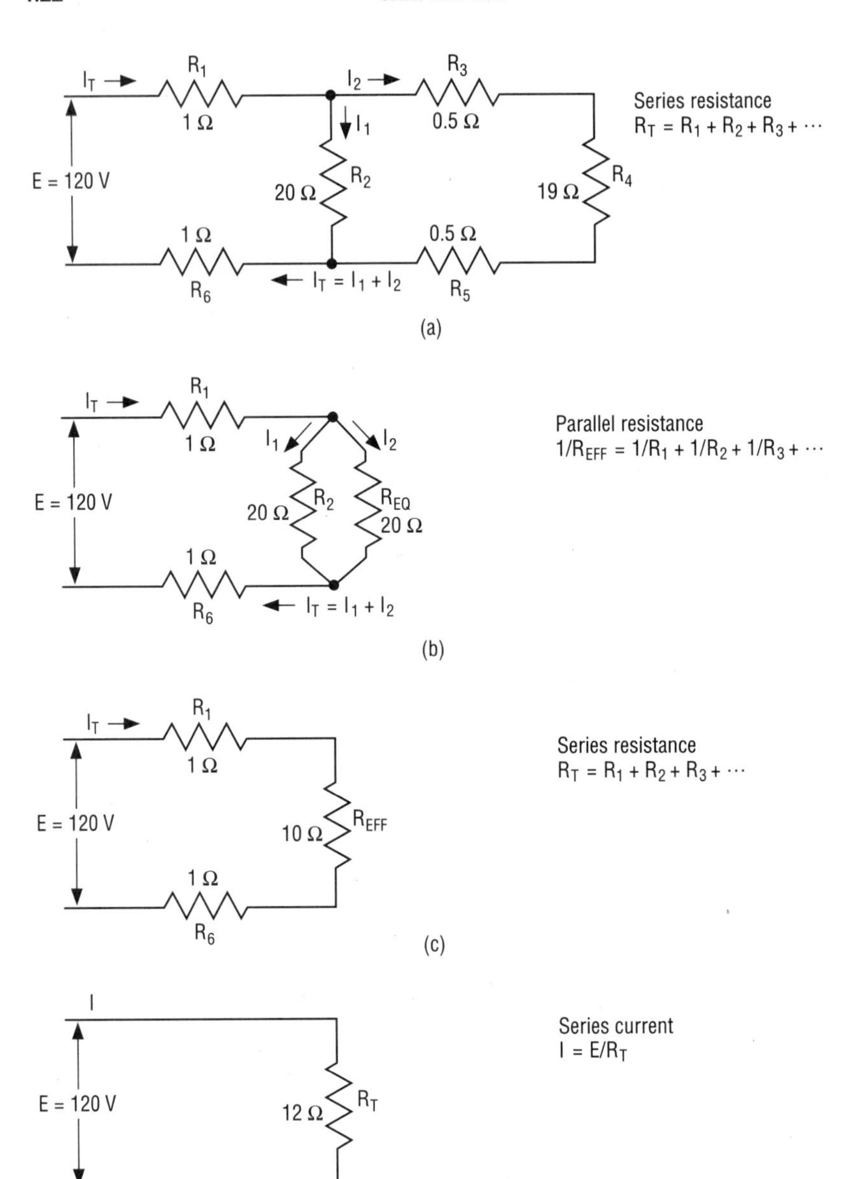

FIG. 1-14 Simplifying a series-parallel circuit.

or, in this case,

$$R_{\text{EFF}} = 20\ \Omega/2$$

$$= 10\ \Omega$$

Regardless of which formula was used, the effective series resistance of these two parallel resistances is 10 Ω. And that value may be substituted in the circuit as a series resistance (see Fig. 1-14*c*). For series circuits, the total resistance R_T is equal to the sum of all resistances:

$$R_T = 1\ \Omega + 10\ \Omega + 1\ \Omega$$

$$= 12\ \Omega$$

We now have a simple series circuit (see Fig. 1-14*d*). And the total current, which is constant in series circuits, can be determined from Ohm's law:

$$I = E/R_T$$

$$= 120\ \text{V}/12\ \Omega$$

$$= 10\ \text{A}$$

Now that the total current drawn by this combination of resistances is known, it is possible to calculate other circuit characteristics. For example, the voltage drops across R_1 and R_6 will reduce the amount of voltage that will be delivered across R_2. And the actual applied voltage will dictate how much current flows through that parallel resistance. As can be seen in Fig. 1-14*a,* the total circuit current flows through both R_1 and R_6. But some lesser value actually flows through the parallel load, R_2. To determine the value of current flowing through R_2, the applied voltage must be calculated. Because voltage drops are additive in series circuits, the voltage applied across R_2 can be found by subtracting the voltage drops E_{VD} at R_1 and R_6 from the supply voltage E. For R_1,

$$E_{\text{VDI}} = I \times R_1$$

$$= 10\text{A} \times 1\ \Omega$$

$$= 10\ \text{V}$$

For R_6,

$$E_{\text{VD6}} = I \times R_6$$

$$= 10\ \text{A} \times 1\ \Omega$$

$$= 10\ \text{V}$$

Because voltage drops are additive, the total dropped across those two series resistances is 20 V (10 V at R_1 plus 10 V at R_6). Therefore, the voltage applied across R_2 is 100 V (120 V − 20 V).

In parallel circuits, only a portion of the total current flows through each parallel load. The amount of current that flows through each parallel element will depend upon the applied voltage and the individual resistance. The lower the resistance, the greater the current flow through that parallel load.

From the simplified parallel circuit shown in Fig. 1-14*b*, it can be seen that both R_2 and the equivalent series resistance of the remainder of the circuit R_{EQ} have a value of 20 Ω. Therefore, one would expect an even split of the 10-A total current between the two parallel resistances. To determine the actual current flow through R_2, divide the applied voltage (100 V) by the load's resistance (20 Ω), as follows:

$$I = E/R$$
$$= 100 \text{ V}/20\ \Omega$$
$$= 5 \text{ A}$$

We now know that 5 A flows through R_2. And because current is additive in parallel circuits ($I_T = I_1 + I_2 + I_3 + \cdots$), we know that 5 A flows through the remainder of the circuit (10 A − 5 A = 5 A). And that is consistent with what was expected.

Although R_4 is in series with R_3 and R_5, it is connected in parallel across the supply circuit conductors. And the applied voltage will be constant, while the current will be additive. For example, if the point of connection were a duplex receptacle and two loads instead of one were plugged in, the voltage at that point in the circuit would be the same for both cord-connected loads.

To determine the value of voltage delivered across R_4, we would establish the voltage dropped across R_3 and R_5—using the remaining 5 A, which is constant throughout the three series loads R_3, R_4, and R_5—and subtract those voltage drops from the 100 V that was applied at R_2.

The voltage drops across R_3 and R_5 will be equal because both have the same resistance and the current is constant at 5 A.

For either R_3 or R_5, the voltage drop E_{VD} will be

$$E_{VD} = I \times R$$
$$= 5 \text{ A} \times 0.5\ \Omega$$
$$= 2.5 \text{ V}$$

Because that voltage will be dropped across each of the two resistances, it is multiplied by 2 and then subtracted from the voltage applied across R_2 (100 V). That is, the voltage applied across R_4 (E_{VD4}) is equal to the voltage at R_2 (100 V) minus that dropped at R_3 and R_5 (2.5 V each, or 5 V total), as follows:

$$E_{VD4} = 100 \text{ V} - 5 \text{ V}$$
$$= 95 \text{ V}$$

To check that calculation, we could multiply the known constant current (5 A) by the resistance value of R_4 (19 Ω), as follows:

$$E_{VD4} = I \times R_4$$
$$= 5 \text{ A} \times 19\ \Omega$$
$$= 95 \text{ V}$$

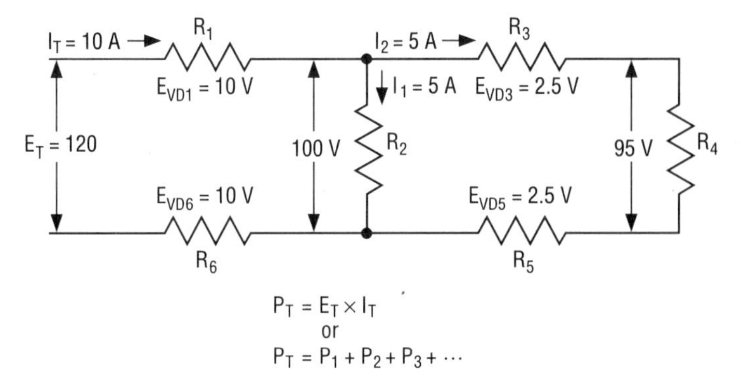

FIG. 1-15 Calculating the power consumption in series-parallel circuits.

The circuit shown in Fig. 1-14*a* is redrawn in Fig. 1-15 with the voltage and current values included. Once the voltage and total current are known, the total circuit power consumption can be calculated from Watt's law:

$$P = E \times I$$

$$= 120 \text{ V} \times 10 \text{ A}$$

$$= 1200 \text{ W}$$

The power consumed by each individual resistance can also be calculated by multiplying the voltage dropped or applied across each resistance by the current flowing through the individual resistance. And the sum of the power consumed by each individual resistance should equal the total power ($P_T = P_1 + P_2 + P_3 + \cdots$). If the power consumed at R_1 through R_6 is calculated and summed, it should equal the total power.

Solve for individual power consumption using the formula $P = E \times I$, where $E =$ the voltage dropped or applied and $I =$ the current flow through the individual resistance.

For R_1:

$$P_1 = E_{\text{VD1}} \times I_T$$

$$= 10 \text{ V} \times 10 \text{ A}$$

$$= 100 \text{ W}$$

For R_2:

$$P_2 = E_{\text{applied}} \times I_1$$

$$= 100 \text{ V} \times 5 \text{ A}$$

$$= 500 \text{ W}$$

For R_3:

$$P_3 = E_{VD3} \times I_2$$
$$= 2.5 \text{ V} \times 5 \text{ A}$$
$$= 12.5 \text{ W}$$

For R_4:

$$P_4 = E_{applied} \times I_2$$
$$= 95 \text{ V} \times 5 \text{ A}$$
$$= 475 \text{ W}$$

For R_5:

$$P_5 = E_{VD5} \times I_2$$
$$= 2.5 \text{ V} \times 5 \text{ A}$$
$$= 12.5 \text{ W}$$

For R_6:

$$P_6 = E_{VD6} \times I_T$$
$$= 10 \text{ V} \times 10 \text{ A}$$
$$= 100 \text{ W}$$

Total:

$$P_1 = 100 \text{ W}$$
$$P_2 = 500 \text{ W}$$
$$P_3 = 12.5 \text{ W}$$
$$P_4 = 475 \text{ W}$$
$$P_5 = 12.5 \text{ W}$$
$$P_6 = 100 \text{ W}$$
$$P_T = 1200 \text{ W}$$

The wattage of non-motor-operated appliances and lamps is established at 120 V in accordance with Underwriters' Laboratories (UL) standards. This is true whether the device is labeled 115 V, 110–120 V, or 115–125 V. The design voltage is 120 V. Where wattage values for loads are given instead of resistance values (see Fig. 1-16), the resistance of the load can be calculated using the design voltage (120 V) and the formula

$$R = E^2/P$$

For the load shown in Fig. 1-16, the resistance of load P_1 is

$$R = (120 \text{ V})^2/40 \text{ W}$$

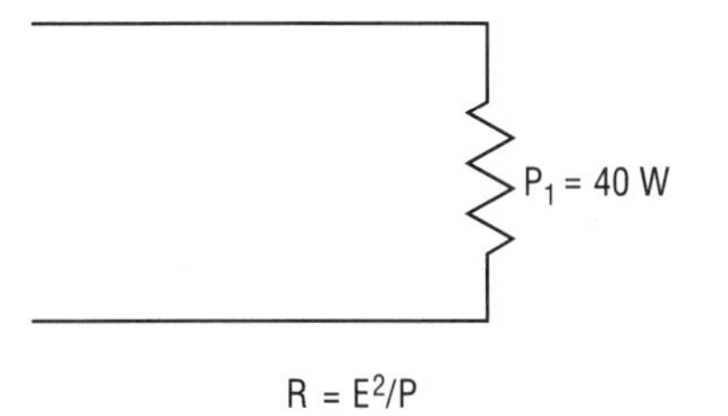

FIG. 1-16 Establishing resistance where wattage value is given in a series circuit. E = design voltage; P = nameplate wattage rating.

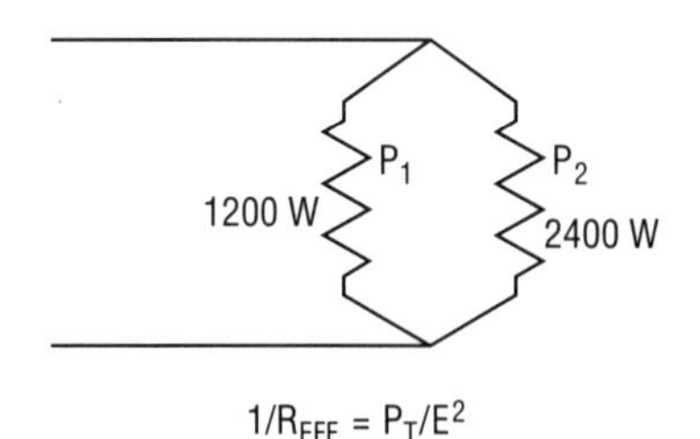

FIG. 1-17 Establishing effective series resistance where wattage values are given for parallel loads. P_T = sum of parallel load's nameplate wattage; E = design voltage.

$$= 14{,}400 \text{ V}/40 \text{ W}$$

$$= 360 \ \Omega$$

Where wattage values are given for parallel loads, the resistance value R_{EFF} can be calculated using the device design voltage, 120 V, and the formula

$$1/R_{\text{EFF}} = P_T/E^2$$

Refer to the parallel loads shown in Fig. 1-17. The effective resistance of those loads is

$$1/R_{\text{EFF}} = (1200 \text{ W} + 2400 \text{ W})/(120 \text{ V})^2$$

$$= 3600 \text{ W}/14{,}400 \text{ V}$$

$$= 0.25 \ \Omega$$

$$= 4 \ \Omega$$

VOLTAGE VALUES

In Art. 100, the National Electric Code (NEC) gives a definition for *nominal voltage.* That definition is reproduced in Fig. 1-18. And those are the values UL uses when evaluating equipment. The NEC, in Sec. 220-2, also calls for the use of nominal voltages when calculating permissible loading of branch circuit, feeder, and service conductors (Fig. 1-19). But, in many instances, the value of voltage actually delivered to the load is some lesser value. That reality is recognized by the second paragraph of the Code definition.

The Fine Print Note (FPN) following the definition of nominal voltage refers the reader to an American National Standards Institute (ANSI) document, specifically ANSI C84.1-1989. That standard shows nominal voltages and *utilization* voltages (Fig. 1-20). Although the NEC *requires* use of nominal system voltages for establishing maximum circuit loading, for practical application, use of the so-called utilization voltages will ensure adequacy for future loading and minimize the possibility of overload. Because the NEC is establishing a *maximum* permissible load, it would be acceptable to

Voltage, Nominal: A nominal value assigned to a circuit or system for the purpose of conveniently designating its voltage class (as 120/240, 480Y/277, 600, etc.).

The actual voltage at which a circuit operates can vary from the nominal within a range that permits satisfactory operation of equipment.

(FPN): See Voltage Ratings for Electric Power Systems and Equipment (60 Hz), ANSI C84.1-1989.

FIG. 1-18 The NEC definition for nominal voltage.

220-2. Voltages. Unless other voltages are specified, for purposes of computing branch-circuit and feeder loads, nominal system voltages of 120, 120/240, 208Y/120, 240, 480Y/277, 480, and 600 volts shall be used.

FIG. 1-19 Section 220-2 of the NEC effectively states that nominal voltages are to be used when calculating permissible circuit loading.

Nominal system voltage	Nameplate (utilization) voltage	
	Three-phase	Single-phase
120	—	115
208	200	—
240	230	230
480	460	—
600	575	—
2400	2300	—
4160	4000	—
4800	4600	—
6900	6600	—
13,800	13,200	—

FIG. 1-20 Reproduction of the table in Appendix C of ANSI C84.1-1989. The right-hand column of this table shows the nameplate rating of motors, motor control equipment, and all heating, refrigeration, and air conditioning equipment, as well as food waste disposals, clothes washers, and dishwashers. All other non-motor-operated equipment, such as lighting fixtures, appliances, water heaters, is rated according to the nominal system voltage shown in the left-hand column.

use the lower utilization voltages, since that would result in an overall greater current value. For example, consider the circuits shown in Fig. 1-21.

In Sec. 220-3(c)(7), the NEC requires that each receptacle strap supplied in nonresidential occupancies be counted as a 180-VA load. To determine the maximum number permitted on a single branch circuit in any nonresidential occupancy, the total voltampere (VA) capacity of the circuit is established and divided by 180 VA. If the voltage value used is 120 V, then the maximum permissible number of receptacles would be

$$\text{Total VA capacity} = \text{V} \times \text{A}$$

$$= 120\text{ V} \times 20\text{ A}$$

$$= 2400\text{ VA}$$

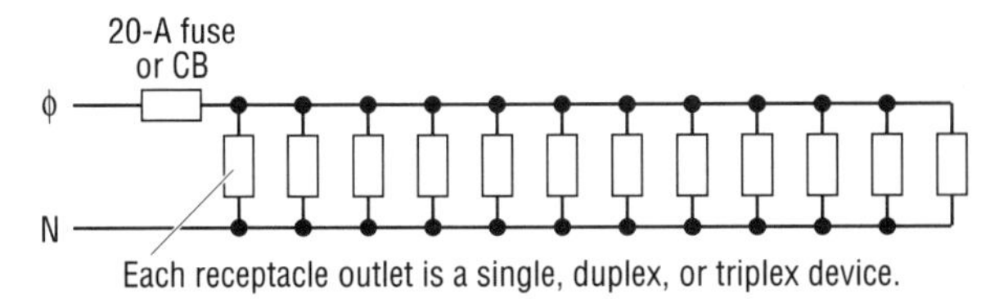

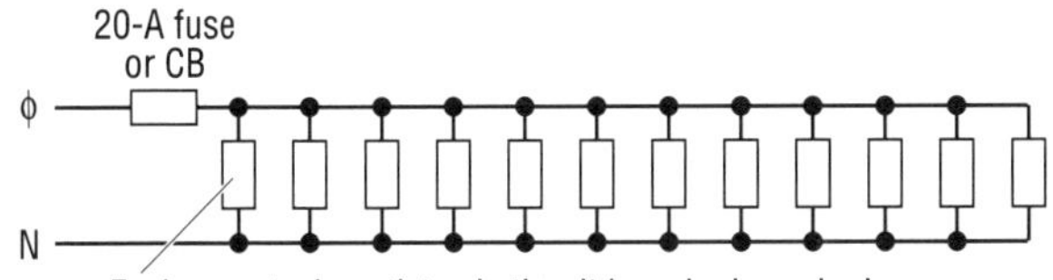

FIG. 1-21 Calculating the maximum number of receptacles permitted on a 20-A branch circuit in a nonresidential occupancy. As can be seen, use of the lower utilization voltage (115 V) instead of the nominal system voltage (120 V) permits less loading and, therefore, would not be a Code violation.

$$\begin{aligned}\text{Maximum no. of receptacles} &= \text{total VA/180 VA}\\ &= \text{2400 VA/180 VA}\\ &= 13.3\end{aligned}$$

If 115 V is used, the maximum number of receptacles would be

$$\begin{aligned}\text{Total VA capacity} &= \text{V} \times \text{A}\\ &= 115\text{ V} \times 20\text{ A}\\ &= 2300\text{ VA}\end{aligned}$$

$$\begin{aligned}\text{Maximum no. of receptacles} &= \text{total VA/180 VA}\\ &= \text{2300 VA/180 VA}\\ &= 12.7\text{ VA} \qquad \text{which is taken as 12}\end{aligned}$$

While use of the lower utilization voltage values will ensure adequacy and minimize circuit overload, the utilization voltage values should be used *only* in real-life applications. It is important to remember that when you take any type of standardized test—for electrician, journeyman, contractor, designer, engineer, or inspector licensing—NEC requirements must be observed. Therefore, on any such test, use the Code-designated voltage value for the particular load being evaluated.

MAXIMUM, INSTANTANEOUS, AND EFFECTIVE AC VOLTAGE AND CURRENT

Because of the comparatively complex nature of alternating-current circuits, there are specific characteristics based on the varying values of current during each cycle. They are electrical degrees, effective current, and effective voltage.

Electrical Degrees

Each cycle of current alternation takes place in a fixed time, depending upon the frequency of the alternating current. For instance, a cycle of 60-Hz ac occurs every $\frac{1}{60}$ s. A half-cycle of 60-Hz ac occurs every $\frac{1}{120}$ s.

There is another way of referring to a single cycle, or portion thereof, that is not related to the specific frequency of the ac source. This is known as *electrical degrees.* The use of electrical degrees can help to facilitate analysis of specific ac characteristics, such as phase relationships.

The basic idea is to represent any point along the alternating current or voltage waveform as a point on a circle. The common ac generator produces a continuously variable voltage and current, the instantaneous values of which, when plotted against degrees of rotation, form a sine wave. One complete cycle is equal to 360°, one-half cycle is equal to 180°, etc. (see Fig. 1-22). The positive alternation or half cycle occurs as the vector

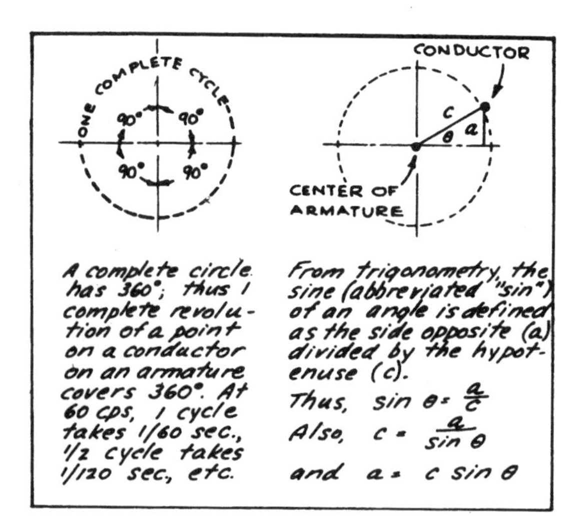

FIG. 1-22 Representation of an ac cycle in electrical degrees.

moves from 0° to 180°. The negative alternation occurs from 180° to 360°. In the diagram, the angle Θ (theta) represents electrical degrees of rotation of a point on a conductor moving through a magnetic field about the center axis of the armature. The *maximum* value of generated voltage or current (E_m or I_m) is proportional to the radius of the circle (the hypotenuse *c* of the triangle shown in Fig. 1-22). The *instantaneous* value of voltage or current (E_i or I_i is proportional to the side *a* opposite, shown in Fig. 1-22. And the instantaneous voltage or current (E_i or I_i) is related to the maximum voltage or current (E_m or I_m) as follows:

$$E_i = E_m \times \sin \Theta$$

and

$$I_i = I_m \times \sin \Theta$$

At 0°, 180°, and 360°, $a = 0$. At 90°, $a = c$, and at 270°, $a = -c$. Thus, at 0°, 180°, and 360°, E_i or $I_i = 0$. And, at 90°, E_i or $I_i = E_m$ or I_m. At 270°, E_i or $I_i = -E_m$ or $-I_m$.

Using the above formula and the table shown in Fig. 1-23, determine the instantaneous voltage for a full cycle at 30° intervals for a generator whose maximum voltage E_m is 170 V.

At 0°: $\sin \Theta = 0.0$

$$E_i = E_m \times \sin \Theta$$
$$= 170 \text{ V} \times 0.0$$
$$= 0.0$$

At 30°: $\sin \Theta = 0.5$

$$E_i = 170 \text{ V} \times 0.5$$
$$= 85 \text{ V}$$

At 60°: $\sin \Theta = 0.866$

$$E_i = 170 \text{ V} \times 0.866$$
$$= 147 \text{ V}$$

TABLE OF SINES

θ	SIN θ
0°	0
30°	0.5
60°	0.866
90°	1.0
120°	0.866
150°	0.5
180°	0
210°	-0.5
240°	-0.866
270°	-1.0
300°	-0.866
330°	-0.5
360°	0

FIG. 1-23 Table of sines.

At 90°: $\sin \Theta = 1.0$ $\quad E_i = 170 \text{ V} \times 1.0$

$= 170 \text{ V}$

At 120°: $\sin \Theta = 0.866$ $\quad E_i = 170 \text{ V} \times 0.866$

$= 147 \text{ V}$

At 150°: $\sin \Theta = 0.5$ $\quad E_i = 170 \text{ V} \times 0.5$

$= 85 \text{ V}$

At 180°: $\sin \Theta = 0.0$ $\quad E_i = 170 \text{ V} \times 0.0$

$= 0.0$

At 210°: $\sin \Theta = -0.5$ $\quad E_i = 170 \text{ V} \times (-0.5)$

$= -85 \text{ V}$

At 240°: $\sin \Theta = -0.866$ $\quad E_i = 170 \text{ V} \times (-0.866)$

$= -147 \text{ V}$

At 270°: $\sin \Theta = -1.0$ $\quad E_i = 170 \text{ V} \times (-1.0)$

$= -170 \text{ V}$

At 300°: $\sin \Theta = -0.866$ $\quad E_i = 170 \text{ V} \times (-0.866)$

$= -147 \text{ V}$

At 330°: $\sin \Theta = -0.5$ $\quad E_i = 170 \text{ V} \times (-0.5)$

$= -85 \text{ V}$

At 360°: $\sin \Theta = 0.0$ $\quad E_i = 170 \text{ V} \times 0.0$

$= 0.0 \text{ V}$

Those individual values are plotted on the graph shown in Fig. 1-24. As can be seen, when the voltage value is plotted against the corresponding electrical degrees, the resultant curve is a sine wave. But there is little practical use for instantaneous voltage or current values in everyday application. The most useful values for ac voltage and current are the *effective* values.

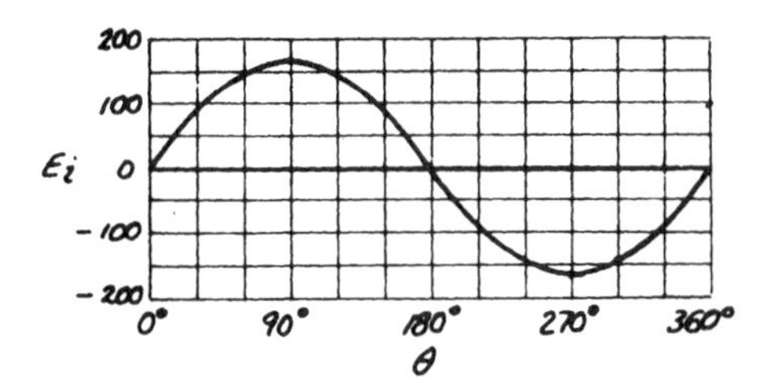

FIG. 1-24 Instantaneous voltage values plotted against the corresponding electrical degrees.

The effective current of an ac circuit is based on the resulting heating effect. That is, a value of 1 A would be assigned to an alternating current that produced the same heat loss as 1 A of direct current.

The value of effective current I_e is given in amperes and is also known as the rms value of current. The letters rms stand for root-mean-square. That expression is derived from the mathematical method used to compare ac and dc heating effects.

A direct current produces heat at a rate equal to I^2R. The heating effect of an alternating-current half cycle is produced at an instantaneous rate equal to i^2R, where i is the instantaneous current value at a given instant. The heating effect of each half cycle depends upon the average of the squares of the instantaneous current values over that half cycle. The rms current value I_e is established by taking the square root of the mean, or average, square of instantaneous currents. That value then represents the heating effect, or rate of heat production during each half cycle. And, as long as current alternations of the same value follow one after the other, the effective or rms value I_e is maintained at a steady rate.

The same relationship exists with effective or rms voltages E_e. That is, the effective voltage will be equal to the square root of the average square of all instantaneous voltages. Both the rms current I_e and rms voltage E_e of a sine wave are equal to the maximum value E_m or I_m times 0.707.

For an ac sine wave with a maximum alternation of 100 A, the rms current value I_e would be

$$I_e = I_m \times 0.707$$

$$= 100 \text{ A} \times 0.707$$

$$= 70.7 \text{ A}$$

Applying the formula for effective voltage (rms) to the voltage sine wave shown in Fig. 1-24, we see that the rms voltage E_e would be

$$E_e = E_m \times 0.707$$

$$= 170 \text{ V} \times 0.707$$

$$= 120.19, \text{ or } 120 \text{ V}$$

INDUCTANCE AND CAPACITANCE

Up to this point, all discussions have focused on resistive ac circuits only. When an ac voltage is applied to a circuit containing a resistance only, the current wave is produced by the voltage wave. In such a circuit, the voltage and current both pass through their zero values and increase to their maximum in the same direction and at the same time. In that case, the voltage and current are said to be *in phase* with each other (see Fig. 1-25). However, in most ac circuits, there are factors other than just resistance that oppose current flow, such as magnetic coils and capacitive effects. In such circuits, the current and voltage waveforms do *not* occur coincidentally. That is, the current and voltage sine waves do *not* pass through their zero and maximum values at the same time. Instead, there is a fixed time interval between the zero and maximum values of current and those of voltage.

The two possible out-of-phase relationships between voltage and current would be

1. The current may pass through zero and increase to maximum at some time later than the voltage passes through zero and increases to maximum. In such a case, the current is said to *lag* the voltage (Fig. 1-26).
2. The current may pass through its zero value and increase to maximum at some time earlier than the voltage waveform. In this case, the current is said to *lead* the voltage (Fig. 1-27).

The first out-of-phase relationship—lagging current—is due to inductance. Inductance is the property of a circuit or circuit element, such as a coil or other electromagnetic device, which opposes any change in the current flowing in that circuit or circuit element. This opposition to current flow is a result of the self-induced voltage, which is due to varying flux caused by the alternating character of the current. Any part of a circuit that produces flux has inductance. This includes all forms of coils wound on magnetic cores—motors, ballasts, transformers, solenoids, etc.

CURRENT FLOW THROUGH A RESISTANCE *encounters no opposing emf; hence circuit voltage and current rise and fall together, or "in phase."*

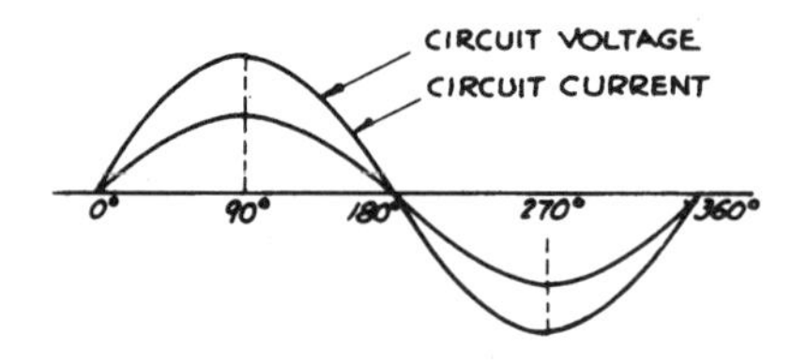

FIG. 1-25 Where an alternating current flows through a resistance only, the zero and maximum values for both current and voltage will occur simultaneously and in the same direction.

CURRENT FLOW THROUGH AN INDUCTANCE,

however, sets up a magnetic field about the coil which induces an opposing emf in the coil and causes the circuit current to lag behind the circuit impressed voltage.

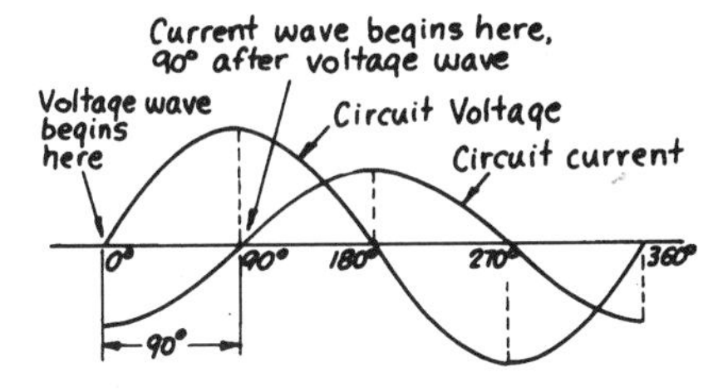

FIG. 1-26 Current and voltage waveforms supplying an inductive load.

CAPACITANCE

Consider the capacitor across an ac generator. Each quarter of one complete cycle will be analyzed:

AS GENERATOR VOLTAGE E_G INCREASES, CAPACITOR C IS CHARGED. WHEN E_G REACHES MAXIMUM VALUE, C IS AT MAXIMUM CHARGE.

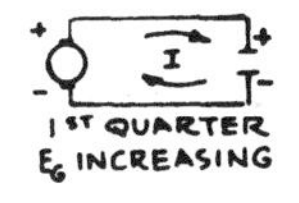

E_G, STILL IN SAME DIRECTION, BEGINS TO DECREASE. C BEGINS TO DISCHARGE CURRENT BACK TOWARD GENERATOR. CURRENT REACHES MAXIMUM WHEN E_G REACHES ZERO.

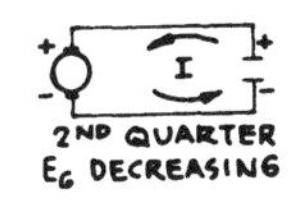

AS E_G BUILDS UP IN OPPOSITE DIRECTION, C CHARGES IN OPPOSITE DIRECTION. CURRENT REACHES ZERO AS E_G REACHES ITS MAXIMUM. AT THIS POINT C AGAIN HAS A MAXIMUM CHARGE BUT OF THE OPPOSITE POLARITY AS IN 2ND QUARTER-CYCLE.

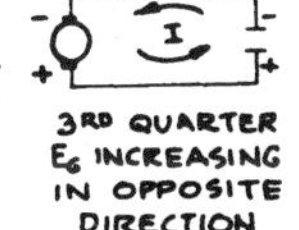

AS E_G DECREASES, C AGAIN DISCHARGES, THIS TIME IN OPPOSITE DIRECTION AS IN 2ND QUARTER-CYCLE.

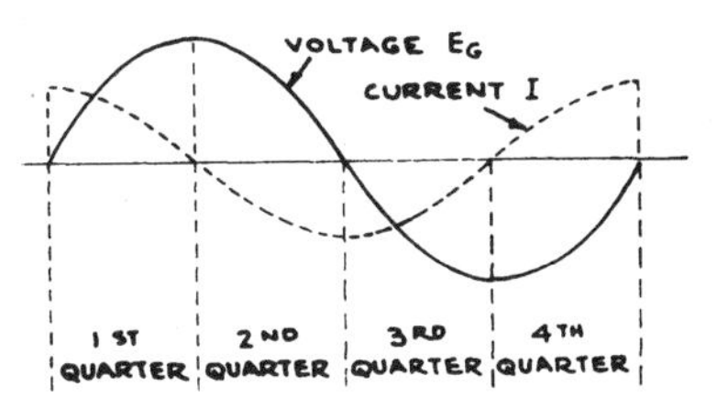

FIG. 1-27 Current and voltage waveforms supplying a capacitive load.

Inductance is measured in units of *henrys* (H) and represented by the symbol L. A circuit is said to have an inductance of one henry if, when the current is changing at a rate of one ampere per second, the average induced counter-emf is one volt.

Determining the total inductance L_T of a given circuit is very similar to establishing the total resistance. That is, in series applications, inductance is additive: $L_T = L_1 + L_2 + L_3 + \cdots$. When connected in parallel, the equivalent inductance L_E is found from the formula

$$1/L_E = 1/L_1 + 1/L_2 + 1/L_3 + \cdots$$

Because the self-induced voltage or counter-emf is constantly opposing the alternations in current, it is in effect opposing the alternating current and can be measured in ohms, just as resistance is. The opposition that inductance presents to the flow of alternating current is known as *inductive reactance* and is represented by the symbol X_L.

In any circuit the amount of inductive reactance depends upon the value of inductance and the frequency of the alternating current. The value of inductive reactance in ohms may be found from the formula

$$X_L = 2\pi f L$$

where $\pi = 3.14$
f = frequency, hertz (Hz)
L = value of inductance, H

When a voltage is applied to a circuit containing only inductance (that is, resistance is negligible), the magnitude of current flow is determined from the following formula:

$$I = E/X_L$$

where I = effective or rms current, A
E = applied rms voltage, V
X_L = inductive reactance, Ω

Consider the circuit shown in Fig. 1-28. Calculation of the effective current through the 30-millihenry (mH) coil supplied from a 120-V rms 60-Hz source would be as follows: First, determine the ohmic value of inductive reactance X_L.

$$X_L = 2' \times \pi \times f \times L$$

$$= 2 \times 3.14 \times 60 \text{ Hz} \times 0.03 \text{ H}$$

$$= 11.3\ \Omega$$

Next, find the value of current.

$$I = E/X_L$$

$$= 120 \text{ V}/11.3\ \Omega$$

$$= 10.6 \text{ A rms}$$

Let's take another example. If 120-V rms 60-Hz ac is applied to a circuit with 20 mH of inductance, what is the amount of current flowing in that circuit? As before, first find the inductive reactance X_L.

INDUCTIVE REACTANCE

is the opposition, expressed in ohms, presented to the flow of ac current due to the inductance in the circuit:

$$X_L = 2\pi f L$$

X_L = inductive reactance (ohms)
π = 3.14
f = frequency (cycles per second)
L = inductance (henrys)

$$I = \frac{E}{X_L}$$

I = effective current (amperes)
E = effective voltage (volts)

EXAMPLE: What effective current will flow through the coil in the following circuit? (Resistance of coil is negligible.)
Note: 1 milli-henry (mh) = $\frac{1}{1000}$ henry.

FIG. 1-28 Calculating inductive reactance and effective current.

$$X_L = 2 \times \pi \times f \times L$$

$$= 2 \times 3.14 \times 60 \text{ Hz} \times 0.02 \text{ H}$$

$$= 7.54\ \Omega$$

Find circuit current as follows:

$$I = E/X_L$$

$$= 120 \text{ V}/7.54\ \Omega$$

$$= 15.9 \text{ A rms}$$

The second out-of-phase relationship—leading current—is caused by capacitance. Capacitance has a reactive load effect somewhat similar to that of inductance but the opposite relationship between current and voltage.

Capacitors are devices used in electric circuits to obtain certain characteristics based on their ability to hold a charge. Typically, capacitor units are made for power-factor correction in power circuits and for many uses in electronics. In any ac circuit, capacitance—from either capacitor units or stray accidental effects—is an important consideration because of its reaction to alternating current.

Every capacitor is rated according to the amount of electrons that are stored in it when a one-volt source is connected across it. The ability to store electrons is called *capacitance* and is expressed in units called *farads* (F). The symbol *C* is used to represent farads of capacitance.

A capacitor has a capacitance of one farad when an applied source of one volt produces a charge of one coulomb. Because the farad is too large a unit for practical use, the unit *microfarad* (one millionth of a farad) is more commonly used (Fig. 1-29).

CAPACITANCE

Remember that current does not flow through a capacitor. The constantly reversing voltage causes a "back-and-forth" movement of electrons to alternately charge the capacitor first in one direction, then the other.

Conventional current is normally represented as moving away from the positive terminal of the source, opposite to the actual direction of electron flow:

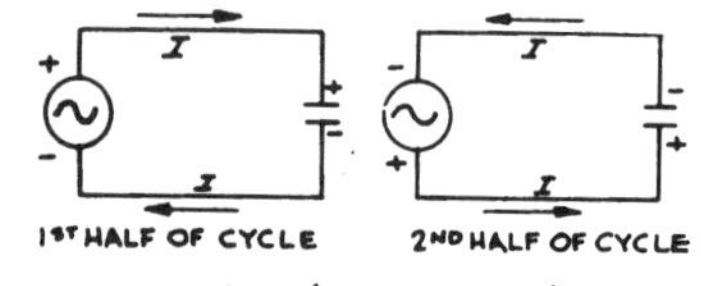

I = effective current

CAPACITANCE = 1 FARAD *when 1 volt produces a charge of 1 coulomb on the capacitor.*

A COULOMB IS A QUANTITY *of electricity, equal to approximately 6 million million million electrons.*

1 FARAD = 1,000,000 MICROFARADS, *abbreviated MFD, mfd, or μfd.*

FIG. 1-29 Capacitance is the measure of a capacitor's ability to store electrons.

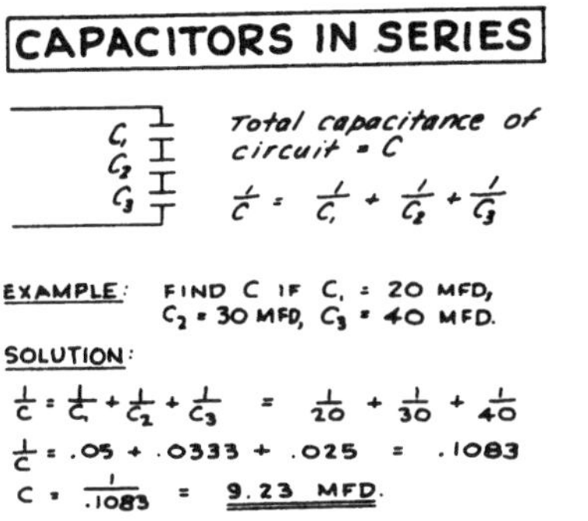

CAPACITORS IN SERIES

C_1, C_2, C_3

Total capacitance of circuit = C

$$\frac{1}{C} = \frac{1}{C_1} + \frac{1}{C_2} + \frac{1}{C_3}$$

EXAMPLE: FIND C IF C_1 = 20 MFD, C_2 = 30 MFD, C_3 = 40 MFD.

SOLUTION:

$$\frac{1}{C} = \frac{1}{C_1} + \frac{1}{C_2} + \frac{1}{C_3} = \frac{1}{20} + \frac{1}{30} + \frac{1}{40}$$

$$\frac{1}{C} = .05 + .0333 + .025 = .1083$$

$$C = \frac{1}{.1083} = \underline{9.23 \text{ MFD}}.$$

FIG. 1-30 Calculation of total capacitance for series-connected capacitors.

CAPACITORS IN PARALLEL

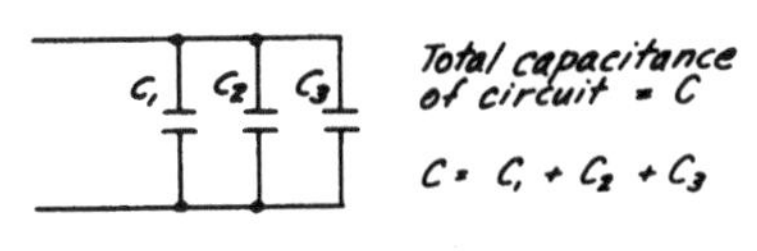

EXAMPLE:

FIND C IF C_1 = 20 MFD, C_2 = 30 MFD, C_3 = 40 MFD.

SOLUTION:

$C = C_1 + C_2 + C_3$

$= 20 + 30 + 40 = \underline{90 \text{ MFD}}.$

FIG. 1-31 Calculation of total capacitance for parallel-connected capacitors.

Unlike resistance and inductance connected in series, capacitance in series is found from the following formula:

$$1/C_T = 1/C_1 + 1/C_2 + 1/C_3 + \cdots$$

As shown in Fig. 1-30, where three capacitors rated at 20 μF, 30 μF, and 40μF are connected in series, the equivalent or total capacitance is equal to 9.23 μF.

When connected in parallel, the total capacitance is additive. That is, the total capacitance for parallel-connected capacitors is equal to the sum of all parallel capacitances (Fig. 1-31). Note that the rules for determining total capacitance are just the opposite of the rules for determining total resistance or total inductance in series and parallel circuits.

As described, a capacitor in an ac circuit permits the current flow by reason of its ability to store current on one direction of current flow, then discharge the current as the current flows in the opposite direction; then store current as the alternating current flows in the original direction again. But capacitors also present an opposition to current flow. This is known as *capacitive reactance.* And, like resistance and inductive reactance, it can be expressed as an ohmic value. The value of capacitive reactance depends on two factors: the value of capacitance and the rate at which the applied voltage is changing (which is the frequency of the ac voltage source). With an ac source connected across a capacitance only, an increase in either capacitance or frequency will decrease the opposition or ohmic value of the capacitive reactance. That is, capacitive reactance is inversely proportional to the capacitance and frequency. The formula used to determine capacitive reactance X_C is

$$X_C = 1/(2 \times \pi \times f \times C)$$

where X_C = capacitive reactance
π = 3.14
f = source frequency, Hz
C = value of capacitance, F

When the value of capacitance C is given in microfarads, the formula becomes

$$X_C = 1{,}000{,}000/(2 \times \pi \times f \times C)$$

CAPACITIVE REACTANCE

The opposition offered by a capacitor to the flow of alternating current is called capacitive reactance and is measured in ohms.

$X_C = \frac{1}{2\pi f C}$ $\quad I = \frac{E}{X_C}$

X_C = capacitive reactance (ohms)
π = 3.14
f = frequency (cycles per sec.)
C = capacitance (farads)
I = effective current (amperes)
E = effective voltage (volts)

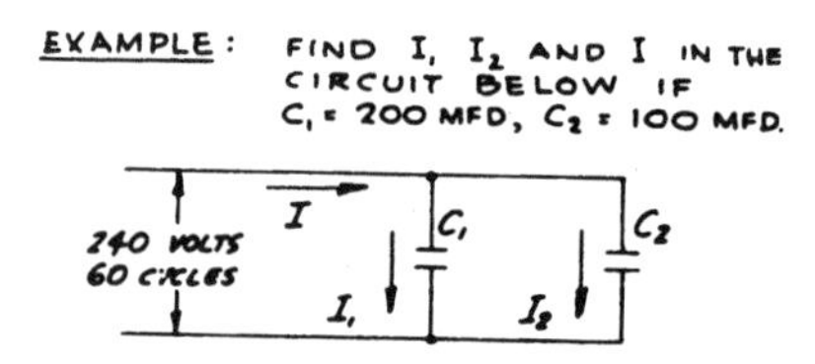

FIG. 1-32 Calculating capacitive reactance and effective current for parallel-connected capacitors.

As is the case with a circuit containing only inductance, a circuit containing only capacitance will have a current flow equal to

$$I = E/X_C$$

where I = effective or rms current
E = applied rms voltage
X_C = capacitive reactance, Ω

Consider the circuit diagram shown in Fig. 1-32. Apply the above formulas and determine the current flow through C_1, C_2, and the total circuit. First establish the capacitive reactance at C_2 (X_{C2}).

$$X_{C2} = 1{,}000{,}000/(2 \times \pi \times f \times C_2)$$
$$= 1{,}000{,}000/(2 \times 3.14 \times 60 \text{ Hz} \times 100 \ \mu\text{F})$$
$$= 1{,}000{,}000/37{,}680$$
$$= 26.5 \ \Omega$$

Effective current through C_2 (I_2) is

$$I_2 = E/X_{C2}$$
$$= 240 \text{ V}/26.5 \ \Omega$$
$$= 9.05 \text{ A}$$

Capacitive reactance at C_1 (X_{C1}) is

$$X_{C1} = 1{,}000{,}000/(2 \times \pi \times f \times C_1)$$
$$= 1{,}000{,}000/(2 \times 3.14 \times 60 \times 200 \ \mu\text{F})$$
$$= 1{,}000{,}000/75{,}360$$
$$= 13.26 \ \Omega$$

Effective current through C_1 (I_1) is

$$I_1 = E/X_{C1}$$
$$= 240\ \text{V}/13.26\ \Omega$$
$$= 18.10\ \text{A}$$

Those results can be checked as follows: Total capacitance in parallel is additive. Therefore,

$$C_T = C_1 + C_2$$
$$= 200\ \mu\text{F} + 100\ \mu\text{F}$$
$$= 300\ \mu\text{F}$$

Total capacitive reactance is

$$X_C = 1{,}000{,}000/(2 \times \pi \times f \times C_T)$$
$$= 1{,}000{,}000/(2 \times 3.14 \times 60 \times 300\ \mu\text{F})$$
$$= 1{,}000{,}000/113{,}040$$
$$= 8.84\ \Omega$$

Total effective or rms current I_T is

$$I_T = E/X_C$$
$$= 240\ \text{V}/8.84\ \Omega$$
$$= 27.15\ \text{A}$$

In parallel circuits, total current is equal to the sum of the individual currents. Therefore, using the individual currents previously calculated (I_1 and I_2), total effective or rms current equals

$$I_T = I_1 + I_2$$
$$= 18.10\ \text{A} + 9.05\ \text{A}$$
$$= 27.15\ \text{A}$$

CIRCUIT IMPEDANCE

Virtually any circuit which consists of an inductance and/or capacitance connected in series will also have resistance in series with the reactance(s). The combined effect of resistance, inductive reactance, and capacitive reactance is known as the circuit *impedance,* represented by the symbol Z. However, determining the combined effect or circuit impedance when resistance and reactance(s) are connected in series is not a simple matter of adding the ohmic values of each component. This is a result of the different phase relationships.

For example, consider the circuit shown in Fig. 1-33. Because of different phase relationships between the resistance, whose current wave is in phase with the voltage wave, and the inductance, whose current wave lags the voltage wave by 90°, the true combined effect or circuit impedance must be summed vectorially. As a result, the circuit impedance is determined from the following:

$$Z^2 = R^2 + X_L^2 \qquad \text{or} \qquad Z = \sqrt{R^2 + X_L^2}$$

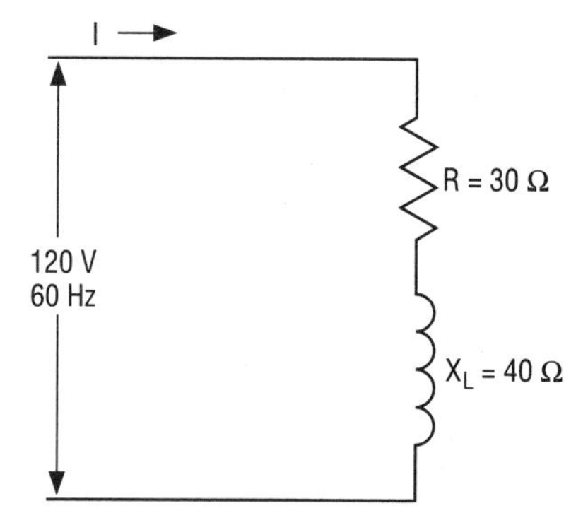

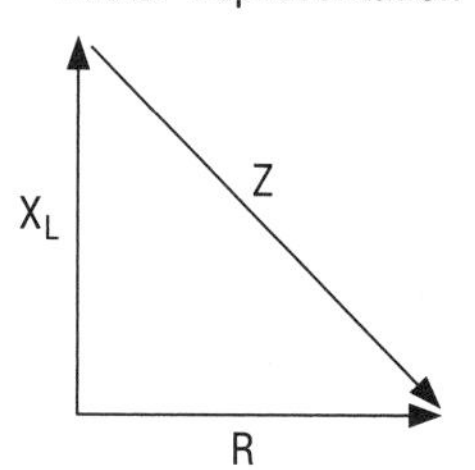

Due to the 90° phase displacement of the current wave through the inductance, it is possible to solve for Z using the Pythagorean theorem for right triangles. That is:

$$Z^2 = R^2 + X_L^2$$
or
$$Z = \sqrt{R^2 + X_L^2}$$

FIG. 1-33 Calculating the circuit impedance Z when a resistance and an inductance are connected in series.

The impedance Z of the circuit shown in Fig. 1-33 is calculated as follows:

$$\begin{aligned} Z &= \sqrt{30^2 + 40^2} \\ &= \sqrt{900 + 1600} \\ &= \sqrt{2500} \\ &= 50\ \Omega \end{aligned}$$

Total current drawn by the circuit shown in Fig. 1-33 is calculated from the equation

$$\begin{aligned} I &= E/Z \\ &= 120\ \text{V}/50\ \Omega \\ &= 2.4\ \text{A} \end{aligned}$$

Now that the total current has been established, the voltage drop across the resistance E_R and the inductance E_L can be determined as follows:

Voltage drop across the resistance E_R:

$$\begin{aligned} E_R &= I \times R \\ &= 2.4\ \text{A} \times 30\ \Omega \\ &= 72\ \text{V} \end{aligned}$$

Voltage drop across the inductance E_L:

$$\begin{aligned} E_L &= I \times X_L \\ &= 2.4 \times 40\ \Omega \\ &= 96\ \text{V} \end{aligned}$$

As is the case with the circuit impedance, the voltage drops across the resistance E_R and the inductance E_L are not purely additive because of their phase relationship. That is, if E_R (72 V) is added to E_L (96 V), which is the method applied to purely resistive ac circuits, the total (168 V) will be greater than the applied voltage (120 V). These voltages must also be added vectorially (Fig. 1-34).

$$\begin{aligned} E_T^2 &= E_R^2 + E_L^2 \\ E_T &= \sqrt{E_R^2 + E_L^2} \\ &= \sqrt{(72\ \text{V})^2 + (96\ \text{V})^2} \\ &= \sqrt{5184\ \text{V} + 9216\ \text{V}} \\ &= \sqrt{14{,}400\ \text{V}} \\ &= 120\ \text{V} \end{aligned}$$

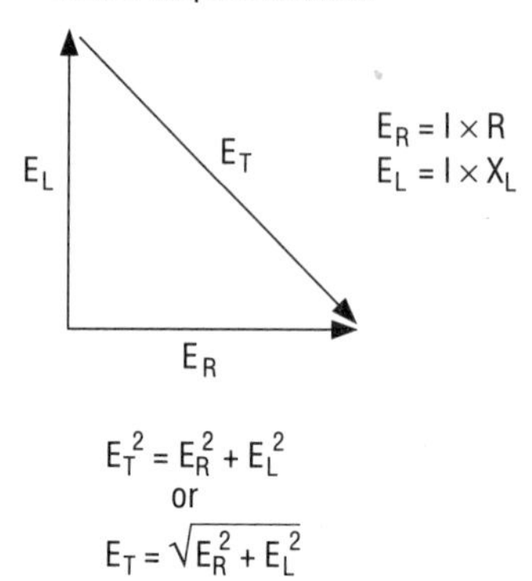

FIG. 1-34 Calculating total voltage E_T when a resistance and an inductance are connected in series.

RC CIRCUIT IMPEDANCE

I

$R = 60\ \Omega$

120 V
60 Hz

$X_C = 80\ \Omega$

Vector Representation

R

X_C

Z

$Z^2 = R^2 + X_C^2$
or
$Z = \sqrt{R^2 + X_C^2}$

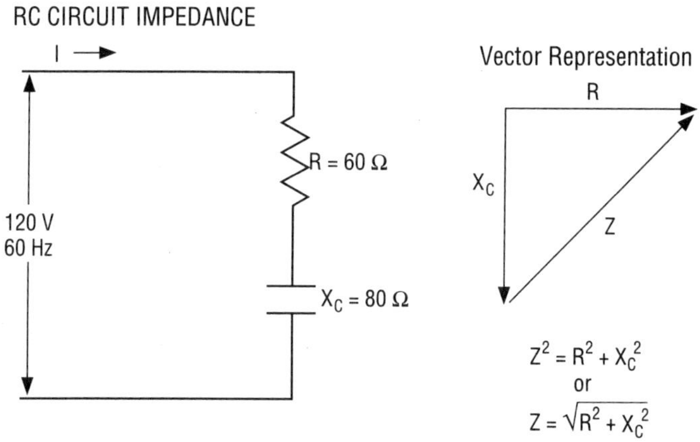

FIG. 1-35 Calculating the circuit impedance Z when a resistance and a capacitance are connected in series.

Circuit impedance when a capacitance is connected in series with a resistance is determined in the same manner. That is, the series capacitance and resistance must be added vectorially. For example, consider the circuit shown in Fig. 1-35.

$$Z^2 = R^2 + X_C^2$$

$$Z = \sqrt{R^2 + X_C^2}$$

$$= \sqrt{(60\ \Omega)^2 + (80\ \Omega)^2}$$

$$= \sqrt{3600\ \Omega + 6400\ \Omega}$$

$$= \sqrt{10{,}000\ \Omega}$$

$$= 100\ \Omega$$

Total circuit current is

$$I = E/Z$$

$$= 120\ \text{V}/100\ \Omega$$

$$= 1.2\ \text{A}$$

Voltage drop across R, denoted by E_R, is

$$E_R = I \times R$$

$$= 1.2\ \text{A} \times 60\ \Omega$$

$$= 72\ \text{V}$$

Voltage drop across C, or E_C, is

$$E_C = I \times X_C$$

$$= 1.2\ \text{A} \times 80\ \Omega$$

$$= 96\ \text{V}$$

Summing those two vectorially (Fig. 1-36), we see that the total circuit voltage is

$$E_T^2 = E_R^2 + E_C^2$$

$$E_T = \sqrt{(E_R)^2 + (E_C)^2}$$

$$= \sqrt{(72\ \text{V})^2 + (96\ \text{V})^2}$$

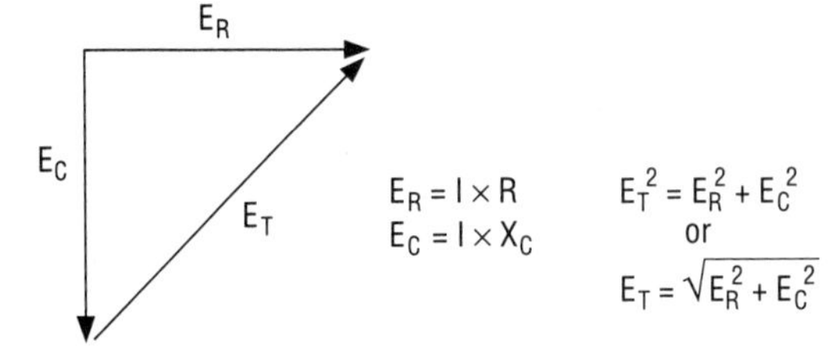

FIG. 1-36 Calculating total voltage E_T when a resistance and a capacitance are connected in series.

$$= \sqrt{5184\text{ V} + 9216\text{ V}}$$

$$= \sqrt{14{,}400\text{ V}}$$

$$= 120\text{ V}$$

As we have seen, the current through a capacitance leads the voltage by 90°. The current wave through an inductance lags the voltage wave by 90°. When they are connected in series, the effect of capacitance is 180° out of phase with the effect of inductance. As a result, the circuit series reactance X will be equal to the difference between the inductive reactance X_L and the capacitive reactance X_C, or $X = X_L - X_C$ (Fig. 1-37). Therefore, the circuit impedance is

$$Z^2 = R^2 + (X_L - X_C)^2$$

or

$$Z = \sqrt{R^2 + (X_L - X_C)^2}$$

$$= \sqrt{R^2 + X^2}$$

RLC SERIES CIRCUITS

Where both inductance and capacitance are included in the circuit, the net reactance X is given by

$$X = X_L - X_C$$

and the circuit impedance Z is given by

$$Z = \sqrt{R^2 + (X_L - X_C)^2} = \sqrt{R^2 + X^2}$$

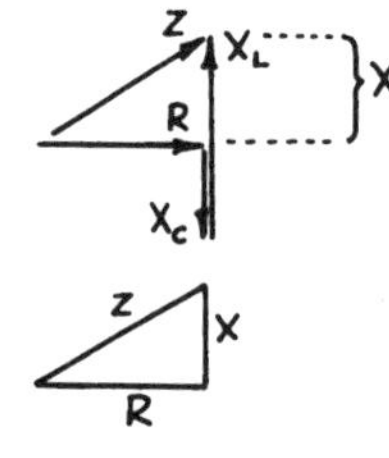

These diagrams illustrate this relation graphically. Resistance is drawn to scale horizontally; X_C and X_L are drawn at right angles to R and 180° out of phase with each other. The net reactance X is shown where X_L is greater than X_C and vice versa.

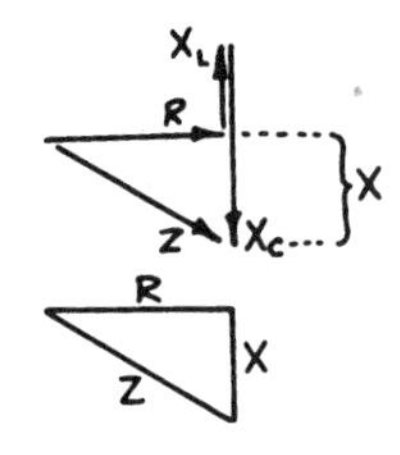

EXAMPLE:

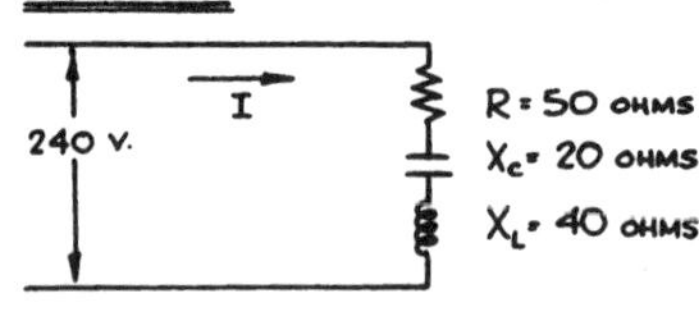

FIND IMPEDANCE Z, EFFECTIVE CURRENT I, AND VOLTAGE DROPS ACROSS R, X_C AND X_L

FIG. 1-37 Calculating circuit impedance Z when a resistance, an inductance, and a capacitance are connected in series.

Consider the *RLC* series circuit shown in Fig. 1-37. If the resistance, inductive reactance, and capacitive reactance have values of 50, 40, and 20 Ω, respectively, what are the circuit impedance, current, and individual voltage drops? Solve for Z as follows:

$$Z = \sqrt{R^2 + (X_L - X_C)^2}$$
$$= \sqrt{(50\ \Omega)^2 + (40\ \Omega - 20\Omega)^2}$$
$$= \sqrt{(50\ \Omega)^2 + (20\ \Omega)^2}$$
$$= \sqrt{2500\ \Omega + 400\ \Omega}$$
$$= \sqrt{2900\ \Omega}$$
$$= 53.85\ \Omega$$

Solve for circuit current:

$$I = E/Z$$
$$= 240\ \text{V}/53.85\ \Omega$$
$$= 4.46\ \text{A}$$

Solve for individual voltage drops:

$$E_R = I \times R = 4.46\ \text{A} \times 50\ \Omega = 223\ \text{V}$$
$$E_C = I \times X_C = 4.46\ \text{A} \times 20\ \Omega = 89.2\ \text{V}$$
$$E_L = I \times X_L = 4.46\ \text{A} \times 40\ \Omega = 178.4\ \text{V}$$

Total voltage E_T of an *RLC* circuit must also be vectorially summed. As can be seen in Fig. 1-38, because E_C is 180° out of phase with E_L, some cancellation will occur. And, as was the case for establishing circuit impedance, the total voltage drop will be equal to the square root of the resistive drop squared ($E_R{}^2$) plus the difference between the inductive drop and the capacitive drop squared [$(E_L - E_C)^2$]. The slight difference between our calculated total voltage and the given voltage can be attributed to rounding of previously calculated values.

It would also be possible to calculate the total voltage E_T based on the use of the total reactance X. As previously stated, the total reactance is equal to the difference between the inductive reactance X_L and the capacitive reactance X_C. For this example, the reactance is 20 Ω (40 Ω −20 Ω). Next, establish the voltage drop across the total reactance E_X:

$$E_X = I \times X$$
$$= 4.46\ \text{A} \times 20\ \Omega$$
$$= 89.2\ \text{V}$$

Now the formula $E_T = \sqrt{E_R{}^2 + E_X{}^2}$ can be used to determine the total circuit voltage. As can be seen in the equation shown at the bottom of Fig. 1-38, $E_T = 240.178$ V, or 240 V.

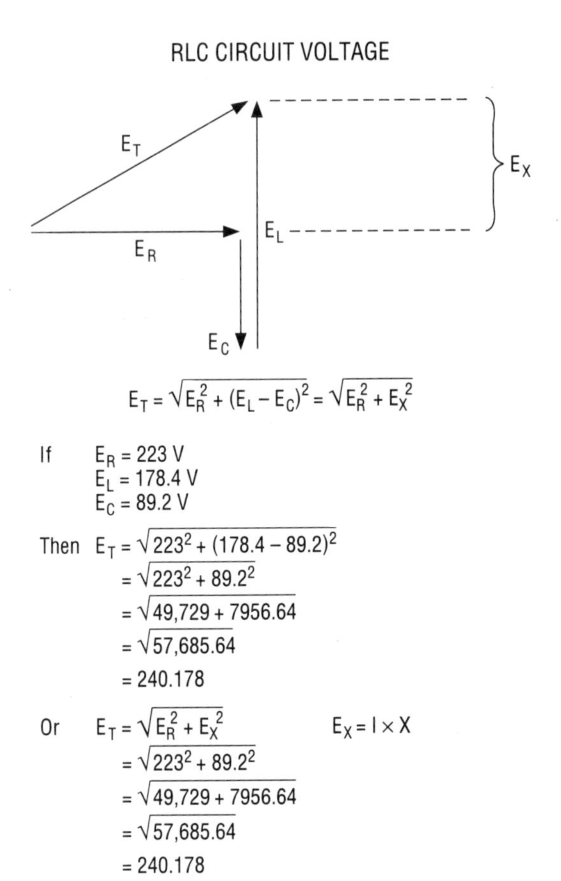

FIG. 1-38 Calculating total voltage E_T when a resistance, an inductance, and a capacitance are connected in series.

POWER IN AC CIRCUITS

The power in any ac circuit is a function of current, voltage, and phase angle, if any, between the current wave and the voltage wave. This differs from dc circuits, in which power is simply equal to the current value I times the voltage value E, with the answer expressed in watts.

As previously indicated, in ac circuits that contain only a resistive load, the current and voltage waves are in phase with each other, and the power is equal to the instantaneous value of current multiplied by the instantaneous value of voltage, with the product expressed as watts. However, an easier way to express the power is to average the power over one-half cycle. Thus, the average power in each half cycle of a purely resistive ac circuit is equal to the rms current value times the rms voltage value, which is the same as the formula for power in dc circuits, or $P = E \times I$ (Fig. 1-39*a*).

When ac circuits contain reactance in addition to resistance, the current and voltage waveforms have a phase difference between them. Whether the reactance is inductive

POWER

The product of the current i and the voltage e at any instant of time t gives the power p at that instant. Note that when both e and i are negative, their product, and thus p, is positive. Therefore power is being expended throughout the cycle.

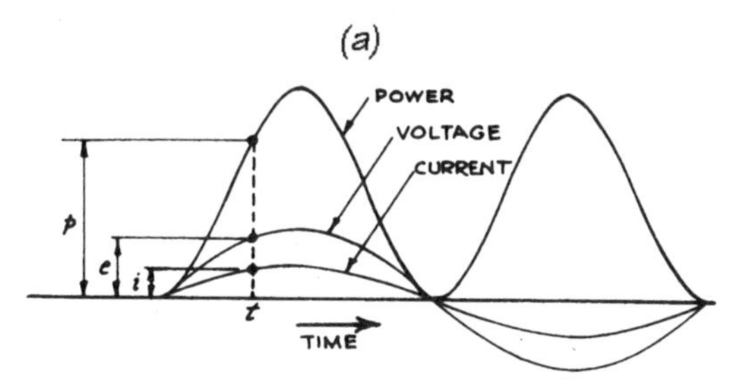

In terms of effective values,

P = EI

when current and voltage are in phase.

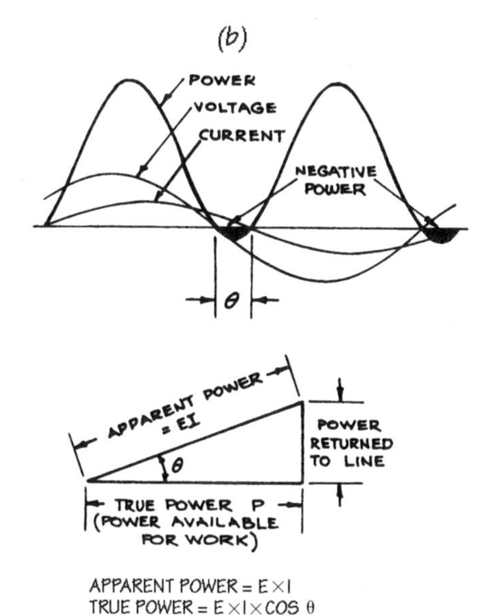

FIG. 1-39 The waveforms in (*a*) show that power P in a purely resistive ac circuit, where voltage and current are in phase with each other, is equal to the rms voltage times the rms current and is expressed in watts, or $P = E \times I$. When an ac circuit contains any reactance, true power is equal to the apparent power (voltage times current, or $E \times I$) times the cosine of the phase angle Θ, or true power $= E \times I \times \cos \Theta$. The value obtained by dividing true power by apparent power is known as the power factor (PF).

only, capacitive only, or the difference between inductive and capacitive reactance, the circuit has a net value of reactance. And the current will either lead or lag the voltage, depending on the nature of the reactance. Under such conditions, the power in the circuit is *not* simply the product of voltage times current.

Refer to the waveforms shown in Fig. 1-39*b*. When the current is out of phase with the voltage, there are regularly recurring periods when the voltage is positive and the current is negative, or vice versa, and the product of current and voltage is negative, indicating that the circuit is delivering power to the source. Because reactive loads essentially exchange power back and forth with the source, the net result is that they consume no power. In ac circuits containing resistance and reactance, the product of voltage and current is referred to as *apparent power* and is expressed in voltamperes, or VA. That is,

$$\text{Apparent power, VA} = E \times I$$

As shown by the vector representation in Fig. 1-39*b,* the actual useful power, or "true" power, in a resistive/reactive ac circuit is related to the apparent power multiplied by the cosine of the angle (Θ). That is,

$$\text{True power, W} = \text{apparent power, VA} \times \cos \Theta$$

$$= E \times I \times \cos \Theta$$

The angle Θ is the angle by which the current either leads or lags the voltage. The cosine of Θ is also referred to as the *power factor* and represented by the letters PF. The cosine of the angle will vary from 1 when the phase angle is 0° to 0 when the phase angle is 90° due to a purely reactive load. Therefore, when Θ is zero, we see that $P = E \times I \times 1$, or $P = E \times I$—the formula for in-phase operation. When Θ is 90°, $P = E \times I \times 0$, which indicates that no power is consumed. Substituting PF for cos Θ in the preceding formula, we see that

$$P = E \times I \times \text{PF} \qquad \text{and} \qquad \text{PF} = P/(E \times I)$$

Cos Θ, which *is* the PF value, either can be taken from trigonometric tables or can be calculated. As we know from trigonometry, the cosine is equal to the ratio of the adjacent side divided by the hypotenuse. If the true power and apparent power of the vector diagram shown in Fig. 1-39 were equal to, say 1200 W and 1440 VA, then cos Θ—and the power factor—would be

$$\cos \Theta = \text{PF} = \text{true power/apparent power}$$

$$= 1200 \text{ W}/1440 \text{ VA}$$

$$= 0.833, \text{ or } 83.3 \text{ percent}$$

Let's apply what we have said to an example. If a current of 7 A lags a voltage of 240 V by 30°, what are the power factor, true power, and apparent power? From trigonometric tables,

$$\cos 30° = 0.866$$

$$\text{PF} = \cos \Theta \qquad \text{which in this case is 0.866, or 87 percent}$$

True power = voltage × current × power factor (cos Θ):

$$\text{True power} = E \times I \times \text{PF}$$
$$= 240\ \text{V} \times 7\ \text{A} \times 0.866$$
$$= 1455\ \text{W}$$

Apparent power = voltage×current:

$$\text{Apparent power} = 240\ \text{V} \times 7\ \text{A}$$
$$= 1680\ \text{VA}$$

What would be the power factor of a 240-V, 9-A rated motor load that consumes 1728 W?

$$\text{True power} = \text{apparent power} \times \text{PF}$$

That can be rewritten as

$$\text{PF} = \text{true power/apparent power}$$
$$= 1728\ \text{W}/(240\ \text{V} \times 9\ \text{A})$$
$$= 1728\ \text{W}/2160\ \text{VA}$$
$$= 0.8, \text{ or } 80 \text{ percent}$$

Unity power factor is another expression for the condition where the PF is 1.0 or 100 percent, that is, where the current and voltage are in phase with each other. In all circuits that contain reactance as well as resistance, the PF will be something other than unity (Fig. 1-40).

Thus far, discussions of resistive/reactive ac circuits have covered single-phase series circuits. Such circuits are common in electrical systems for power, light, heat, signals, and communications. Two wires feeding an incandescent lamp constitute a resistive series circuit. Two wires supplying a motor or solenoid coil make up a series circuit combining both resistance and inductance. But practical electrical systems are always made up of both series *and* parallel circuits.

Note that there is a set of opposite relationships between series and parallel circuits. As has been seen, in a series circuit containing resistive and reactive components, there is *one value of current* through all loads. The voltage across any one load is equal to the total current multiplied by the load's impedance. And the total voltage is equal to the vector sum of the individual voltage drops. However, in a parallel circuit, *the voltage value* is constant across all loads. The current drawn by any one load is equal to the applied voltage divided by the impedance of that load. And the total current is equal to the vector sum of the individual load currents.

Consider first a voltage impressed across a resistance connected in parallel with an inductance. The current through the resistance will be in phase and equal to the voltage E divided by the load's resistance R, or $I_R = E/R$. The current through the inductance will be lagging the voltage wave by 90° and will be equal to the voltage E divided by

POWER FACTOR

In a series circuit containing resistance and inductance, the current lags the voltage by an angle θ. During part of each cycle, the voltage is negative while the current is positive; therefore their product is negative. This "negative" power is not available for work; it is actually power returned to the line. This may be represented by a right triangle, as shown by the drawing at the left. From trigonometry,

TRUE POWER P = EI COS θ,

where all values are effective values. The factor Cos θ is also known as the "power factor" and is equal to

$$\frac{\text{TRUE POWER}}{\text{APPARENT POWER}}$$

EXAMPLE: A motor rated 240 volts, 9 amps draws 1728 watts at full load. What is its power factor?

$$P = EI \cos \theta$$

$$1728 = (240)(9)(\cos \theta)$$

$$\cos \theta = \frac{1728}{(240)(9)} = \frac{1728}{2160}$$

POWER FACTOR = 0.8 OR 80%

FIG. 1-40 The relationship between true power and apparent power is a function of the cosine of the phase angle.

the inductive reactance X_L, or $I_L = E/X_L$. Those two current values must then be vectorially summed as follows:

$$I_T = \sqrt{I_R^2 + I_L^2}$$

where I_T = total current of the parallel loads
I_R = current drawn by resistance
I_L = current through inductance

If a resistance and a capacitance are connected in parallel, the current through the capacitance will lead the voltage waveform and the current through the resistance by 90°. But total current also is found by vectorially summing those individual currents as follows:

$$I_T = \sqrt{I_R^2 + I_C^2}$$

where I_C is the current drawn by the capacitance.

Notice that this is essentially the same method used to determine total *voltage* in a reactive *series* circuit with a 90° phase difference.

Another example of the parallel circuit consists of resistance, inductance, and capacitance in parallel (Fig. 1-41). In such a circuit, the current through each of the parallel loads is established. Then, because the current through the inductance is lagging and the current through the capacitance is leading, those two currents are 180° out of phase and oppose each other. The total current drawn will be the arithmetic difference between them. The difference is then combined with the resistive current I_R to obtain the total current as follows:

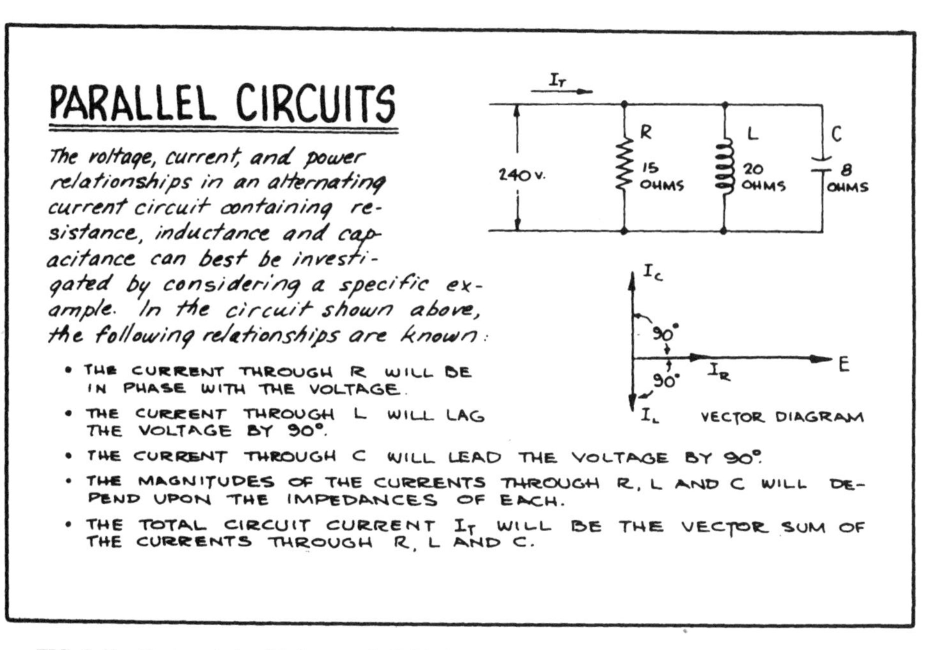

FIG. 1-41 Vector relationship between individual currents in parallel resistive/reactive circuits.

$$I_T = \sqrt{I_R^2 + (I_L - I_C)^2}$$

In parallel circuits which include reactance of either type, the total or resultant current drawn will not be in phase with the voltage. The resultant current will either lead or lag the voltage depending upon the type and amount of net reactance—inductive or capacitive—and the value of resistance. As a result, the power consumed in such a circuit will *not* be equal to the voltage times the current. Instead, just as was the case for series circuits, the power is found from the formula

$$P = E \times I \times \cos \Theta$$

where P = true power, W
E = circuit voltage
I = total or resultant current
Θ = phase angle of the circuit

The cos Θ, which is the power factor of the circuit, is equal to the current through the resistive load divided by the resultant or total current, or $\cos \Theta = \text{PF} = I_R/I_T$ (see Fig. 1-41).

Apply the foregoing information and determine total current, power factor, true power, and apparent power (see Fig. 1-42).

First, find the current drawn by each individual load.

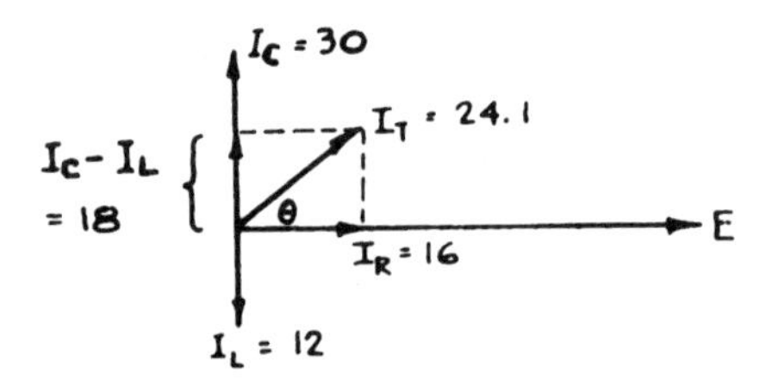

FIG. 1-42 Vector representation of individual currents through the circuit shown in Fig. 1-41.

Current through *R:*

$$I_R = E/R$$
$$= 240 \text{ V}/15\ \Omega$$
$$= 16 \text{ A}$$

Current through *L:*

$$I_L = E/X_L$$
$$= 240 \text{ V}/20\ \Omega$$
$$= 12 \text{ A}$$

Current through *C:*

$$I_C = E/X_C$$
$$= 240 \text{ V}/8\ \Omega$$
$$= 30 \text{ A}$$

Next, determine the total current:

$$I_T^2 = I_R^2 + (I_L - I_C)^2$$
$$I_T = \sqrt{T_R^2 + (T_L - T_C)^2}$$
$$= \sqrt{(16 \text{ A})^2 + (30 \text{ A} - 12 \text{ A})^2}$$
$$= \sqrt{(16 \text{ A})^2 + (18 \text{ A})^2}$$
$$= \sqrt{256 \text{ A} + 324 \text{ A}}$$
$$= \sqrt{580 \text{ A}}$$
$$= 24.1 \text{ A}$$

Find cos Θ or the power factor of the circuit:

$$\text{PF} = I_R/I_T$$
$$= 16\ \text{A}/24.1\ \text{A}$$
$$= 0.664, \text{ or } 66.4 \text{ percent}$$

Establish the apparent power of the circuit:

$$\text{Apparent power} = E \times I_T$$
$$= 240\ \text{V} \times 24.1\ \text{A}$$
$$= 5784\ \text{VA}$$

And the true power equals the apparent power times the power factor:

$$\text{True power, W} = \text{apparent power, VA} \times \text{PF}$$
$$= E \times I \times \text{PF}$$
$$= 240\ \text{V} \times 24.1\ \text{A} \times 0.664$$
$$= 3841\ \text{W}$$

COMPLEX CIRCUITS

Up to this point, all single-phase ac circuits that have been considered were series or parallel connections of resistance, inductance, and/or capacitance. In the discussion of parallel circuits, idealistic conditions were assumed in which a pure resistance was connected in parallel with a pure inductance and/or a pure capacitance. Although such conditions are possible in practical applications, it is much more likely to find, say, a purely resistive load connected in parallel with a load that is actually an inductance and resistance in series with each other. Or, a capacitance might be connected in parallel with a series resistive-inductive load. Such circuits are actually combinations of series and parallel circuits.

Basic analysis of a complex circuit with a purely resistive load and series resistive-inductive load connected in parallel begins with determining current flow through the individual loads by using the known voltage. For example, consider the circuit shown in Fig. 1-43.

Current through the resistive load (the incandescent lamp) is found by using the basic formula $I = E/R_1$. If the resistance of that lamp were 50 Ω and the circuit voltage were 120 V, the current I_1 would be

$$I_1 = 120\ \text{V}/50\ \Omega$$
$$= 2.4\ \text{A}$$

Before the current I flowing through the series resistive-inductive load (the fan motor) is found using the basic formula $I = E/Z$, the inductive reactance X_L must be found from the formula $X_L = 2 \times \pi \times f \times L$, where $\pi = 3.14$, $f =$ the frequency in hertz, and $L =$ the inductance in henrys. Then the motor's impedance Z is found from $Z = \sqrt{R^2 + X_L^2}$. If the inductive reactance X_L is 30 Ω and the resistance is 40 Ω, the impedance Z is

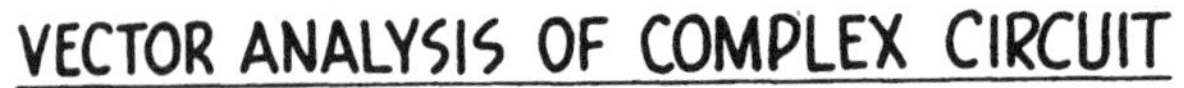

In lamp portion of the circuit shown, current through R_1 is in phase with the circuit voltage E; thus the power factor of this portion of the circuit is 1.0.

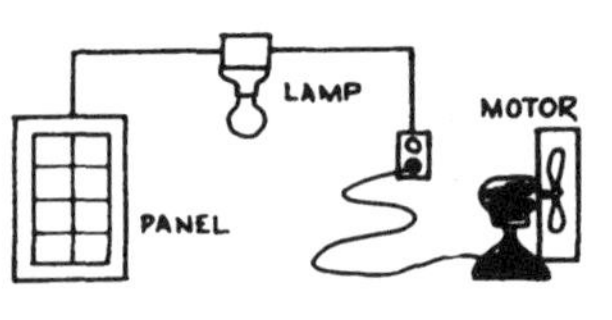

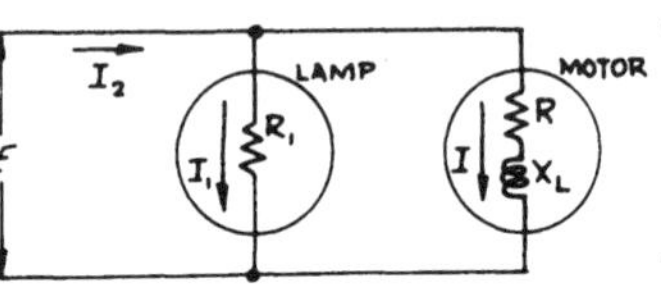

In motor portion of the circuit, current through R is in phase with E, and current through X_L lags E by 90°. Resultant current I lags E by an angle ϕ. The cosine of ϕ is the power factor of the motor.

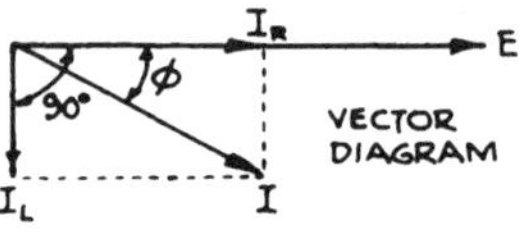

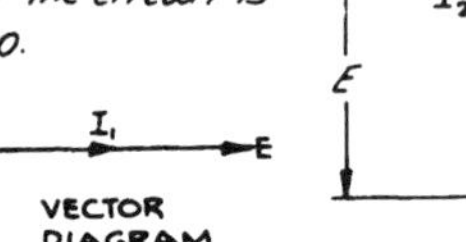

If the two current vector diagrams above are added together, the result is the vector diagram of the complete series-parallel circuit.

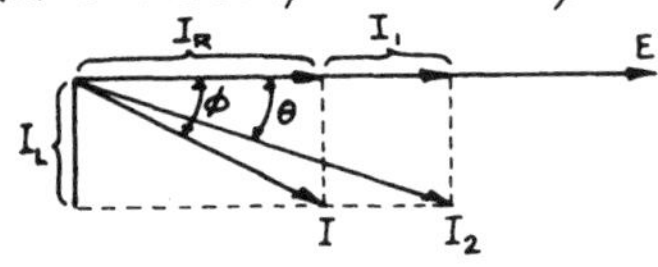

The total current I_2 is the resultant of I, I_L, and I_1. It will lag the voltage E by an angle θ. The cosine of θ is the power factor of the complete circuit.

APPLICABLE FORMULAS

$I_1 = \frac{E}{R_1}$ $\quad X_L = 2\pi f L$ $\quad I = \frac{E}{Z}$

$I_R = \frac{IR}{Z}$ $\quad Z = \sqrt{R^2 + X_L^2}$

$I_L = \frac{IX_L}{Z}$ $\quad I_2 = \sqrt{(I_1 + I_R)^2 + (I_L)^2}$

$\cos\phi = \frac{I_R}{I} = \frac{R}{Z}$ = MOTOR PF

$\cos\theta = \frac{I_1 + I_R}{I_2}$ = PF OF COMPLETE CIRCUIT

FIG. 1-43 Vector representation of a complex circuit.

$$
\begin{aligned}
Z &= \sqrt{R^2 + X_L{}^2 \Omega} \\
&= \sqrt{(40\ \Omega)^2 + (30\ \Omega)^2} \\
&= \sqrt{1600\ \Omega + 900\ \Omega} \\
&= \sqrt{2500\ \Omega} \\
&= 50\ \Omega
\end{aligned}
$$

Now, the motor current I can be found:

$$
\begin{aligned}
I &= E/Z \\
&= 120\ \text{V}/50\ \Omega \\
&= 2.4\ \text{A}
\end{aligned}
$$

The power factor, or cosine of the angle ϕ, for this motor is equal to the resistance R divided by the impedance Z, or

$$\begin{aligned} PF &= R/Z \\ &= 40\ \Omega/50\ \Omega \\ &= 0.8, \text{ or } 80 \text{ percent} \end{aligned}$$

As is shown by the vector diagram of the motor current in Fig. 1-43, the total motor current I is the resultant of the resistive current I_R and the inductive current I_L, which lags the resistive current by 90°. Additionally, the total current I_2 in the entire circuit is *not* the simple sum of I and I_1. The two vectors are combined to find the resultant current I_2 drawn by this parallel combination of a purely resistive and resistive-inductive load. Total current I_2 is found from the equation

$$I_2 = \sqrt{(I_1 + I_R)^2 + I_L^2}$$

But the values of I_R and I_L are not yet known. They can be found by substitution. That is, because the voltage drop across the resistance E_R is equal to the motor current I times the ohmic value of resistance R, or $E_R = I \times R$, the current drawn by the resistive portion of the motor load can be found by substituting the known values of motor current I and resistance R for the voltage drop across the resistance E_R in the equation $I_R = E_R/Z$:

$$\begin{aligned} I_R &= I \times R/Z \\ &= 2.4\ \text{A} \times 40\ \Omega/50\ \Omega \\ &= 96\ \text{V}/50\ \Omega \\ &= 1.92\ \text{A} \end{aligned}$$

The same approach is applied to the inductive portion of that series combination. That is,

$$\begin{aligned} I_L &= E_L/Z \\ &= I \times X_L/Z \\ &= 2.4\ \text{A} \times 30\ \Omega/50\ \Omega \\ &= 72\ \text{V}/50\ \Omega \\ &= 1.44\ \text{A} \end{aligned}$$

Substitute those values as follows:

$$\begin{aligned} I_2 &= \sqrt{(I_1 + I_R)^2 + I_L^2} \\ &= \sqrt{(2.4\ \text{A} + 1.92\ \text{A})^2 + (1.44\ \text{A})^2} \\ &= \sqrt{(4.32\ \text{A})^2 + (1.44\ \text{A})^2} \\ &= \sqrt{18.6624\ \text{A} + 2.0736\ \text{A}} \end{aligned}$$

$$= \sqrt{20.736\ \text{A}}$$

$$= 4.5537\ \text{A}$$

As previously indicated, the power factor PF of the motor load is equal to the ratio of resistance R to impedance Z, or PF = R/Z. The motor's power factor is also a function of the resistive current I_R in the motor, in this case 1.92 A, and total motor current I, or PF = I_R/I_T. Substituting the calculated values for I_R and I, we see that

$$\text{PF} = \frac{I_R}{I}$$

$$= \frac{1.92\ \text{A}}{2.4\ \text{A}}$$

$$= 0.8, \text{ or } 80 \text{ percent}$$

The power factor for the entire circuit is the ratio of all resistive currents to total current. In this case,

$$\text{PF} = (I_1 + I_R)/I_2$$

$$= (2.4\ \text{A} + 1.92\ \text{A})/4.5537\ \text{A}$$

$$= 4.32\ \text{A}/4.5537\ \text{A}$$

$$= 0.9486, \text{ or } 94.86 \text{ percent}$$

In another example of complex circuits, a power factor correction capacitor is installed in parallel with a series resistive-inductive load. When a load consisting of a resistance R and an inductance L, such as a motor, is connected across a circuit's conductors, the current will lag the voltage by some value, depending on the R-to-L ratio. That is, the power factor will be something less than unity, and the product of volts and amperes (voltamperes, or VA) will be greater than the wattage of the load (W). That's because part of the current drawn is magnetizing current and not useful current. Because a capacitor draws leading current, the right size capacitor can be used in parallel with the series resistive-inductive load to provide the magnetizing current. The reactive component of the series load and the purely reactive current of the capacitor will be 180° out of phase with each other and will effectively cancel. The net result is that the current supplied by the system up to the point where the capacitor is installed, is reduced. And where the power factor is corrected to unity, the voltamperes supplied will be equal to the load's wattage. Remember, even though the amount of current supplied by the system is reduced, the current flow through the series resistive-inductive load will still be the same. The difference is that the capacitor is now supplying the nonuseful reactive current, and the system is supplying the useful working current (Fig. 1-44).

If the motor load diagrammed in Fig. 1-44 drew 143 A at full load and had a power factor of 0.7, or 70 percent, the resistive current I_R can be found from

$$I_R = I \times \cos \Theta$$

$$= 143\ \text{A} \times 0.7$$

$$= 100\ \text{A}$$

POWER FACTOR CORRECTION

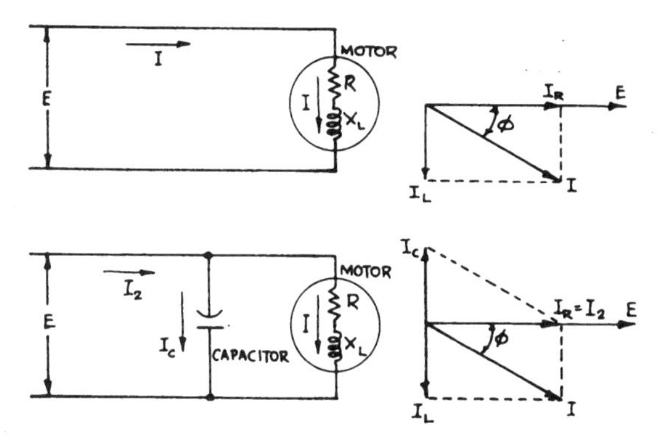

FIG. 1-44 Vector representation of power factor correction.

If a capacitor connected across the motor as shown in Fig. 1-44 raises the PF to unity (1.0 or 100 percent), the current furnished by the system I_2 will equal the resistive current I_R. If the power factor is corrected to other than unity, the current supplied by the system I_2 will be

$$I_2 = \sqrt{I_R^2 + (I_L - I_C)^2}$$

3-PHASE SYSTEMS

Although single-phase ac circuits are widely used and are involved in just about every electrical system, electrical generation and distribution systems are most commonly 3-phase systems.

Three-phase power is derived from a generator with three separate windings. These windings are spaced around a rotating cylinder—the generator rotor—and each winding occupies one-third of the rotor's 360° circumference. This means that each winding covers 120°. Thus, any one winding is physically displaced by 120° from either of the other two windings, and the sinusoidal voltage waveforms generated are displaced one from the other by 120° (Fig. 1-45). In other words, they are 120° out of phase with each other. (Each phase is designated *A, B,* or *C.)*

There are two basic connection arrangements used for 3-phase systems: the wye-connected configuration and the delta-connected configuration. Voltage sources—generators and transformers—and load devices may be connected in either a wye or delta configuration.

The wye connection is shown in Fig. 1-46. As can be seen, one end of each winding or coil is connected at a common or, more correctly, neutral point. The net result is that two voltage levels are available; that is, there is one value of voltage between any phase and the neutral point and another value of voltage between any two phases. The relationship between these two voltage levels is expressed mathematically as

$$E_1 = E_p \times \sqrt{3}$$

where E_1 = phase-to-phase voltage and E_p = phase-to-neutral voltage.

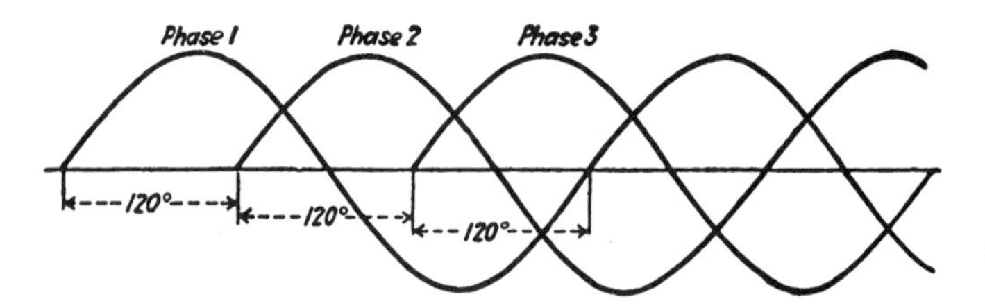

FIG. 1-45 Representation of an oscilloscope trace shows that individual voltage waveforms in 3-phase systems are displaced by 120°.

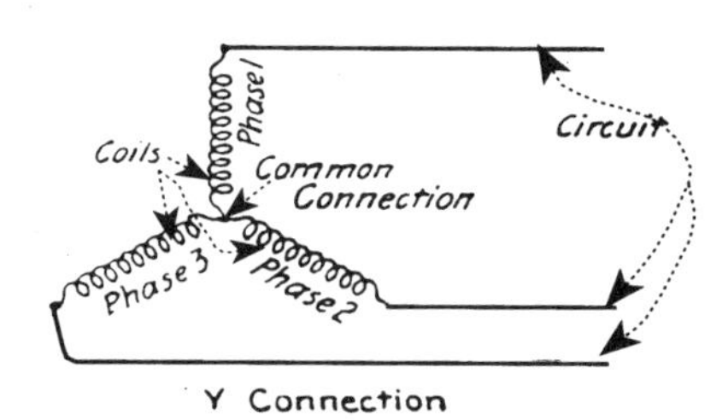

FIG. 1-46 The wye-connected configuration takes its name from the vector representation of each coil's voltage. When all coils are connected at a common point, they take on the appearance of the letter Y.

In a wye-connected source, the coil or winding voltage is the phase-to-neutral voltage. If, for example, each winding in a wye-connected generator or transformer produced 120 V, that voltage value would be measured between any phase conductor and the neutral. But the phase-to-phase voltage would be equal to 120 V times the square root of 3 (1.73), or 208 V. This latter voltage is the result of vectorially combining the voltages of two phases that are 120° out of phase with each other.

Consider the following: If a wye-connected source has a phase-to-phase voltage of 13,840 V, what is the phase-to-neutral voltage?

$$E_1 = E_p \times \sqrt{3} \quad \text{or} \quad E_1/1.73 = E_p$$

$$13{,}840\ \text{V}/1.73 = E_p$$

$$8000\ \text{V} = E_p$$

The phase winding, or phase coil, current I_p in a wye-connected source is equal to the amount of current flowing in the connected conductor I_l. Thus, for wye-connected system, $I_p = I_l$. The most common wye-connected secondary distribution and utilization systems are the 3-phase, 4-wire, 208 Y/120 V and 480 Y/277 V rated systems.

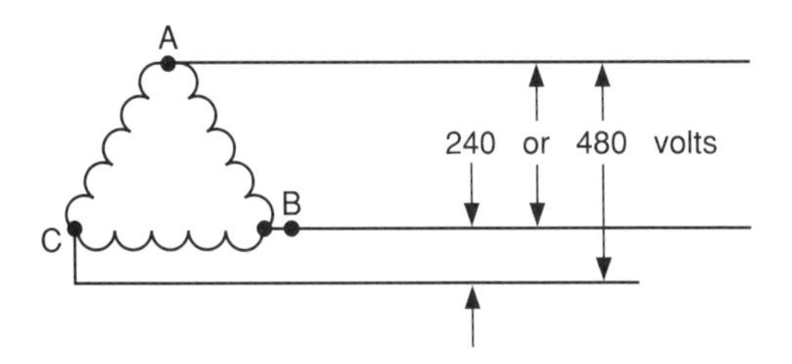

FIG. 1-47 If the vectors representing the magnitude and direction of the voltage coils are connected end to end, they form a triangle, which is the Greek letter delta. This is why sources connected in this manner are called *delta*-connected.

When a source's windings are connected end to end, it is said to be *delta-connected* (Fig. 1-47). Although the windings or coils are still 120° displaced from each other, the coil or winding voltage E_p is equal to the phase-to-phase line voltage E_l, or

$$E_p = E_l$$

In the wye-connected system, there are two voltage levels. But in the delta-connected configuration there are two current values: the line current I_l and the coil current I_p. These currents are related to each other in a manner similar to the voltages in a wye-connected source. That is, the line current I_l is equal to the coil current I_p times the square root of 3 (1.73), or

$$I_l = I_p \times 1.73$$

where I_l = current in the connected conductor and I_p = current in the source coil current.

The voltage and current relationship in 3-phase sources is summarized in Fig. 1-48.

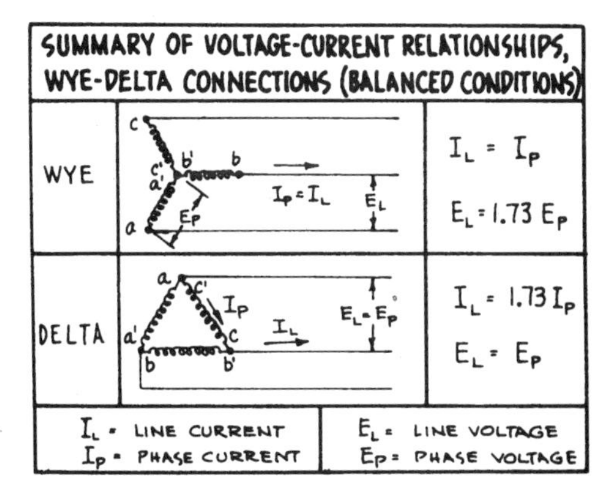

FIG. 1-48 Summary of the relationship between voltage and current in 3-phase sources.

3-PHASE POWER

The relationships between voltage and current in 3-phase systems slightly alter calculation of total power for such systems. For example, the power P_p, in watts, delivered by each phase winding in a wye-connected source operating at unity power factor is equal to the phase winding current I_p times the phase winding voltage E_p, or

$$P_p = E_p \times I_p$$

The total power P_t delivered by all three windings is equal to 3 times the phase winding voltage E_p times the phase winding current I_p, or

$$P_t = 3 \times E_p \times I_p$$

Because the phase winding current I_p is equal to the line current I_l and the phase winding voltage E_p is equal to the line-to-line voltage E_l divided by the square root of 3, the total power rating P_t of a wye-connected source can be expressed as

$$P_t = 3 \times E_l/\sqrt{3} \times I_l$$

or

$$P_t = \sqrt{3} \times E_l \times I_l$$

If the power factor of the 3-phase load is something other than unity, then the angle by which the current leads or lags the voltage must be considered. That is, because the power factor is the cosine of the angle between phase current and phase voltage, the formula for power in watts becomes

$$P_t = \sqrt{3} \times E_l \times I_l \times \text{PF}$$

For other than unity power factor, the formula $P_t = \sqrt{3} \times E_l \times I_l$ gives the voltampere rating of sources and loads. Division of that value by 1000 will give the kilovoltampere (kVA) rating of the load or system.

In a delta-connected system, the same formulas are used to obtain the true power (in watts) and apparent power (in kilovoltamperes) rating of sources and loads. Basically stated, the power of a balanced 3-phase delta system is equal to

$$P_t = 3 \times E_p \times I_p \times \text{PF}$$

But since phase winding voltage E_p is equal to line-to-line voltage E_l and phase winding current I_p is equal to the line current I_l divided by the square root of 3, the formula can be rewritten as

$$P_t = 3 \times E_l \times I_l/\sqrt{3} \times \text{PF}$$

or

$$P_t = \sqrt{3} \times E_l \times I_l \times \text{PF}$$

The foregoing assumes that all the phase windings of the source have the same current rating and the same voltage rating. The same formulas can be used to determine the power consumed by a 3-phase load of balanced current and voltage rating.

For example, how many kilowatts, voltamperes, and kilovoltamperes are drawn from a 3-phase source when it is delivering 40 A at 240 V to a motor with a power factor of 86 percent?

Power in kilowatts:

$$\begin{aligned} \text{kW} &= \sqrt{3} \times E_l \times I_l \times \text{PF}/1000 \\ &= 1.73 \times 240 \text{ V} \times 40 \text{ A} \times 0.86/1000 \\ &= 14{,}283/1000 \\ &= 14.283 \text{ kW} \end{aligned}$$

Voltamperes:

$$\begin{aligned} \text{VA} &= \sqrt{3} \times E_l \times I_l \\ &= 1.73 \times 240 \text{ V} \times 40 \text{ A} \\ &= 16{,}608 \text{ VA} \end{aligned}$$

Kilovoltamperes:

$$\begin{aligned} \text{kVA} &= \text{VA}/1000 \\ &= 16{,}608 \text{ VA}/1000 \\ &= 16.608 \text{ kVA} \end{aligned}$$

What would be the secondary current of a 75-kVA, 3-phase, 4-wire transformer with a secondary rated at 208Y/120 V?

$$\text{kVA} = \sqrt{3} \times E_l \times I_l/1000$$

Substituting known values gives

$$75 \text{ kVA} = 1.73 \times 208 \text{ V} \times I_l/1000$$

Multiplying both sides of the equation by $1000/1.73 \times 208$ V gives us

$$\begin{aligned} 1000 \times 75 \text{ kVA}/(1.73 \times 208 \text{ V}) &= I_l \\ 75{,}000/359.84 &= I_l \\ 208 \text{ A} &= I_l \end{aligned}$$

NEUTRAL CURRENTS IN 3-PHASE, 4-WIRE, WYE-CONNECTED SYSTEMS

In common applications of today's electrical distribution systems, wye-connected sources generally supply both single-phase loads (phase-to-neutral connected equipment such as incandescent and fluorescent lighting, appliances, business equipment, single-phase motors) and 3-phase loads (3-phase motors and transformers). When 3-phase loads are fed from a wye-connected source, current flow in all three phase conductors is the same and no current is carried by the neutral; this is called a *balanced* load. But when a wye source supplies a number of single-phase loads, these loads may or may not represent balanced loading (that is, the same current in each phase conductor and no current in the neutral).

To determine the current carried by any conductor of a 3-phase, 4-wire, wye-connected system, each phase conductor can be treated as a separate single-phase circuit. But the neutral conductor current is the vector sum of the phase-to-neutral loads supplied by each of the phase conductors; that is—with careful attention to magnitude, direction, and scale—it is possible to establish the value of current that will flow in the neutral using vector analysis. However, it is worth noting that complete, accurate solution of neutral current using vector analysis is often cumbersome and subject to inaccuracies. Therefore, it is recommended that vector analysis be used only to provide a graphical representation of the phase relationships between the individual currents. An accurate solution can be quickly established using the actual current values and the angles between them in a basic trigonometric formula.

Another point that needs to be considered is the power factor of each phase; that is, when the loads are operating at other than unity power factor, a more detailed analysis is required. However, the basic calculations described here still apply after the vector quantity of each phase is appropriately adjusted according to the decimal value indicating the displacement from its position at unity power factor. The following discussions assume unity power factor (see Fig. 1-49).

As previously stated, the voltages of any 3-phase system are alternating with a displacement of 120° between any two phases. The currents in the phase conductors are also displaced by 120°. Therefore, these currents can be represented vectorially as three individual vectors with a common origin and equiangular displacement between them. By representing the individual currents in some scale, a resultant indicative of the neutral current can be obtained.

For instance, on a 208Y/120 V, 3-phase, 4-wire system, two fixtures might be connected between phase *A* and neutral, with five fixtures connected between phase *B* and neutral and four fixtures connected between phase *C* and neutral (Fig. 1-50). If all the fixtures have the same rating and each draws 2 A, the current in phase *A* will be 4 A, in phase *B* it will be 10 A, and in phase *C* it will be 8 A. These values are then plotted according to scale and angular displacement. As shown in Fig. 1-50, the resultant vector is equal to 5.29 A. Therefore, the neutral conductor in this case carries 5.29 A.

What if phases *A, B,* and *C* were each supplying 10 A of phase-to-neutral load? To determine the neutral current, the values of each phase current can be represented as shown in Fig. 1-51. The vector diagram in Fig. 1-51 shows a completed parallelogram

POWER IN 3-PHASE CIRCUITS

For both Y- and delta-connected systems:

$$P = \sqrt{3}\, E_L\, I_L\, PF$$

where P = TOTAL POWER (WATTS)
$\sqrt{3}$ = 1.732
E_L = LINE-TO-LINE VOLTAGE (VOLTS)
I_L = LINE CURRENT (AMPERES)
PF = POWER FACTOR (EXPRESSED AS DECIMAL)

NOTE: Omitting the power factor in the power formula gives volt-amperes:

$$\text{VOLT-AMPERES} = \sqrt{3}\, E_L\, I_L$$

FIG. 1-49 Summary of formulas for power and kilovoltampere ratings of 3-phase sources and loads.

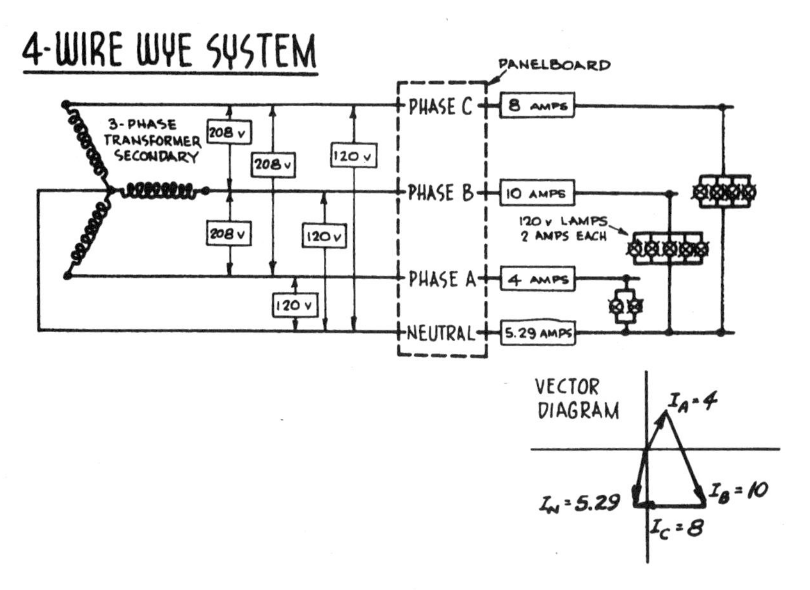

FIG. 1-50 Unbalanced loading on 3-phase, 4-wire circuits causes current to flow in the neutral conductor. The magnitude of this current is the vector sum of the 3-phase currents.

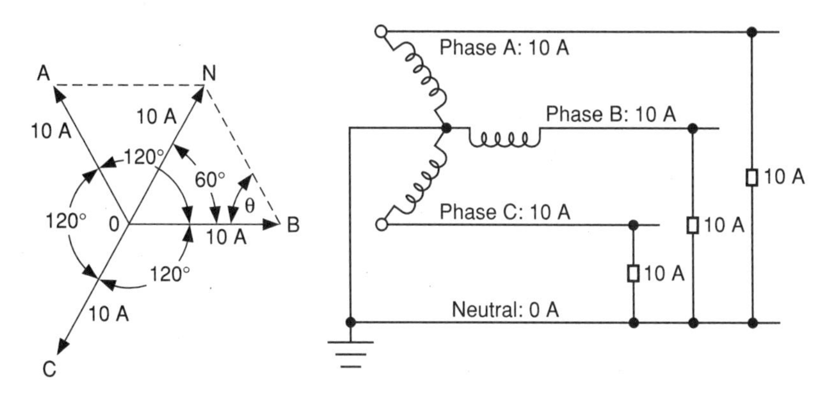

FIG. 1-51 Vector analysis of phase currents where equal amounts of phase-to-neutral loads are supplied by each phase in a 3-phase, 4-wire circuit.

on the *A* and *B* phase-current vectors (*OA* and *OB*). When they are measured, we see that the resultant *ON* of these two vectors is equal in length to *OA* and *OB*. Therefore, because *OA* is a scalar quantity representing 10 A, if the resultant *ON* is the same length, it is also equal to 10 A.

Now that the vector sum of phase *A* and phase *B* has been determined, the resultant *ON* is summed vectorially with the current drawn by phase *C* (vector *OC*) to establish the value of neutral current. As can be seen, the resultant vector *ON* is 180° out of phase with the phase *C* current vector. From the basic concept of vector analysis, it is known that the resultant of two diametrically opposed vectors is the difference between them. Therefore, because the vectors we are considering both have the same magnitude (10 A) and are 180° out of phase with each other, the vector sum is zero (10 A − 10 A = 0).

The preceding example illustrates a basic advantage associated with 3-phase, 4-wire circuits: When the current in all three phase conductors is equal (when the loading is balanced), the neutral current will be zero.

Determining the value of neutral current in a 3-phase, 4-wire circuit that supplies phase-to-neutral loads can also be calculated by using the known vector quantities, their angular relationships, and a basic trigonometric formula. That is, the square of the resultant *ON* is equal to the square of the vector *BN* connecting the phase-current and resultant vectors (*BN* will be equal to the current vector of the other phase-current vector selected for summing) plus the square of the phase-current vector *OB* minus 2 times the product of the phase-current vector times the connecting vector times the cosine of angle Θ. Applying this to the example shown in Fig. 1-51 and substituting the appropriate vector designations, we see that the equation for calculating the value of *ON* is as follows:

$$ON^2 = BN^2 + OB^2 - 2 \times (BN \times OB \times \cos \Theta)$$

To establish the value of angle Θ, refer to the triangle shown in the vector diagram in Fig. 1-51. Sides *OB* and *ON* are equal by construction; therefore, the triangle is an isosceles triangle, and base angles *ONB* and *BON* are equal. Additionally, angle *ONB* is equal to angle *AON* because they are alternate interior angles of two parallel lines. Also angles *AON* and *BON* are equal and total 120° by construction—remember that the two phases are 120° displaced. Therefore, each of the three angles in the triangle is equal to 60°, angle Θ is equal to 60°, and the cosine of 60° is 0.5 (Fig. 1-52). Now, the value of *ON* can be calculated:

$$\begin{aligned} ON^2 &= 10^2 + 10^2 - 2 \times (10 \times 10 \times 0.5) \\ &= 100 + 100 - 2 \times (50) \\ &= 100 + 100 - 100 \\ &= 100 \\ ON &= 10 \end{aligned}$$

In some instances, it may not be possible to determine the value of angle Θ through analysis of the vector quantities and angular relationships. In these cases, the known displacement between any two phase currents (120°) can be used in conjunction with another trigonometric formula to establish the resultant's magnitude and direction, which are then vectorially combined with the remaining phase-current vector to establish actual neutral loading.

Consider three equally loaded phase conductors as in Fig. 1-51. The angular displacement between phase *A* and phase *B* in a 3-phase, 4-wire system is 120°. This angular value can be used as angle Θ and combined with the known vector quantities (10 A on phase *A* and 10 A on phase *B*) in a trigonometric formula:

$$ON^2 = OA^2 + OB^2 + 2(OA \times OB \times \cos 120°)$$

A	cos A	A	cos A
0	1.000		
1	1.000	46	.695
2	.999	47	.682
3	.999	48	.669
4	.998	49	.656
5	.996	50	.643
6	.994	51	.629
7	.992	52	.616
8	.990	53	.602
9	.988	54	.588
10	.985	55	.574
11	.982	56	.559
12	.978	57	.545
13	.974	58	.530
14	.970	59	.515
15	.966	60	.500
16	.961	61	.485
17	.956	62	.470
18	.951	63	.454
19	.946	64	.438
20	.940	65	.423
21	.934	66	.407
22	.927	67	.391
23	.920	68	.375
24	.914	69	.358
25	.906	70	.342
26	.899	71	.326
27	.891	72	.309
28	.883	73	.292
29	.875	74	.276
30	.866	75	.259
31	.857	76	.242
32	.848	77	.225
33	.839	78	.208
34	.829	79	.191
35	.819	80	.174
36	.809	81	.156
37	.799	82	.139
38	.788	83	.122
39	.777	84	.104
40	.766	85	.087
41	.755	86	.070
42	.743	87	.052
43	.731	88	.035
44	.719	89	.018
45	.707	90	.000

FIG. 1-52 Table of cosines.

Note that instead of *subtracting* 2 times the product of *OA, OB,* and cos Θ, this value is *added.* Additionally, it can be seen that the table of cosines (Fig. 1-52) does not give a value for 120°. Because the angular displacement is greater than 90°, cos Θ will be a negative number equal to the cosine of the angle established by subtracting the angle of displacement from 90° and then moving this number of degrees from 90° toward zero in the table of cosines. For our example,

$$90^\circ - 120^\circ = -30^\circ$$

Then, moving 30° from 90° toward zero, we arrive at 60°. Therefore,

$$\cos 120^\circ = -\cos 60^\circ = -0.5$$

Substitute −0.5 for cos 120° and 10 A for both *OA* and *OB* in the formula above and solve, as follows:

$$ON^2 = 10^2 + 10^2 + 2[10 \times 10 \times (-0.5)]$$

$$= 100 + 100 + (-100)$$

$$= 100$$

$$ON = 10 \text{ A}$$

As was the case in the first calculation of neutral current where each phase was identically loaded, *ON* is 180° out of phase with *OC*; therefore, the neutral current is zero.

Let's look at another example, one where the loads are not balanced. Consider a 3-phase, 4-wire circuit supplying phase-to-neutral loads where phase *A* carries 5 A, phase *B* carries 15 A, and phase *C* carries 10 A. What would be the value of current carried by the neutral conductor?

First, diagram vectorially the magnitude and direction of the current on each of the phases (Fig. 1-53). Next, determine the value of *ON* from the following:

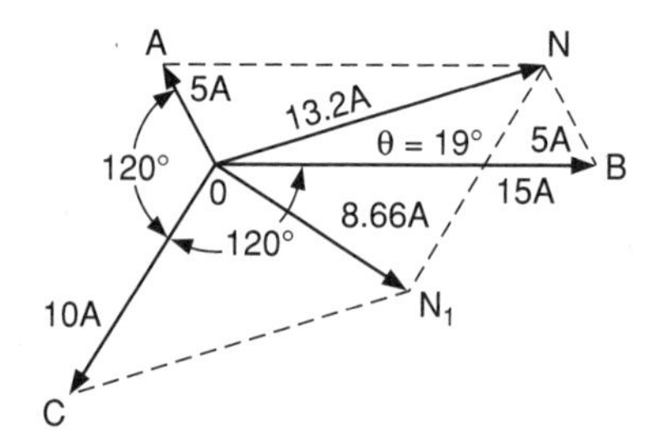

FIG. 1-53 Vector analysis of the neutral current in an unbalanced 3-phase, 4-wire circuit supplying phase-to-neutral loads of 5, 15, and 10 A on phases *A, B,* and *C,* respectively.

$$ON^2 = OA^2 + OB^2 + 2(OA \times OB \times \cos 120°)$$

$$= 5^2 + 15^2 + 2[5 \times 15 \times (-0.5)]$$

$$= 5 + 225 + 2[75 \times (-0.5)]$$

$$= 250 + 2(-37.5)$$

$$= 50 + (-75)$$

$$= 175$$

$$ON = 13.2 \text{ A}$$

Next, find angle Θ as follows:

$$NB^2 = OB^2 + ON^2 - 2(OB \times ON \times \cos \Theta)$$

$$5^2 = 15^2 + 13.2^2 - 2(15 \times 13.2 \times \cos \Theta)$$

$$25 = 225 + 175 - 2(198 \times \cos \Theta)$$

$$25 - 400 = 400 - 2(198 \times \cos \Theta) - 400$$

$$-375/(-2 \times 198) = -2 \times 198 \times \cos \Theta/(-2 \times 198)$$

$$(-375)/(-396) = \cos \Theta$$

$$0.946 = \cos \Theta$$

From the table of cosines (Fig. 1-52), we see that 0.946 is the cosine of 19°. Therefore, angle Θ is equal to 19°. Because we know that *OB* is displaced from *OC* by 120°, and because we know that *ON* is displaced from *OB* by 19°, it can be said that *ON* is displaced from *OC* by 139° (120 + 19). The value of ON_1 is determined from the following:

$$ON_1^2 = OC^2 + ON^2 + 2(OC \times ON \times \cos 139°)$$

$$= 10^2 + 13.2^2 + 2[10 \times 13.2 \times (-0.755)]$$

$$= 100 + 175 + 2(-99.6)$$

$$= 275 + (-199.32)$$

$$= 75.6, \text{ or } 75$$

$$ON_1 = 8.66 \text{ A}$$

CHAPTER 2
WIRING METHODS

EQUIPMENT AND CONDUCTOR TEMPERATURE LIMITATIONS

The primary enemy to virtually all electrical equipment and conductors is the I^2R heat generated during use. Although this heating of equipment and conductors is normal and predictable, if the amount of heating is not limited, it can adversely affect the insulating material used and can result in premature failure of the equipment or conductors. As a result, when equipment or conductors are evaluated by UL, or other third-party testing laboratories, for listing, the insulation system used is also examined to determine and establish the maximum temperature at which the equipment or conductor can be safely operated without causing damage to the insulation.

Note that the magnitude of overload influences only the amount of time before failure occurs. That is, a slight overload will take a long time to cause insulation breakdown and failure, and a great overload will cause insulation breakdown and failure more quickly. But in either case, the equipment and conductors will not have as long a service life as would normally be expected.

In addition to a shortened service life, failure to observe the temperature limitations of equipment and conductors is a violation of the NEC. In Secs. 110-3(b) and 110-14(c)(1), (2), and (3) of the 1996 edition, the NEC requires all listed and labeled products be used in accordance with their listing or labeling instructions. This requirement has been repeatedly interpreted—by inspectors in the field, as well as judges in the courts—to mandate strict adherence to all instructions given by the listing laboratory. And among the many thousands of installation details given in the UL's "General Information Directory" or "White Book," as it is often called, there are general and specific temperature limitations that must be observed when using listed equipment and conductors.

General Instructions

On pages 12, 13, 14, and 15 of the 1996 edition of the UL's White Book and in Sec. 110-14(c)(1), (2), and (3) of the 1996 edition of the NEC, a variety of general instructions are given. Most important to this discussion are the instructions given for "Appliances and Utilization Equipment Terminations" and "Distribution and Control Equipment Terminations." Unless a specific equipment listing instruction states otherwise, these two provisions establish maximum temperature ratings for terminations in *all* equipment that is to be interconnected using Code-recognized conductors and cables to form the building's or facility's electrical system.

Although these two provisions address the same concern—temperature limitations of terminations—each defines the limitation in a slightly different way. However, the net effect is nearly the same.

Appliances and Utilization Equipment

Under "Appliances and Utilization Equipment," the White Book states that, unless otherwise marked, all field terminations in such equipment are based on the use of 60°C insulated conductors for circuits rated 100 A or less. And for circuits rated over 100 A, the termination provisions are based on the use of 75°C insulated conductors. These statements are intended to indicate that, depending on the rating of the supply circuit, the terminations within the equipment provided by the manufacturer for connection to branch circuit conductors are evaluated using either a 60°C or 75°C rated conductor at their 60 or 75°C ampacity, as given in Table 310-16. Additionally, UL is indicating that when the appliance or utilization equipment is installed, unless the equipment is marked otherwise, the ampacity of the conductors for supply circuits rated 100 A or less is to be based on the ampacity given for a conductor of the same size with 60°C rated insulation, as shown in Table 310-16. And for conductors of supply circuits rated over 100 A, the ampacity is to be based on the ampacity given for the 75°C rated insulations, as shown in Table 310-16 for a given conductor size.

Notice that the UL statement relates this requirement to the circuit rating and not the nameplate rating. Because UL takes requirements of the NEC into consideration when evaluating appliances or utilization equipment, if a piece of equipment is to be supplied by a circuit that is required by an NEC rule to be rated at a greater amperage than the equipment's nameplate amperage rating, the equipment will be evaluated accordingly. That is, when a piece of equipment has a nameplate rating that is 100 A or less and the NEC would require a circuit rated over 100 A to supply the equipment, the equipment is evaluated using 75°C insulated conductors at their full 75°C ampacity. And, therefore, when installed, such equipment is permitted to be supplied by 75°C insulated conductors at the full 75°C ampacity.

An additional statement makes clear that higher-temperature rated insulations may be used provided the conductor ampacity is based on the 60°C value for circuits rated up to 100 A and on the 75°C value for circuits over 100 A. That is, the amount of current carried by a conductor with a higher-temperature rated insulation must not be greater than the value of current shown in Table 310-16 for the same size conductor with either 60°C or 75°C rated insulation, depending on the rating of the supply circuit.

Application of these requirements would show that when the NEC requires a circuit rated at 100 A to supply an appliance or piece of utilization equipment, and the appliance or utilization equipment is not marked with a specific temperature rating or temperature rating and conductor size, conductors connected to such equipment would have to be at least a No. 1 AWG, copper with 60°C rated insulation. Additionally, conductors with either 75°C or 90°C rated insulations may also be used; *but* where 75°C rated (for example, THW) or 90°C rated (for example, THHN) insulation is used, the ampacity must be no more than that given for a 60°C rated insulated wire. Therefore, in this case, the THW- or THHN-insulated conductors must carry not more than 110 A, after any derating that may be required because of an elevated ambient and/or number of conductors, as may be necessary.

If conductors are to be connected to an appliance or utilization equipment where the supply circuit would be required to be rated at more than 100 A, and there is no specific temperature rating or temperature rating and conductor size marked on the equipment, the UL information tells us that conductors with 75°C rated insulation may be used at the full 75°C ampacity shown in Table 310-16. Conductors with 90°C rated insulations

(RHH, THHN, XHHW, etc.) may be used, *but* the ampacity of these conductors must be taken to be that of a 75°C insulated conductor of the same AWG or kcmil size, after any derating for ambient temperature or number of conductors is applied. Additionally, if the equipment is not marked with a specific temperature rating or a temperature rating and conductor size, then it is also permissible to use a conductor with 60°C rated insulation, provided the conductor selected has sufficient ampacity as described within applicable Code rules for the load being supplied.

Guidance is also provided for equipment whose termination provisions are based on the use of conductors with higher-temperature rated insulation. When an appliance or piece of utilization equipment is intended for use with conductors of higher-temperature ratings than described above, such equipment will be marked to show the temperature rating and size of conductor necessary, or just a temperature rating.

If the equipment is marked to show only a higher-temperature rating, then the insulation of the conductor used must be rated as indicated, *but* the conductor ampacity must be not more than the value given in Table 310-16 for the same size conductor with 60°C rated insulation for applications up to 100 A or with 75°C rated insulation for applications over 100 A. Additionally, any derating required may be applied against the ampacity value given for the higher-temperature rated insulation in Table 310-16, but the final value of ampacity must not exceed that shown for a 60 or 75°C insulated conductor of the same size, where the equipment is supplied by a circuit rated 100 A or less, or over 100 A, respectively.

If the equipment is marked with both a temperature rating and a conductor size, only a conductor of the indicated size and insulation temperature rating may be used.

Distribution and Control Equipment

As indicated above, the UL temperature limitations for distribution and control equipment are defined on a different basis from those for appliances and utilization equipment. Instead of considering the ampere rating of the supply circuit conductors, for distribution and control equipment, temperature limitations of the terminations are based on the size of the circuit conductors used, but the net effect is almost identical.

The general guideline for terminations within distribution and control equipment states that, unless marked otherwise, when the terminations are intended for use with conductors sized from No. 14 to No. 1 AWG, the ampacity must be based on the ampacity given in Table 310-16 for a conductor of the same size with 60°C rated insulation. And when terminating devices are intended for use with conductors Nos. 1/0 AWG and larger, the ampacity of the conductor used must be based on the 75°C value for a conductor of the same size, as shown in Table 310-16. With this differentiation in mind, it can be seen in Table 310-16 that the ampere value at the point where the UL changes to the higher-temperature rated conductors when evaluating such equipment is about 100 A, which is the value used as the dividing line between 60° and 75°C rated conductors supplying appliance and utilization equipment.

The White Book also states that a marking of "75C only" or "60/75C" on a switch, circuit breaker (CB), or lug does not necessarily mean that 75°C insulated conductors may be used at their full 75°C ampacity. In addition to such marking on the switch, CB, or lug, the enclosure or equipment in which the switch, breaker, or lug is installed must be marked (that is, "75C only" or "60/75C"), independently of the marking on the switch, CB, or lug.

While this is the general requirement, an additional statement clarifies that when a switch or CB is to be used by itself, such as in a separate enclosure, if the switch or breaker is marked for use with 75°C insulated conductors, then the enclosure is not required to be independently marked, and it is permissible to use such conductors at

Listed equipment must be used in accordance with all instructions given by the third-party testing laboratory. Even though this lug is marked AL9CU, which indicates it is for use with either copper or aluminum conductors at their full 90°C ampacity (arrow), a general instruction given in the UL's "General Information Directory," the so-called White Book, states that such a lug may be used with 90°C rated conductors at their full 90°C ampacity *only* if the enclosure or equipment is also marked to indicate that 90°C rated conductors may be used at their full 90°C ampacity.

Distribution and control equipment intended for use with conductors in sizes from No. 14 to No. 1 AWG are to be used with 60°C rated conductors at their 60°C ampacities, unless otherwise marked. The label on this CB states: "60/75°C WIRE." If this CB is mounted in an individual enclosure, or if it is mounted in a panelboard that is also marked for use with 75°C rated wires, then 75°C rated conductors may be used at the full 75°C ampacity.

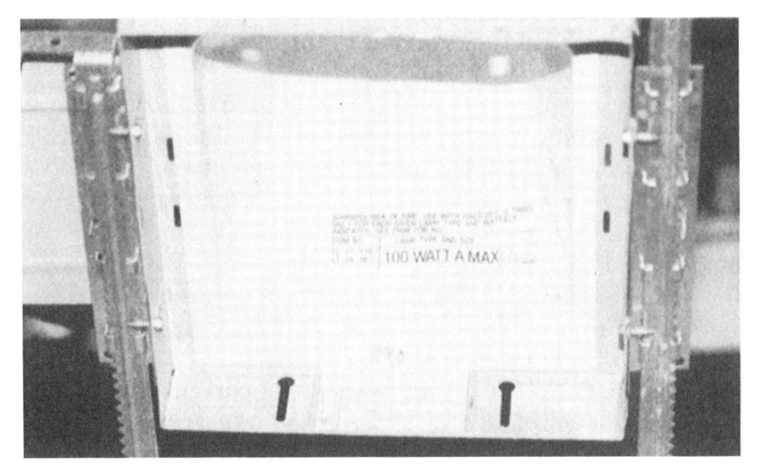

Appliances and utilization equipment supplied by circuits rated 100 A or less are evaluated using 60°C rated conductors, unless the appliance or equipment is otherwise marked. Additional restriction will also be given either on labels or in the manufacturer's installation instructions. The label inside this fixture warns that the fixture is suitable for use with an incandescent lamp rated at not more than 100 W. Use of any lamp that is rated over 100 W would be in violation of the manufacturer's labeling instruction, and would therefore be a violation of Sec. 110-3(b) in the National Electric Code.

their full 75°C ampacity. *This permission applies to switches and CBs installed in individual enclosures, only!* For lugs, only when the enclosure or equipment is also marked is it permissible to use 75°C insulated conductors at their full 75°C.

The UL general listing instruction for distribution and control equipment also states that higher-temperature rated conductors may be used, provided that the conductor's ampacity is not greater than that given in the 60°C column for size No. 14 to No. 1 AWG, and the 75°C column for No. 1/0 AWG and larger, as given in Table 310-16.

Circuit Breakers

Circuit breakers are categorized in a different manner. As indicated on page 24 of the 1996 UL White Book, CBs rated 125 A or less are assumed to have 60°C temperature rating unless they are otherwise marked. And those rated over 125 A are assumed to be suited for 75°C insulated conductors at their full 75°C ampacity, as given in Table 310-16.

As we have seen, for distribution and control equipment, the determination of the maximum temperature rating of the connected conductors is based on the size of conductor to which it is intended to be connected. And for appliances and utilization equipment, the maximum temperature rating is based on the ampere rating of the circuit. However, although a CB is a piece of distribution and control equipment, its suitability for use at greater than 60°C is based on the *ampere* value and *not* the size of the conductor that is intended to be connected, which is the case with other distribution and control equipment. Instead, the CB is classified according to its ampere rating, in a similar manner to appliances and utilization equipment. And because this listing data appear at the beginning of the product category for molded-case CBs, they take precedence over the general rules put forth at the beginning of the White Book. That is what

UL means by the statement "except as noted in the information at the beginning of some product categories...," which is given in the general information for both appliances and utilization equipment and distribution and control equipment. Therefore, where the information at the beginning of a specific product category permits or requires different temperature ratings than those required by the general information at the beginning of the White Book, the permission or requirement given in the specific product category should be followed.

Conductors

As we have seen, these UL requirements are all based on the use of conductors with an ampacity not more than that shown for certain insulation ratings as given in Table 310-16. But what do these ratings actually mean and how are they related to the overheating of equipment terminations?

Table 310-16 in the NEC gives the Code-recognized ampacities for a variety of copper, aluminum, and copper-clad aluminum conductors when the conditions of application are as described at the top of this table—that is, where there are not more than three current-carrying conductors in raceway or cable and the ambient temperature is not more than 86°F (30°C). When these conditions exist, the ampacity value given in Table 310-16 is the amount of current in amperes that a conductor of a given size, material, and insulation can carry continuously.

As can be seen in Table 310-16, the amount of current that a conductor of a given material and size is permitted to carry varies with the type of insulation used. For example, where there are not more than three current-carrying conductors in a raceway or cable and the ambient temperature is not over 86°F (30°C), the ampacity of a No. 1 AWG copper conductor with TW insulation is 110 A. Under the same conditions of application, a No. 1 AWG copper conductor with THW insulation has an ampacity of

In addition to temperature ratings, equipment labels and instructions will give the designer and installer a variety of requirements regarding the application and installation of the listed equipment. Always take the time to review these instructions and labels to assure that the equipment is being used in accordance with the manufacturer's listing instructions.

As is the case with many conductors manufactured today, this cable is marked THHN or THWN. This marking is intended to indicate that when used in a dry location, the ampacity of this conductor may be taken to be that of a THHN-insulated (90°C) conductor, as given in Table 310-16. But when it is used in a wet location, as defined by the NEC, the ampacity of this conductor must be based on that given in the 75°C column of Table 310-16, which gives the ampacity of THWN-insulated conductors. Regardless of whether this conductor is used in a dry or wet location, the temperature rating of the equipment terminations must be carefully correlated with the conductor ampacity to prevent overheating and premature failure.

130 A. And a No. 1 AWG copper conductor with THHN insulation has an ampacity of 150 A under the same conditions of use.

The reason for the difference in ampacity for the different insulation types has to do with the maximum temperature that the various insulations can withstand. With TW-insulated wire, the maximum acceptable operating temperature is 60°C. For THW insulation, the maximum operating temperature is 75°C. And THHN has a maximum operating temperature of 90°C ampacity.

When the conditions of use are as described in the heading to Table 310-16, the values given in Table 310-16 represent that amount of current that will cause the copper, where it is in contact with the insulation, to reach and stabilize at the temperature indicated for the particular insulation type. That is, for a No. 1 AWG copper conductor with type TW insulation, where there are not more than three current-carrying conductors in a raceway and the ambient temperature is not over 86°F, 110 A is the amount of current that will cause the copper to reach and stabilize at a temperature of 60°C. Any additional current will cause the copper to "run hotter" than 60°C and will contribute to insulation breakdown. The same is true for the 75 and the 90°C rated insulations.

When conductors are used to interconnect the various pieces of distribution and control equipment, as well as appliance and utilization equipment, the UL rules discussed above warn that the terminations within such equipment were evaluated using conductors that were operated at a specific temperature. Because these terminations were tested at a specific temperature, proper operation and normal service life can be expected only if the equipment is used as it was tested. And failure to do so is to violate the equipment's listing instruction.

For example, with a panelboard whose main lugs or CB is intended for use with No. 1 AWG conductors, unless the terminating device and enclosure are otherwise marked, the No. 1 AWG conductors must have an ampacity no greater than would be permitted in the 60°C column of Table 310-16 for a No. 1 AWG conductor. Use of 75 or 90°C insulated wire is permitted, but the ampacity or amount of current, which is proportional to the amount of heat generated, must not be greater than the ampere value given for conductors of the same size with 60°C rated insulations.

In addition to violating the listing instructions and, therefore, Sec. 110-3(b), use of 75 or 90°C rated wire at its full 75 or 90°C ampacity will cause the terminations to run hot. This is completely undesirable, as it can contribute to equipment failure and such failure may be violent and destructive. It must also be remembered that if the terminations within the panelboard were evaluated at a 60°C maximum, the amount of wiring space in the enclosure, which is related to the enclosure's ability to dissipate heat, was also evaluated on the basis of a 60°C maximum. Higher-rated conductors used at their higher ampacities can also become overloaded because the enclosure, which was evaluated using conductors operating at 60°C, may not have sufficient capacity to permit proper dissipation of the additional heat caused by the additional current. This is an important point. Although conductors with higher-rated insulations may still have any derating for number of conductors and/or elevated ambient applied against their higher ampere ratings, as given in Table 310-16, if the final value of ampacity is greater than that shown for the same size conductor in either the 60 or 75°C column, depending on the conductor size, this "additional" current will generate additional heat. Remember, this additional heating, no matter how slight, can eventually cause the conductor insulating material to break down and can result in premature conductor failure.

In addition to these general instructions regarding temperature limitations, there are other limitations described in the general instructions, as well as many specific listing instructions dealing with the installation and use of listed equipment. Take the time to read and understand every such instruction before you design systems with, or install listed equipment and conductors to ensure strict adherence to, all listing instructions given by the third-party testing laboratory.

AMPACITY AND OVERCURRENT PROTECTION FOR NONCONTINUOUS LOADS

The NEC approach to establishing the ampacity of any conductor is aimed at designating the current that will cause the conductor insulation to reach and stabilize at its thermal limit, the temperature beyond which it will be damaged. The Fine Print Note (FPN) following Sec. 240-1 verifies this assessment. The wording of the first sentence here, which has been in the Code and virtually unchanged for over 38 years, reads:

> Overcurrent protection for conductors is provided to open the circuit if the current reaches a value that will cause dangerous temperature in conductors or conductor insulation.

As described in the definition of *ampacity* given in article 100, ampacity is the amount of current, in amperes, that the conductor can carry continuously under specified conditions of use without developing a temperature in excess of the value that represents the maximum temperature that the conductor insulation can withstand. NEC Table 310-16, for instance, specifies ampacities for conductors where not more than three (current-carrying) conductors are contained in a single raceway or cable or

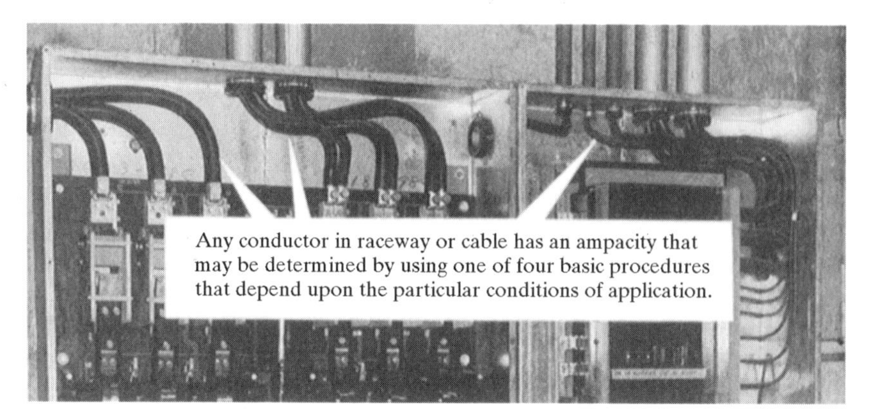

Any conductor in a raceway or cable has an ampacity that may be determined by using one of four basic procedures that depend upon the particular conditions of application.

directly buried in the earth, provided that the ambient temperature is not in excess of 30°C (86°F).

In the 1984 NEC, Tables 310-16 through 310-19 and their accompanying notes produced a clear, logical, coherent, and comprehensive approach to sizing conductors for the load current. However, in the 1987 edition of the Code, there was considerable change in the approach for determining conductor ampacity. This change was intended to recognize the use of the Neher-McGrath method of calculating conductor ampacity. The net result was a procedure that ranged from very difficult to comprehend and apply, at best, to totally incomprehensible and unenforceable, at worst. While the 1996 Code still permits the use (if that's possible) of this method [as now covered in Sec. 310-15(b)], this discussion and analysis is based on the procedure presented by the rule of Sec. 310-15(a) in the 1996 NEC, which states:

> Ampacities for conductors rated 0 through 2000 volts shall be as specified in Tables 310-16 through 310-19 and their accompanying notes.

The Basics

When all NEC rules on conductor ampacity are considered and correlated, it can be clearly established that any given conductor in a raceway, in a cable, or directly buried in the ground must have its ampacity determined in accordance with one of four possible conditions of use. Summarized, they are as follows:

1. When there are *not more than three* current-carrying conductors in a raceway or cable or directly buried, and the ambient temperature is *not over 30°C (86°F),* the conductor has an *ampacity equal to* the current value shown in Table 310-16 for the particular size and insulation of the conductor.
2. When there are *not more than three* current-carrying conductors in a raceway or cable or directly buried, but the ambient temperature *exceeds 30°C (86°F),* the conductor has an *ampacity that is calculated* by multiplying the current value shown in

The conductors in these conduits that are run across the roof of a building will be baking in the sun on hot summer days and, therefore, require ampacity derating for elevated ambient temperature.

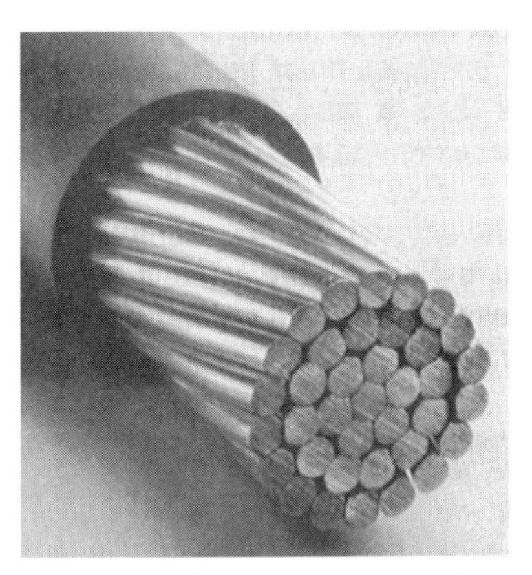

Ampacity: The current in amperes a conductor can carry continuously under the conditions of use without exceeding its temperature rating.

Table 310-16 (for the particular size and insulation of the conductor) by the ampacity correction factor selected from the bottom part of Table 310-16, based on the actual ambient temperature and the particular size and insulation of the conductor.

3. When there are *more than three* current-carrying conductors in a raceway or cable or directly buried, and the ambient temperature is *not over 30°C (86°F),* the conductor has an *ampacity that is calculated* by multiplying the current value shown in Table 310-16 (for the particular size and insulation of the conductor) by the percentage shown in the table to Note 8 to Table 310-16, based on the actual number of current-carrying conductors.
4. When there are *more than three* current-carrying conductors in a raceway or cable or directly buried, and the ambient temperature *exceeds 30°C (86°F),* the conductor has an *ampacity that is calculated* by multiplying the current value shown in Table 310-16 (for the particular size and insulation of the conductor) by the ampacity correction factor from the bottom part of Table 310-16 (based on the actual ambient temperature and the particular size and insulation of the conductor) *and then* times the percentage shown in the table to Note 8 to Table 310-16 (based on the actual number of conductors).

After the ampacity of a given conductor is established by one of the four above procedures, a suitable overcurrent protective device must be selected on the basis of that established ampacity. In accordance with all the rules of Sec. 240-3, the maximum current rating of the protective device (fuse or CB pole) must be based on the ampacity of the circuit conductor to be protected.

Clearly, the rule wants overcurrent devices to prevent conductors from being subjected to currents in excess of the ampacity values for which the conductors are rated by Tables 310-16 through 310-19 *including all notes.*

That last phrase in the first sentence of this section (about "all applicable notes to these tables") is important because it points out that when conductors have their ampacities derated because of conduit fill (Note 8 to the tables) or because of elevated ambient temperature, the conductors must be protected at the *derated* ampacities and *not* at the values given in the tables.

The basic rule of Sec. 240-3 (Fig. 2-1) requires that every conductor be protected in accordance with its ampacity, as determined above. That is, the fuse or CB pole protecting a conductor must ideally be rated at not more than the conductor ampacity. But because ampacity values as calculated above often do not exactly match the rating of standard available fuses or CBs (as given in Sec. 240-6), Sec. 240-3(b) says that, in such

240-3. Protection of Conductors. Conductors, other than flexible cords and fixture wires, shall be protected against overcurrent in accordance with their ampacities as specified in Section 310-15, unless otherwise permitted in (a) through (m) below.

(b) Devices Rated 800 Amperes or Less. Conductors not part of a multioutlet branch circuit supplying receptacles for cord- and plug-connected portable loads and where the ampacity of the conductors does not correspond with the standard ampere rating of a fuse or a circuit breaker without overload trip adjustments above its rating (but that may have other trip or rating adjustments), the next higher standard device rating shall be permitted only if this rating does not exceed 800 amperes.

FIG. 2-1

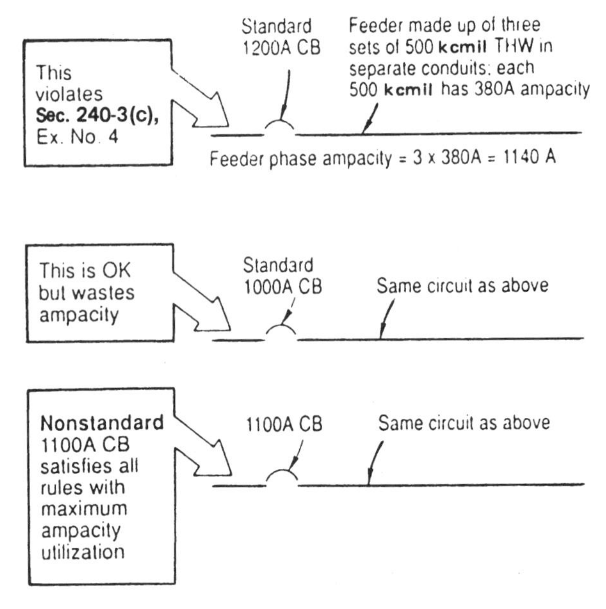

FIG. 2-2 Protection in accordance with the NEC may use standard or non-standard rated fuses or circuit breakers. (Secs. 240-3 and 240-6.)

cases, the conductor may be considered to be protected by a fuse or CB that has a current rating that is the next higher standard rating of protection above the ampacity value of the conductor—*but only up to 800 A* (Fig. 2-2). When a conductor has an ampacity greater than 800 A [covered in Sec. 240-3(c)] and there is no standard rating of fuse or circuit breaker that exactly corresponds to the ampacity value, the conductor *may not* be protected by a fuse or CB with a rating above its ampacity. The maximum rating of fuse or CB that may be used to protect such a conductor is either the next lower *standard* rating *or* (as covered in the FPN in Sec. 240-6) any listed *nonstandard* rated overcurrent protective device whose rating does not exceed the conductor's ampacity. (Remember, fuses or CBs for protecting motor-circuit conductors are permitted in Secs. 430-52 and 430-62 to be substantially higher than the conductor ampacity.)

For example, if a circuit conductor of, say, 500-kcmil THW copper (not more than three in a conduit at not over 86°F ambient) satisfies design requirements and NEC rules for a particular load current not in excess of the conductor's table ampacity of 380 A, then the conductor *may* be protected by a 400-A rated fuse or CB.

Section 240-6, which gives the standard ampere ratings of protective devices to correspond to the word *standard* in Sec. 240-3(b), shows a device rated at 350 and 400 A, but none at 380 A. In such a case, the NEC accepts a 400-A rated device as the next higher standard device rating above the conductor ampacity of 380 A.

But such a 400-A device would permit load increase above the 380 A that is the safe maximum limit for the conductor. It would be more effective practice to use 350-A rated protection and prevent such overload.

In the 1987 edition, the NEC required that a circuit breaker with an adjustable or changeable long-time trip rating have load-circuit conductors of an ampacity at least equal to the highest rating at which the adjustable trip might be set. But the exception to

Sec. 240-6 recognizes the use of such a CB with a conductor of an ampacity less than the highest possible trip setting *if* the CB is equipped with a lockable or sealable cover, or is in a locked enclosure, or is in a locked room. In such an application, however, the actual trip setting must protect the conductor in accordance with its ampacity, as required by the basic rule of Sec. 240-3. This is more consistent with the Code-recognized treatment of fusible switches in that fusible switches still can be, and always could be, cabled for the actual rating of the fuse, and not the maximum value of fuse that might be installed.

Section 240-3(b) requires that the rating of overcurrent protection *never* exceed the ampacity of circuit conductors supplying one or more receptacle outlets on a branch circuit with more than one outlet. This wording coordinates with the rules described under Sec. 210-19(a) on conductor ampacity. The effect of Sec. 240-3(b) is to require that the rating of the overcurrent protection not exceed the Code table ampacity (NEC Table 310-16) or the derated ampacity dictated by Note 8 to the tables for any conductor of a multioutlet branch circuit supplying any receptacles for cord- and plug-connected portable loads. If a standard rating of fuse or CB does not match the ampacity (or derated ampacity) of such a circuit, the next lower standard rating of protective device must be used. *But* where branch-circuit conductors of an individual circuit to a single load or a multioutlet circuit supply *only* fixed connected (hard-wired) loads, such as lighting outlets or permanently connected appliances, the next larger standard rating of protective device *may* be used in those cases where the ampacity (or derated ampacity) of the conductor does not correspond to a standard rating of protective device—but, again, that is permitted only up to 800 A, above which the next lower rating of fuse or CB must be used, as described under Sec. 210-19(a).

From the definition of *ampacity*, it is clear that heating of the conductor is the factor that determines ampacity ("without exceeding its temperature rating"). It is also clear that ampacity is the amount of current that the conductor can carry under *a specific set of* "conditions of use." And finally, it is also clear that when the ampacity of a conductor is determined in accordance with all applicable Code rules, it is a *continuous* rating. The NEC makes no differentiation between ampacity for continuous load and for noncontinuous load. Any determined value of ampacity represents the amount of current the conductor is rated to carry continuously or noncontinuously. *But* because of limitations related to non-100 percent rated CBs, an additional evaluation is necessary where the load to be supplied is either all continuous or a combination of continuous and noncontinuous loads. The following examples show the method for determining a conductor's ampacity where the load to be supplied is noncontinuous.

Typical Example of Applying These Rules

Any given size of conductor and type of insulation used in raceway or cable has only one of four possible procedures, depending on conditions of use, to determine its ampacity. The following exercise will examine these different procedures as they apply to a conductor of specific size and insulation type and provide a basis for developing a thorough understanding of this subject. The procedures and concepts put forth in this exercise can then be applied to a conductor of any size or insulation type to determine its ampacity. Give this Code change your fullest attention and highest priority.

Question. What is the ampacity of a No. 6 AWG THW copper conductor?

Answer. It can have any one of hundreds, even thousands, of different ampacities. The ampacity will be determined by the conditions of use.

This is the basic table to use for over 99% of all conductor applications.

Table 310-16. Allowable Ampacities of Insulated Conductors Rated 0-2000 Volts, 60° to 90°C (140° to 194°F) Not More Than Three Conductors in Raceway or Cable or Earth (Directly Buried), Based on Ambient Temperature of 30°C (86°F)

Size	Temperature Rating of Conductor. See Table 310-13.						Size
	60°C (140°F)	75°C (167°F)	90°C (194°F)	60°C (140°F)	75°C (167°F)	90°C (194°F)	
AWG kcmil	TYPES TW†, UF†	TYPES FEPW†, RH†, RHW†, THHW†, THW†, THWN†, XHHW† USE†, ZW†	TYPES TA, TBS, SA SIS, FEP†, FEPB†, MI RHH†, RHW-2, THHN†, THHW†, THW-2, THWN-2, USE-2, XHH, XHHW† XHHW-2, ZW-2	TYPES TW†, UF†	TYPES RH†, RHW†, THHW†, THW†, THWN†, XHHW†, USE†	TYPES TA, TBS, SA, SIS, THHN†, THHW†, THW-2, THWN-2, RHH†, RHW-2 USE-2 XHH, XHHW XHHW-2, ZW-2	AWG kcmil
	COPPER			ALUMINUM OR COPPER-CLAD ALUMINUM			
18	....	....	14	....	....	....	....
16	....	....	18	....	....	....	....
14	20†	20†	25†	....	....	....	....
12	25†	25†	30†	20†	20†	25†	12
10	30	35†	40†	25	30†	35†	10
8	40	50	55	30	40	45	8
6	55	65	75	40	50	60	6
4	70	85	95		65	75	4
3	85	100	110		75	85	3
	95	115				100	
	110	130				115	

FIG. 2-3 NEC Table 310-16 gives ampacities under two conditions: that the raceway or cable containing the conductors is operating in an ambient temperature not over 30°C (86°F) and that there are not more than three current-carrying conductors in the raceway or cable. Under those conditions, the ampacities shown correspond to the thermal limit of each particular insulation, and the value shown *is* the ampacity of each conductor.

Condition 1. When the conditions of use are as described by the heading of one of the ampacity tables, the ampacity of the conductor can be taken directly from that ampacity table. (See Fig. 2-3.)

As a result of the changes in the 1990 NEC, there are now only four tables covering ampacities for copper, aluminum, and copper-clad aluminum conductors rated 0 to 2000 V. Tables 310-16 and 310-17 are for conductors with insulation rated from 60 to 90°C in an ambient temperature no greater than 86°F (30°C), and Tables 310-18 and 310-19 cover conductor insulations rated from 150 to 250°C in an ambient temperature of 104°F (40°C). Both Tables 310-16 and 310-18 are for use where there are not more than three current-carrying conductors in a raceway, cable, or directly buried; and Tables 310-17 and 310-19 are for "single insulated conductors…in free air" or, in other words, spaced open conductors. Inasmuch as the vast majority of electric circuits incorporate 60 to 90°C rated insulated conductors in raceway, cable, or directly buried, it quickly becomes clear that Table 310-16 is going to be used for the vast majority of installations, including our example.

Although Tables 310-16 through 310-19 refer to "*three conductors* in raceway…," this should read "three *current-carrying* conductors." In some applications more than three conductors are installed in a raceway, or cable, or directly buried, *but not* all the conductors are counted because the Code does not consider them to be current-carrying conductors. For example, a 3-phase, 4-wire branch circuit with an equipment-grounding conductor consists of five conductors, but in many instances, not more than three of the conductors are considered to be current-carrying conductors. As covered by Note 10, the neutral in such a circuit "shall not be counted" as one of the three conductors referred to in the tables' headings if less than half the load is electric discharge lighting, data processing equipment, or other "nonlinear" loads. And Note 11 to the ampacity tables states that a grounding or bonding conductor also "shall not be counted." Therefore, even though there are actually five conductors, only three "count" as current-carrying conductors, and the ampacity can be taken directly from the appropriate table for the particular size and insulation of the conductor.

As shown in Fig. 2-4, the ampacity of each No. 6 AWG THW copper conductor [where there are not more than three current-carrying conductors in the raceway and the ambient temperature is not in excess of 86°F (30°C), that is, the conditions set forth in the heading of Table 310-16] is taken directly from the table. Under these conditions of use, the ampacity shown in the table corresponds to the thermal limit of this particular insulation, and the ampacity of each No. 6 AWG THW copper conductor is 65 A.

Rating of Overcurrent Protection. Because there is not a standard rated fuse or circuit breaker of 65-A rating, each No. 6 AWG THW copper conductor may be protected by the next standard rating of protective device above 65 A [Sec. 240-3(b)]. Section 240-6 shows that to be a 70-A pole or fuse.

The maximum Code-recognized loading permitted on each No. 6 THW conductor is 65 A—if the load current does not persist for a period of 3 h or more. That is, if the loading does *not* constitute "continuous load" (where the current flows for 3 h or more, as defined in NEC Article 100, "Definitions"), the circuit may be properly loaded up to the conductor ampacity of 65 A—*but* not over 65 A, even though the 70-A protection would permit current higher than 65 A.

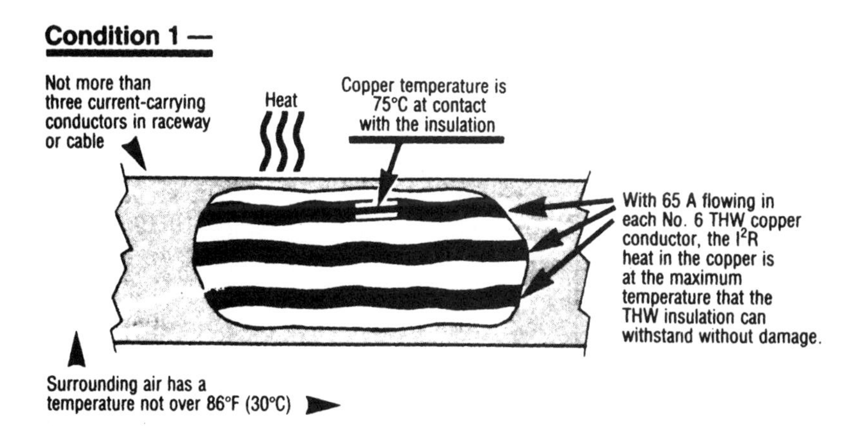

FIG. 2-4

Condition 2. NEC Table 310-16 gives ampacities under two conditions: that the raceway or cable containing the conductors is operating in a surrounding ambient temperature not over 86°F (30°C) and that there are not more than three current-carrying conductors in the raceway or cable. Under those conditions, the ampacities shown correspond to the thermal limit of each particular insulation, and the value shown *in the table is* the ampacity of each conductor. But in any case where either (or both) of the two conditions is (are) not satisfied, the ampacity of the conductors must be reduced (and protection must be based upon or provided at the reduced ampacity!) to ensure that the temperature limit of the insulation is not exceeded.

If the ambient temperature is above 86°F (30°C), the current value given in Table 310-16 for the given size and insulation of the conductor must be reduced by the appropriate correction factor from the bottom part of Table 310-16, for the particular temperature that exists. (See Fig. 2-5.) However, while that requirement is easy to understand, the Code provides no guidance on how to determine an appropriate adjusted "ambient temperature." Should it be the maximum temperature reached? An average maximum? The Code does not say.

In the absence of any direct guidance from the NEC, one approach that should produce satisfactory results in actual application is to determine a 3-h "average maximum" over the course of a week. Ideally (but realistically improbable), hourly temperature readings would be taken over an entire day to determine that 3-h period when the ambient temperature is greatest. Then for the next week, take hourly readings over this 3-h

Use these reduction factors to calculate ampacity when ambient temperature is above 86°F.

CORRECTION FACTORS

Ambient Temp. °C	For ambient temperatures other than 30°C (86°F), multiply the allowable ampacities shown above by the appropriate factor shown below.						Ambient Temp. °F
21-25	1.08	1.05	1.04	1.08	1.05	1.04	70-77
26-30	1.00	1.00	1.00	1.00	1.00	1.00	78-86
31-35	.91	.94	.96	.91	.94	.96	87-95
36-40	.82	.88	.91	.82	.88	.91	96-104
41-45	.71	.82	.87	.71	.82	.87	105-113
46-50	.58	.75	.82	.58	.75	.82	114-122
51-55	.41	.67	.76	.41	.67	.76	123-131
56-60	....	.58	.71	....	.58	.71	132-140
61-70	....	.33	.58	....	.33	.58	141-158
71-80	....	....	.41	....	....	.41	159-176

†Unless otherwise specifically permitted elsewhere in this Code, the overcurrent protection for conductor types marked with an obelisk (†) shall not exceed 15 amperes for No. 14, 20 amperes for No. 12, and 30 amperes for No. 10 copper; or 15 amperes for No. 12 and 25 amperes for No. 10 aluminum and copper-clad aluminum after any correction factors for ambient temperature and number of conductors have been applied.

FIG. 2-5 The ampacity correction factors, which must be applied to the value given in the ampacity table when the conductors are to be installed where the ambient temperature is in excess of 86°F (30°C), appear at the bottom of each of the ampacity tables. Derating of the table values is required to compensate for the diminished heat-dissipating capability of the conductors and prevent them from reaching an operating temperature in excess of that for which the conductor insulation material is rated.

period. Average these readings and use that temperature value for selecting the appropriate derating factor from the ampacity correction factors (Fig. 2-5) for ambient temperature that appear at the bottom of Tables 310-16 through 310-19. While such an approach is not always practicable, especially at the design stage, the only other method available to determine what the adjusted ambient temperature should be is to take an "educated guess" at an average maximum based on previous experience.

In Fig. 2-6, while there are still only three current-carrying conductors within the raceway, each No. 6 AWG THW copper conductor now has an ampacity of 53 A. This is the product of multiplying the ampacity value from Table 310-16 (65 A) by the factor (0.82) from the ampacity correction table for an ambient temperature of 108°F, which expressed mathematically is 65 A × 0.82 = 53 A. Because these conductors are in an elevated ambient temperature, they are less able to dissipate heat. Under these conditions, 53 A now represents the conductor ampacity, that is, the amount of current that will produce an I^2R heat generation, establishing thermal equilibrium at a temperature of 75°C in the copper, which is the maximum temperature that THW insulation can withstand without damage.

Rating of Overcurrent Protection. Here, each No. 6 AWG THW copper conductor, with an ampacity of 53 A, may be protected by a fuse of breaker rated at 60 A, which is the next standard rating of protective device above the conductor's 53-A ampacity—again, as permitted by Sec. 240-3(b), which permits protection by the next standard rating of protective device above the conductor's ampacity when there is not a standard rated protective device of the conductor's ampacity.

The maximum Code-recognized loading permitted on each No. 6 conductor is 53 A—that is, up to the conductor's ampacity—if the loading is "noncontinuous load."

Condition 3. When there are more than three current-carrying conductors in a raceway or cable, their current-carrying capacities must be decreased to compensate for proximity heating effects and reduced heat dissipation due to reduced ventilation of individual

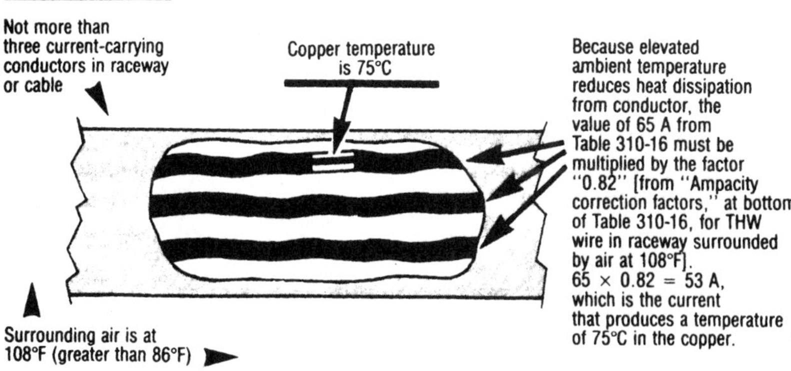

FIG. 2-6

conductors, which are bunched or which form an enclosed group of closely placed conductors. In such cases, the ampacity must be reduced from the Table 310-16 value, as required in the table of Note 8, which appears after Tables 310-16 through 310-19.

Part (a) of Note 8 to Table 310-16 says: "Where the number of current-carrying conductors in a raceway or cable exceeds three, the ampacities given shall be reduced as shown in the following table."

Figure 2-7 shows that each No. 6 AWG THW copper conductor now has an ampacity of 52 A. This is the product of multiplying the ampacity value from Table 310-16 (65 A) by the factor (80 percent) for four through six conductors given in the table to Note 8 (65 A × 0.80 = 52 A) (Fig. 2-8). Because more than three current-carrying conductors are used in a single raceway or cable, their heat-dissipating capability is reduced. Under these conditions, 52 A now represents the conductor ampacity—that is, the amount of current that will produce an I^2R heat generation, establishing thermal equilibrium at a temperature of 75°C in the copper, which is the maximum temperature that THW insulation can withstand without damage.

Rating of Overcurrent Protection. As in condition 2, a 60-A fuse or CB pole may be used to protect each No. 6 AWG THW copper conductor with its ampacity of 52 A, for the same reasons explained under condition 2. And the maximum Code-recognized loading is 52 A—the conductor ampacity—for a noncontinuous load.

Condition 4. Where more than three current-carrying conductors are used in a single raceway or cable *and* the conduit or cable containing the conductors is in an ambient air temperature higher than 86°F (30°C), the ampacity of the contained conductors must be reduced in accordance with the factors of Note 8 *and* in accordance with the ampacity correction factor table for higher ambient temperatures given at the bottom of Table 310-16. This is made clear by the heading in the table to Note 8. The heading requires that any ampacity derating for elevated ambient be made *in addition* to the one for number of conductors.

As shown in Fig. 2-9, each No. 6 AWG THW copper conductor now has an ampacity of 39 A. The heading for the table to Note 8 requires that the ampacity first be

Condition 3 —

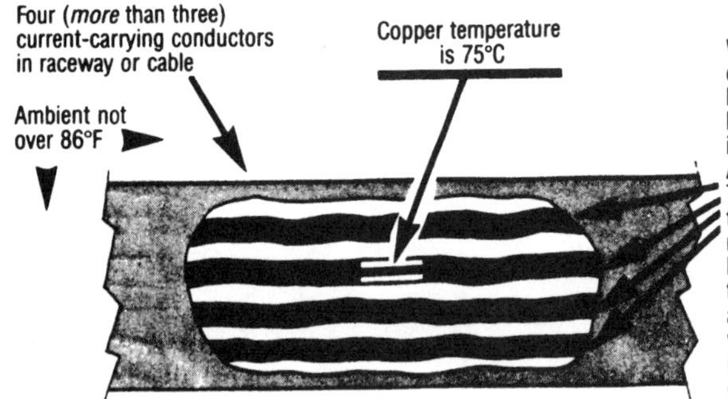

When more than three current-carrying conductors are used, heat dissipation is impeded because some conductors are in the middle of the bundle. As required by Note 8 to Table 310-16, the value of 65 A from Table 310-16 must be multiplied by 80% [from Note 8, for 4 to 6 conductors] to determine conductor ampacity. 65 × 0.8 = 52 A, which is the current that produces a temperature of 75°C in the copper.

Under these conditions, No. 6 THW copper conductor has an ampacity value of 52 *A*—not 65 A.

FIG. 2-7

This data must be applied to determine conductor ampacity when more than three current-carrying conductors are installed in a single raceway or cable.

Number of current-carrying conductors	Percent of values in tables as adjusted for ambient temperature if necessary
4 through 6	80
7 through 9	70
10 through 20	50
21 through 30	45
31 through 40	40
41 and above	35

FIG. 2-8 Whenever there are more than three current-carrying conductors in a raceway, a cable, or directly buried, the basic rule of Note 8 requires derating in accordance with the percentages given in this table.

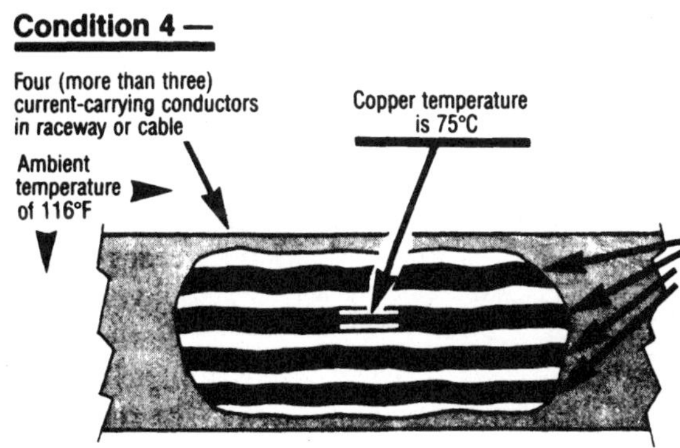

FIG. 2-9

adjusted for ambient temperature. As was done under condition 2, the base ampacity of the No. 6 AWG THW copper conductor is multiplied by the ampacity correction factor for an elevated ambient—(over 86°F)—65 A × 0.75 = 49 A. Then this product is multiplied by the factor given for four through six conductors—49 A × 0.80 = 39 A. Therefore, under these conditions of use, each No. 6 AWG THW copper conductor has an ampacity of 39 A.

Rating of Overcurrent Protection. Now, the same No. 6 AWG THW copper conductors have an ampacity of only 39 A, which calls for a maximum 40-A rated protective device, with a loading of 39 A maximum of noncontinuous load.

AMPACITY AND OVERCURRENT PROTECTION FOR CONTINUOUS LOADS

Sections 210-22, 220-3, and 220-10 put forth the NEC rules for minimum circuit rating and conductor size. Generally, for noncontinuous loading, they must be adequate for the load to be served. Parts (c), (a), and (b), respectively, of those sections also present requirements for the minimum rating of overcurrent devices and the minimum size for conductors where supplying continuous or continuous and noncontinuous loads. Collectively, those rules cover *all* types of circuits—service, feeder, and branch circuits. And for continuous loading or continuous and noncontinuous loading, the overcurrent device and circuit conductors for services, feeders, and branch circuits must be selected in accordance with those limitations.

As indicated, the wording only impacts applications where the load is either all continuous or a combination of continuous and noncontinuous loads. For circuits supplying noncontinuous loads, only, there is no change. The overcurrent device rating and conductor's ampacity must simply be equal to or greater than the total of load to be served, as calculated in accordance with Art. 220. However, the requirement for continuous loading or combination (continuous and noncontinuous) loading is a bit more involved. Therefore, to establish which applications are, or are not, covered, the first thing that must be established is the difference between continuous and noncontinuous loads.

For that we refer to the definition of continuous load in Art. 100. As can be seen, simply put, a *continuous load* is a load that is energized for 3 h or more. And, by inference, the term *noncontinuous load* is a load that is energized for less than 3 h. Circuits supplying store or office lighting are common examples of continuous loads. And where those or similarly energized loads are supplied, the conductors that serve those loads—branch-circuit, feeder, and service conductors—will typically require greater capacity than in the 1993 Code.

The actual wording used differs only slightly in that Secs. 220-22(c) and 220-3(a) apply to branch circuits and only specifically mention branch circuits. Section 220-10(b) only specifically identifies feeders, which include service conductors. But the requirement is the same for both. The pertinent part of all sections reads as follows:

> The minimum...conductor size, without application of any adjustment or correction factors, shall have allowable ampacity equal to or greater than the noncontinuous load plus 125 percent of the continuous load.

Although similar to what the Code requires for overcurrent devices, the qualifying phrase *without application of any adjustment or correction factors* alters the application slightly. The conductor selected must have a Table 310-16 ampacity that is equal to or greater than the sum of noncontinuous and 125 percent of the continuous load, before determining the conductor's final ampacity, which may be less than the sum of noncontinuous and 125 percent of continuous load. That is, for the procedure described in Secs. 210-22(c), 220-3(a), and 220-10-(b), the conductor is only evaluated based on its ampacity as given in the ampacity Tables (310-16 through 310-19). And if required, the *adjustment* factors of Note 8 and the correction factors for ambient temperature at the bottom of Table 310-16 must be applied to establish the conductor's actual ampacity.

It must be clearly understood that the qualifying phrase *without application of any adjustment or correction factors* does *not* mean derating is *not* required. As indicated above, it is intended to direct that evaluation of the conductors for compliance with Secs. 210-22(c), 220-3(a), and 220-10(b) may be done before or, as the rule states, *without* application of any derating, which may otherwise be required. The rules covering protective device ratings and conductor sizing where supplying continuous or continuous and noncontinuous loads are completely separate from the rules for determining conductor ampacity—even though the rules are closely related. The application of derating factors is regulated by Sec. 310-15(a), which calls for the use of Tables 310-16 through 310-19 and their accompanying notes to establish ampacity. If derating is needed, Sec. 310-15(a) essentially says it must be applied. Section 220-3(a) says it can be applied *after* establishing the continuous-loading capacity. The point is that both requirements can be and must be in all ways fully satisfied.

The following examples are based on 75°C terminations. But what if the terminations on the connected equipment were rated for 60°C? Would the minimum conductor size have to be based on the 60°C ampacity before any other consideration? That is, must the conductor selected have a 60°C ampacity equal to, or greater than, the noncontinuous plus 125 percent of the continuous load? Although not entirely clear, it appears that the answer is yes.

Examples

As indicated, these new requirements must be correlated with the rules for ampacity (Sec. 310-15), rating of the overcurrent device (Sec. 240-3), and termination temperature limitations (Sec. 110-14) described in the previous discussion.

The steps necessary to thoroughly evaluate conductors where supplying continuous or continuous and noncontinuous loads have been arranged in the flowchart shown in Fig. 2-10. The flowchart indicates a sequence of evaluation that will simplify the entire process. Note that the literal wording used in the Code only requires that the conductor have sufficient capacity for the sum of noncontinuous load plus 125 percent of the continuous load. The same is required for the protective device. But in those cases where the calculated value does not correspond to a standard rating of protective device, the next *larger* size will have to be selected in order to satisfy the *minimum* rating requirement. This is true even for circuits rated over 800 A, which will automatically require larger conductors according to Sec. 240-3(c). By basing the conductor size on the rating of the overcurrent device from the beginning, the likelihood of underprotected conductors, which will necessitate a reevaluation of the conductors, is greatly reduced.

The following examples will cover four different applications:

1. Continuous loading only, without derating
2. Continuous loading only, with derating
3. Combination loading, without derating
4. Combination loading, with derating

All are evaluated by using the flowchart in Fig. 2-10.

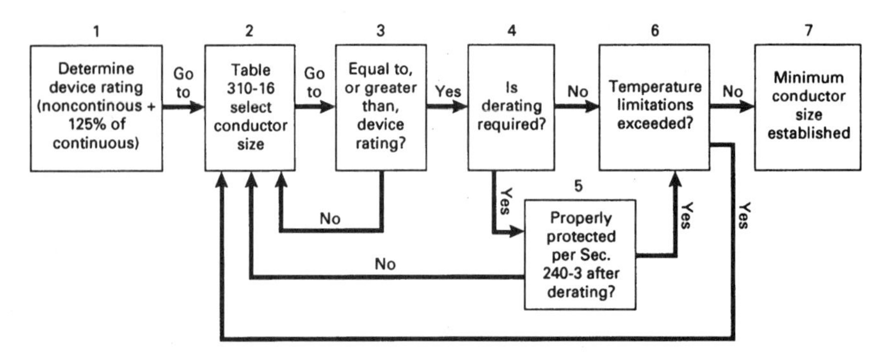

FIG. 2-10 Flowchart showing steps involved for determining the minimum size conductor permitted by the 1996 NEC where supplying continuous loads or a combination of continuous and noncontinuous loads.

Continuous Loading without Derating. Determine the minimum overcurrent (OC) device rating and minimum conductor size for a 3-phase, 4-wire, 208Y/120 V circuit supplying 100 A of incandescent floodlighting. Assume all equipment terminations suitable for 75°C operation.

Step 1. Establish the OC device rating by multiplying the continuous load by 1.25 and adding the product to the value of noncontinuous load.

$$\text{Device rating} = \text{noncontinuous load} + (1.25 \times \text{continuous load})$$

$$= 0\text{ A} + (1.25 \times 100\text{ A})$$

$$= 125\text{ A}$$

The OC device rating must be not less than 125 A. Note that if the minimum rating for the OC device were, say, 132 A, then the next larger standard rating above that value, which is 150 A, would have to be used because the device rating may not be less than the calculated value of noncontinuous plus 125 percent of the continuous load. But if the calculated minimum OC device rating is 132 A, the conductor size may be based on the 132-A value, provided that the conductors are properly protected by the 150-A rated device after any derating that may be required.

Steps 2 and 3. These two steps are conducted simultaneously. Using the 125-A rating of the OC device, a No. 2 copper THHN, with an ampacity of 130 A at 90°C, is initially selected. But even though it is permissible to use 90°C value for derating purposes, such permission is *not* extended to this determination. As previously indicated, the 75°C value must be used, because that's what was used by the manufacturers when the OC devices were tested. As can be seen in Table 310-16, No. 2 copper only has a 115-A underated 75°C ampacity. Therefore, No. 1 copper with either THW (75°C) insulation or THHN (90°C) insulation may be used because the 130-A value from the 75°C column for No. 1s in Table 310-16 is "equal to, or greater than" the 125-A OC device rating. A No. 1 copper conductor with either a 75°C rated insulation (for example, THW) or a 90°C rated insulation (for example, THHN) may be used.

Step 4. Where derating is required for ambient temperature or for number of conductors, we go to step 5 and evaluate the conductor's derated ampacity to ensure it is properly protected in accordance with Sec. 240-3. Here, no derating is required and, because the selected conductors are already properly protected by the 125-A rated OC device, just proceed to step 6.

Step 6. The next consideration is the temperature limitations of the equipment terminations. Here, use of either 75°C rated or 90°C rated No. 1 copper conductor is acceptable because the equipment is rated for 75°C rated conductors used at their full 75°C ampacity. The 75°C value is greater than the actual load, which is 100 A.

Step 7. Once all the other steps have been completed, the minimum conductor size is established. To serve the incandescent lighting load, a No. 1 copper with either 75°C rated or 90°C rated insulation may be used.

Continuous Loading with Derating. Determine the minimum OC device rating and minimum conductor size for a 3-phase, 4-wire 208Y/120-V circuit supplying 100 A of fluorescent lighting. Assume all equipment terminations are suitable for 75°C operation.

Step 1. Again, establish the minimum OC device rating by multiplying the continuous load by 1.25 and adding the product to the value of noncontinuous load.

$$\text{Device rating} = \text{noncontinuous load} + (1.25 \times \text{continuous load})$$

$$= 0\text{ A} + (1.25 \times 100\text{ A})$$

$$= 125\text{ A}$$

The OC device rating must be not less than 125 A.

Steps 2 and 3. Select a conductor with a 75°C ampacity value not less than 125 A. From Table 310-16, it can be seen that a No. 1 with either 75°C rated or 90°C rated insulation may be used.

Step 4. Derating *is* required here because Note 10 to the ampacity tables indicates that where nonlinear loads, such as electric discharge lighting, are supplied from a grounded-wye system using a 3-phase, 4-wire circuit, the neutral conductor will carry current that approximates the phase current under balanced loading conditions. As a result, the table ampacity must be derated according to the table given in Note 8 to the ampacity tables. In that table, it can be seen that where the number of current-carrying conductors is four to six, the ampacity of the No. 1 copper conductors must be derated to 80 percent of the table value, or

$$\text{Usable ampacity} = \text{table ampacity} \times 0.80$$

- For the 75°C insulated conductors:

$$\text{Usable ampacity} = 130\text{ A} \times 0.80$$

$$= 104\text{ A}$$

- For the 90°C insulated conductors:

$$\text{Usable ampacity} = 150\text{ A} \times 0.80$$

$$= 120\text{ A}$$

Step 5. Reference to Sec. 240-6 shows that the 125-A OC device does not properly protect the 75°C insulated No. 1s at their derated value of 104 A. The next-higher rated device above the conductors' ampacity is 110 A, and that is the maximum rating of OC device that may be used with these derated conductors. But the 90°C insulated No. 1s, with their derated value of 120 A, are properly protected.

Step 6. The 90°C insulated No. 1s, with their 120-A ampacity, may be used because the total load, which is only 100 A, does not exceed the 75°C table value of 130 A. Therefore, the termination temperature limitations are satisfied. Notice that the 104-A derated ampacity of the 75°C insulated conductors is also greater than the total load of 100 A, *but* because the usable ampacity of 104 A is not properly protected by a 125-A rated OC device, only 90°C insulated No. 1s may be used.

Step 7. To supply 100 A of fluorescent lighting, the circuit conductors' minimum size is No. 1 copper with 90°C insulation, or the next-larger size (No. 1/0) with 75 or 90°C rated insulation.

Continuous and Noncontinuous Loading without Derating. Assume a combination load in which the continuous load is 48 A and the noncontinuous load is 90 A and less than half of the total load is nonlinear load. Determine the minimum rating for OC protection and the minimum conductor size permitted for a 3-phase, 4-wire, 208Y/120 V circuit supplying these loads. All terminations are suited for 75°C.

Step 1. Again, establish the minimum OC device rating by multiplying the continuous load by 1.25 and adding the product to the value of noncontinuous load.

$$\begin{aligned}\text{Device rating} &= \text{noncontinuous load} + (1.25 \times \text{continuous load})\\ &= 90\text{ A} + (1.25 \times 48\text{ A})\\ &= 90\text{ A} + 60\text{ A}\\ &= 150\text{ A}\end{aligned}$$

The OC device rating must be not less than 150 A.

Steps 2 and 3. Select a conductor with a 75°C ampacity of at least 150 A. From Table 310-16, a No. 1/0 copper with either 75°C insulation (150 A) or 90°C insulation (170 A) may be used.

Step 4. Where derating is required for ambient temperature or for number of conductors, we go to step 5 and evaluate the conductor's derated ampacity to ensure it is properly protected in accordance with Sec. 240-3. Here no derating is required, and because the selected conductors are already properly protected by the 150-A rated OC device, we proceed to step 6.

Step 6. The next consideration is the temperature limitations of the equipment terminations. Here, use of either 75 or 90°C rated No. 1/0 copper conductors is acceptable because the equipment is rated for 75°C insulated conductors used at their full 75°C ampacity, and the 75°C value (150 A) is greater than the actual load, which is 138 A (90 A + 48 A). Notice that 75°C insulated No. 1s do not have adequate table ampacity (130 A) to serve the total 138-A load.

Step 7. Again, once all the other steps have been completed, the minimum conductor size is established. To serve this combination load, a No. 1/0 copper with either 75 or 90°C rated insulation may be used.

Continuous and Noncontinuous Loading with Derating. Assume a combination load in which the continuous load is 48 A and the noncontinuous load is 90 A and more than half of the total load is nonlinear load. Determine the minimum rating of OC protection and minimum conductor size for a 3-phase, 4-wire, 208Y/120 V circuit to supply these loads. All terminations are suited for 75°C.

Step 1. Again, establish the minimum OC device rating by multiplying the continuous load by 1.25 and adding the product to the value of noncontinuous load.

$$\text{Device rating} = \text{noncontinuous load} + (1.25 \times \text{continuous load})$$

$$= 90 \text{ A} + (1.25 \times 48 \text{ A})$$

$$= 90 \text{ A} + 60 \text{ A}$$

$$= 150 \text{ A}$$

The OC device rating must be not less than 150 A.

Steps 2 and 3. Select a conductor with a 75°C ampacity of at least 150 A. From Table 310-16, a No. 1/0 copper with either 75°C insulation (150 A) or 90°C insulation (170 A) may be used.

Step 4. Derating is required here because Note 10 to the ampacity tables indicates that whereas more than half of the loads are nonlinear loads, and are supplied from a grounded-wye system using a 3-phase, 4-wire circuit, the neutral conductor must be counted as a current-carrying conductor. As a result, the table ampacity must be derated according to the table given in Note 8 to the ampacity tables. In that table, it can be seen that where the number of current-carrying conductors is four to six, the ampacity of the No. 1/0 copper conductors must be derated to 80 percent of the table value, or

$$\text{Usable ampacity} = \text{table ampacity} \times 0.80$$

- For the 75°C insulated conductors:

$$\text{Usable ampacity} = 150 \text{ A} \times 0.80$$

$$= 120 \text{ A}$$

- For the 90°C insulated conductors:

$$\text{Usable ampacity} = 170 \text{ A} \times 0.80$$

$$= 136 \text{ A}$$

Step 5. Reference to Sec. 240-6 shows that the 150-A OC device does not properly protect the 75°C insulated No. 1/0s at their derated value of 120 A. The next-higher rated device above the conductors' ampacity is 125 A, and that is the maximum rating of OC device that may be used with these derated conductors. Additionally, although the 90°C insulated No. 1/0s, with their derated value of 136 A, are properly protected, the total load of 138 A is greater than the conductors' derated ampacity. Therefore, the next larger size conductor, No. 2/0 copper, must be considered.

A 75°C insulated No. 2/0 copper has a table value of 175 A, and a 90°C insulated No. 2/0 copper has a 195-A table value. When these conductors are derated to 80 percent, the 75°C insulated No. 2/0s have an ampacity of 140 A and the 90°C insulated No. 2/0s have an ampacity of 156 A. Both are properly protected by the 150-A rated OC device and are adequate to carry the actual total load of 138 A.

Step 6. The next consideration is the temperature limitations of the equipment terminations. Here, use of either 75 or 90°C rated No. 2/0 copper conductors is acceptable. That is, the equipment is rated for 75°C rated conductors used at their full 75°C ampacity, and the actual load, which is 138 A, is less than the 75°C table value for No. 2/0s of 175 A.

Step 7. To supply this combination load of 138 A, minimum rating of OC device is 150 A, and the circuit conductors' minimum size is No. 2/0 copper with either 75 or 90°C insulation.

Notes. It should be clearly understood that any reduced ampacity, required because of higher ambient and/or conductor bundling, has the same meaning as the value shown in Table 310-16: each represents a current value above which excessive heating would occur under the particular conditions. And if two separate conditions individually diminish the ability of the conductor to dissipate heat, then more reduction of current is required than if only one such condition existed.

In conductor sizes Nos. 14, 12, and 10 AWG, Table 310-16 clearly indicates that 90°C rated conductors do, in fact, have higher ampacities than those given for the corresponding sizes of 60 and 75°C conductors. As shown, for copper, No. 12 TW and No. 12 THW are both assigned an ampacity of 25 A under the basic conditions of the table. But a No. 12 THHN, RHH, or XHHW (dry location) has an ampacity of 30 A. However, the obelisk (dagger) note to Table 310-16 requires that overcurrent protection for No. 14, No. 12, and No. 10 be taken as 15, 20, and 30 A, respectively, regardless of the type and temperature rating of the insulation on the conductors. And the obelisk note says that these limitations apply "after any correction factors for ambient temperature and number of conductors have been applied."

When applied to the selection of branch-circuit wires in cases where conductor ampacity derating is required by Note 8 of Tables 310-16 through 310-19 for conduit fill (over three wires in a raceway), the obelisk note to Table 310-16 affords advantageous use of the higher current ratings of 90°C wires for branch-circuit makeup. The reason is that, as stated in Note 8, the derating of ampacity is based on taking a *percentage of the actual current value shown in the table,* and the table current values for 90°C conductors *are higher* than those for 60 and 75°C conductors. The derated ampacity must be properly protected by the 15-, 20-, and 30-A protective devices.

The following examples illustrate these concepts.

As shown in Fig. 2-11, the makeup of a branch circuit consists of first selecting the correct size of wire for the particular load current (based on the number of wires in the raceway, ambient temperature, and ampacity deratings) and then relating the rating of the overcurrent protective device to all the conditions.

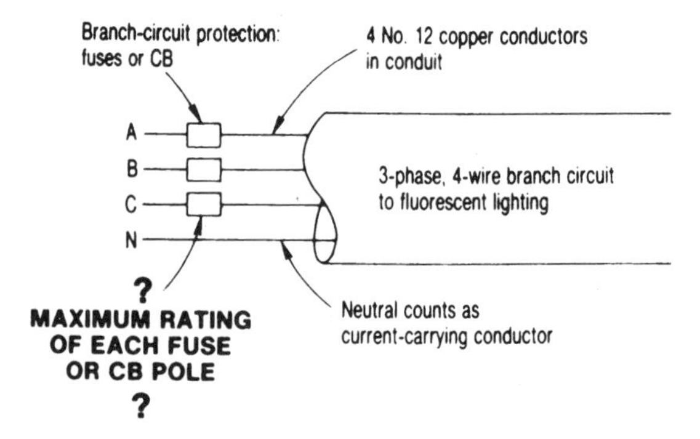

FIG. 2-11 Loading and protection of branch circuit wires must account for derating of conductor "ampacity."

In Table 310-16, which applies to conductors in raceways and in cables and covers the vast majority of conductors used in electrical systems for power and light, the ampacities for Nos. 14, 12, and 10 conductors are particularly significant because copper conductors of those sizes are involved in the vast majority of branch circuits in modern electrical systems. Number 14 wire has an ampacity of 20 A; No. 12 has an ampacity of 25 A; and No. 10 has an ampacity of 30 A. The typical impact on circuit makeup and loading is as follows.

1. Number 12 TW or THW copper is shown to have an ampacity of 25 A instead of 20 A; based on the general UL requirement that equipment terminals be limited to use with 60°C conductors up to a 100-A rating, No. 12 THHN or XHHW copper conductors must also be treated as having only a 25-A continuous rating and not a 30-A rating, as shown in Table 310-16. But the footnote to Table 310-16 limits all No. 12 copper wires to a maximum load of 20 A and requires that they be protected at not more than 20 A.

2. The ampacity of 25 A for No. 12 TW and THW copper wires interacts with Note 8 of Tables 310-16 to 310-19 where there are, say, six No. 12 TW current-carrying wires for the phase legs of two 3-phase, 4-wire branch circuits in one conduit supplying, say, receptacle loads. In such a case, the two neutrals of the branch circuits do not count in applying Note 8, and only the six phase legs are counted in determining the extent to which all circuit conductors must have their ampacities derated to the percent of values in Tables 310-16 and 310-18, given in the table in Note 8. In the case described here, that literally means that each No. 12 phase leg may be used at an ampacity of 0.8 × 25 A, or 20 A. And the footnote to Table 310-16 would require the use of a fuse or CB rated not over 20 A to protect each No. 12 phase leg. Each No. 12 would be protected at the current value that represents the maximum I^2R heat input that the conductor insulation can withstand.

3. If two 3-phase, 4-wire branch circuits of No. 12 TW or THW copper conductors were installed in a single conduit or EMT run to supply, say, fluorescent or other electric-discharge lighting, the two neutrals of the branch circuits would carry harmonic current even under balanced conditions and would have to be counted (Note 10c to Tables 310-16 through 310-19), along with the six phase legs, as current-carrying wires for ampacity derating in accordance with Note 8. In such a case, as the table of Note 8 shows for seven to nine conductors, their ampacities would have to be derated to 70 percent of the 25-A value shown in Table 310-16, which gives each conductor an ampacity of 17.5 A (0.7 × 25 = 17.5). If 20-A overcurrent protection were used for each No. 12 phase leg and the load were limited to no more than 17.5 A (or 16 A, which is 0.8 × 20 A, for a continuous load), the application would satisfy Note 8 in the NEC. And Sec. 210-19(a) permits the use of 20-A protection as the next-higher standard rating of protective device above the conductor ampacity of 17.5 A, provided that the circuit supplies only lighting outlets—and not readily accessible receptacle outlets that might allow the conductors to be overloaded above 17.5 A.

Use of the 20-A protection on conductors rated at 17.5 A is recognized only for fixed circuit loading (like lighting-fixture outlets) and not for the variable loading of receptacle outlets, because any increase in load current over 17.5 A would produce excessive heat input to the eight bundled No. 12 conductors in the conduit, which would damage and ultimately break down conductor insulation. Of course, the question arises: Over the operating life of the electrical system, how can excessive current be prevented? The practical, realistic answer is: It can't! It would be better to use 15-A protection on the No. 12 conductors or use conductors rated at 90°C.

4. Number 12 THHN or XHHW conductors, with their 90°C rating and consequently greater resistance to thermal damage, could be used for the two 3-phase, 4-wire,

20 A circuits to the electric-discharge lighting load; they would satisfy all NEC rules and would not be subject to insulation damage. With eight current-carrying wires in the conduit, the ampacity derating to 70 percent as required by Note 8 would be applied to the 30-A value shown in Table 310-16 as the ampacity of No. 12 THHN, RHH, or XHHW (dry locations). Then, because 0.7×30 A = 21 A, the maximum of 20-A protection required by the footnote to Table 310-16 would ensure that the conductors could never be subjected to excessive current and its damaging heat. And if the original loading on the conductors is set at 16 A (the 80 percent load limitation) for continuous operation of the lighting, any subsequent increase in load—even up to the full 20-A capacity, which would be a violation of Secs. 210-22(c) and 220-3(a)—would not reach the 21-A maximum ampacity set by Note 8.

Figure 2-12 summarizes the application described in items 3 and 4 above.

Advantage of 90°C Wires

If the four circuit wires in Fig. 2-11 are rated at 90°C, as are THHN, RHH, or XHHW conductors, then the loading and protection of the circuit must be related to the

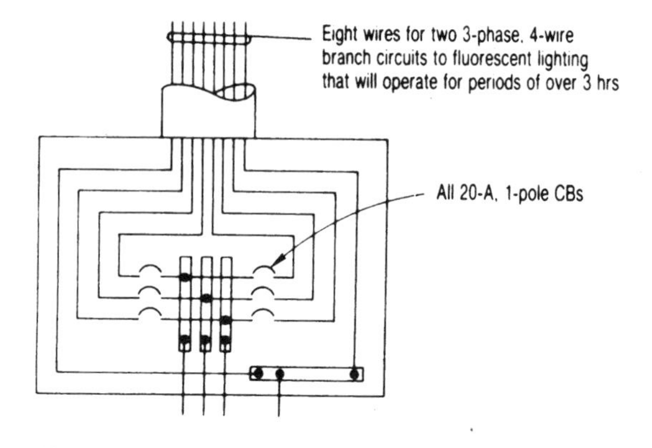

CASE 1
With TW or THW No. 12 copper conductors:
No. 12 ampacity = 25 A, from Table 310-16
From Note 8, max ampacity = 0.7 × 25 = 17.5 A

CASE 2
With THHN, XHHW or RHH No. 12 copper conductors:
No. 12 ampacity = 30 A, from Table 310-16
From Note 8, max ampacity = 0.7 × 30 = 21 A
Max load = 0.8 × 20 = 16 A

IN CASE 1, CONDUCTORS ARE NOT PROTECTED IN ACCORDANCE WITH THE 17.5-A MAXIMUM ALLOWABLE LOAD CURRENT. IN CASE 2, THEY *ARE* PROTECTED AGAINST EXCESSIVE LOAD CURRENT.

FIG. 2-12 Conductors with 90°C insulation eliminate the chance of conductor damage due to overload.

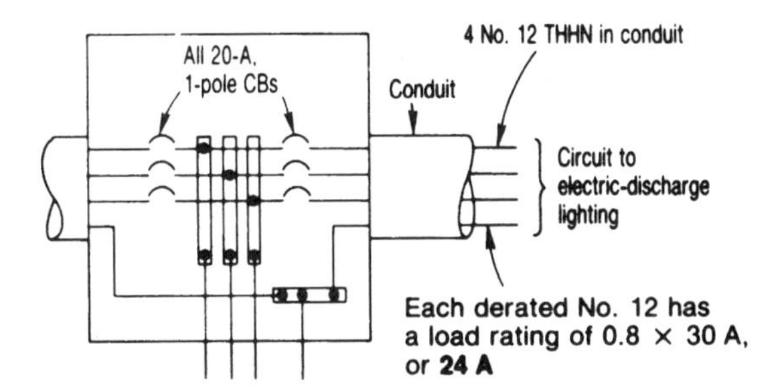

NOTE: TW or THW wire would have to be derated from 25A to 20 A.

FIG. 2-13 Derating of 90°C branch-circuit wires (No. 14, No. 12, and No. 10) is based on higher ampacities.

required ampacity derating as shown in Fig. 2-13. The application is based on these considerations:

1. As indicated, each No. 12 THHN has an ampacity of 30 A from NEC Table 310-16, but the footnote to the table limits the load and overcurrent protection on any No. 12 THHN to no more than 20 A.
2. Because the neutral of the 3-phase, 4-wire circuit must be counted as a current-carrying wire, there are four conductors in the conduit; each conductor must therefore have its ampacity derated to 80 percent of its table ampacity value, as required by Note 8 of Tables 310-16 to 310-19. Each No. 12 then has a new (derated) ampacity of 0.8 × 30 A, or 24 A.
3. By using a 20-A, single-pole protective device (fuse, single-pole CB, or one pole of a three-pole CB), which is the maximum protection permitted by the footnote to Table 310-16, each No. 12 THHN is easily made to comply with the requirement of Sec. 210-19 that the branch-circuit wire have an ampacity "not less than" the maximum load current being supplied.
4. If the lighting load on the circuit is noncontinuous, that is, does not operate for any period of 3 h or more, the circuit may be loaded up to its 20-A maximum rating.
5. If the load is continuous—full-load current flows for 3 h or more—the load on the circuit must be limited to 80 percent of the rating of each 20-A fuse or CB pole.

Figure 2-14 shows the use of two 3-phase, 4-wire circuits of THHN conductors in a single conduit. The 90°C wires offer distinct advantages (substantial economies) over the use of either 60°C (TW) or 75°C (THW) wires for the same application, as follows.

Resistive Load. If the circuit shown feeds only incandescent lighting or other resistive loads (or electric-discharge lighting does not make up "a major portion of the load"), then Note 10(c) of Tables 310-16 to 310-19 does not require the neutral conductor to be counted as a current-carrying conductor. In such cases, circuit makeup and loading could include these considerations:

1. With the six phase legs as current-carrying wires in the conduit, Note 8 requires that the ampacity of each No. 12 be derated from its basic table value of 30 A to 80 percent of that value, or 24 A.

2. Then each No. 12 is properly protected by a 20-A CB or fuse, satisfying both Sec. 210-19 and the footnote to Table 310-16.
3. If the circuit load is not continuous, each phase leg may be loaded to 20 A.
4. If the load is continuous, a maximum of 16 A (80 percent) must be observed to satisfy Sec. 384-16(c).

Electric-Discharge Load. If the two circuits of Fig. 2-14 supply electric-discharge lighting (fluorescent, mercury vapor, metal halide, high-pressure sodium, or low-pressure sodium), the makeup and loading must be as shown in case 2 of Fig. 2-12. If that same makeup of circuits supplies noncontinuous loads, the circuit conductors may be loaded up to 20 A per pole.

In case 2 of Fig. 2-12, the only difference between such circuit makeups using XHHW or RHH conductors and those using THHNs is the need for ¾-in conduit instead of ½-in conduit because of the larger cross-sectional area of RHH and XHHW (see tables 3A and 3B in NEC Chap. 9).

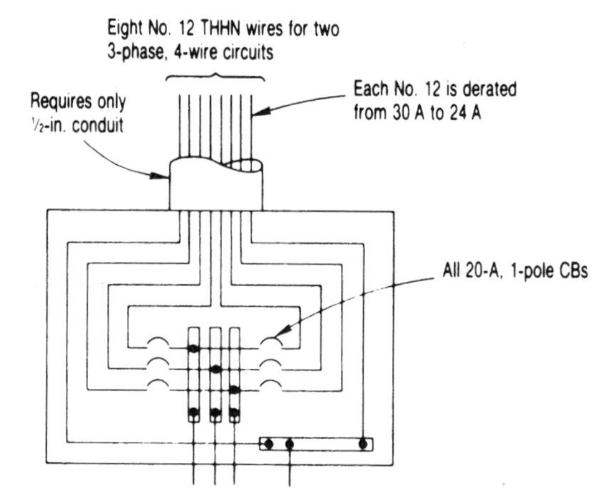

FIG. 2-14 The 90°C conductors can take derating without losing the full circuit load-current rating.

VOLTAGE DROP

Branch-Circuit Voltage Drop

In laying out circuits, the loading and lengths of homeruns and runs between outlets must be related to voltage drop and the need for spare capacity in the circuits for possible future increases in load. Each lamp, appliance, or other utilization device on the circuit is designed for best performance at a particular operating voltage. Although such devices will operate at voltages on either side of the design value, adverse effects will generally result from operation at voltages lower than the specified value.

A 1 percent drop in voltage to an incandescent lamp produces about a 3 percent decrease in light output; a 10 percent voltage drop will decrease the output about 30 percent. In heating devices of the resistance-element type, voltage drop has a similar effect on heat output. In motor-operated appliances, low voltage to the device will affect the starting and pullout torques, and the current drawn from the line will increase with drops in voltage. As a result, the heat rise in the motor windings will be above normal.

Voltage drop in the conductors is due to the resistance of the conductors plus, in ac circuits, reactance. And the heat developed by the dissipation of power (the I^2R losses) in the wiring, which itself costs money, deteriorates the conductor insulation. In good design, the voltage drop is held to no more than 1 percent.

Figure 2-15 shows the basic formulas used to calculate voltage drop, with examples of their application. Figure 2-16 shows how copper loss (the I^2R loss) is calculated for circuits with power factors of 100 percent and less than 100 percent.

FPN No. 4 to Sec. 210-19(a) of the NEC comments (but does not make a mandatory rule) on branch-circuit voltage drop, as follows:

> Conductors for branch circuits as defined in Article 100, sized to prevent a voltage drop exceeding 3 percent at the farthest outlet of power, heating, and lighting loads, or combinations of such loads and where the maximum total voltage drop on both feeders and branch circuits to the farthest outlet does not exceed 5 percent, will provide reasonable efficiency of operation. See Section 215-2 for voltage drop on feeder conductors.

That wording is only a recommendation, and the voltage drop should be held to the above-recommended 1 percent in lighting branch-circuit conductors to minimize energy losses in the conductors and ensure optimal operation and efficiency for load devices. Figure 2-17 is a tabulation of circuit loads and the maximum circuit lengths from the panelboard to the first outlet (on a balanced 3-wire, single-phase circuit or a 4-wire, 3-phase circuit) that will hold the voltage drop to a 1 percent maximum. Figure 2-18 contains a graph that may be used to directly select the wire size required to hold the voltage drop to 1 percent. Figure 2-19 gives examples of applying those data.

Although 50 percent loading of circuits is recommended for substantial protection against excessive voltage drop, the sizes of conductors for long runs should always be carefully selected to ensure that if spare capacity is put to use in the future, it will not increase the voltage drop beyond the limit. For this reason, when the design intent is to use 50 percent loading to provide spare capacity in the circuit, the conductors used in long runs should be sized for voltage drop on the basis of the maximum possible loading.

For lighting and appliance branch circuits, calculations of voltage drop should be made on a single-phase, 2-wire basis. That is, the voltage drop in the wires should be limited to some percentage of the voltage from the hot leg to the neutral—whether the circuit is 2-wire or 3-wire, single-phase or 4-wire, 3-phase. Since most branch-circuit loads are connected hot leg to neutral, the major concern should be with the voltage delivered to the loads, rather than with the phase-to-phase voltage when that voltage is not supplied to any load device. Of course, 2-wire circuits made up of ungrounded legs of single-phase or 3-phase systems should be designed for a voltage drop that is some percentage of the voltage between them when they supply loads at other than phase-to-neutral connection.

The voltage drop of any 2-wire circuit (either phase to neutral or phase to phase) can be taken simply as the *IR* drop of the conductors, using the total dc resistance of the two circuit wires. For conductors up to No. 3 AWG, which are used in the vast majority of branch circuits, the ac resistance of the wire is equal to the dc resistance; the latter is readily obtained from the right-hand columns of Table 8 in NEC Chap. 9. In such wire sizes, the inductive reactance of the circuit is negligible. Figure 2-20 shows an example of this approach.

1 Two-wire, single-phase circuits (inductance negligible):

$$V = \frac{2k \times L \times I}{d^2} = 2R \times L \times I$$

$$d^2 = \frac{2k \times I \times L}{V}$$

V = drop in circuit voltage (volts)
R = resistance per ft of conductor (ohms/ft)
I = current in conductor (amps)
L = one-way length of circuit (ft)
d^2 = cross section area of conductor (circular mils)
k = resistivity of conductor metal (cir mil-ohms/ft) (k = 12 for circuits loaded to more than 50% of allowable carrying capacity; k = 11 for circuits loaded less than 50%; k = 18 for aluminum conductors) at 30C

EXAMPLE:

What is the voltage drop in this circuit?

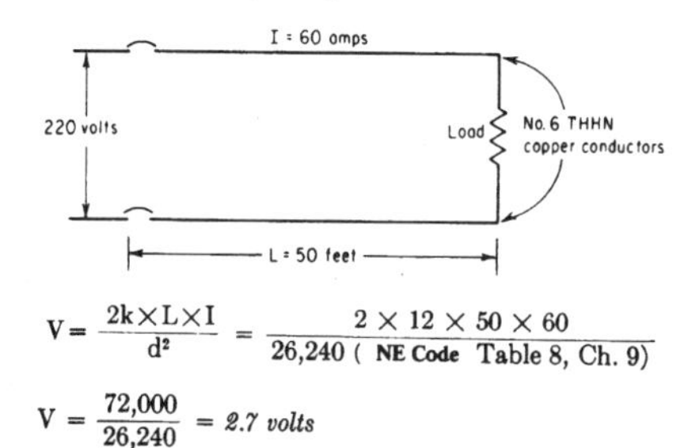

$$V = \frac{2k \times L \times I}{d^2} = \frac{2 \times 12 \times 50 \times 60}{26{,}240 \text{ (NE Code Table 8, Ch. 9)}}$$

$$V = \frac{72{,}000}{26{,}240} = \textit{2.7 volts}$$

2 Three-wire, single-phase circuits (inductance negligible):

$$V = \frac{2k \times L \times I}{d^2}$$

V = drop between outside conductors (volts)
I = current in more heavily loaded outside conductor (amps)

EXAMPLE:

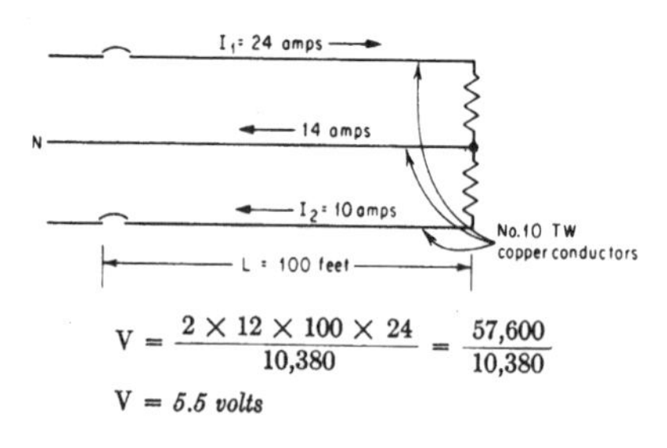

$$V = \frac{2 \times 12 \times 100 \times 24}{10{,}380} = \frac{57{,}600}{10{,}380}$$

V = *5.5 volts*

3 Three-wire, 3-phase circuits (inductance negligible):

$$V = \frac{2k \times I \times L}{d^2} \times 0.866$$

V = voltage drop of 3-phase circuit

EXAMPLE:

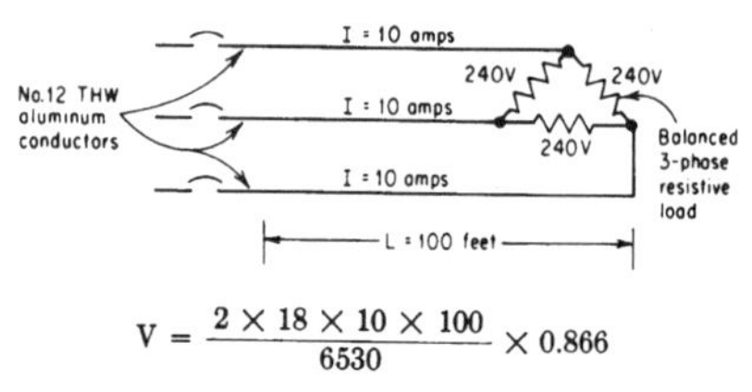

$$V = \frac{2 \times 18 \times 10 \times 100}{6530} \times 0.866$$

$$V = \frac{36{,}000}{6530} \times 0.866 = \textit{4.76 volts}$$

4 Four-wire, 3-phase balanced circuits (inductance negligible):

For lighting loads: Voltage drop between one outside conductor and neutral equals one-half of drop calculated by formula for 2-wire circuits.

For motor loads: Voltage drop between any two outside conductors equals 0.866 times drop determined by formula for 2-wire circuits.

EXAMPLE:

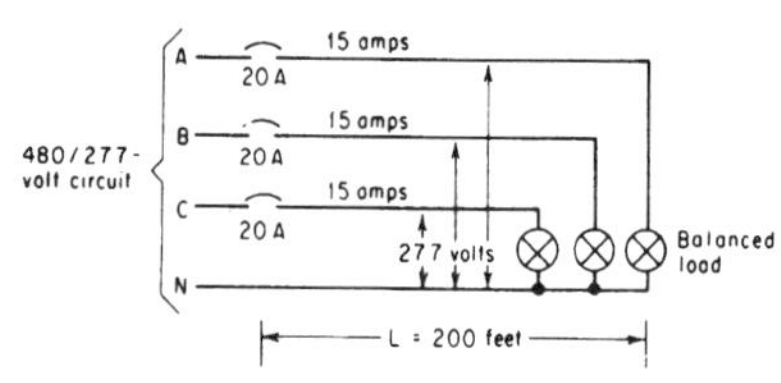

What size of copper wire must be used here to keep voltage drop to 1% under balanced conditions (i.e., 1% of 277 or 2.8 volts)?

For a 2-wire circuit, $d^2 = \frac{2k \times I \times L}{V}$

For a 3-ϕ, 4-wire circuit, $d^2 = \frac{2k \times I \times L}{V} \times \frac{1}{2}$

$$d^2 = \frac{2 \times 12 \times 15 \times 200}{2.8} \times \frac{1}{2} = \frac{72{,}000}{5.6}$$

= 12,857 circular mils

From Table 8, Ch. 9, NE Code —
No. 10 = 10,380 CM No. 8 = 16,510 CM
No. 8 must be used.

FIG. 2-15 Basic formulas for branch-circuit voltage drop.

With 100% load power factor and negligible conductor reactance

When load power factor is 100% and conductor reactance due to self-induction is negligible, calculation of voltage drop in the circuit and calculation of copper loss due to heating effect of current in the circuit conductors follows the standard relations between current and voltage. The supply voltage is equal to the arithmetic sum of the voltage drop in the conductors and the voltage across the load. Copper loss—energy wasted as heat produced in the circuit conductors—is a simple "I^2R" loss.

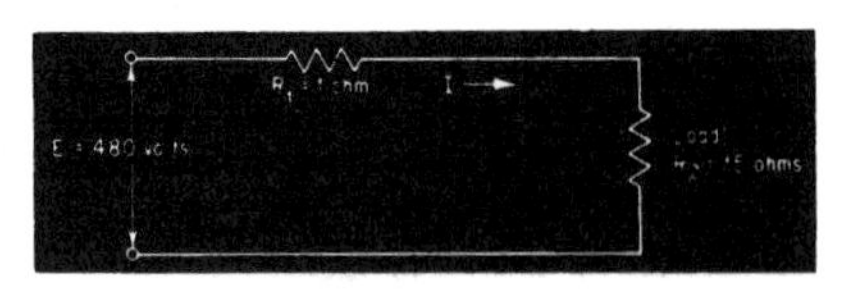

where E = circuit voltage (volts)
R_1 = total resistance of circuit conductors (ohms)
R_2 = resistance of noninductive load (ohms)
I = current flowing in circuit (amps)

$$I = \frac{E}{R_1 + R_2} = \frac{480}{1 + 15} = 30 \text{ amps}$$

Voltage drop in conductors $= I \times R_1 = 30 \times 1 = 30$ *volts*

Copper loss in conductors $= I^2 \times R_1 = (30)^2 \times 1 = 900$ *watts*

$$\text{Percent voltage drop} = \frac{30}{480} \times 100 = 6\%$$

With less than 100% load power factor and negligible conductor reactance

When the power factor of the load is less than 100% and the conductor reactance is negligible, the sum of the voltage drops across the load and the resistance of the circuit conductors is no longer equal to the circuit supply voltage. In the following example, note that although the voltage drop in the circuit conductors is 30.4 volts and might appear to leave only 449.6 volts for the load (480 − 30.4), there is actually 456 volts across the load (30.4 amps x 15 ohms of load impedance). The real loss in voltage to the load is, therefore, only 24 volts (480 − 456). The 30.4-volt drop in the conductors is only the voltage across the resistive conductor load, which differs in phase from the voltage across the reactive load device fed by the circuit. Although the arithmetic sum of the two voltages (across the conductor resistance and across the load) is equal to 486.4 (30.4 + 456), correct vector addition of the two voltages gives the 480-volt value of the supply. And note that the significant percent voltage drop is different from the apparent value. Copper loss, however, remains a simple "I^2R" value.

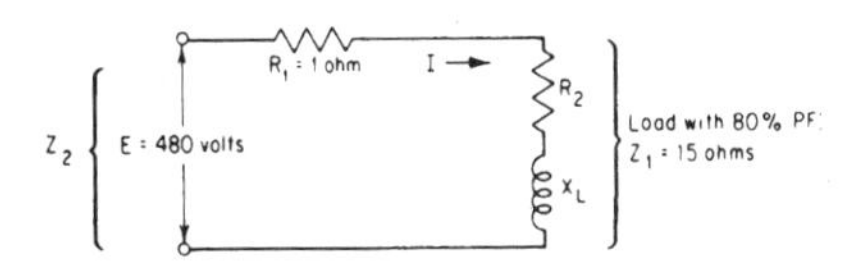

where X_L = inductive reactance component of load (ohms)
Z_1 = impedance of load (ohms)
Z_2 = total impedance of circuit (ohms)

$Z_1 = \sqrt{(R_2)^2 + (X_L)^2}$ $(Z_1)^2 = (R_2)^2 + (X_L)^2$

then, $(15)^2 = (R_2)^2 + (X_L)^2$

but, POWER FACTOR $= 0.8 = \frac{R_2}{Z_1} = \frac{R_2}{15}$

thus, $R_2 = 0.8 \times 15 = 12$ *ohms*

then from $(X_L)^2 = (Z_1)^2 - (R_2)^2$
$= (15)^2 - (12)^2$
$= 225 - 144 = 81$

$X_L = \sqrt{81} = 9$ *ohms*

The total impedance of the circuit, Z_2, includes conductor resistance and must be calculated in order to find circuit current.

$$Z_2 = \sqrt{(R_1 + R_2)^2 + (X_L)^2} = \sqrt{(1 + 12)^2 + (9)^2} = \sqrt{169 + 81} = \sqrt{250} = 15.8 \text{ ohms}$$

Then $I = \frac{480}{15.8} = 30.4$ *amps*

Voltage drop in conductors $= I \times R_1 = 30.4 \times 1 = 30.4$ *volts*

Copper loss in conductors $= I^2 \times R_1 = (30.4)^2 \times 1 = 924$ *watts*

$$\text{Apparent \% voltage drop} = \frac{30.4}{480} \times 100 = 6.3\%$$

Voltage delivered to load $= I \times Z_1 = 30.4 \times 15 = 456$ *volts*

Significant % voltage drop

$$= \frac{480 - 456}{480} \times 100 = \frac{24}{480} \times 100 = 5\%$$

FIG. 120 Calculating voltage drop and copper loss.

FIG. 2-16 Calculating voltage drop and copper loss.

Loads and lengths in feet for 1% drop on 3- and 4-wire, 115-volt circuits

Amp load	#10 wire	#12 wire	#14 wire
1	946	596	374
2	474	298	188
3	316	198	124
4	236	148	94
5	190	120	76
6	158	100	62
7	136	86	54
8	118	74	46
9	106	66	42
10	94	60	38
11	86	54	34
12	78	50	32
13	72	46	28
14	68	42	26
15	64	40	24
16	60	38	
17	56	36	
18	52	34	
19	50	32	
20	48	30	
21	46		
22	44		
23	42		
24	40		
25	38		
26	36		
27	36		
28	34		
29	32		
30	32		

Calculations based on copper resistance of 13 ohms per CM-ft at 60C (140F).

For a circuit made up of two phase wires and the neutral of a 3-phase, 4-wire wye system, multiply given lengths by 0.67.

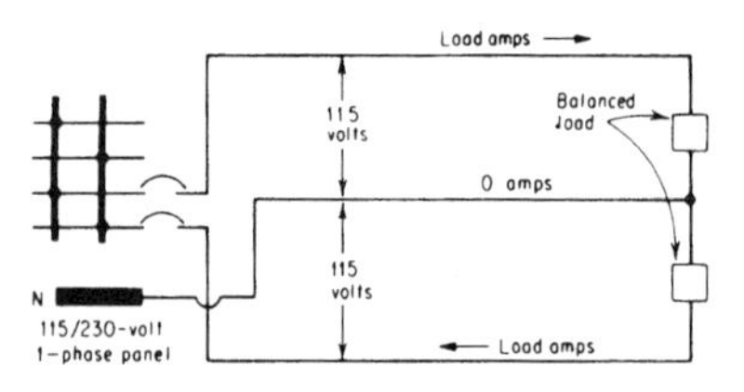

Balanced 3-wire single-phase circuit

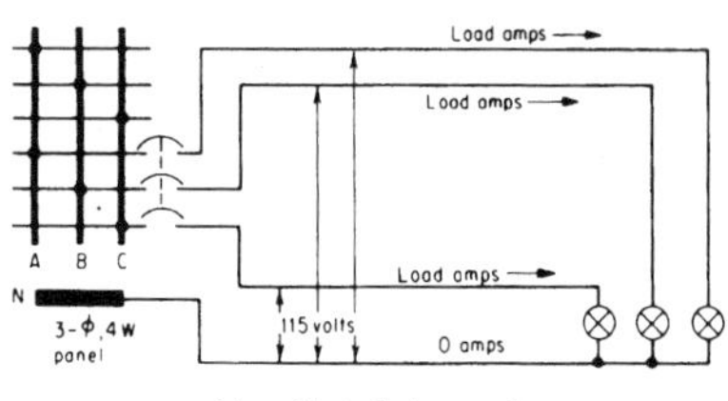

Balanced 4-wire 3-phase circuit

NOTE:
This table is based on a 1% voltage drop in each 115-volt circuit which is part of a 3-wire, single-phase circuit or part of a 4-wire, 3-phase circuit—provided the circuit is operating under balanced conditions. This is important because under balanced conditions, the neutral associated with each 115-volt circuit does not carry current. As a result, each 115-volt circuit is, in effect, fed by only one conductor. This produces half the voltage drop that would occur under unbalanced conditions, with both the phase leg and neutral carrying current. The distances given are, therefore, twice what they would be for the same voltage drop under unbalanced conditions.

EXAMPLE:

Q. In making up a 115/230-volt, 3-wire circuit to carry 15 amps on each hot leg for a total circuit length of 57 ft, what size wire must be used to keep the voltage drop to 1% with a balanced load?

A. Come down the left hand column, "Amp load," to the circuit load of 15 amps. Move across and note maximum distances for each wire size.

It can be seen that No. 12 wire reaches a drop of 1% (1.15 volts) at a length of 40 ft. No. 10 wire can carry a 15-amp circuit for up to 64 ft before reaching the voltage drop level.

Because the required circuit length of 57 ft is greater than 40, No. 12 cannot be used. But because it is less than 64 ft, *No. 10* wire will keep voltage drop within limits.

FIG. 2-17 Sizing branch-circuit wires for 1 percent voltage drop in the homerun (the wires between the panelboard and the first outlet).

The maximum tolerable voltage drop in a motor circuit usually depends upon the speed and torque requirements of the motor and how they would be affected by low voltage. In Fig. 2-21, for example, the characteristics of the dc motor and its load-torque requirements indicate that the motor needs at least 223 V for proper performance. For operation at normal ambient temperature, the branch-circuit conductors would be sized as follows to ensure at least the minimum allowable voltage at the motor.

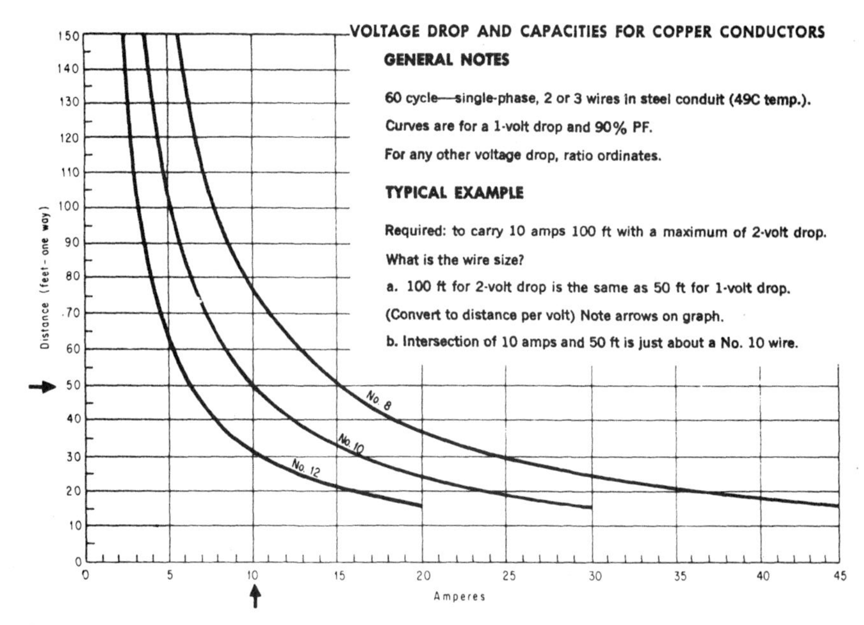

FIG. 2-18 Graph shows branch-circuit voltage drop and minimum wire size for a given load current.

The branch-circuit conductors must provide the voltage drop along with the required carrying capacity. NEC Sec. 430-22 requires branch-circuit conductors supplying a single motor to have an ampacity not less than 125 percent of the motor full-load current rating:

$$1.25 \times 20 \text{ A} = 25 \text{ A}$$

From NEC Table 310-16, a No. 12 TW copper conductor has an ampacity of 25 A; this conductor would therefore satisfy Sec. 430-22 as regards the minimum ampacity for the motor circuit. And because the actual full-load current of the motor is 20 A, there is no conflict with the footnote to Table 310-16 requiring that the load current rating of No. 12 copper not exceed 20 A.

Although the same footnote also sets 20 A as the maximum rating of overcurrent protection for No. 12 copper conductors, Sec. 240-3(f) exempts "motor-circuit conductors" from the basic requirement of Sec. 240-3 that "conductors...shall be protected against overcurrent in accordance with their ampacities as specified in Sec. 310-15...."

To ensure a voltage of at least 223 V at the motor, the maximum permitted voltage drop in the 250-ft circuit run from the panelboard to the motor must be

$$228 \text{ V} - 223 \text{ V} = 5 \text{ V}$$

Table 8 in NEC Chap. 9 shows that a No. 12 copper conductor has a dc resistance of 1.93 Ω per 1000 ft at a temperature of 77°F. Because resistance is the only cause of voltage drop in a dc circuit (there is no inductive reactance as in ac circuits), the voltage drop of this circuit, using No. 12 copper wire, can be readily calculated.

Balanced 3-wire single-phase circuit

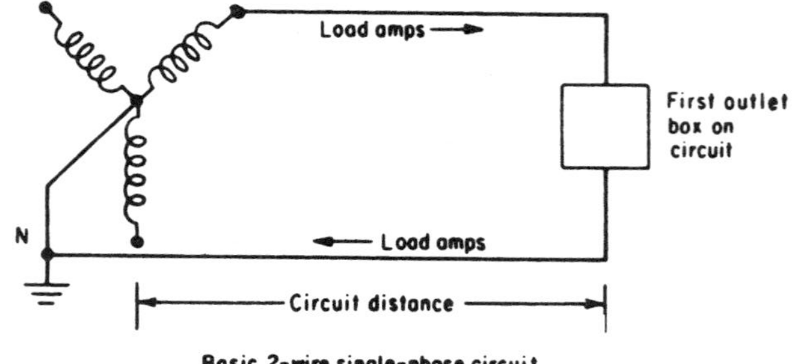

Basic 2-wire single-phase circuit

NOTE:

In both of the types of circuits shown, there are two conductors carrying current—one out to the load and one back from the load. For the single-phase circuit derived from the 3-phase, 4-wire wye, it is best to select conductors for a 3-phase, 4-wire circuit on the basis of the voltage drop which would exist if only one phase was operating. This gives the worst condition for voltage drop. When such a circuit is truly balanced, there is no current in the neutral, and the phase-to-neutral connected load is being served by one conductor—which produces only half the voltage drop that will be present under conditions of maximum unbalance.

EXAMPLES:

1. **Q.** What wire size should be used to make up a 2-wire, 480-volt circuit to carry 15 amps a distance of 100 ft with a drop of 5 volts?

A. A drop of 5 volts over a 100-ft circuit is equal to a 1-volt drop for each 20 ft ($100 \div 5$). By converting to this value of distance per volt, we can use the table directly. On the horizontal scale, locate the load of *15 amps*. On the vertical scale, locate the distance of *20 ft*. The intersection of coordinates shows that *No. 12* conductors would be O.K.

2. **Q.** What wire size should be used for a 3-phase, 4-wire, 480/277-volt circuit that supplies a balanced lighting load of 16 amps per phase—to keep voltage drop at not more than 3 volts over the circuit length of 120 ft, even under conditions of maximum unbalance?

A. A drop of 3 volts over 120 ft is the same as 1 volt for 40 ft. Using the coordinates of 16 amps and 40 ft indicates that No. 10 conductors would do the job.

FIG. 2-19

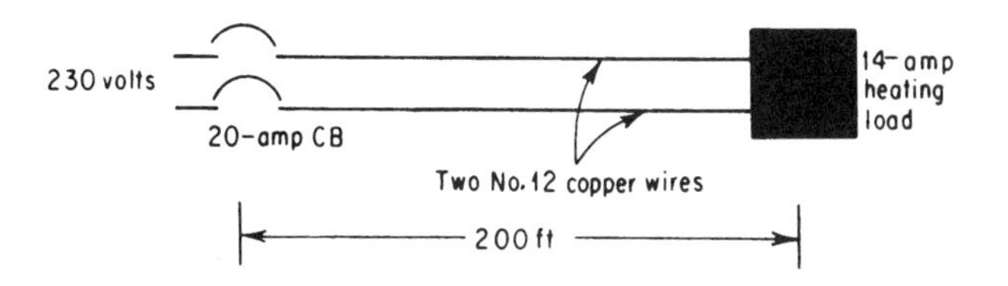

Total length of No. 12 conductor:

$$200 \times 2 = 400 \text{ ft}$$

From Table 8, Chapter 9, **NE Code**, No. 12 solid copper conductor has a dc resistance of 1.93 ohms per 1000 ft.

Then the 400-ft circuit length has a total resistance of

$$1.93 \times \frac{400}{1000} = 0.772 \text{ ohm}$$

$$\text{Voltage drop} = \text{I} \times \text{R} = 14 \text{ amps} \times 0.772 \text{ ohm} = 10.808 \text{ volts.}$$

In the preceding problem, the conductors were assumed to be exposed to an ambient temperature of 25C (77F). The resistance of the conductors, and thus the voltage drop, would increase under conditions of higher ambient temperature.

FIG. 2-20 Calculating voltage drop from NEC table of wire resistances.

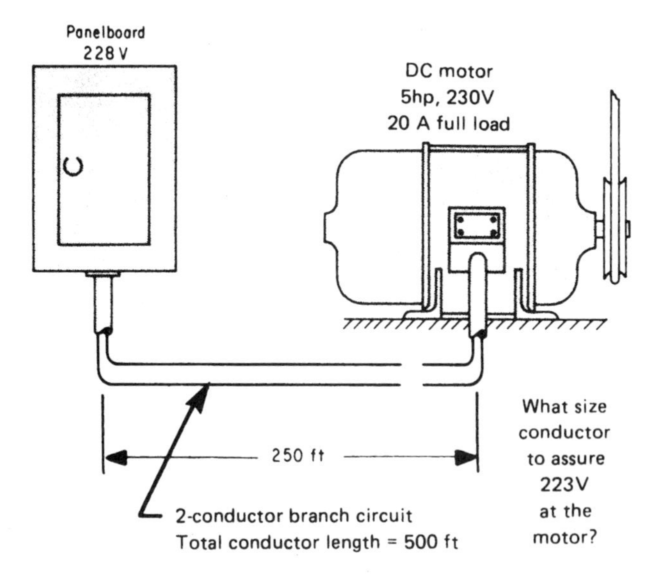

FIG. 2-21 Voltage drop for dc motor branch circuit involves only resistance and current.

With two conductors for the 250-ft-long circuit, the total length of No. 12 wire is 2 × 250 ft, or 500 ft. At 1.93 Ω per 1000 ft, the total resistance of the circuit would be half of 1.93 Ω, or 0.965 Ω.

At a full-load current of 20 A, the drop in voltage from the panel to the motor would be

$$V_d = I \times R = 20 \text{ A} \times 0.965 \ \Omega = 19.3 \text{ V}$$

That would produce a motor voltage of

$$228 \text{ V} - 19.3 \text{ V} = 208.7 \text{ V}$$

Obviously, this would not satisfy the requirement of 223 V at the motor. A larger conductor, with less resistance, must be used.

Because the drop of 19.3 V with No. 12 wire is almost 4 times the maximum Code-recommended drop of 5 V, the required conductor should have a resistance about one-fourth that of No. 12 wire. Reference to the table of dc resistances shows that No. 6 wire has a resistance of 0.491 Ω per 1000 ft, which is about 1.93 Ω/4. So a No. 6 looks like it might do.

The calculation is repeated for the No. 6 wire: At a resistance of 0.491 Ω per 1000 ft, the required 500 ft of No. 6 wire for the 250-ft-long 2-wire circuit would have a total resistance equal to half of 0.491 Ω, or 0.2455 Ω.

$$V_d = I \times R = 20 \text{ A} \times 0.2455 \ \Omega = 4.91 \text{ V}$$

The motor supply voltage then would be

$$228 \text{ V} - 4.91 = 223.09 \text{ V}$$

This satisfies the need for at least 223 V at the motor, and No. 6 copper conductor (TW, THW, RHH, THHN, or XHHW) must be used.

As shown in Table 9 in NEC Chap. 9, conductors smaller than No. 2 AWG have negligible self-inductance, and therefore there is no difference between the resistance of such conductors to direct current and their resistance to alternating current. In larger conductors, the *skin effect* (self-inductance forcing current to flow in the outer part of the conductor cross-sectional area) produces an ac resistance that is higher than the dc resistance. From the vast majority of branch-circuit designs, the dc resistance of wire is entirely satisfactory for use in voltage-drop calculations.

In the smaller sizes of conductors, the inductive reactance produced under ac conditions is also generally negligible. However, it is important to distinguish between the increase in ac resistance due to self-inductance and that due to inductive reactance, which is also produced by the self-inductance of the conductor.

As noted above, ac flow in conductors is subject to the skin effect, which produces an apparent increase in resistance over the resistance value which would obtain for dc flow. This is due to a reduction in the effective conductor cross section resulting from the tendency of ac to flow close to the surface (or "skin") of the conductor. Generally, this increase in resistance to alternating current is of little consequence in conductors smaller than No. 4/0 AWG.

Reactance in conductors carrying ac power depends upon the size of the conductor, the spacing between it and other conductors carrying current, the position of the conductor with respect to conductors close to it, the frequency of the alternating current, and the presence of magnetic materials close to the conductor. In an ac circuit, the reactance of the conductors may be reduced by placing the conductors close together and/or by placing them in nonmagnetic raceway instead of steel conduit or raceway. In many large-size or long ac circuits, the voltage drop due to impedance is often far greater than the drop due simply to the resistance of the conductors.

When current flows in a conductor in which the reactance due to self-induction is negligible, the voltage drop is equal to the product of the current in amperes and the total resistance of the conductor in ohms. But when the reactance of the conductor is not negligible, the voltage drop is equal to the product of the current in amperes and the total impedance of the conductor, which is determined from the formula

$$Z = \sqrt{R^2 + X^2}$$

where Z = total impedance, Ω
R = total ac resistance of conductor, Ω
X = reactance of conductor, Ω

The voltage drop in such a conductor is

$$V_d = I \times Z$$

where V_d = voltage drop, V
I = current flowing in conductor, A
Z = total impedance of conductor, Ω

Resistance and reactance data on wires and cables are given in literature made available by the manufacturers. Tables and graphs are also available for quickly and easily computing the voltage drops in large, heavily loaded feeders operating at less than unity power factor and with considerable conductor reactance. But for branch-circuit calculations of voltage drop, inductive reactance is not nearly as significant as it is for the larger conductors used for feeders.

Feeder Voltage Drop

Voltage drop must be carefully considered in sizing feeder conductors, and calculations should be made for peak load conditions. The voltage drop must be calculated on the basis that full system voltage (for example, 120, 240, 480 V) is available at the service entrance or transformer secondary supplying the feeder. The voltage drop is the amount of reduction from the nominal supply voltage. Feeder conductors should be sized such that the voltage drop up to the branch-circuit panelboards or the point at which the branch circuits originate is not more than 1 percent for lighting loads or combined lighting, heating, and power loads, and not more than 2 percent for power or heating loads. Local codes may impose lower limits on voltage drop. Voltage drop limitations are shown in Fig. 2-22, as follows:

1. For combinations of lighting and power loads on feeders and branch circuits, use the voltage drop percentages for lighting loads (at the left in Fig. 2-22).
2. The word *feeder* here refers to the overall run of conductors carrying power from the source (the service entrance, load center substation, or local transformer at which the full value of system voltage is available) to the point of final branch-circuit distribution, including feeders, subfeeders, sub-subfeeders, etc.
3. The voltage drop percentages are based on nominal circuit voltage being available at the load terminals of the source of each voltage level in the distribution system.

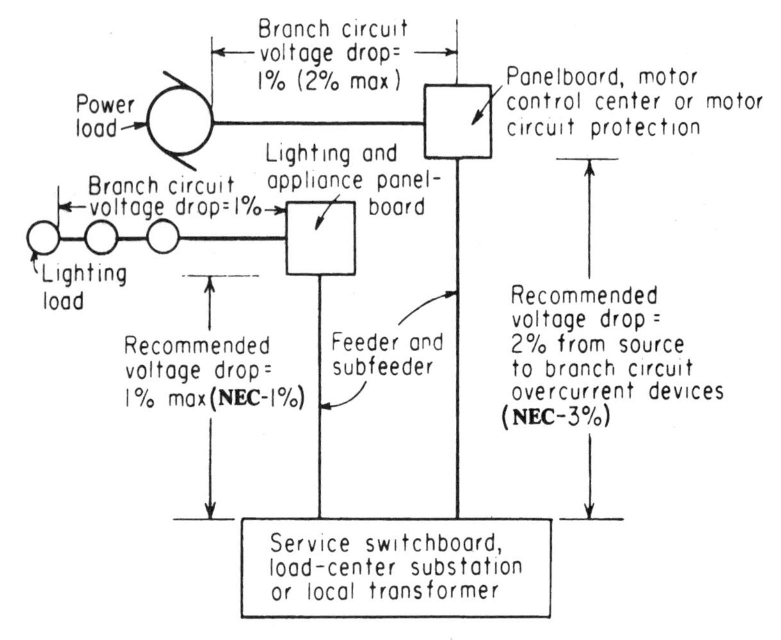

FIG. 2-22 Feeder voltage drop must be carefully calculated to be within the limits of good design.

The voltage drop in any set of feeders can be calculated with the formulas given in Fig. 2-23. From this calculation, it can be determined if the conductor size initially selected to handle the load will be adequate to maintain the voltage drop within given limits. If it is not, the size of the conductors must be increased (or other steps taken where conductor reactance is not negligible) until the voltage drop is within the prescribed limits.

In application, the loadings and lengths of feeders can be adjusted to accommodate voltage drop requirements, with additional advantages as shown in Fig. 2-24 for a modern office building. A total of nine risers supply lighting and receptacle circuits in all the building's office space—floors 2 through 22. Each of these floors has three electric closets. At each feeder riser location, one feeder supplies the bottom eight floors, the second feeder supplies seven floors in the middle of the building's height, and the third feeder supplies the top six floors. With the design demand load the same for each of the electric closets, the feeder serving the bottom eight closets has a heavier load than the one serving the middle seven closets, which in turn is more heavily loaded than the feeder to the top six closets. Because each feeder is designed to deliver its kilovoltampere capacity with only a 2 percent drop in voltage from the switchboard to any panel, the unbalanced loading of the feeders made possible the use of a single makeup—four 500-kcmil THW conductors in $3\frac{1}{2}$-in conduit for all the feeders. This one size of feeder can supply a total load of eight closets with a 2 percent voltage drop for the distance from the switchboard to the lowest load, but it can only supply a total of six closets with a 2 percent drop when the feeder has to run to the top of the building.

Two-Wire, Single-Phase Circuits (Inductance Negligible):

$$V = \frac{2k \times L \times I}{d^2} = 2R \times L \times I \qquad d^2 = \frac{2k \times I \times L}{V}$$

V = drop in circuit voltage (volts)
R = resistance per ft of conductor (ohms/ft)
I = current in conductor (amps)

Three-Wire, Single-Phase Circuits (Inductance Negligible):

$$V = \frac{2k \times L \times I}{d^2}$$

V = drop between outside conductors (volts)
I = current in more heavily loaded outside conductor (amperes)

Three-Wire, Three-Phase Circuits (Inductance Negligible):

$$V = \frac{2k \times I \times L}{d^2} \times 0.866$$

V = voltage drop of 3-phase circuit

Four-Wire, Three-Phase Balanced Circuits (Inductance Negligible):

For lighting loads: Voltage drop between one outside conductor and neutral equals one-half of drop calculated by formula for 2-wire circuits.

For motor loads: Voltage drop between any two outside conductors equals 0.866 times drop determined by formula for 2-wire circuits.

In above formulas

L = one-way length of circuit (ft)
d^2 = cross-section area of conductor (circular mils)
k = resistivity of conductor metal (cir mil-ohms/ft) (k = 12 for circuits loaded to more than 50% of allowable carrying capacity; k = 11 for circuits loaded less than 50%; k = 18 for aluminum conductor)

FIG. 2-23 These basic formulas indicate resistive voltage drop in feeders.

The use of a single basic feeder makeup for all feeders provides two distinct advantages. First, four 500-kcmil conductors in 3½-in conduit is a highly economical and efficient makeup, based on apparent power delivered, voltage drop, and cable power loss and measured against the installed cost of the feeder. Second, the use of a single feeder makeup minimizes construction costs. The use of only 500-kcmil conductors in 3½-in conduit permitted a high degree of standardization and mechanization in installing the conduit, pulling the conductors, coupling to panel enclosures, making taps, installing lugs, etc.

There are many cases in which the above-mentioned limits of voltage drop (1 percent for lighting feeders, etc.) should be relaxed in the interest of reducing the prohibitive costs of conductors and conduits that may be required by such low drops. In many installations, a 5 percent drop in feeders is not critical or unsafe. A recommended plan for real-

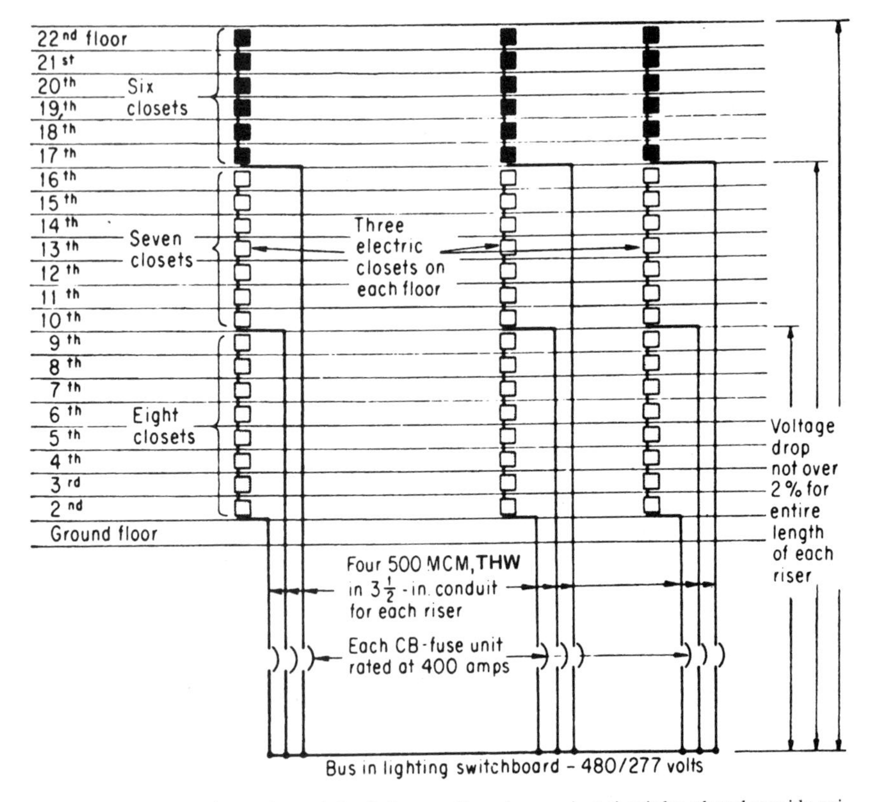

FIG. 2-24 Feeder loads may be varied to balance voltage drop against circuit length and provide uniform circuit makeup.

istic voltage drop in apartment houses is shown in Fig. 2-25, where the slightly greater drop is not considered detrimental to the lighting and appliance loads in the apartments.

Figure 2-26 shows how load center distribution improves voltage level and regulation by deriving full-voltage feeders from transformers located at the centers of load groupings, thereby shortening the secondary feeders and reducing the voltage drop.

Figures 2-27 and 2-28 show sets of curves correlating size of conductor, ampere load, and voltage drop for 3-phase circuits in steel conduit. The curves are for a 1-V drop and 90 percent power factor, based on ampacities for types RHW and THW (75°C) 600-V insulation. Many such graphs and tabulated data on voltage drop are available in handbooks and from manufacturers.

Figures 2-29 and 2-30 are handy tables for determining maximum feeder circuit lengths for 1 percent voltage drop on single-phase and 3-phase feeders.

The table in Fig. 2-29 gives the one-way lengths for 115/230-V circuits under balanced loading. With a balanced 3-wire load, the drop will be 1.15 V for each 115-V circuit, that is, for each side of the 115/230-V circuit. For a 230-V, 2-wire circuit, the given lengths produce a voltage drop of 2.3 V. An example is shown below the table.

The table in Fig. 2-30 gives the one-way lengths for 3-phase, 230-V, delta ac load circuits with 85 percent power factor. For balanced load conditions, the conductors and lengths shown produce voltage drops of 2.3 V (1 percent of 230 V). For other voltage drop percentages (2 percent, 3 percent, etc.), the lengths shown can be doubled, tripled,

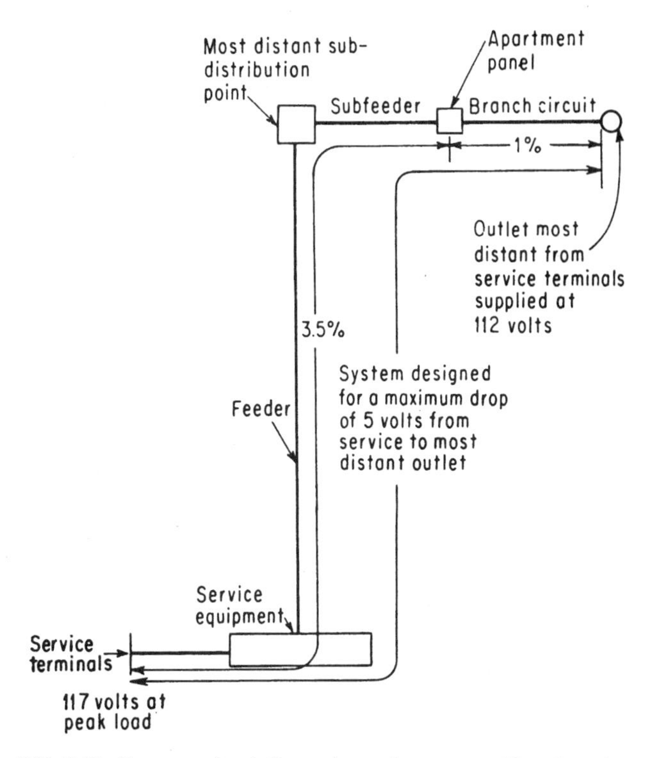

FIG. 2-25 Economy often indicates the maximum acceptable voltage drop for noncritical loads.

etc. Thus, if, as shown, a 90-A-loaded 3-phase circuit can be run 113 ft using No. 1/0 conductors for a 2.3-V drop, then the same load could be carried twice as far (226 ft) if double the voltage drop (4.6 V) were acceptable. The table may be used for calculating other feeder lengths for 1 percent voltage drop:

- For 208-V, 4-wire wye feeders, multiply the given lengths by 0.9.
- For 230-V, single-phase feeders, multiply the given lengths by 0.85.
- For 460-V, 3- or 4-wire feeders, multiply the given lengths by 2.
- For aluminum wire, multiply the given lengths by 0.7 or use the length of copper wire which is two sizes smaller than the aluminum under consideration.

Short-Circuit Calculations

The selection of circuit overcurrent devices must include careful consideration of the ability of the devices to operate properly and safely on short-circuit faults—phase-to-ground and phase-to-phase faults in grounded wiring systems and phase-to-phase faults in ungrounded wiring systems. In all cases, the following points are important:

1. Under normal operation, a circuit draws current in proportion to the voltage applied and the impedance of the load. When a short circuit occurs, the source voltage no longer encounters the opposition to current flow that the normal load had presented.

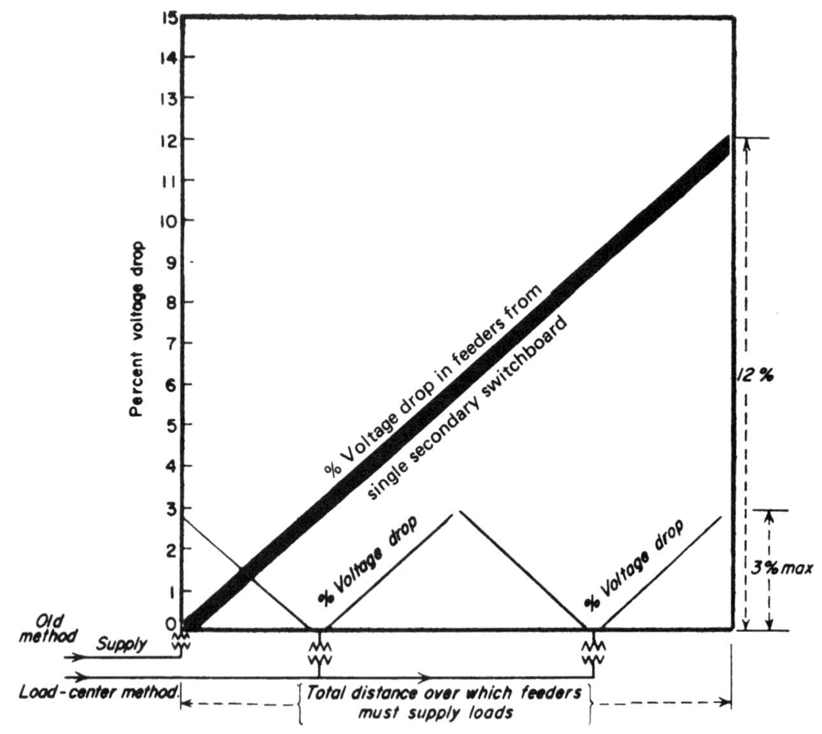

FIG. 2-26 Voltage drop can be greatly reduced by deriving feeders from load center substations or transformer distribution centers.

Instead, the voltage is applied across a load of much lower impedance, made up of the impedance of the conductors from the source of the voltage to the short-circuit fault, the impedance of the transformer from which the circuit is derived, and any other impedances due to equipment interposed in the circuit between the transformer and the fault.

In the top part of Fig. 2-31, the amount of current which will flow through the short circuit is determined from the system voltage and the total impedance connected in the path of current flow from the source to the fault. This impedance includes resistance and reactance in the conductors, in one or more transformers (going from the fault back to the source), and in any equipment connected in the path of current flow. If, as shown, a source of infinite capacity is assumed on the primary side of the transformer—as if it were a source of voltage with no internal impedance—then the impedance that determines the amount of short-circuit current on the secondary of the transformer consists of the impedance of the transformer itself and the impedance of the secondary conductors up to the fault.

In a transformer circuit, the "impedance" of the unit is involved in the ability of the transformer to supply short-circuit currents to faults on the load side of the transformer.

VOLTAGE DROP AND CAPACITIES

60 cycle—3-phase in steel conduit (49°C temp.)

GENERAL NOTES

Curves are for a 1-volt drop and 90% PF.
For any other voltage drop, ratio ordinates.
Capacities are for Type RHW 600-volt insulation.

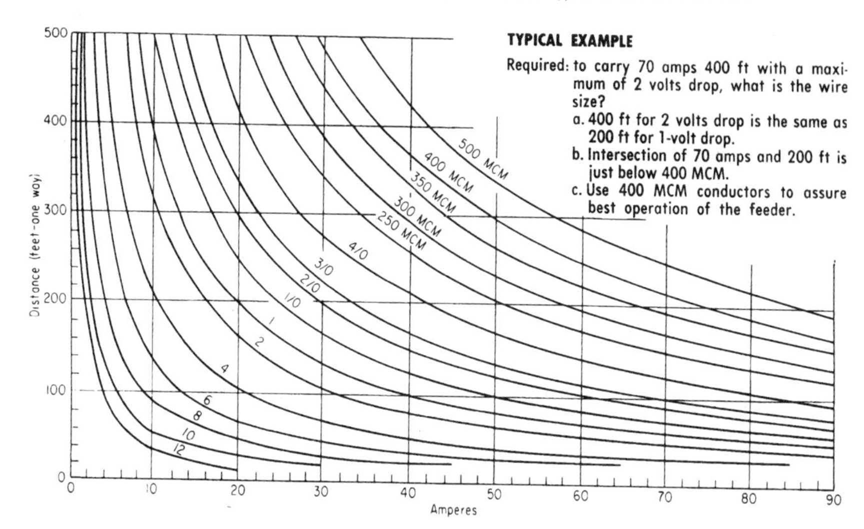

FIG. 2-27 Maximum lengths of 3-phase feeder circuits for a 1-V drop at loads up to 90 A.

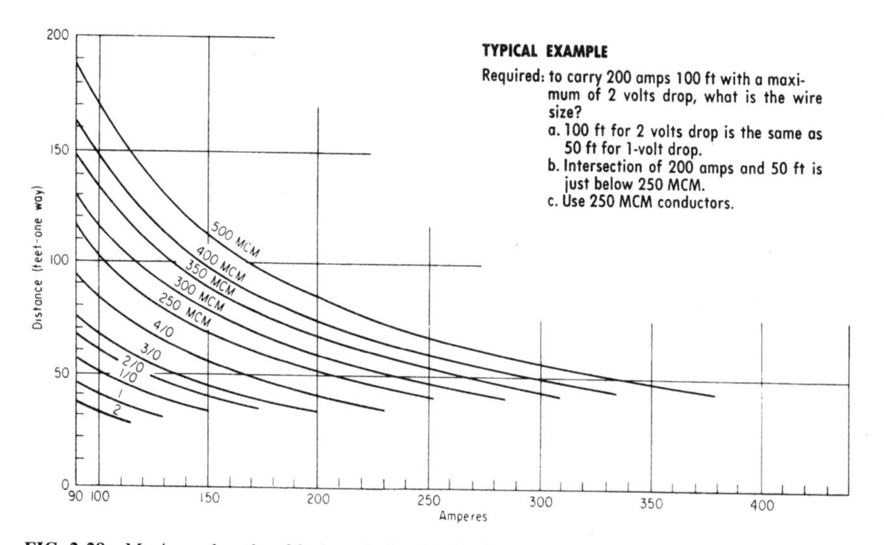

FIG. 2-28 Maximum lengths of 3-phase feeder circuits for a 1-V drop at loads from 90 to 380 A.

SINGLE-PHASE AC LOADS

115/230 volts, 60 cycles, 100% PF

Amp load	Wire size—circular mils					Wire size—B & S or AWG									
	500	400	350	300	250	4/0	3/0	2/0	1/0	1	2	3	4	6	8
40	1106	898	788	669	558	475	378	299	239	188	150	119	94	59	38
50	885	719	630	535	447	380	303	240	191	150	120	91	75	47	30
60	737	599	525	446	372	317	252	200	159	125	100	79	62	39	
70	632	513	450	382	319	271	216	171	136	107	86	68	53	34	
80	553	449	394	334	279	238	189	150	119	94	75	59	47		
90	491	399	350	297	248	211	168	133	106	83	67	53	42		
100	442	359	315	267	223	190	151	120	95	75	60	47			
110	402	327	286	243	203	173	138	109	87	68	55				
120	369	299	263	223	186	158	126	100	79	63					
130	340	276	242	206	172	146	116	92	73	58					
140	316	257	225	191	159	136	108	86	68						
150	295	240	210	178	149	127	101	80	64						
160	276	225	197	167	140	119	95	75	60						
170	260	211	185	157	131	112	89	70							
180	246	200	175	148	124	106	84	66							
190	233	189	166	140	117	100	80								
200	221	180	157	134	112	95	76								
210	211	171	150	127	106	90									
220	201	163	143	122	101	86									
230	192	156	137	116	97	83									
240	184	150	131	111	93										
250	177	144	126	107	89										
260	170	138	121	103	80										
270	164	133	117	99											
280	158	128	112	96											
290	152	124	109	92											
300	147	120	105												
310	143	116	102												
320	138	112													
330	134	109													
340	130	106													

Calculations based on copper resistance of 12.5 ohms per CM-ft at 50C (122F).

Reactance and impedance losses calculated for each wire.

Conductors closely grouped in metallic conduit.

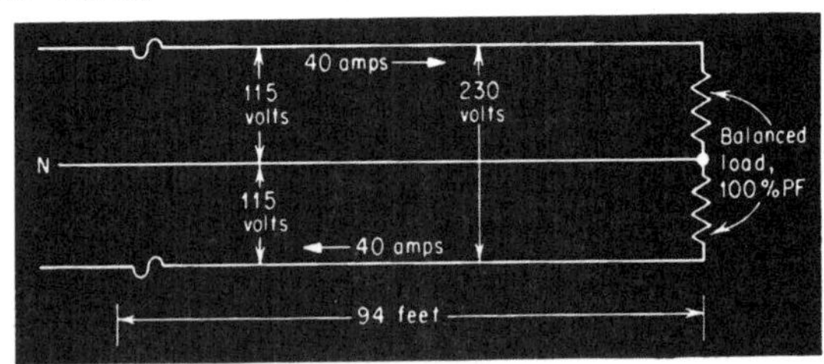

REFER TO TABLE: For a load of 40 amps, it shows that No. 4 conductors must be used if the circuit length is between 60 and 94 ft. Under the condition shown in the diagram, there will be a 1.15-volt drop in each 115-volt circuit.

FIG. 2-29 Circuit lengths (in feet) for a 1 percent voltage drop in single-phase feeders.

3-PHASE DELTA AC LOADS

230 volts, 60 cycles, 85% PF

Amp load	Wire size—circular mils					Wire size—B & S or AWG									
	500	400	350	300	250	4/0	3/0	2/0	1/0	1	2	3	4	6	8
40	710	625	584	530	475	429	364	303	253	208	173	139	113	75	49
50	568	500	467	424	380	343	291	242	203	167	139	111	90	60	39
60	473	417	389	353	317	286	243	202	169	139	115	93	75	50	
70	406	357	333	303	271	245	208	173	145	119	99	79	64	43	
80	355	312	292	265	238	214	182	151	127	104	87	69	56		
90	316	278	259	235	211	191	162	134	113	93	77	62	45		
100	284	250	233	212	190	172	146	121	101	83	69	55			
110	258	227	212	193	173	156	132	110	92	76	63				
120	237	208	195	177	158	143	121	101	84	69	58				
130	218	192	180	163	146	132	112	93	78	64					
140	203	179	167	151	136	123	104	86	72						
150	189	168	156	141	127	114	97	81	67						
160	177	156	146	132	119	107	91	76							
170	167	147	137	125	112	101	86	71							
180	158	139	130	118	106	95	81	67							
190	149	132	123	112	100	90	77								
200	142	125	117	106	95	86	73								
210	135	119	111	101	90	82									
220	129	114	106	96	86	78									
230	123	109	101	92	83	75									
240	118	104	97	88	79										
250	114	100	93	85	76										
260	109	96	90	81	73										
270	105	93	86	78											
280	101	89	83	76											
290	98	86	80	73											
300	95	83	78												
310	92	81	75												
320	89	78													
330	86	76													
340	83	73													
350	81														
360	79														
370	77														
380	75														

Calculations based on copper resistance of 12.5 ohms per CM-ft at 50C (122F).

Reactance and impedance losses calculated for each wire.

Conductors closely grouped in metallic conduit.

FIG. 2-30 Circuit lengths (in feet) for a 1 percent voltage drop in 3-phase, 3-wire feeders.

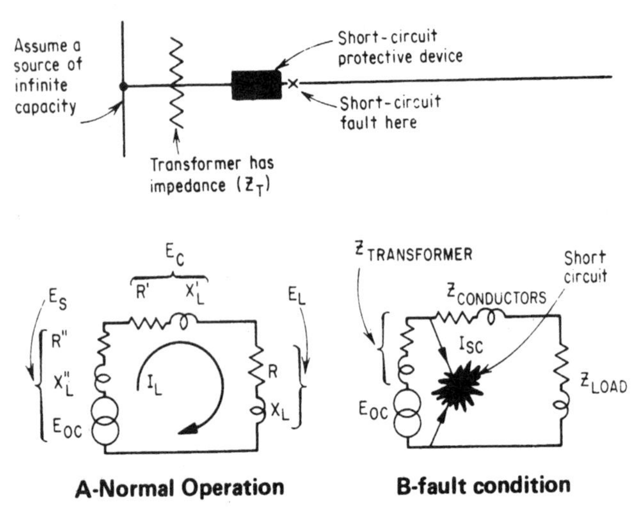

FIG. 2-31 Elements of a short-circuit condition.

The impedance of a transformer is the opposition which the transformer presents to the flow of short-circuit current through it.

As shown at the bottom of Fig. 2-31, under normal conditions of operation (*a*), a transformer winding can be considered to be a source of voltage E_s. This source is made up of a generator producing open-circuit voltage E_{oc} and having internal resistance R' and inductive reactance X_L''. Under normal conditions, load current I_L flows, determined by the transformer impedance, the impedance of the circuit conductors, and the impedance of the load. When a short occurs (*b*), the transformer open-circuit voltage E_{oc} is connected across a total load made up only of the transformer impedance and whatever part of the circuit conductor impedance is in the short circuit. If the supply to the transformer primary can deliver the necessary primary current, the secondary short-circuit current is equal to E/Z. Thus the transformer impedance is the limiting factor for short-circuit current.

All transformers have impedance, which generally is expressed as a percentage of the normal rated primary voltage. That percentage of the primary voltage must be applied to the transformer to cause full-load rated current to flow in the short-circuited secondary. For instance, if a 480- to 120-V transformer has an impedance of 5 percent, then 5 percent of 480 V, or 24 V, applied to the primary will cause rated load current to flow in the short-circuited secondary. If 5 percent of the primary voltage will cause such current, then 100 percent of the primary voltage will cause 100/5 (or 20) times the full-load rated secondary current to flow through a solid short circuit on the secondary terminals.

From the foregoing, it can be seen that the lower the impedance of a transformer of given kilovoltampere rating, the higher the short-circuit current which it can deliver. Take two transformers, each rated at 500 kVA. Assume the rated secondary load current is the same for both transformers. If one transformer is rated at 10 percent impedance, it can supply 100/10 (or 10) times the rated secondary current into a short circuit on its secondary terminals. If the other transformer is rated at 2 percent impedance, it can supply 100/2 (or 50) times the rated secondary current into a short circuit on its secondary terminals. Thus, the second transformer can supply 5 times as much short-circuit current as the first, even though they have the same load-handling ability.

Figure 2-32 presents a simple calculation that yields the symmetric, single-phase short-circuit current. The short-circuit protective device must be capable of safely interrupting this value of current and the asymmetric current value that is obtained by applying some multiplier (such as 1.25) to the symmetric value. In a more rigorous analysis, this value would be reduced by all impedances in the circuit.

2. Every overcurrent protective device, in a circuit at any voltage level, must be capable of interrupting the maximum possible short-circuit current that might be delivered by the system into a solid short on the load terminals of the device—without destroying itself in the process. This is shown in Fig. 2-33, and the short-circuit protective device in Fig. 2-32 must be rated for at least the 16,680-A calculated current.

3. Coordinated selective protection for modern circuits provides fast, effective isolation of any faulted section of a system but does not interrupt service to any other section. A careful study of the time-current characteristics of protective devices along with correct application can ensure clearing of a fault by the device nearest to the fault on its supply side, as shown in Fig. 2-34.

Selective coordination of overcurrent protection, in which every fault is cleared by the closest protective device on the line side of the fault, is widely considered to be a particular capability of modern CB equipment—either CBs alone or CBs in conjunction

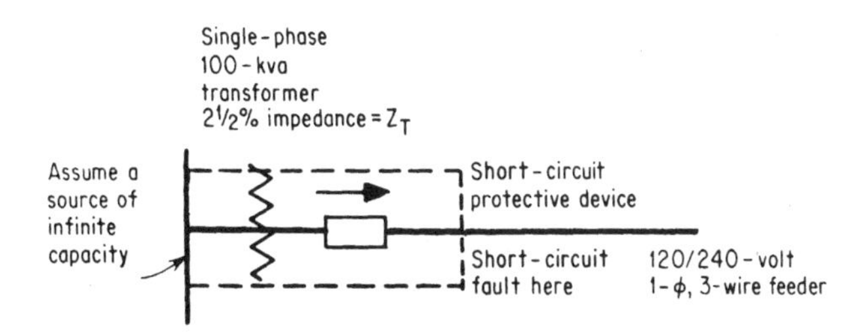

Assuming negligible line and other impedances between the transformer and the fault:

1. Transformer full-load secondary current $= \dfrac{100{,}000 \text{ va}}{240 \text{ v}} = \textit{417 amps}$

2. Maximum short-circuit current based on transformer impedance =

$$\frac{100\%}{\%Z_T} \times \text{secondary current}$$

$$\text{Max. } I_{sc} = \frac{100}{2.5} \times 417 = \textit{16,680 amps symmetrical}$$

FIG. 2-32 Basic idea behind short-circuit calculations.

All short-circuit protective devices. . .

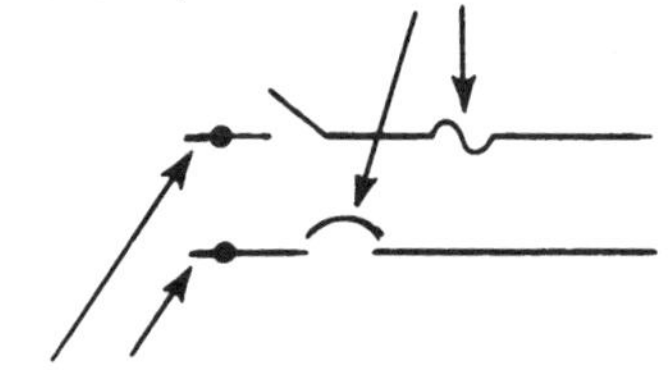

. . . must have an interrupting rating at least equal to the maximum fault current that the circuit could deliver into a short circuit on the *line side* of the device.

NOTE: That means that the fault current "available" at the line terminals of all fuses and circuit breakers *must be known* in order to assure that the device has a rating sufficient for the level of fault current.

FIG. 2-33 All feeder fuses and circuit breakers must have adequate interrupting capacity.

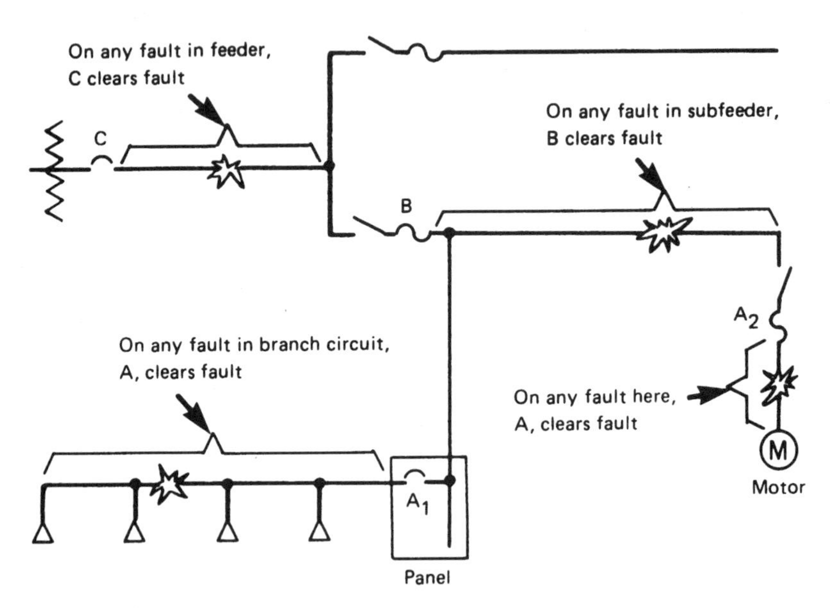

NOTE: Any fault is cleared by nearest device on supply side of fault, and all other devices in series with the fault do not operate.

FIG. 2-34 Selection coordination of overcurrent devices minimizes the extent of an outage due to a fault.

with load-side fusible equipment. Selective coordination ensures fault-initiated opening at only one point in the system, such as on a branch circuit or a feeder; it eliminates cascade operation, where, say, both a branch device and a feeder device open on a branch-circuit fault.

Curve-shaping of the time-current trip characteristics of both molded-case and air CBs by means of long-time-delay, short-time-delay, and/or instantaneous-trip elements is the technique by which sophisticated selective coordination can be achieved, even though there are a number of voltage transformations. Only the faulted circuit is opened, and the extent of the outage is minimized. But, again, such coordination must be carefully and precisely related to all other design, installation, operational, and economic factors.

4. Let-through current is another factor in the effective application of protective devices for short circuits. A given protective device may be able to interrupt the maximum short-circuit current at its point of installation safely, but it may take so much time to open the faulted circuit that damage will occur to equipment or devices connected in series with the fault. Unless the device operates quickly enough, the tremendous rupturing stresses created by short circuits can cause severe damage. The system components must be related to the let-through current—the current which flows from the time the fault develops until the circuit is opened. That is required by NEC Sec. 110-10, as shown in Fig. 2-35. Current-limiting fuses, for instance, open a short circuit in much less than half a cycle, thereby squelching let-through current.

Because a short-circuit fault is an accidental condition that occurs at random with respect to the alternating voltage wave of a circuit, the fault can occur at any point in the waveform of a voltage cycle. The two extremes of the waveform occur where the voltage wave passes through the "zero" axis and where the voltage is a maximum. Figure 2-36 shows these two extreme conditions at which short-circuit current can start and continue

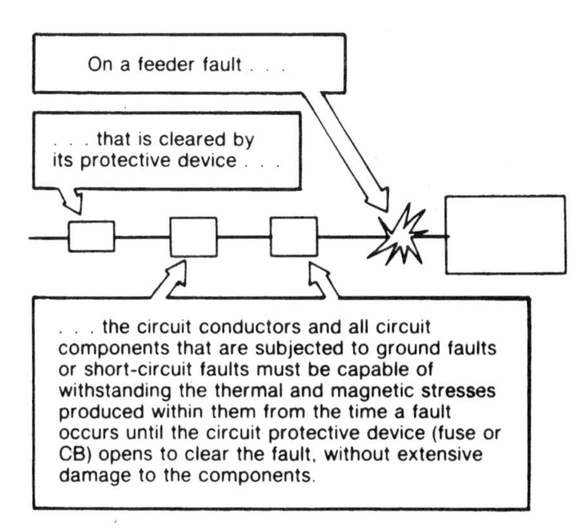

FIG. 2-35 Operating speed of short-circuit protection must be related to equipment *withstandability* ratings.

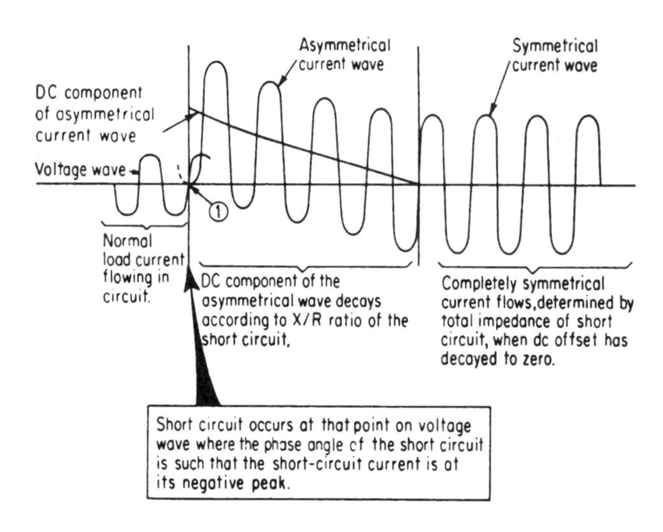

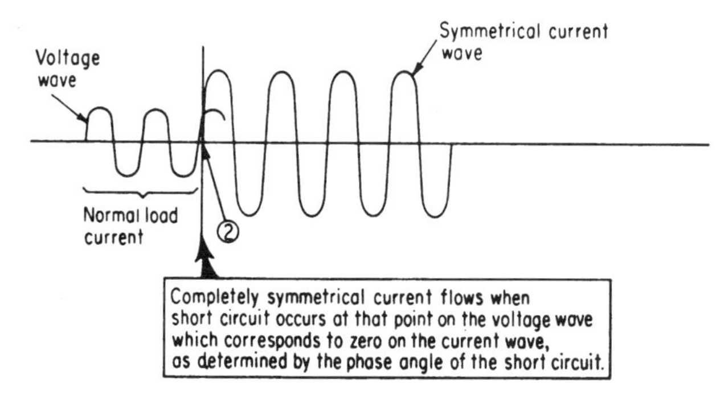

FIG. 2-36 How short-circuit currents could flow if they were not opened by overcurrent protection.

to flow if the circuit is not opened by fuses or breakers. Maximum asymmetry in the flow of short-circuit current is produced when the short circuit occurs at that point in the voltage wave for which the short-circuit current is exactly at its positive or negative peak value, as determined by the ratio of reactance to resistance in the short circuit; this ratio establishes the phase-angle difference between the voltage wave and the short-circuit current wave. In the case shown in Fig. 2-36*a,* with the short circuit occurring as shown on the voltage wave, the phase angle is such that the short-circuit current is exactly at its negative peak. Maximum asymmetry is thus produced. The instantaneous change in phase relation between current and voltage (from the normal condition to fault condition) produces an offset in the ac wave which acts as a dc transient.

From the immediate condition of asymmetry, the dc component of the short-circuit current wave decays to zero at a rate determined by the ratio of reactance to resistance in the short circuit. For a purely resistive short circuit (zero ratio of reactance to resistance), which is a completely theoretical condition that can never be achieved in an ac circuit, the decay to zero of the dc component will be instantaneous. If the short circuit is purely reactive (infinite ratio of reactance to resistance), which is also a practical impossibility, the dc component never decays to zero, the current remains offset, and the asymmetric rms current is equal to 1.73 times the symmetric rms current. Of course, actual circuits have ratios of reactance to resistance up to about 20, although the vast majority of cases have ratios of less than 7. This means the dc offset will decay to zero in several cycles.

In Fig. 2-36*b,* a short-circuit current wave of complete symmetry flows when the short occurs at that point in the voltage wave for which the instantaneously initiated short-circuit current is at its zero value. Again the exact point in the voltage wave which will correspond in time to a zero value for the short-circuit current wave is determined by the phase angle (or power factor, or ratio of reactance to resistance) of the short circuit.

Although short circuits can start at the two extremes shown in Fig. 2-36, chances are greatest that a fault will occur at some intermediate degree of asymmetry. But because design must account for the worst possibility, short-circuit calculations should be performed to determine the asymmetric current which represents the most destructive current that the protective device might be called upon to break.

From the character of the asymmetry, it can be seen that the ratio of reactance to resistance is important in applying protective devices because it indicates how fast the direct current will decay and how much current the device will be called upon to handle in an attempt to open the circuit in the first half cycle, or in the second cycle, or the third, etc.—since the current decreases with each succeeding cycle. A fast-acting fuse such as a current-limiting fuse, which opens the short circuit in the first half cycle, will break almost the maximum value of the asymmetric wave and must be rated for that current value. On the other hand, a circuit breaker which will not act until the fourth cycle of the short-circuit current wave will have to interrupt only the level of current to which the asymmetric wave has decayed.

On a fully asymmetric short circuit, the maximum instantaneous current—the peak asymmetric current—occurs at the end of the first half cycle of the current wave. Owing to decay of the dc component, subsequent peaks are lower in value. Various values for this half-cycle asymmetric current have been determined from study of actual circuits and are shown in Fig. 2-37.

The foregoing discussion explains the use of multipliers to determine the asymmetric current from the calculated symmetric value, such as that arrived at in Fig. 2-37. For fuses, which operate relatively fast, a multiplier of 1.4 is commonly applied to the rms symmetric short-circuit current to get the rms asymmetric value. For circuit breakers which operate almost as fast as some fuses, the same multiplier would be used in determining the possible asymmetric current when the breaker opens the circuit. But for breakers which operate more slowly, say two or three cycles after the fault starts, a multiplier of only 1.25 or 1.1 will give the value to which the asymmetric current has decayed when the circuit is opened.

The value of 1.4, mentioned above as the multiplier for determining the first-half-cycle value of the rms asymmetric current, has been established for a ratio of reactance to resistance which is not exceeded in the majority of cases. Thus, this multiplier yields the value to which the asymmetric current has decayed in the first half cycle. But in cir-

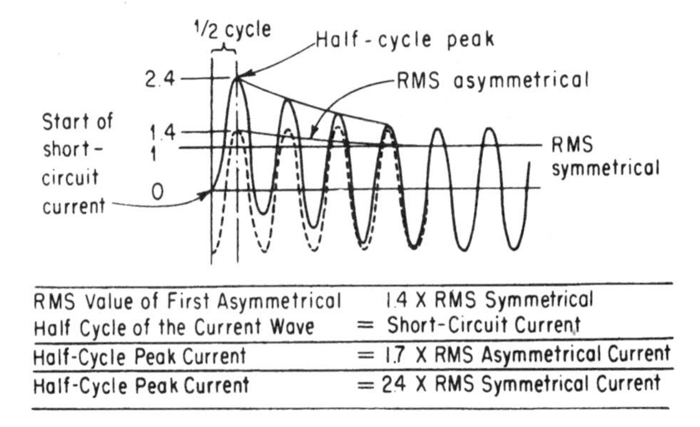

FIG. 2-37 Fast-acting fuses and CBs must be rated for the current value in the first half cycle.

cuits with the higher-than-normal ratio of reactance to resistance, the decay will not be as great in the first half cycle. A multiplier of, say, 1.5 or 1.6 would be required to compute the possible value of the first-half-cycle rms asymmetric current in such circuits. In all applications, when multipliers are used, the *X/R* ratio to which the multiplier applies should be known, as well as the *X/R* ratio of the circuit being protected.

Figure 2-38 shows the application of current-limiting fuses to quickly clear ground and short-circuit faults and to limit the energy let-through on circuits which could deliver fault currents over 100,000 or 200,000 A. On a given short circuit, a current-limiting fuse may be called upon to interrupt either of the two extreme conditions shown here, or some intermediate condition. In the fully asymmetric case, the fuse would limit both rms current and peak available current if the melting time of the fuse were anything less than one-half cycle. But within its current-limiting range, a true current-limiting fuse must clear the circuit in a time interval that is not greater than the duration of the first symmetric current loop, and it must limit the peak let-through current to a value less than the peak available current. Thus, it is the action of a fuse in clearing a symmetric short circuit that determines its current-limiting ability, as shown in the lower sketch. From the upper sketch, it can be seen that a fuse which has a melting time almost up to the peak of the asymmetric first half cycle will limit current let-through on an asymmetric fault but will not limit let-through if the short happens to occur as a completely symmetric wave.

The rms value of the triangular wave of let-through current through the fuse is

$$I_{LT} = \frac{I_p}{1.7}$$

where I_{LT} = rms let-through current, A
I_p = peak let-through current, A

The application of standard thermomagnetic or fully magnetic circuit breakers should be based on a correlation of their time-delay tripping curves and instantaneous trip setting with the requirements for overload protection, the current characteristics of the circuit being protected, and the need for coordination in the operation of overcurrent devices connected in series (such as a feeder device in series with a subfeeder device in

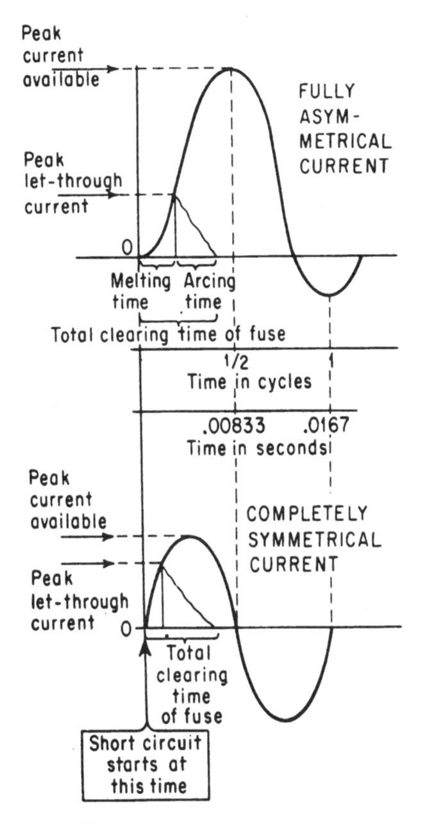

FIG. 2-38 Current-limiting operation of fuse protection.

series with a branch-circuit device). Breaker operation is readily analyzed in relation to a particular circuit by reference to the manufacturer's curves on the breaker.

Figure 2-39 shows a typical set of curves for a molded-case CB. The curves show trip time versus current for a 225-A breaker with interchangeable trip units. Each trip unit consists of a fixed thermal trip element and an independent adjustable magnetic trip element. The various trip units give the breaker its particular continuous current rating and are available in standard settings: 70, 100, 125, 150, 175, 200, and 225 A. Then the adjustable magnetic element associated with each thermal trip can be adjusted to 3 to 10 times the current rating of the thermal trip. The curves are shown as bands to cover the tolerances to which the elements are manufactured and within which they will operate. A band can be taken to be a large number of individual curves laid one against another. The curves show time plotted against multiples of any current rating (that is, multiples of 70 A, or 150 A, etc.). The two operating bands shown below 100 s represent the two extreme conditions of setting of the adjustable magnetic trip—3 times the CB rating at left and 10 times the CB rating at right. Intermediate settings of the magnetic trip would produce bands between the two shown here.

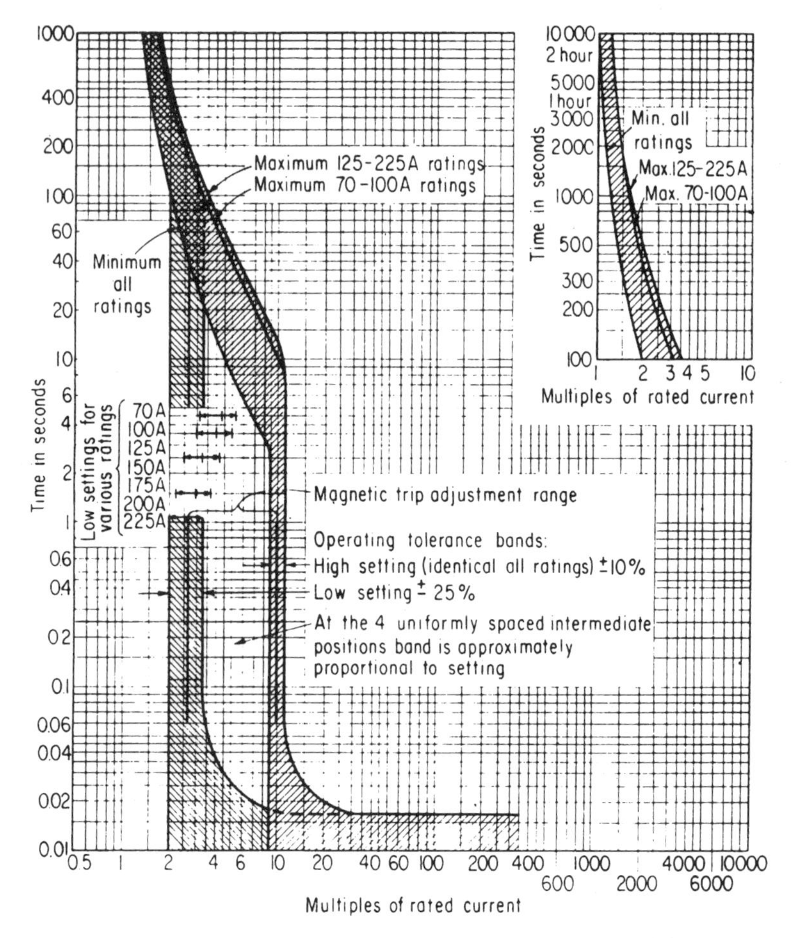

FIG. 2-39 CB time-current curves reveal overload and short-circuit characteristics.

Figure 2-40 shows the application of current-limiting fuses in combination with a power circuit breaker. Current-limiting fuses can be combined with both molded-case and power circuit breakers. A typical application is represented by the curves shown. The fuses extend the maximum interrupting capacity of the CB to values above 200,000 rms symmetric A.

The three-pole CB is an electrically operated low-voltage power circuit breaker with a continuous rating of 3000 A and an interrupting capacity of 75,000 rms asymmetric A at 480 V. The CB has three magnetic tripping characteristics in a 2000-A trip device. The first, the long-time-delay trip, is adjusted to pick up at 2000 A. The second, the short-time-delay trip, is adjusted to pick up at 6000 A. The third trip is the instantaneous trip, set for 12 times the coil rating, or 24,000 A. The tolerance on pickup settings is ±15 percent.

The time-delay trips are factory adjusted for required delays. The fuses used are rated at 3000 A continuous carrying capacity with interrupting capacity of 200,000 A.

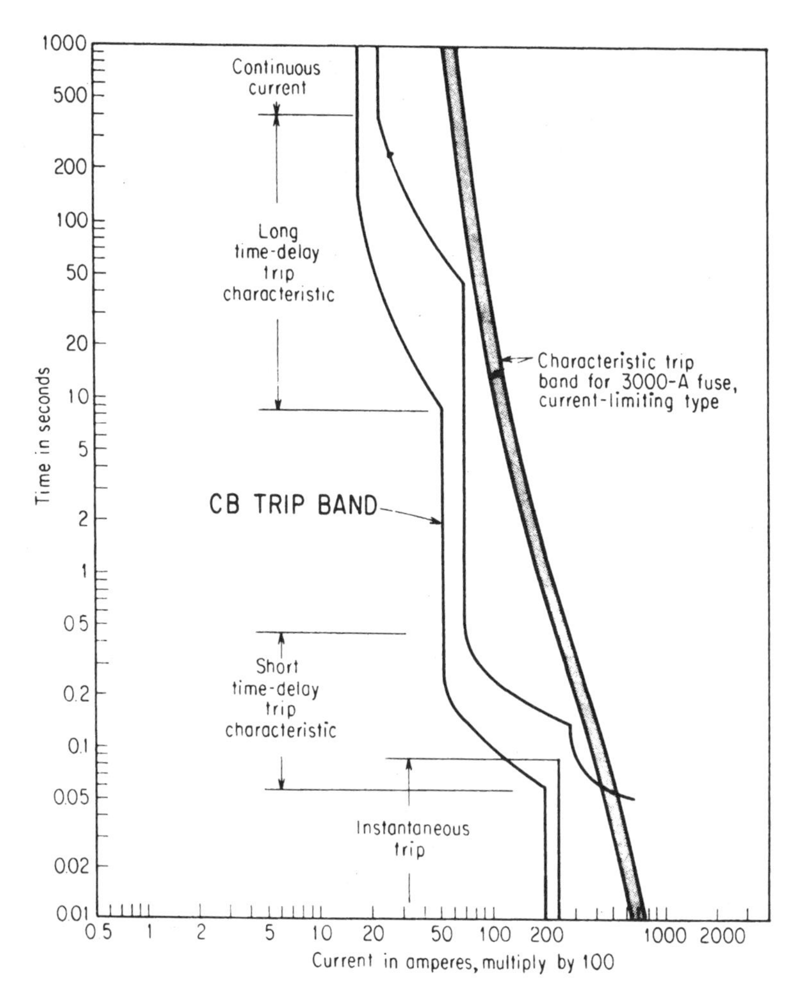

FIG. 2-40 Fuse-CB combinations depend upon coordinated operation.

The coordination between the fuse operating band and the breaker tolerance band is shown in the graph. The combination provides CB operation on faults up to about 60,000 A. Both devices operate above that.

Estimated Short Circuit at End of Low-Voltage Feeder

(See Figs. 2-41 through 2-70.)

Power-system maximum estimated short-circuit currents, as functions of distance along feeder conductors fed from standard 3-phase radial secondary unit substations, can be read directly in rms symmetric amperes from a series of curves, Figs. 2-41 through 2-70). The one-line diagram shows the typical radial circuit investigated.

Calculation Of Short-Circuit Currents—Point-To-Point Method.
Adequate interrupting capacity and protection of electrical components are two essential aspects required by the '—— **National Electrical Code** in Sections 110-9, 110-10, and the FPN to 240-1. The first step to assure that system protective devices have the proper interrupting rating and provide component protection is to determine the available short-circuit currents. The application of the point-to-point method permits the determination of available short-circuit currents with a reasonable degree of accuracy at various points for either 3ϕ or 1ϕ electrical distribution systems. This method assumes unlimited primary short-circuit current (infinite bus).

Basic Short-Circuit Calculation Procedure.

Procedure		Formulae	
Step 1	Determine transf. full-load amperes from either: a) Name plate b) Table SC-2 c) Formula	3ϕ transf.	$I_{FLA} = \frac{KVA \times 1000}{E_{L\text{-}L} \times 1.73}$
		1ϕ transf.	$I_{FLA} = \frac{KVA \times 1000}{E_{L\text{-}L}}$
Step 2	Find transf. multiplier.	—	$\text{Multiplier} = \frac{100}{\text{Transf. } \% Z}$
Step 3	Determine transf. let-thru short-circuit current (Table SC-4 or formula).	—	$\dagger I_{SCA} = \text{Transf.}_{FLA} \times \text{multiplier}$
Step 4	Calculate "f" factor.	3ϕ faults	$f = \frac{1.73 \times L \times I}{C \times E_{L\text{-}L}}$
		1ϕ line-to-line (L-L) faults on 1ϕ, center-tapped transformers	$f = \frac{2 \times L \times I}{C \times E_{L\text{-}L}}$
		1ϕ line-to-neutral (L-N) faults on 1ϕ, center-tapped transformers	$f = \frac{2 \times L \times I^{*}}{C \times E_{L\text{-}N}}$
		L = length (feet) of circuit to the fault. **C** = constant from Table SC-1. For parallel runs, multiply C values by the number of conductors per phase. **I** = available short-circuit current in amperes at beginning of circuit.	
Step 5	Calculate "M" (multiplier) or take from Table SC-3.	$M = \frac{1}{1 + f}$	
Step 6	Compute the available short-circuit current (symmetrical) at the fault.	I_{SCA} at fault $= I_{SCA}$ at beginning of crk. $\times M$	

†**Note 1.** Motor short-circuit contribution, if significant, may be added to the transformer secondary short-circuit current value as determined in Step 3. Proceed with this adjusted figure through Steps 4, 5, and 6. A practical estimate of motor short-circuit contribution is to multiply the total load current in amperes by 4.

Reprinted, with permission, from the "Electrical Protection Handbook" of the Bussman Division, McGraw-Edison Company, St. Louis, Missouri.

Example Of Short-Circuit Calculation.

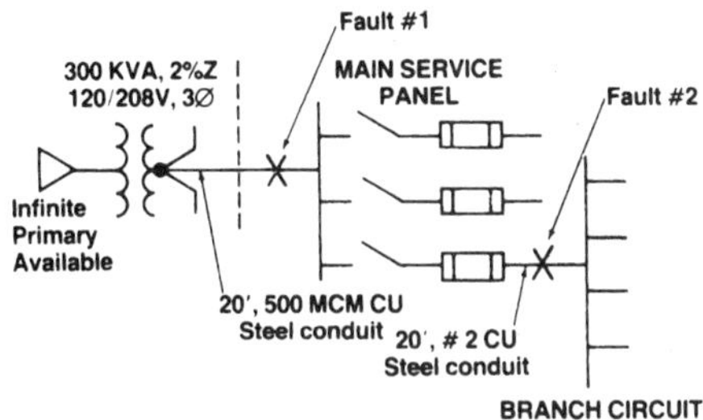

FAULT #1

Step 1 $I_{FLA} = \frac{KVA \times 1000}{E_{L\text{-}L} \times 1.73} = \frac{300 \times 1000}{208 \times 1.73} = 834A$

Step 2 $\text{Multiplier} = \frac{100}{\text{Trans. } \% Z} = \frac{100}{2} = 50$

Step 3 $I_{SCA} = 834 \times 50 = 41{,}700A$
At Transformer Secondary

Step 4 $f = \frac{1.73 \times L \times I}{C \times E_{L\text{-}L}} = \frac{1.73 \times 20 \times 41{,}700}{18{,}100 \times 208} = .383$

Step 5 $M = \frac{1}{1 + f} = \frac{1}{1 + 0.383} = .723$ (See Table SC-3)

Step 6 $I_{SCA} = 41{,}700 \times .723 = 30{,}150A$
Fault #1

FAULT #2

Step 4 Use I_{SCA} @ Fault #1 to calculate

$f = \frac{1.73 \times 20 \times 30{,}150}{4760 \times 208} = 1.05$

Step 5 $M = \frac{1}{1 + f} = \frac{1}{1 + 1.05} = 0.49$ (See Table SC-3)

Step 6 $I_{SCA} = 30{,}150 \times 0.49 = 14{,}770A$
Fault #2

Note: For simplicity, the motor contribution was not included.

***Note 2.** The L-N fault current is higher than the L-L fault current at the secondary terminals of a single-phase center-tapped transformer. The short-circuit current available (I) for this case in Step 4 should be adjusted at the transformer terminals as follows:

At L-N center tapped transformer terminals
I = 1.5 × L-L Short-Circuit Amperes at Transformer Terminals

At some distance from the terminals, depending upon wire size, the L-N fault current is lower than the L-L fault current. The 1.5 multiplier is an approximation and will theoretically vary from 1.33 to 1.67. These figures are based on change in turns ratio between primary and secondary, infinite source available, zero feet from terminals of transformer, and 1.2 × %X and 1.5 × %R for L-N vs. L-L resistance and reactance values. Begin L-N calculations at transformer secondary terminals, then proceed point-to-point.

Calculation Of Short-Circuit Currents At Second Transformer In System.

Use the following procedure to calculate the level of fault current at the secondary of a second, downstream transformer in a system when the level of fault current at the transformer primary is known.

MAIN TRANSFORMER — KNOWN FAULT CURRENT — $I_{P(SCA)}$ — $I_{S(SCA)}$

H.V. UTILITY CONNECTION — KNOWN FAULT CURRENT — $I_{P(SCA)}$ — $I_{S(SCA)}$

Procedure For Second Transformer In System.

Procedure		Formulae	
Step 1	Calculate "f" ($I_{P(SCA)}$, known).	3φ transformer ($I_{P(SCA)}$ and $I_{S(SCA)}$ are 3φ fault values.)	$f = \frac{I_{P(SCA)} \times V_P \times 1.73\ (\%Z)}{100{,}000 \times KVA_{TRANS.}}$
		1φ transformer ($I_{P(SCA)}$ and $I_{S(SCA)}$ are 1φ fault values; $I_{S(SCA)}$ is L-L.)	$f = \frac{I_{P(SCA)} \times V_P \times (\%Z)}{100{,}000 \times KVA_{TRANS.}}$
Step 2	Calculate "M" (multiplier) or take from Table SC-3.	—	$M = \frac{1}{1 + f}$
Step 3	Calculate short-circuit current at secondary of transformer. (See Note 1 under "Basic Procedure")	—	$I_{S(SCA)} = \frac{V_P}{V_S} \times M \times I_{P(SCA)}$

††$I_{P(SCA)}$ = Available fault current at transformer primary
$I_{S(SCA)}$ = Available fault current at transformer secondary
V_p = Primary voltage L-L.
V_s = Secondary voltage L-L.

KVA_{TRANS} = KVA rating of transformer.
%Z = Percent empedance of transformer.
Note — To calculate fault level at the end of a conductor run, follow Steps 4, 5, and 6 of Basic Procedure.

Table SC-1A. "C" Values For Conductors and *Busway.

AWG Or MCM	Copper Three Single Conductors				Copper Three Conductor Cable		Aluminum-Three Single Conductors Or Three Conductor Cables	
	Magnetic Duct		Nonmagnetic Duct		Magnetic Duct	Nonmagnetic Duct	Magnetic	Nonmagnetic Duct
	600V And 5KV Nonshielded	5KV Shielded And 15KV	600V And 5KV Nonshielded	5KV Shielded And 15KV	600V And 5KV Nonshielded	600V And 5KV Nonshielded	600V And 5KV Nonshielded	600V And 5 Nonshielded
12	617	—	—	—	—	—	—	—
10	982	—	—	—	—	—	—	—
8	1230	1230	1230	1230	1230	1230	—	—
6	1940	1940	1950	1940	1950	1950	1180	1180
4	3060	3040	3080	3070	3080	3090	1870	1870
3	3860	3830	3880	3870	3880	3900	2360	2360
2	4760	4670	4830	4780	4830	4850	2960	2970
1	5880	5750	6020	5920	6020	6100	3720	3750
1/0	7190	6990	7460	7250	7410	7580	4670	4690
2/0	8700	8260	9090	8770	9090	9350	5800	5880
3/0	10400	9900	11500	10700	11100	11900	7190	7300
4/0	12300	10800	13400	12600	13400	14000	8850	9170
250	13500	12500	14900	14000	14900	15800	10300	10600
300	14800	13600	16700	15500	16700	17900	11900	12400
350	16200	14700	18700	17000	18600	20300	13500	14200
400	16500	15200	19200	17900	19500	21100	14800	15800
450	17300	15900	20400	18800	20700	22700	—	—
500	18100	16500	21500	19700	21900	24000	17200	18700
600	18900	17200	22700	20900	23300	25700	18900	21000
700	—	—	—	—	—	—	20500	23100
750	20200	18300	24700	22500	25600	28200	21500	24300
1000	—	—	—	—	—	—	23600	27600

*Note—See next page for Busway

Table SC-1B. **"C" Values For Busway.**

Ampacity	Plug-In Busway		Feeder Busway		High Imped. Busway
	Copper	Aluminum	Copper	Aluminum	Copper
225	28700	23000	18700	12000	—
400	38900	34700	23900	21300	—
600	41000	38300	36500	31300	—
800	46100	57500	49300	44100	—
1000	69400	89300	62900	56200	15600
1200	94300	97100	76900	69900	16100
1350	119000	104200	90100	84000	17500
1600	129900	120500	101000	90900	19200
2000	142900	135100	134200	125000	20400
2500	143800	156300	180500	166700	21700
3000	144900	175400	204100	188700	23800
4000	—	—	277800	256400	—

Table SC-2A. **Three-Phase Transformer—Full-Load Current Rating (In Amperes).**

Voltage (Line-To-Line)	Transformer KVA Rating								
	150	**167**	**225**	**300**	**500**	**750**	**1000**	**1500**	**2000**
208	417	464	625	834	1388	2080	2776	4164	5552
220	394	439	592	788	1315	1970	2630	3940	5260
240	362	402	542	722	1203	1804	2406	3609	4812
440	197	219	296	394	657	985	1315	1970	2630
460	189	209	284	378	630	945	1260	1890	2520
480	181	201	271	361	601	902	1203	1804	2406
600	144	161	216	289	481	722	962	1444	1924

Table SC-2B. **Single-Phase Transformer—Full-Load Current Rating (In Amperes).**

Voltage	Transformer KVA Rating									
	25	**50**	**75**	**100**	**150**	**167**	**200**	**250**	**333**	**500**
115/230	109	217	326	435	652	726	870	1087	1448	2174
120/240	104	208	313	416	625	696	833	1042	1388	2083
230/460	54	109	163	217	326	363	435	544	724	1087
240/480	52	104	156	208	313	348	416	521	694	1042

Table SC-3. **"M" (Multiplier).***

f	M	f	M
0.01	0.99	1.50	0.40
0.02	0.98	1.75	0.36
0.03	0.97	2.00	0.33
0.04	0.96	2.50	0.29
0.05	0.95	3.00	0.25
0.06	0.94	3.50	0.22
0.07	0.93	4.00	0.20
0.08	0.93	5.00	0.17
0.09	0.92	6.00	0.14
0.10	0.91	7.00	0.13
0.15	0.87	8.00	0.11
0.20	0.83	9.00	0.10
0.25	0.80	10.00	0.09
0.30	0.77	15.00	0.06
0.35	0.74	20.00	0.05
0.40	0.71	30.00	0.03
0.50	0.67	40.00	0.02
0.60	0.63	50.00	0.02
0.70	0.59	60.00	0.02
0.80	0.55	70.00	0.01
0.90	0.53	80.00	0.01
1.00	0.50	90.00	0.01
1.20	0.45	100.00	0.01

* $M = \frac{1}{1 + F}$

Table SC-4. Short-Circuit Currents Available from Various Size Transformers.

Voltage And Phase	KVA	Full Load Amps	% Impedance	†Short Circuit Amps
120/240 1 ph.	25	104	1.6	10,300
	37½	156	1.6	15,280
	50	209	1.7	19,050
	75	313	1.6	29,540
	100	417	1.6	38,540
	167	695	1.8	54,900
120/208 3 ph.	150	417	2.0	20,850
	225	625	2.0	31,250
	300	834	2.0	41,700
	500	1388	2.0	69,400
	750	2080	5.0	41,600
	1000	2776	5.0	55,520
	1500	4164	5.0	83,280
	2000	5552	5.0	111,040
	2500	6950	5.0	139,000
277/480 3 ph.	150	181	2.0	9,050
	225	271	2.0	13,550
	300	361	2.0	18,050
	500	601	2.0	30,059
	750	902	5.0	18,040
	1000	1203	5.0	24,060
	1500	1804	5.0	36,060
	2000	2406	5.0	48,120
	2500	3007	5.0	60,140

†Three-phase short-circuit currents based on "infinite" primary. Single-phase short-circuit currents on 100,000 KVA primary.

The conditions on which the curves are based were as follows:

1. The fault was a bolted 3-phase short circuit.
2. The primary 3-phase short-circuit duty was 500 MVA (60 cycles) for all curves. A typical supply system *X/R* at the low-voltage bus was used in calculating the curves for each case.
3. Motor contributions through the bus to the point of short circuit were included in the calculations on the basis of 100 percent contribution for the 240-, 480-, and 600-V systems and 50 percent contributions for the 208-V systems.
4. The feeder-conductor impedance values used in the calculations are indicated for various conductor sizes.

These curves can also be used to select feeder conductor sizes and lengths needed to reduce short-circuit duties to desired smaller values. Note that conductors thus selected must be further checked to ensure adequate load and short-circuit capabilities and acceptable voltage drop.

Coordinated ratings are based on two protective devices operating in series with all short-circuit current flowing through the upstream device. If any current bypasses the upstream device (such as motor contribution fed in on load side of upstream device), a fully rated system, not a coordinated rated system, should be used.

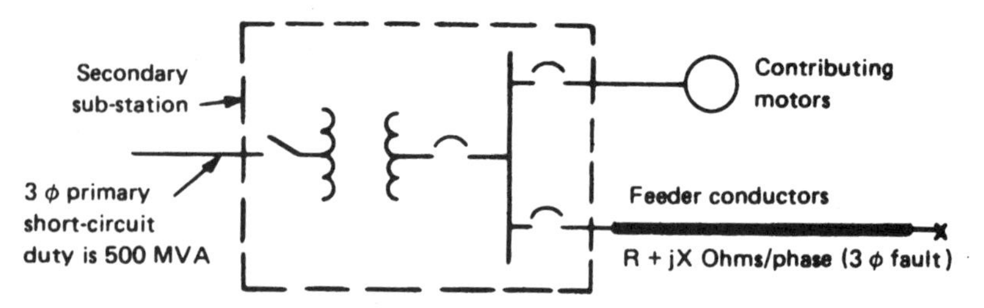

Typical circuit investigated to show effect on short-circuit duty as point of fault is moved away from the low-voltage bus along the feeder conductors.

Feeder Impedance Values Used in Investigation

Feeder Conductor Size/Phase	Resistance (R) Ohms/Phase/1000 Ft	60-cycle Inductive Reactance (X) Ohms/Phase/1000 Ft
#4	0.3114	0.0492
#1/0	.1231	.057
250 MCM	.0534	.0428
2-500 MCM	.0144	.0201
4-750 MCM	.0053	.0099

Reprinted, with permission, from "Short-circuit Current Calculations," publication EESG II-AP-1, of the Apparatus Distribution Sales Division, General Electric Company.

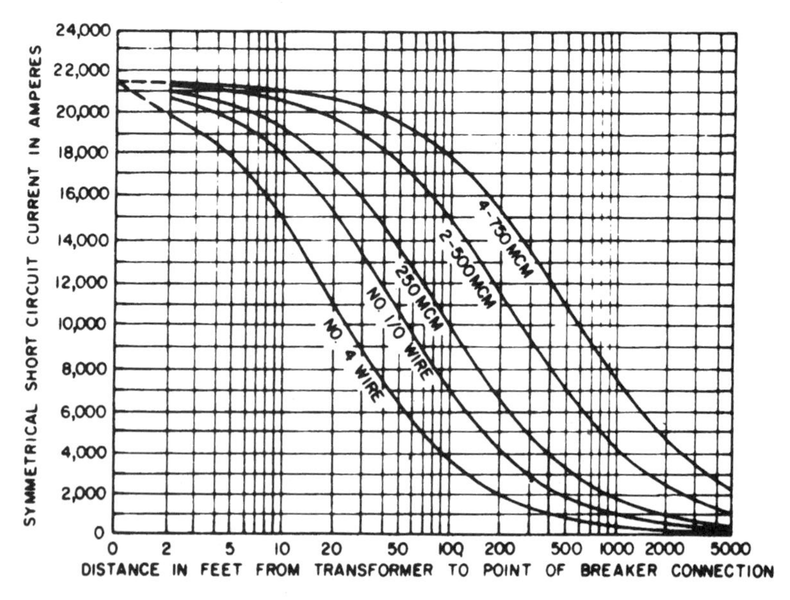

FIG. 2-41 Transf: 150 kVA, 208 V, 2.0% Z.

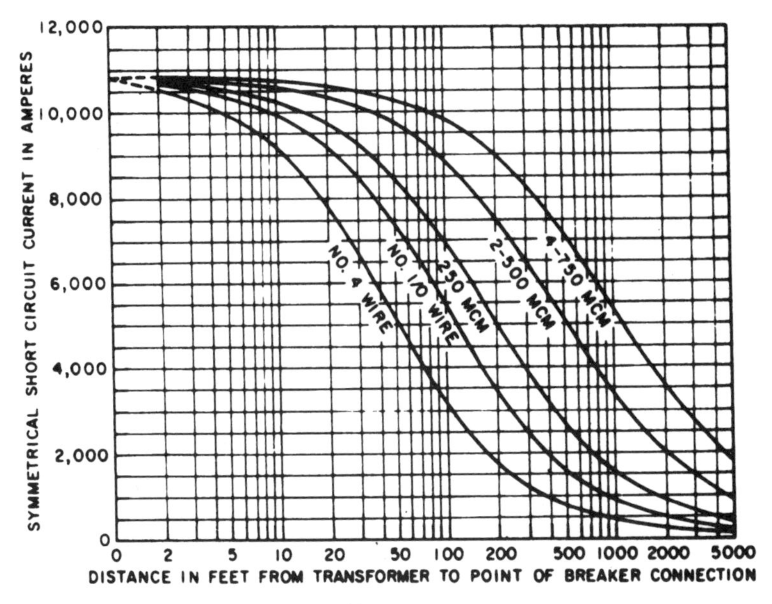

FIG. 2-42 Transf: 150 kVA, 208 V, 4.5% Z.

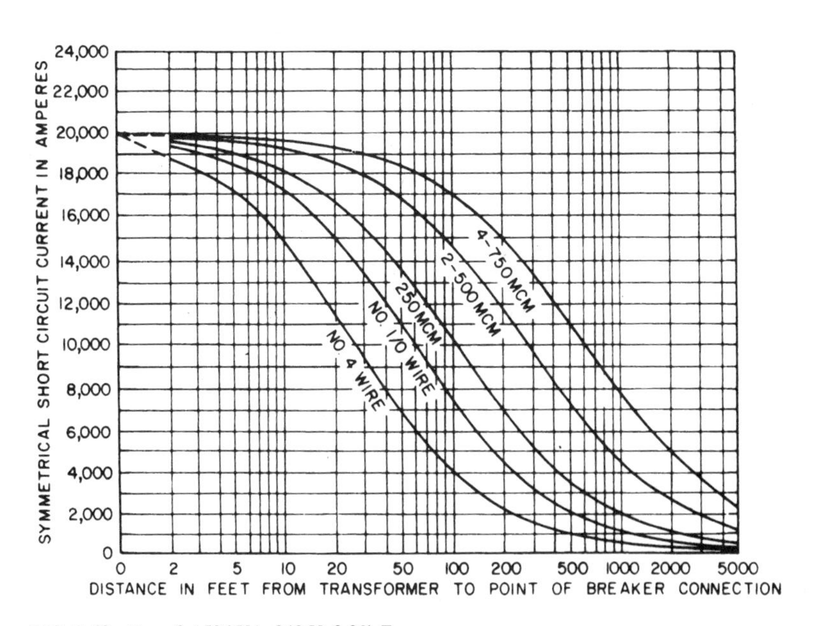

FIG. 2-43 Transf: 150 kVA, 240 V, 2.0% Z.

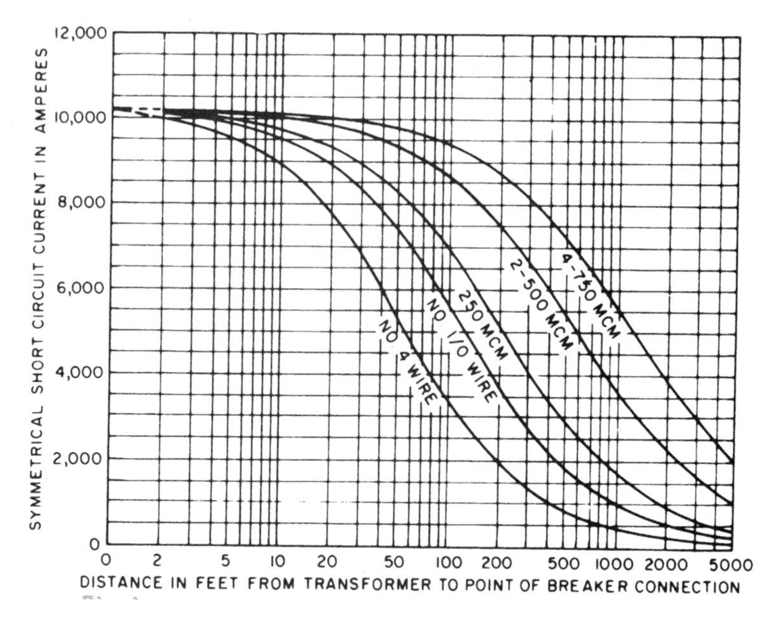

FIG. 2-44 Transf: 150 kVA, 240 V, 4.5% Z.

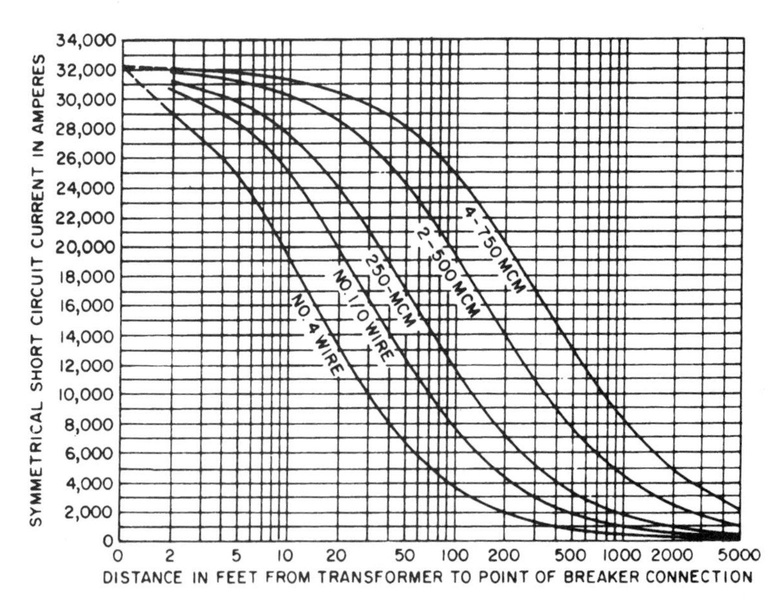

FIG. 2-45 Transf: 225 kVA, 208 V, 2.0% Z.

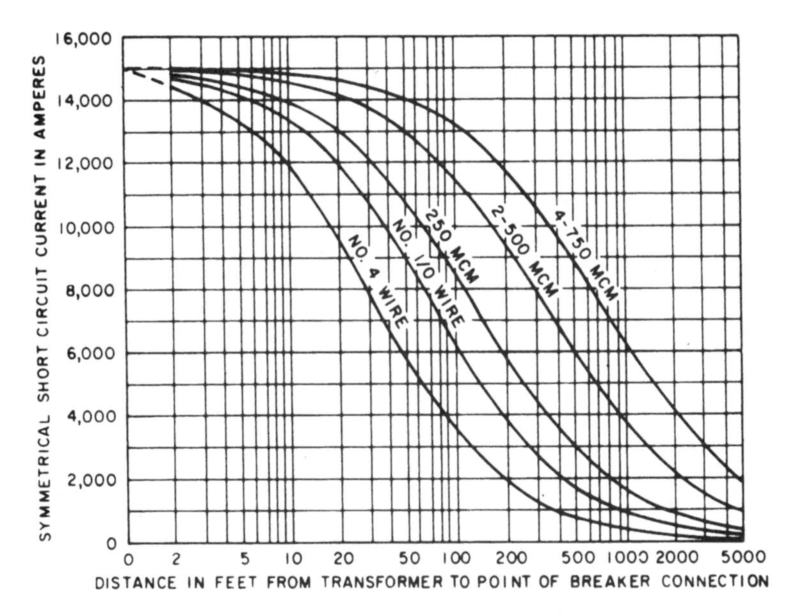

FIG. 2-46 Transf: 225 kVA, 208 V, 4.5% Z.

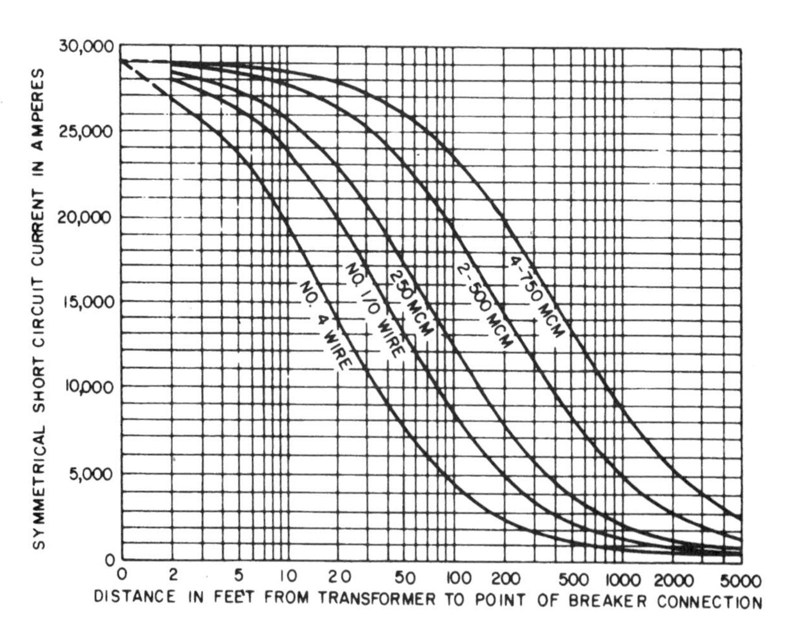

FIG. 2-47 Transf: 225 kVA, 240 V, 2.0% Z.

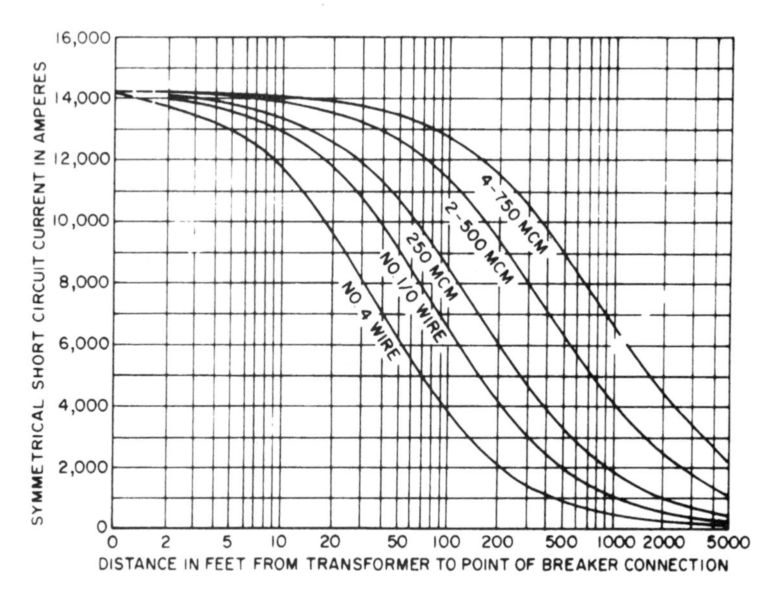

FIG. 2-48 Transf: 225 kVA, 240 V, 4.5% Z.

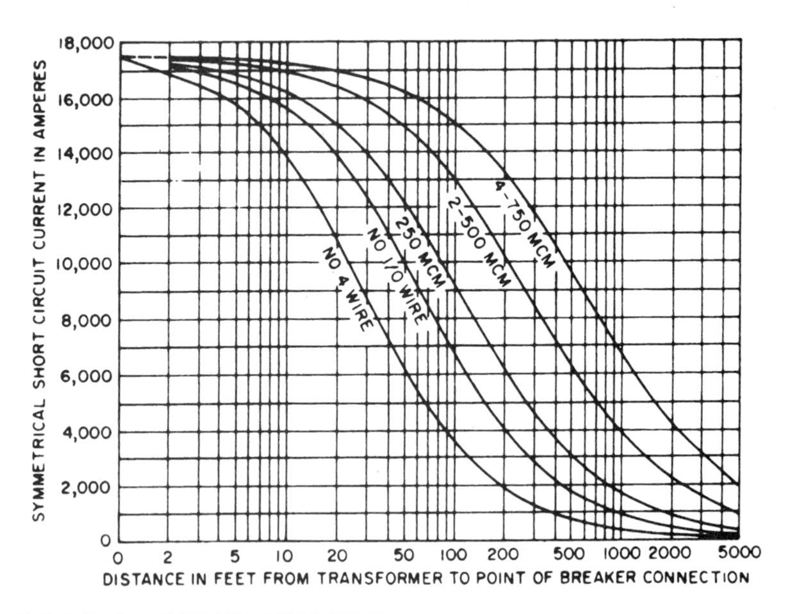

FIG. 2-49 Transf: 300 kVA, 208 V, 4.5% Z.

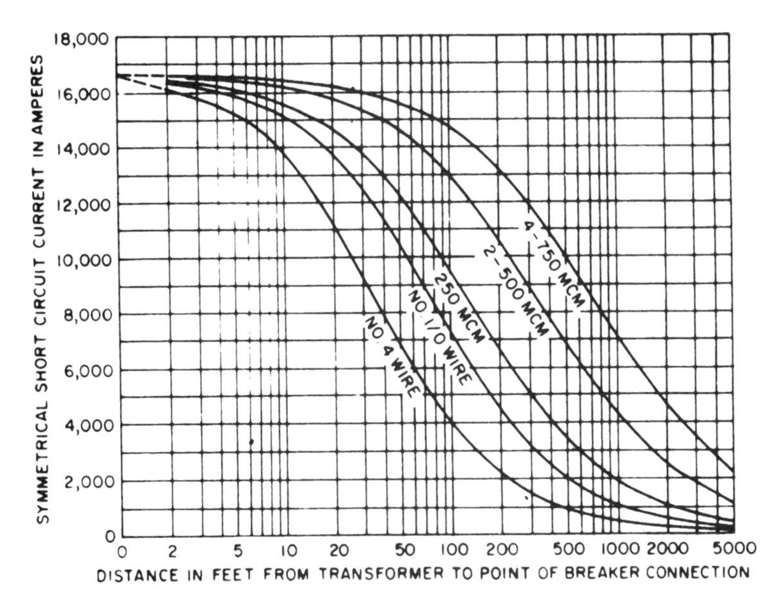

FIG. 2-50 Transf: 300 kVA, 240 V, 4.5% Z.

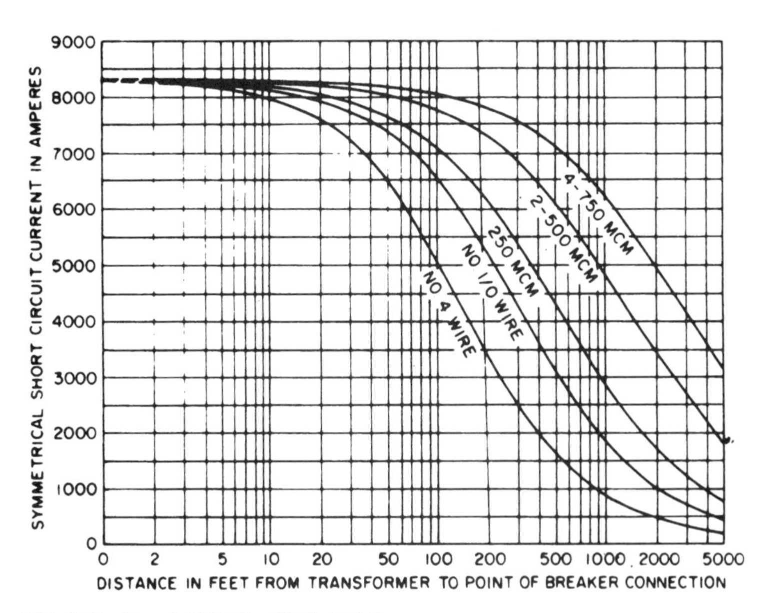

FIG. 2-51 Transf: 300 kVA, 480 V, 4.5% Z.

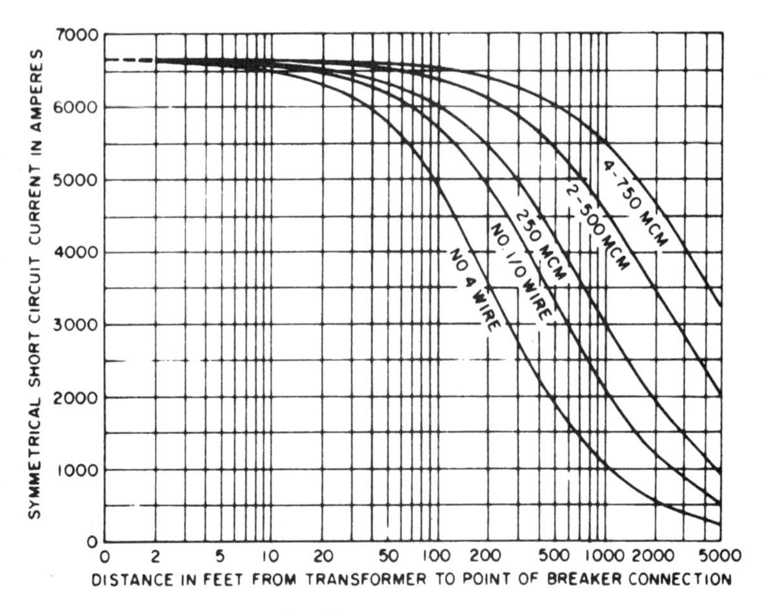

FIG. 2-52 Transf: 300 kVA, 600 V, 4.5% Z.

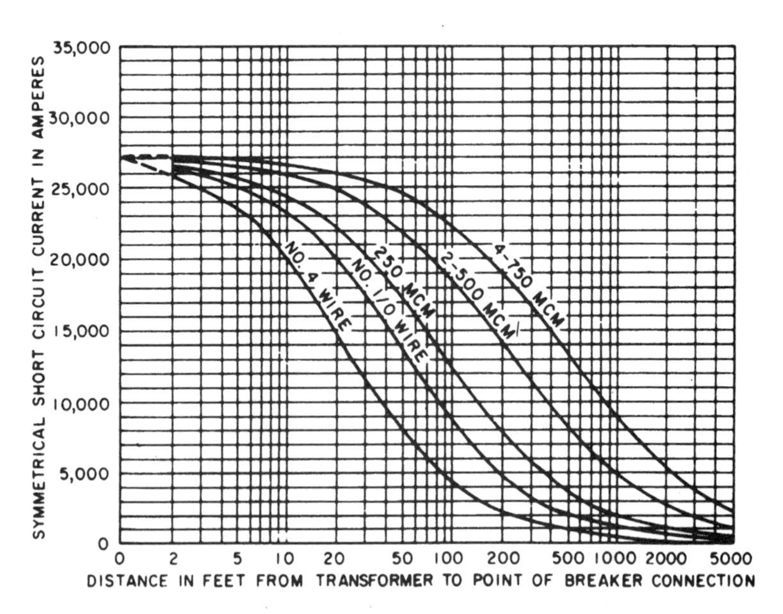

FIG. 2-53 Transf: 500 kVA, 208 V, 4.5% Z.

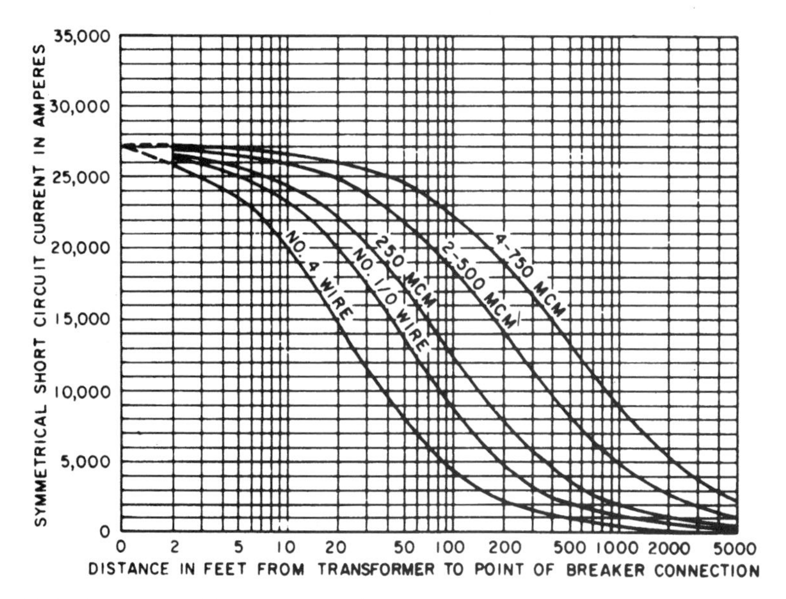

FIG. 2-54 Transf: 500 kVA, 240 V, 4.5% Z.

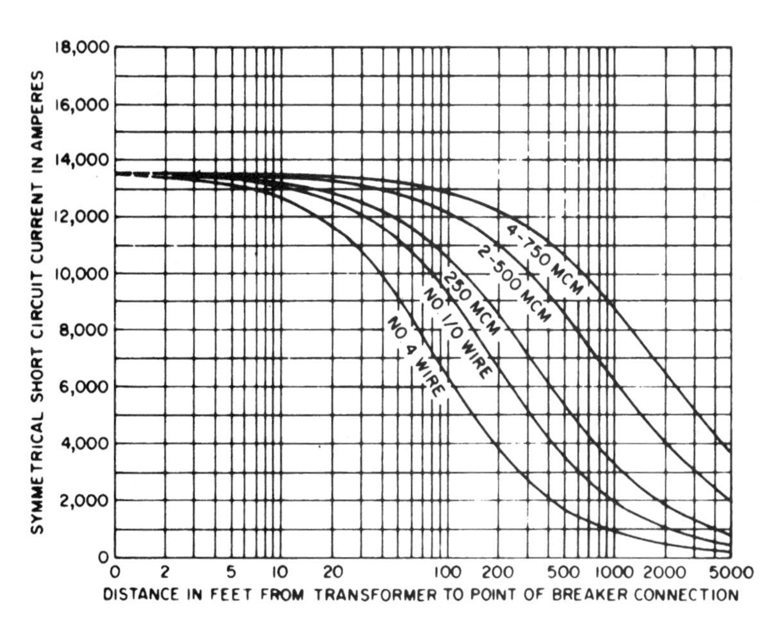

FIG. 2-55 Transf: 500 kVA, 480 V, 4.5% Z.

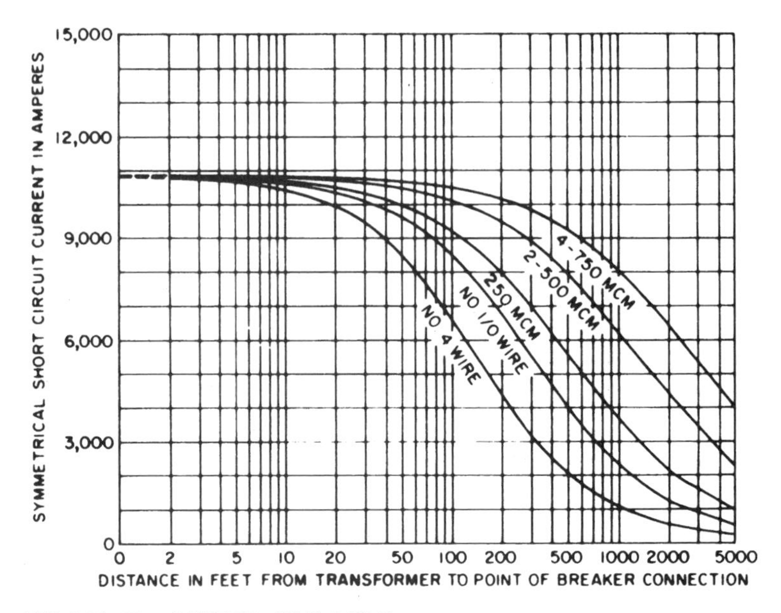

FIG. 2-56 Transf: 500 kVA, 600 V, 4.5% Z.

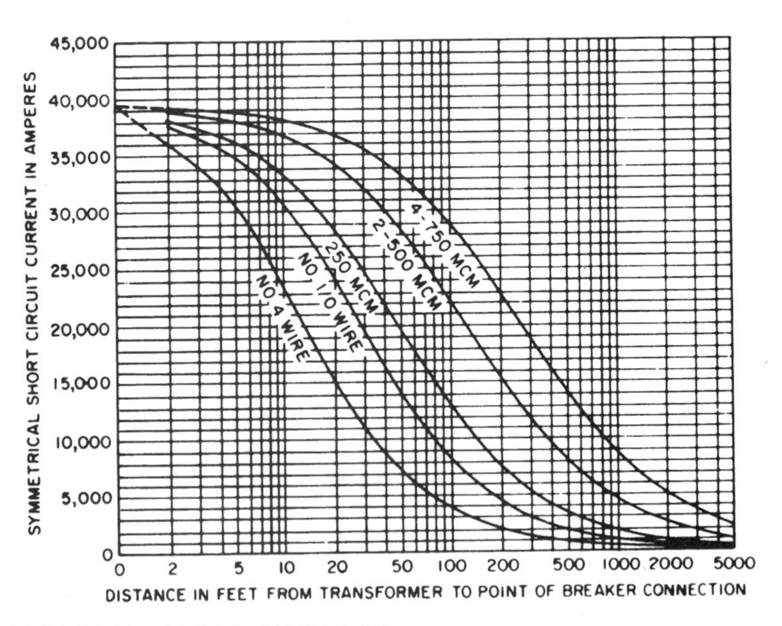

FIG. 2-57 Transf: 750 kVA, 208 V, 5.75% Z.

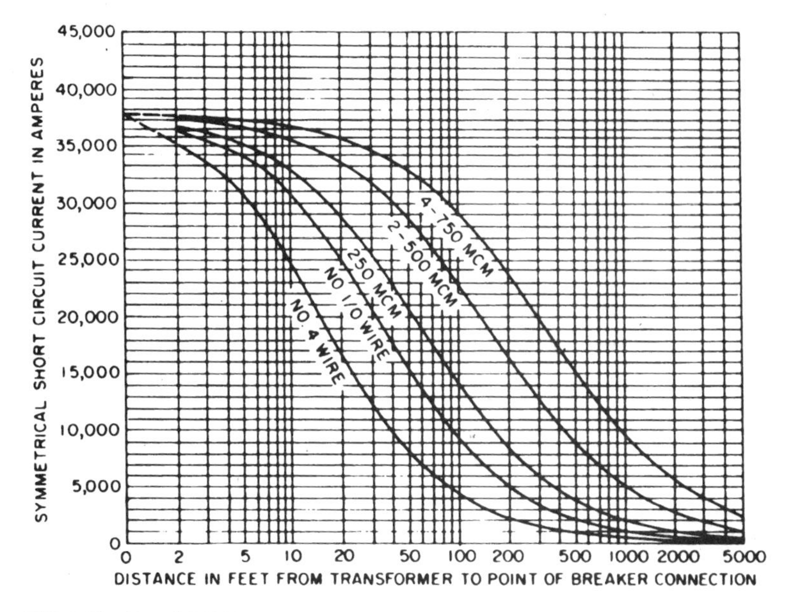

FIG. 2-58 Transf: 750 kVA, 240 V, 5.75% *Z*.

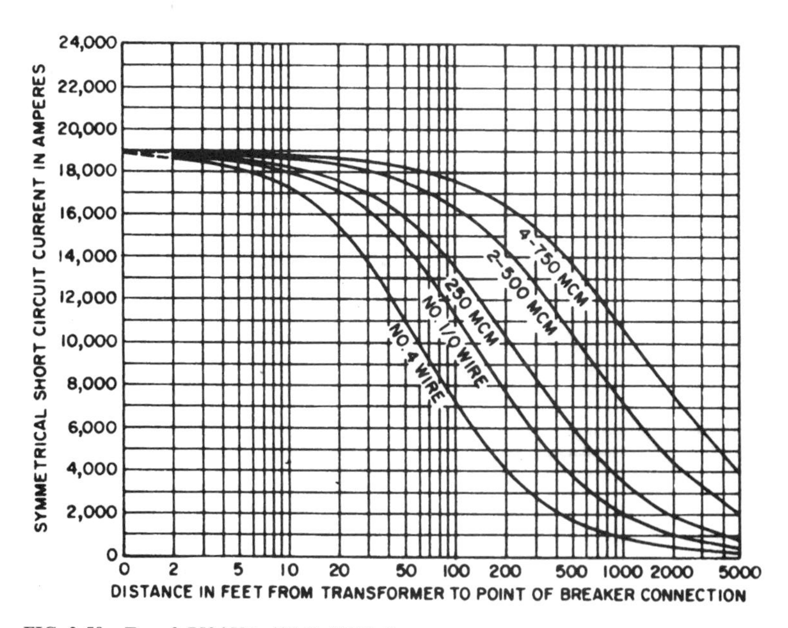

FIG. 2-59 Transf: 750 kVA, 480 V, 5.75% *Z*.

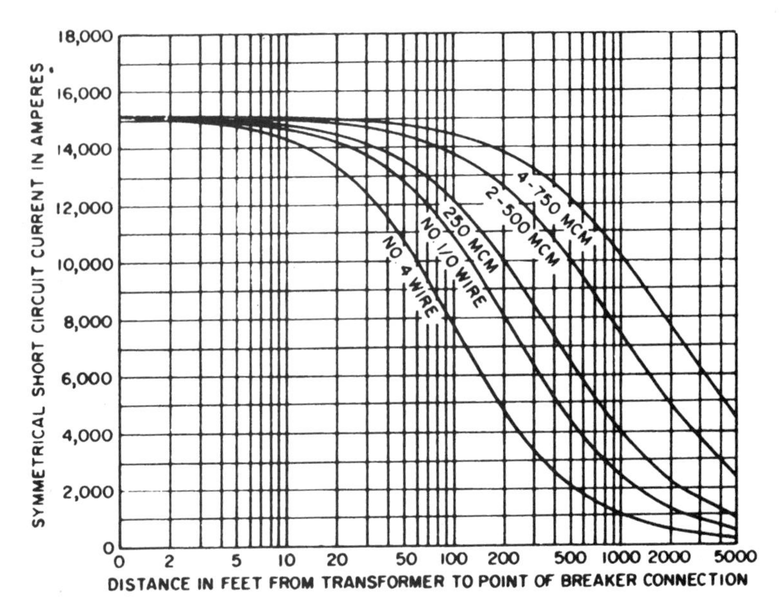

FIG. 2-60 Transf: 750 kVA, 600 V, 5.75% Z.

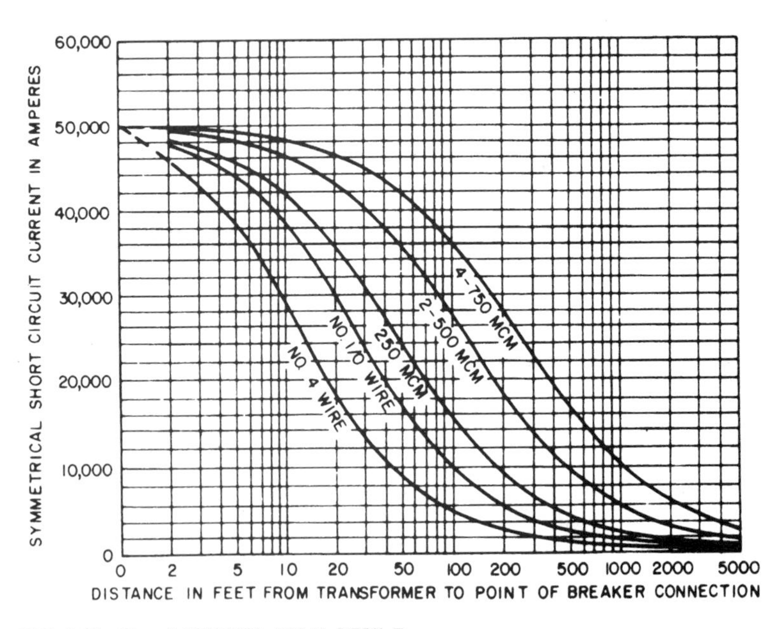

FIG. 2-61 Transf: 1000 kVA, 208 V, 5.75% Z.

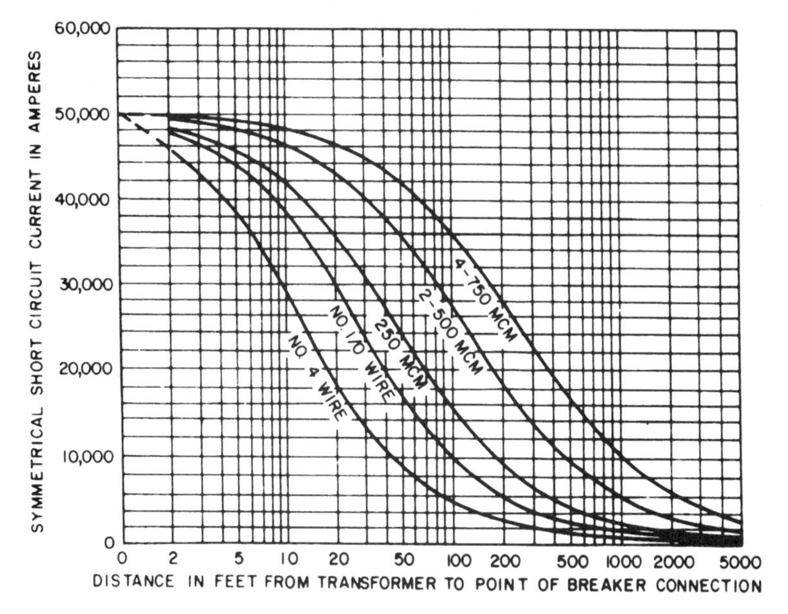

FIG. 2-62 Transf: 1000 kVA, 240 V, 5.75% Z.

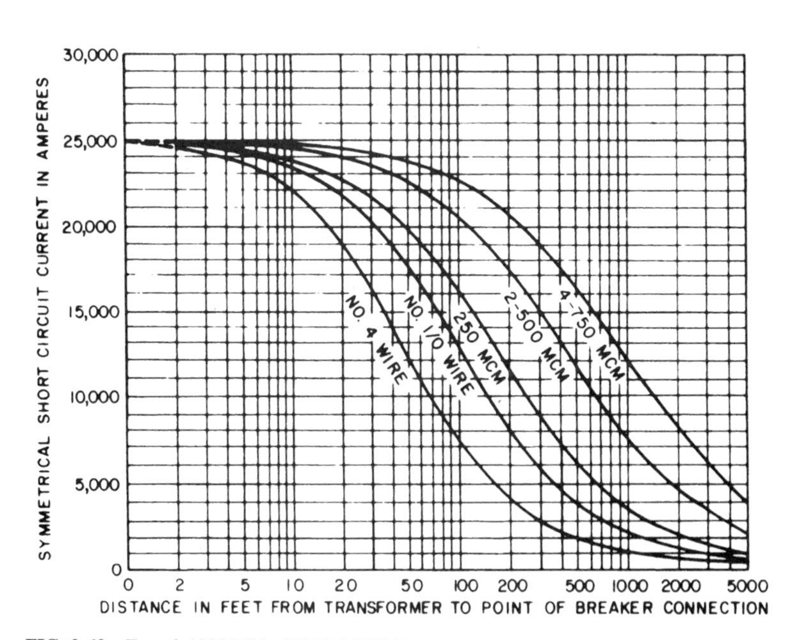

FIG. 2-63 Transf: 1000 kVA, 480 V, 5.75% Z.

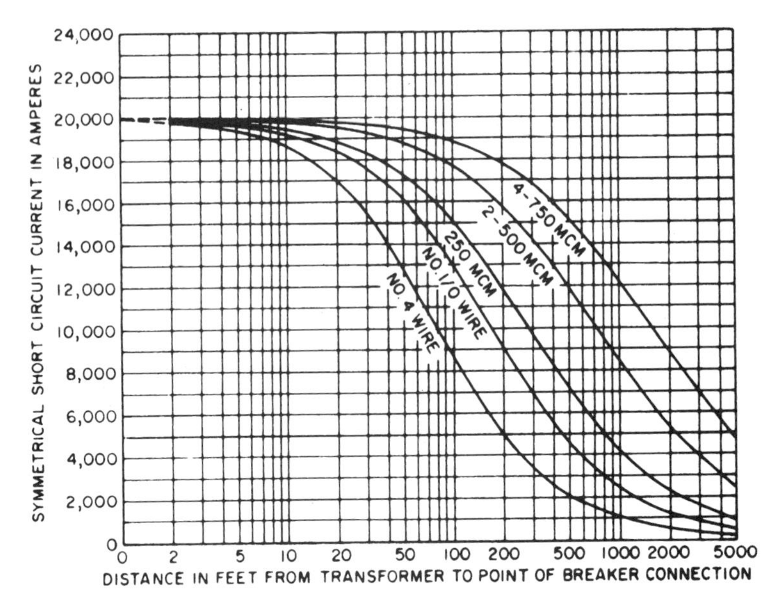

FIG. 2-64 Transf: 1000 kVA, 600 V, 5.75% Z.

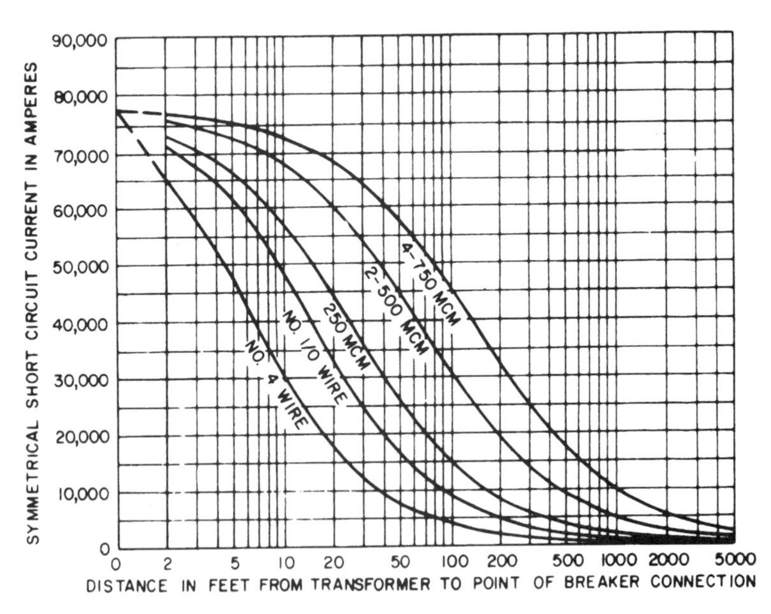

FIG. 2-65 Transf: 1500 kVA, 208 V, 5.75% Z.

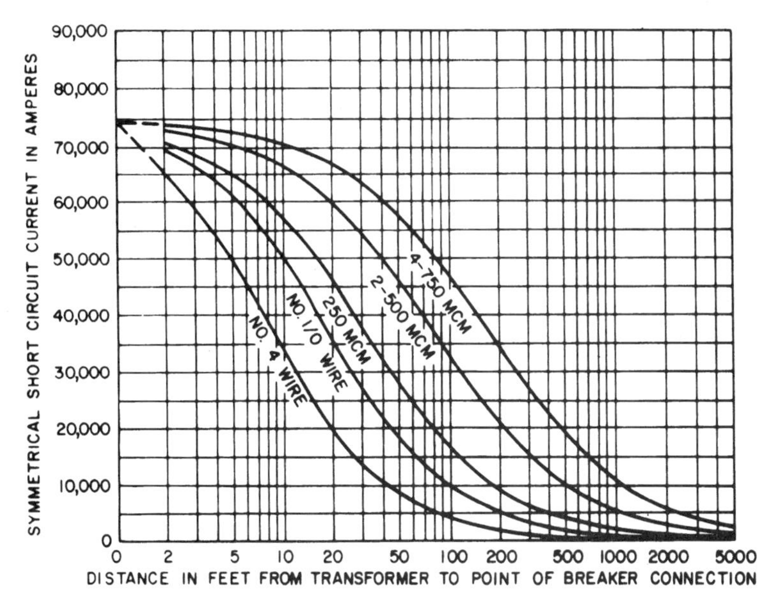

FIG. 2-66 Transf: 1500 kVA, 240 V, 5.75% Z.

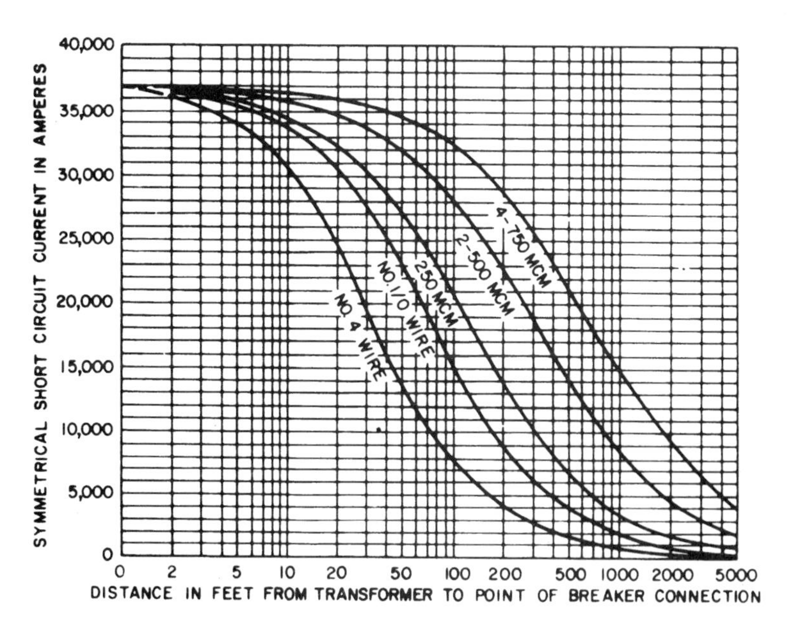

FIG. 2-67 Transf: 1500 kVA, 480 V, 5.75% Z.

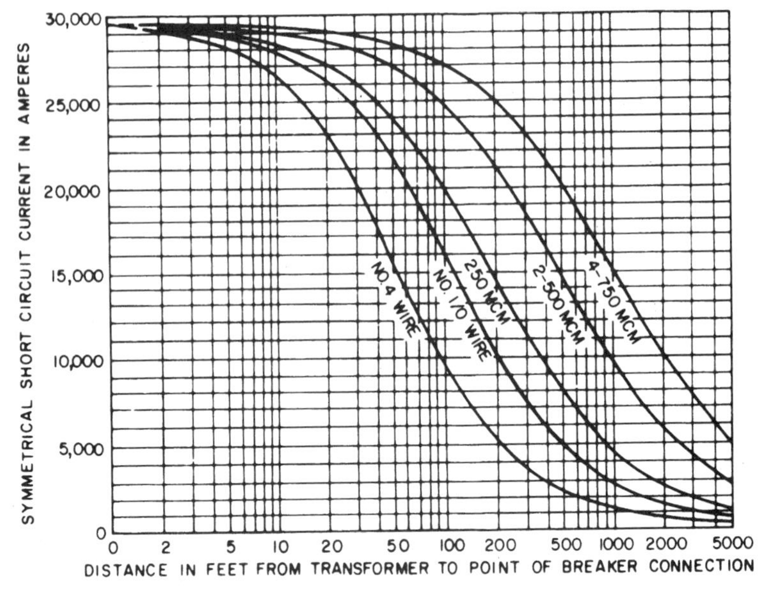

FIG. 2-68 Transf: 1500 kVA, 600 V, 5.75% Z.

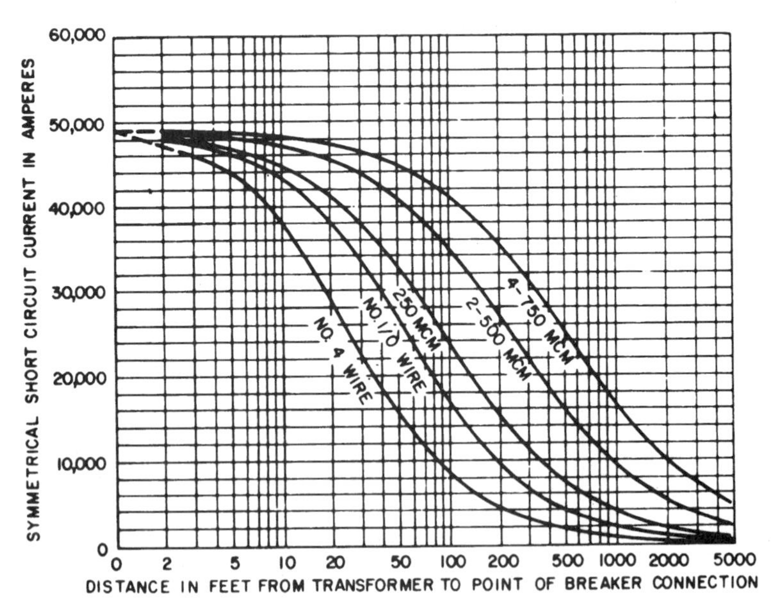

FIG. 2-69 Transf: 2000 kVA, 480 V, 5.75% Z.

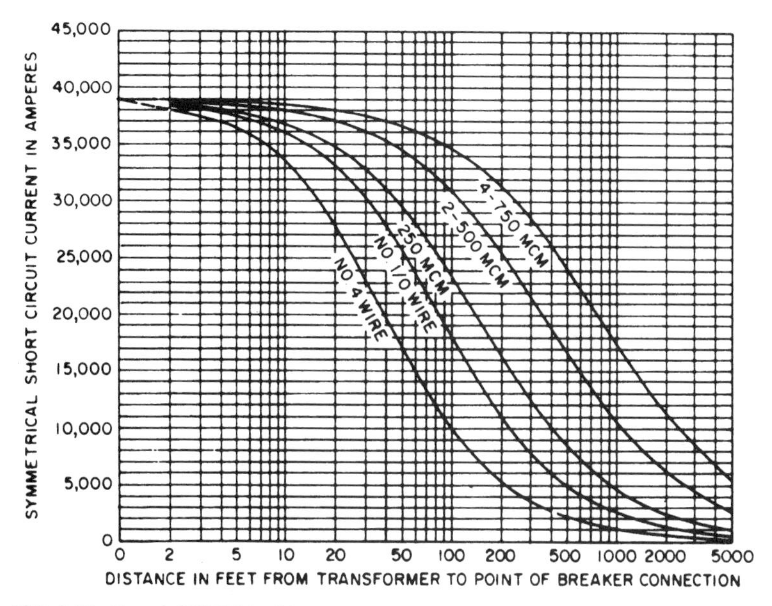

FIG. 2-70 Transf: 2000 kVA, 600 V, 5.75% Z.

CONDUCTORS IN PARALLEL

Conductors that are permitted to be used in parallel (in multiple) include *phase* conductors, *neutral* conductors, and *grounded-circuit* conductors. In the places where this section describes parallel makeup of circuits, a grounded-circuit conductor is identified along with phase and neutral conductors to extend the same permission for paralleling to grounded legs of corner-grounded delta systems.

NEC Sec. 310-4 recogn izes the use of conductors in sizes 1/0 and larger for use in parallel under the conditions stated, to allow a practical means of installing large-capacity feeders and services. Paralleling of conductors relies on a number of factors to ensure equal division of current, and thus all these factors must be satisfied in order to ensure that none of the individual conductors will become overloaded.

When conductors are used in parallel, *all* the conductors making up *each phase, neutral,* or *grounded-circuit* conductor must satisfy the five conditions of the second paragraph in this section. Those characteristics—same length, same conductor material (copper or aluminum), same size, same insulation, and same terminating device—apply only to the paralleled conductor making up each phase or neutral of a parallel-makeup circuit. All the conductors of any phase or the neutral must satisfy the rule, but phase *A* conductors (all of which must be the same length, same size, etc.) may be different in length, material, size, etc., from the conductors making up phase *B* or phase *C* or the neutral. But all phase *B* conductors must be the same length, same size, etc.; phase *C* conductors must all be the same; and neutral conductors must all be alike (Fig. 2-71). As the last sentence in the fine-print note explains, it is not the intent of this Code rule to require that conductors of one phase be the same as those of another phase or of the neutral. The only concern for safe operation of a parallel-makeup circuit is that all the conductors in parallel per phase leg (neutral, or grounded conductor) evenly divide the load current and thereby prevent overloading of any one of the conductors. Of course,

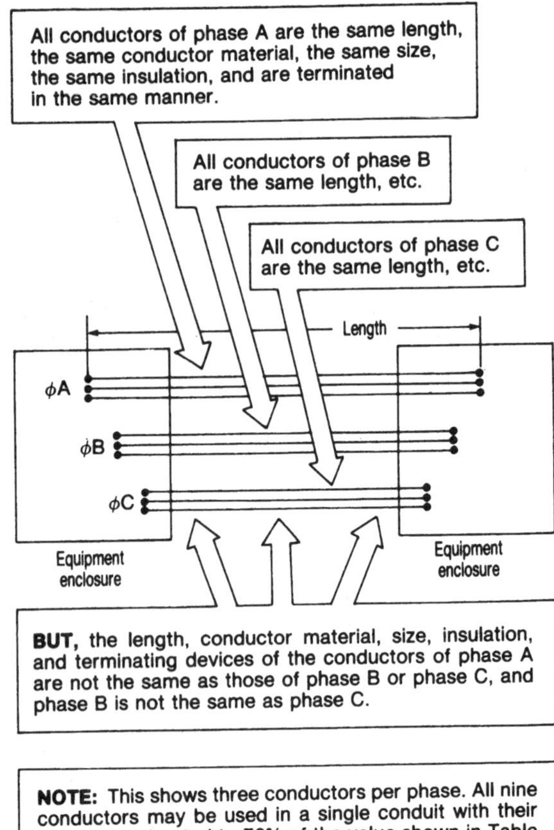

FIG. 2-71 This is the basic rule on conductors used for parallel circuit makeup.

the realities of material purchase and application and good design practice dictate that *all* the conductors of all phases and neutral use the same conductor material, have the same insulation, have as nearly the same length as possible to prevent voltage drop from causing objectionable voltage unbalance on the phases, and are terminated in the same way. The size of conductors may vary from phase to phase or in the neutral, depending upon load currents.

Figure 2-72 shows two examples of parallel conductor circuit makeup. The photograph at bottom shows six conductors used per phase and neutral to obtain 2000-A capacity per phase, which simply could not be done without parallel conductors per phase leg. The CB or fuses on such circuits are rated much higher than the ampacity of each conductor.

Where large currents are involved, it is particularly important that the separate phase conductors be located close together to avoid excessive voltage drop and ensure equal

FIG. 2-72 Examples of circuit makeup using conductors in parallel. (*a*) Multiple conductors (two in parallel for each phase leg) are used for normal and emergency feeder through this automatic transfer switch. (*b*) Six conductors in parallel make up each phase leg and the neutral of this feeder. Fusible limiter lug on each conductor, although not required by Code on other than transformer tie circuits, is sized for the conductor to protect against division of current among the six conductors that would put excessive current on any conductor.

division of current. It is also essential that each phase and the neutral, and grounding wires, if any, be run in each conduit even where the conduit is of nonmetallic material.

The sentence just before the FPN in this section calls for the same type of raceway or enclosure for conductors in parallel in separate raceways or cables. The impedance of the circuit in a nonferrous raceway will be different from that of the same circuit in a ferrous raceway or enclosure. See Sec. 300-20.

From the Code tables of current-carrying capacities of various sizes of conductors, it can be seen that small conductor sizes carry more current per circular mil of cross section than do large conductors. This results from rating conductor capacity according to temperature rise. The larger a cable, the less radiating surface per circular mil of cross section. Loss due to *skin effect* (apparent higher resistance of conductors to alternating current than to direct current) is also higher in the larger conductor sizes. And larger conductors cost more per ampere than smaller conductors.

All the foregoing factors point to the advisability of using a number of smaller conductors in multiple to get a particular carrying capacity, rather than a single conductor of that capacity. In many cases, multiple conductors for feeders provide distinct operating advantages and are more economical than the equivalent-capacity single-conductor makeup of a feeder. But, it should be noted, the reduced overall cross section of conductor resulting from multiple conductors instead of a single conductor per leg produces higher resistance and greater voltage drop than the same length as a single conductor per leg. Voltage drop may be a limitation.

Figure 2-73 shows a typical application of copper conductors in multiple, with the advantages of such use. Where more than three conductors are installed in a single conduit, the ampacity of each conductor must be derated from the ampacity value shown in NEC Table 310-16. The four circuit makeups show:

1. Without ampacity derating because there are more than three conductors in the conduit, circuit 2 would be equivalent to circuit 1.
2. A circuit of six 400-kcmil conductors can be made equivalent in ampacity to a circuit of three 2000-kcmil conductors by dividing the 400s between two conduits (3 conductors/3-in conduit). If three different phases are used in each of two 3-in conduits for this circuit, the multiple circuit will not require ampacity derating to 80 percent, and its 670-A rating will exceed the 665-A rating of circuit 1.
3. Circuit 2 is almost equivalent to circuit 3 in ampacity.
4. Circuit 4 is equivalent to circuit 1 in ampacity, but uses less conductor copper and a smaller conduit. And the advantages are obtained even with the ampacity derating for conduit fill.

Except where the conductor size is governed by conditions of voltage drop, it is seldom economical to use conductors of sizes larger than 1000 kcmil, because above this size the increase in ampacity is very small in proportion to the increase in the size of the conductor. Thus, for a 50 percent increase in the conductor size, that is, from 1,000,000 to 1,500,000 cmil, the ampacity of a type THW conductor increases only 80 A, or less than 15 percent; and for an increase in size from 1,000,000 to 2,000,000 cmil, a 100 percent increase, the ampacity increases only 120 A, or about 20 percent. In any case where single conductors larger than 500,000 cmil would be required, it is worthwhile to compute the total installation cost using single conductors and the cost using two (or more) conductors in parallel.

The next-to-last paragraph of Sec. 310-4 warns that when multiple conductors are used per circuit phase leg, it may require more space at equipment terminals to bend and install the conductors. Refer to Sec. 373-6.

1.  6″

A 3-phase circuit of three 2,000 kcmil type THW conductors in a 6-in. conduit.
Current rating of each phase = 665 amps.
Cross-section area per phase = 2.9013 sq. in.

2. 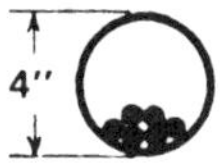 4″

A 3-phase circuit of six 400 kcmil type THW conductors (two per phase) in a 4-in. conduit.
Current rating of each phase might appear to be = 2 X 335 = 670 amps.
But, because of the 80% derating required by Note 8 to Tables 310-16/19:
Current rating of each phase = 670 X 80% = 536 amps.
Cross-section area per phase = 1.3938 sq. in. (two conductors).

3. 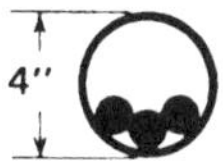 4″

A 3-phase circuit of three 1,000 kcmil type THW conductors in a 4-in. conduit.
Current rating of each phase = 545 amps.
Cross-section area per phase = 1.5482 sq. in.

4. 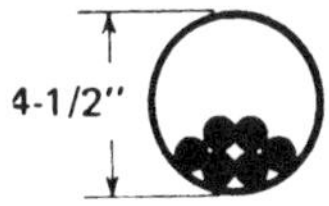 4-1/2″

A 3-phase circuit of six 600 kcmil type THW conductors in a 5-in. conduit
Current rating of each phase might appear to be = 2 X 420 = 840 amps.
But 80% derating must be applied because of the number of conductors in the conduit:
Current rating of each phase = 840 X 80% = 672 amps.
Cross-section area per phase = 2.0522 sq. in. (two conductors).

FIG. 2-73 The above circuit makeups represent typical considerations in the application of multiple-conductor circuits.

Figure 2-74 shows an interesting application of parallel conductors. A 1200-A riser is made up of three conduits, each carrying three phases and a neutral. At the basement switchboard, the 1200-A circuit of three conductors per phase plus three conductors for the neutral originates in a bolted-pressure switch with a 1200-A fuse in each of the 3-phase poles. Because the total of 12 conductors make up a *single* 3-phase, 4-wire circuit, a 400-A, 3-phase, 4-wire tap-off must tap all the conductors in the junction box at top. That is, the three phase *A* legs (one from each conduit) must be skinned and bugged together, and then the phase *A* tap made from that common point to one of the lugs on

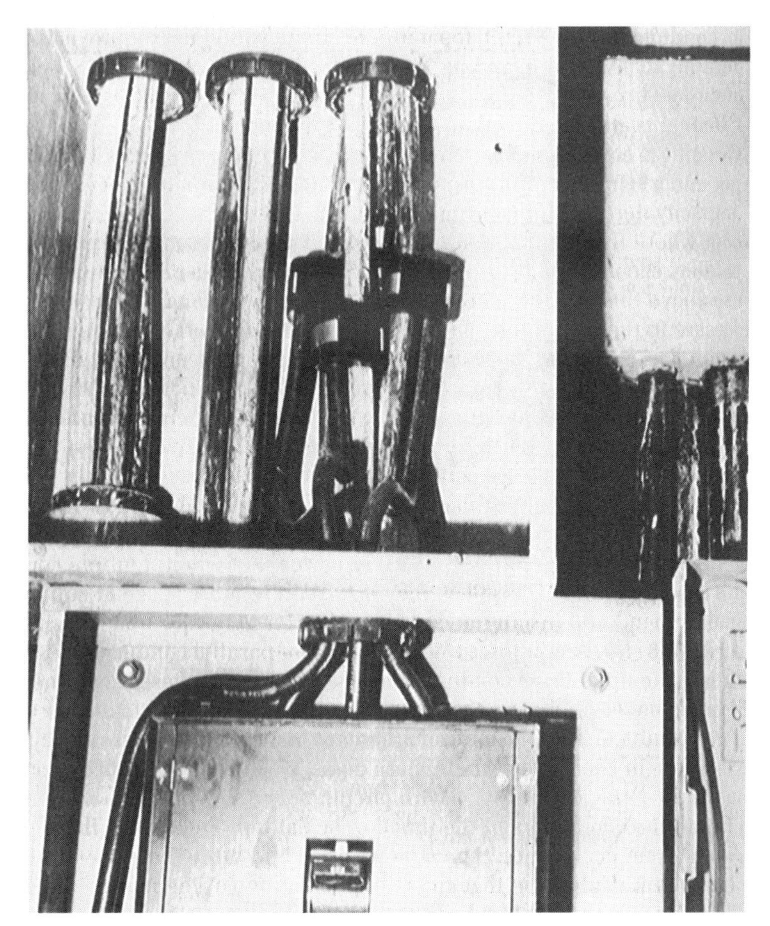

FIG. 2-74 A 1200-A circuit of three sets of four 500-kcmil conductors (top) is tapped by a single set of 500-kcmil conductors to a 400-A CB (bottom) that feeds an adjacent meter center in an apartment house. This was ruled a violation because the tap must be made from all the conductors of the 1200-A circuit. (*Note:* The conduits feeding the splice box at top are behind the CB enclosure at bottom.)

the 400-A CB. Phase *B* and phase *C* must be treated in the same way—as well as the neutral. The method shown in the photo was selected by the installer on the basis that the conductors in the right-hand conduit are tapped on this floor, the center-conduit conductors tapped to a 400-A CB on the floor above, and the left-conduit conductors tapped to a 400-A CB on the floor above that. But such a hookup can produce excessive current on some of the 500-kcmil conductors. Because it does not have the parallel conductors of equal length at points of load tap, the currents will not divide equally, and this is a violation of the second paragraph of Sec. 310-4, which calls for parallel conductors to "be the same length."

Exception 1 of this section clearly indicates long-time Code acceptance of paralleling conductors smaller than No. 1/0 for use in traveling cables of elevators, dumbwaiters, and similar equipment.

Exception 2 of Sec. 310-4 permits parallel-circuit makeup using conductors smaller than 1/0—*but all the conditions given must be observed.* This exception permits use of smaller conductors in parallel for circuit applications where it is necessary to reduce conductor capacitance effect or to reduce voltage drop over long circuit runs. As it was argued in the proposal for this exception,

> If a No. 14 conductor, for example, is adequate to carry some load of not more than the 15-amp rating of the wire, there can be no reduction in safety by using two No. 14 wires per circuit leg to reduce voltage drop to acceptable limits—with a 15-amp fuse or CB pole protecting each pair of No. 14s making up each leg of the circuit.

Where conductors are used in parallel in accordance with this exception, the rule requires that *all* the conductors be installed in the same raceway or cable. And that will dictate application of the last sentence of Sec. 310-4: "Conductors installed in parallel *shall* comply with the provisions of Article 310 Note 8, Notes to Ampacity Tables of 0 to 2000 Volts." Thus a single-phase, 2-wire control circuit made up of two No. 14s for each of the two legs of the circuit would have to be considered as four conductors in a conduit, and the "ampacity" of each No. 14 would be reduced to 80 percent of the value shown in Table 310-16. If TW wires are used for the circuit described, the ampacity of each is no longer the value of 20 A, as shown in Table 310-16. With four of them in a conduit, each has an ampacity of 0.8 × 20, or 16 A. Then using a 15-A fuse or CB pole for each pair of No. 14s would properly protect the conductors and would also comply with the dagger footnote of Table 310-16, which says that No. 14 must not have overcurrent protection greater than 15 A. See Fig. 2-75.

Parallel makeup with smaller than 1/0 conductors

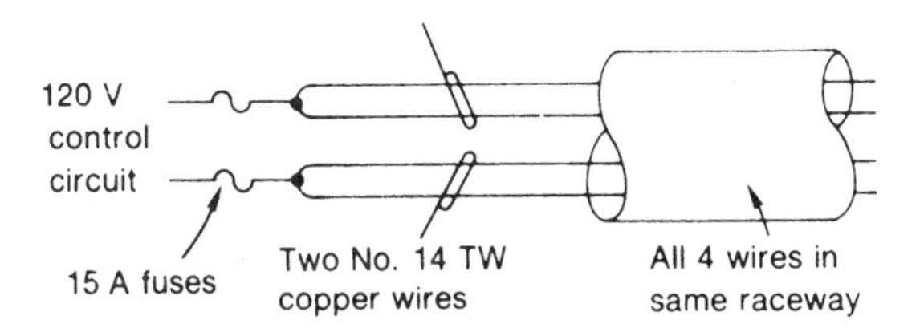

From Table 310-16, No. 14 TW copper has ampacity of 20.

"Ampacity reduction" from Note 8 to table:
20 amps × 0.8 = 16 amps.

15-AMP FUSES PROPERLY PROTECT THE NO. 14s AND SATISFY FOOTNOTE TO TABLE 310-16.

FIG. 2-75 Overcurrent protection must be rated not in excess of the ampacity of one conductor when conductors smaller than No. 1/0 are used in parallel. (Sec. 310-4.)

Exception 3 permits circuits operating at frequencies of 360 Hz or higher to use conductors smaller than 1/0 in parallel. Exception 3 permits parallel use of conductors smaller than 1/0 for circuits operating at 360 Hz or higher frequencies, provided that all the wires are in the same conduit, the ampacity of each wire is adequate to carry the entire current that is divided among the parallel wires, and the rating of the circuit protective device does not exceed the ampacity of any one of the wires. Such use of small conductors in parallel is very effective in reducing inductive reactance and "skin effect" in high-frequency circuits. Interweaving of the multiple wires per phase and neutral produces greater mutual cancellation of the magnetic fields around the wires and thereby lowers inductance and skin effect. Typical application of such usage is made for the 400-Hz circuits that are standard in the aerospace and aircraft industry.

Exception 4 to Sec. 310-4 permits the paralleling of neutral (grounded) conductors smaller than No. 1/0, but not smaller than No. 2. Such permission only applies to neutral conductors in existing facilities and, therefore, may *not* be used in new design. That makes little sense, but is strictly mandated by the literal wording used in Exception 4.

The purpose of this permission is illustrated in Fig. 2-76. As can be seen, although the single 3/0 has approximately 25 percent greater cross-sectional area, the two No. 2s have approximately 25 percent greater surface area. And where nonlinear loads are supplied, the greater surface area will reduce the heating caused by the skin effect. That is, because nonlinear loads generally produce higher-frequency harmonic currents, which tend to flow on the surface of a conductor, the use of two smaller conductors instead of an equivalent single larger conductor will serve to prevent overheating of the neutral due to additive harmonics.

	Cross-section Area, in^2	*Diameter*, in
3/0:	0.173	0.470
No. 2:	0.067	0.292

There are two lengths of No. 2s, so the total cross-sectional area would be twice the value shown in Table 8, or: 0.134 in^2 (2 × 0.067). Now we can see that the single 3/0 has a greater cross section than do two No. 2s.

Surface Area (skin area)

From the formula (πd) × L, we can determine the surface area of each conductor. But, as before, the value calculated for a single No. 2 is multiplied by two because there are two lengths of No. 2. For the sake of simplicity, we'll use 5 in as the length. When L = 5 in

3/0: Surface area $= \pi \times d \times L$
$= 3.14 \times 0.470$ in $\times$ 5 in
$= 7.379$

No. 2s: Surface area $= \pi \times d \times L \times 2$
$= 3.14 \times 0.292$ in $\times$ 5 in $\times$ 2
$= 9.1688$ in^2

Dividing 9.1688/7.379, we see that the two No. 2s provide about 25% *more* surface area and about 25% *less* cross section!

FIG. 2-76

SIZING EQUIPMENT GROUNDING CONDUCTORS AND EQUIPMENT BONDING JUMPERS

Equipment Ground

This is the permanent and continuous bonding (that is, connecting) of all non-current-carrying metal parts of equipment enclosures—conduit, boxes, cabinets, housings, frames of motors, and lighting fixtures—and the connection of this interconnected system of enclosures to the system grounding electrode, either at the service equipment of a premises or on the secondary side of a transformer within a system (Fig. 2-77). All metal enclosures must be interconnected to provide a low-impedance path for fault-current flow along the enclosures, to ensure the operation of overcurrent devices that will open a circuit in the event of a fault. By opening a faulted circuit, these devices prevent dangerous voltages from being present on equipment enclosures that could be touched by personnel, with consequent electric shock to such personnel.

Simply stated, the grounding of all metal enclosures containing electric wires and equipment prevents the occurrence of any potential above ground on the enclosures.

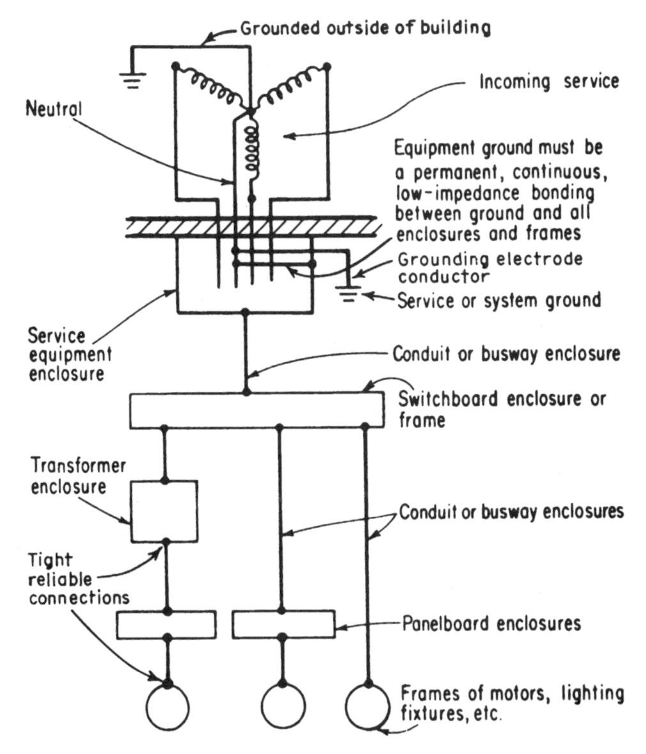

FIG. 2-77 For either a grounded or an ungrounded electrical system, all metallic enclosures of conductors and other operating components must be interconnected and grounded at the service system source.

Such bonding and grounding of all metal enclosures is required for both grounded electrical systems (those systems in which one of the circuit conductors is intentionally grounded) and ungrounded electrical systems (systems with none of the circuit wires intentionally grounded).

But effective equipment grounding is extremely important in grounded electrical systems, to provide the automatic fault clearing that is one of the important advantages of grounded systems. A low-impedance path for fault current is necessary to permit enough current to flow to operate the fuses or circuit breaker protecting the circuit, and low impedance is specifically required by NEC Sec. 250-51.

In a grounded electrical system with a high-impedance equipment ground-return path, if one of the phase conductors of the system (that is, one of the ungrounded conductors of the wiring system) should accidentally come in contact with one of the metal enclosures in which the wires are run, it is possible that not enough fault current would flow to operate the overcurrent devices. In such a case, the faulted circuit would not automatically open, and a dangerous voltage would be present on the conduit and other metal enclosures. This voltage would present a shock hazard and a fire hazard through possible arcing or sparking from the energized conduit to some grounded pipe or other grounded metal.

In a grounded system with a high-impedance equipment ground-return system, a ground fault will not open the faulted circuit; a phase-to-phase fault must develop to operate the overcurrent device. Such a condition is quite hazardous.

For effective protection against common ground faults in grounded electrical systems, therefore, low impedance for the equipment bonding system is even more important than low impedance for the grounding-electrode connection to earth. And in long runs of magnetic-material conductor enclosures, the ground-circuit impedance should be taken as 10 times the dc resistance to allow for the many variables.

When an individual equipment grounding conductor is used in a raceway—either in a nonmetallic raceway, as required by Sec. 347-4, or in a metal raceway where such a conductor is used for grounding reliability even though Sec. 250-91(b) accepts metal raceways as a suitable grounding conductor—the grounding conductor must have a minimum size as shown in Table 250-95. The minimum acceptable size of an equipment grounding conductor is based on the rating of the overcurrent device (fuse or CB) protecting the circuit, run in the same raceway, for which the equipment grounding conductor is intended to provide a path of ground-fault current return (Fig. 2-78). Each size of grounding conductor in the table is adequate to carry enough current to blow the fuse or trip the CB of the rating indicated beside it in the left-hand column. In Fig. 2-78, if the fuses are rated at 60 A, Table 250-95 shows that the grounding electrode conductor used with that circuit must be at least a No. 10 copper or a No. X Aluminum or copper-clad aluminum.

Whenever an equipment grounding conductor is used for a circuit that consists of only one conductor for each hot leg (or phase leg), the grounding conductor is sized simply and directly from Table 250-95, as described. When a circuit is made up of parallel conductors per phase, say an 800-A circuit with two conductors per phase, an equipment grounding conductor is also sized in the same way and would, in that case, have to be at least a No. 1/0 copper or No. 3/0 aluminum. *But* if such a circuit is made up using two conduits, that is, three phase legs and a neutral in each conduit, Sec. 250-95 requires that an individual grounding conductor be run in each of the conduits *and* each of the two grounding conductors be at least No. 1/0 copper or No. 3/0 aluminum (Fig. 2-79). Another example is shown in Fig. 2-80, where a 1200-A protective device on a parallel circuit calls for No. 3/0 copper or 250-kcmil aluminum grounding conductor.

[Note in that example that each 500-kcmil XHHW circuit conductor has an ampacity of 430 A (Table 310-16), and three per phase gives a circuit ampacity of $3 \times 430 =$

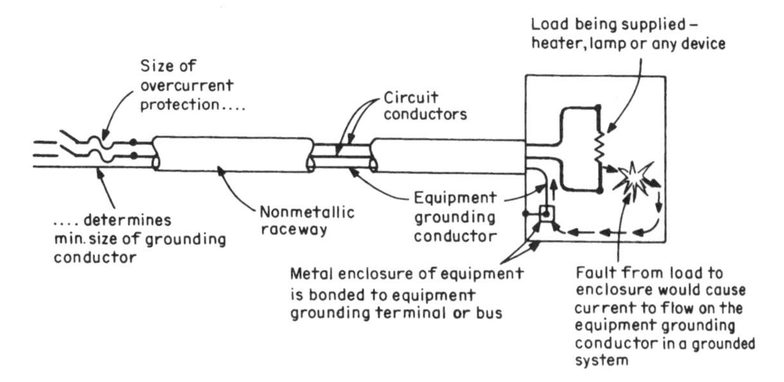

FIG. 2-78 Size of grounding conductor must carry enough current to operate circuit overcurrent device.

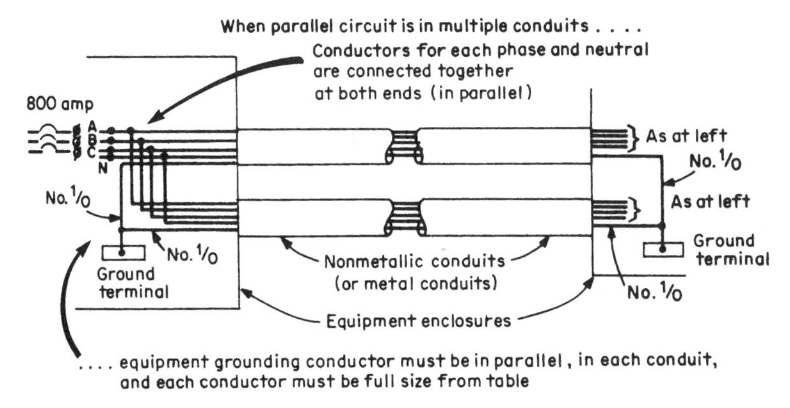

FIG. 2-79 Grounding conductor must be used in each conduit for parallel conductor circuits. (Sec. 250-95.)

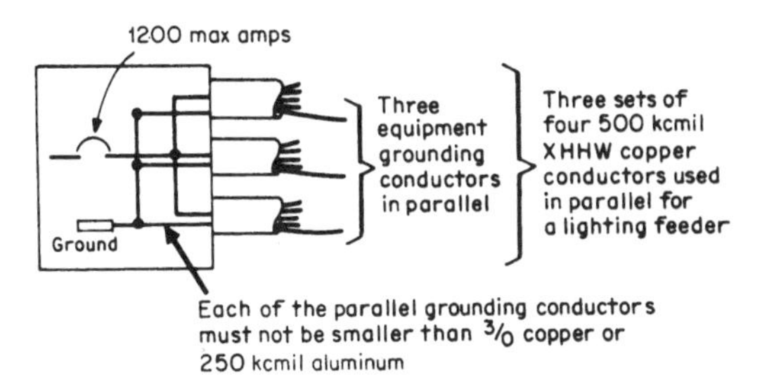

FIG. 2-80 Using equipment grounding conductors in parallel. (Sec. 250-95.)

1290 A. Use of a 1200-A protective device satisfies the basic rule of Sec. 240-3, protecting each phase leg within its ampacity. Because the load on the circuit is continuous (over 3 h), the circuit is loaded to not over 960 A—satisfying Sec. 220-10(b), which requires a continuous load to be limited to no more than 80 percent of the circuit protection rating. Each circuit conductor is actually made up of three 500-kcmil XHHW conductors with a total-per-phase ampacity of 3 × 430 = 1290 A. But load is limited to 0.8 × 1200 = 960 A per phase. Each 500-kcmil conductor is then carrying 960 ÷ 3 = 320 A. Because that value is less than 380 A, which is the ampacity of a 500-kcmil THW copper, the use of XHHW conductors *does* comply with the UL requirement that size 1/0 and larger conductors connected to the equipment be rated at not over 75°C (such as THW) or, if 90°C conductors are used (such as XHHW), they must be used at no more than the ampacity of 75°C conductors of the same size. However, some authorities object to that usage on the grounds that the use of 1200-A protection would not be acceptable to Sec. 240-3(c) if 75°C (THW) conductors were used, with an ampacity of only 3 × 380 A, or 1140 A, per phase, because the rating is over the 800-A level and it is not permitted to go to the next larger size of protective device. Therefore, they note, the XHHW conductors are not actually being used as 75°C conductors; the load current could later be increased above the 75°C ampacity; and the application might be taken to violate the letter and intent of the UL rule, thereby violating Sec. 110-3(b) of the NEC. If such application is acceptable to the local inspector, unless the terminating device and the enclosure are marked as suitable for conductors loaded to their 90°C ampacity, then the load would have to be limited to 1140 A (380 × 3), which is the 75°C ampacity for 500-kcmil copper conductors.]

The fifth paragraph of Sec. 250-95 covers a similar concern for unnecessarily oversizing equipment grounding conductors. Because the minimum acceptable size of an equipment grounding conductor is based on the rating of the overcurrent protective device (fuse or CB) protecting the circuit for which the equipment grounding conductor is intended to provide a path of ground-fault return, a problem arises when a motor circuit is protected by a magnetic-only (a so-called instantaneous) circuit breaker. Because Sec. 430-52 and Table 430-152 permit an instantaneous-trip CB with a setting of 700 percent of (7 times) the motor full-load running current—and even up to 1300 percent for an instantaneous CB or MSCP (motor short-circuit protector), if needed to handle motor inrush current—use of those high values of current rating permitted in Table 430-152 would result in excessively large equipment grounding conductors. Because such large sizing is unreasonable and not necessary, the rule says when sizing an equipment grounding conductor from Table 250-95 for a circuit protected by an instantaneous-only circuit breaker or by an MSCP, the rating of the motor running overload device must be used in the left-hand column of Table 250-95 (Fig. 2-81).

Exception 2 states that the equipment grounding conductor *need not* be larger than the circuit conductors. The main application for this exception is for motor circuits where short-circuit protective devices are usually considerably larger than the motor branch circuit conductor ampacity, as permitted in Sec. 430-52 (up to 1300 percent) to permit starting of a motor without opening on inrush current. In such cases, literal use of Table 250-95 could result in grounding conductors larger than the circuit conductors.

Exception 3 points out that metal raceways and cable armor are recognized as equipment grounding conductors and Table 250-95 does not apply to them.

Figure 2-81 shows details of a controversy that often arises about Sec. 250-95 and Sec. 250-57(a). When two or more circuits are used in the same conduit, it is logical to conclude that a single equipment grounding conductor within the conduit may serve as the required grounding conductor for each circuit if it satisfies Table 250-95 for the circuit with the highest rated overcurrent protection. The common contention is that if a single metal conduit is adequate as the equipment grounding conductor for all the contained circuits, a single grounding conductor can serve the same purpose when installed

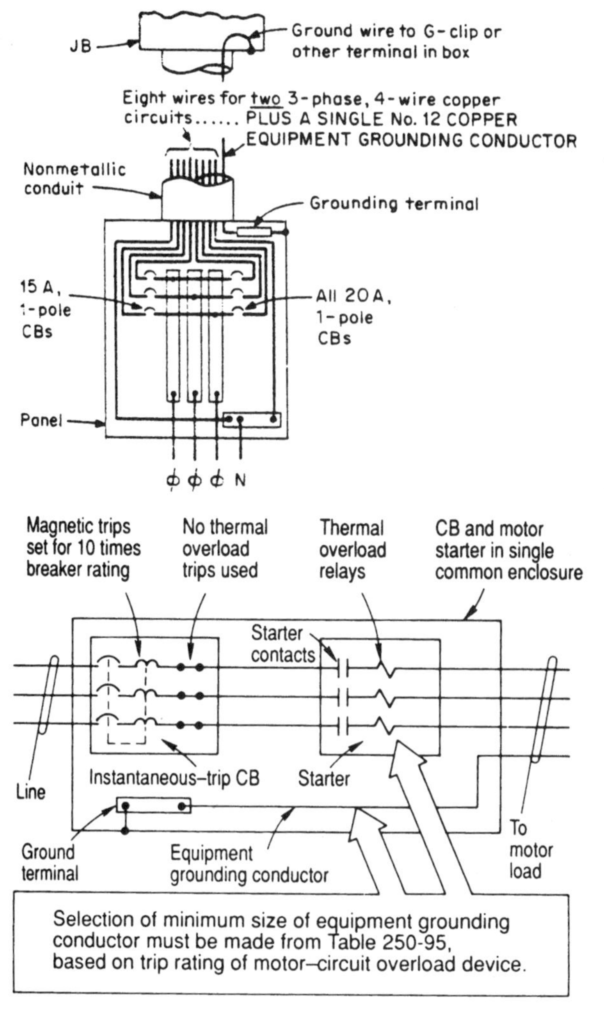

FIG. 2-81 These applications are covered by the fourth and fifth paragraphs of Sec. 250-95. (Sec. 250-95.)

in a nonmetallic conduit that connects two metal enclosures (such as a panel and a homerun junction box) where both circuits are within both enclosures. As shown, a No. 12 copper conductor satisfies Table 250-95 as an equipment grounding conductor for the circuit protected at 20 A. The same No. 12 also may serve for the circuit protected at 15 A, for which a grounding conductor must not be smaller than No. 14 copper. Such application is specifically permitted by the fourth paragraph of Sec. 250-95, just before the exceptions. Although this will have primary application with PVC conduit where an equipment grounding conductor is required, it may also apply to circuits in EMT, IMC, or rigid metal conduit when an equipment grounding conductor is run with the circuit conductors to supplement the metal raceway as an equipment grounding return path.

SIZING EQUIPMENT GROUNDING CONDUCTORS WHERE HIGH FAULT CURRENTS ARE AVAILABLE

Over the years, most designers and installers have considered the minimum size of equipment grounding conductor given in Table 250-95 to be adequate regardless of the short-circuit current available at that point in the distribution system. That is, whether there was 5000 or 50,000 A of available fault current, generally the same size of equipment grounding conductor would be selected. While this may not have been an issue years ago—very few installations had available short-circuit currents greater than 5000 or 10,000 A—today, with much larger distribution systems and lower-impedance transformers (some below 1 percent), it is no longer unusual to see available fault currents in excess of 100,000 and, in some cases, over 200,000 A. It is those installations where high levels of short-circuit current are available that pose the real problem and require additional consideration.

The concern for providing protection for the equipment grounding conductor was recognized a number of years ago, and was explained by Eustace Soares in his now-famous work entitled "Grounding Electrical Distribution Systems for Safety." In that book, Soares indicates that the "validity" of a conductor as a fault return path is maintained only where loaded not in excess of a specific current value for a specified amount of time based on the conductor's cross-sectional circular mils. That is, for a copper conductor, its integrity or "validity" is maintained where the amount of current for each 30 cmils of cross-sectional area is not greater than 1 A and does not persist for more than 5 s. That validity is related to the amount of energy that would cause the copper conductor to become loosened at its point of attachment after the copper cools to ambient. That validity rating is based on raising the temperature of the copper conductor from 75 to 250°C.

Another recognized method looks at the amount of energy that will produce damage to the conductor's insulating material. The method to which reference is made is covered in the International Cable Engineers Association publication P-32-382. That method calculates the energy required to raise the conductor's temperature from 75 to 150°C, which will cause damage to the conductor's insulation. And a third method promoted by Onderdonk calculates the amount of energy necessary to cause the temperature of the conductor material to rise from 75 to 1083°C, which will cause the copper to melt. That is, the Onderdonk method calculates the conductor melting point. Values of short-circuit current and withstand ratings calculated in accordance with each of those recognized methods are shown in Fig. 2-82.

But does the NEC *require* that equipment grounding conductors be evaluated with regard to the available fault currents? In answer to that question, consider the following.

The NEC in Sec. 250-51 requires that each and every equipment grounding conductor be capable of carrying any fault currents likely to be imposed upon it. And compliance with that requirement would necessitate some evaluation of the selected size of grounding conductor with respect to the available fault current. That is, in addition to ensuring compliance with Sec. 250-95 on the minimum sizing, the conductor selected in accordance with that Code section must have sufficient cross-sectional area "to conduct safely" the fault current it will be required to carry.

Although the phrase "to conduct safely" is not very clear, use of one of the three preceding methods—the Soares validity-of-connection method, the ICEA insulation damage method, or the Onderdonk melt-point method—should serve to satisfy the wording of Sec. 250-51. Of course, if the authority having jurisdiction prefers one method over the others, then the method preferred by the local inspector should be used.

Which is the more desirable and realistic approach? That depends on the application. For example, in most applications it would seem that Soares' validity-of-connection

COMPARISON OF EQUIPMENT GROUNDING CONDUCTOR SHORT CIRCUIT WITHSTAND RATINGS.

CONDUCTOR SIZE	5 SEC RATING (AMPS)			I^2t RATING X10^6 (AMPERE Squared Seconds)		
	ICEA P32-382 INSULATION DAMAGE 150 C	SOARES 1 AMP/30 cm VALIDITY 250 C	ONDERDONK MELTING POINT 1083 C	ICEA P32-382 INSULATION DAMAGE 150 C	SOARES 1 AMP/30 cm VALIDITY 250 C	ONDERDONK MELTING POINT 1083 C
14	97	137	253	.047	.094	.320
12	155	218	401	.120	.238	.804
10	246	346	638	.303	.599	2.03
8	391	550	1015	.764	1.51	5.15
6	621	875	1613	1.93	3.83	13.0
4	988	1391	2565	4.88	9.67	32.9
3	1246	1754	3234	7.76	15.4	52.3
2	1571	2212	4078	12.3	24.5	83.1
1	1981	2790	5144	19.6	38.9	132.
1/0	2500	3520	6490	31.2	61.9	210.
2/0	3150	4437	8180	49.6	98.4	331.
3/0	3972	5593	10313	78.9	156.	532.
4/0	5009	7053	13005	125.	248.	845.
250	5918	8333	15365	175.	347.	1180.
300	7101	10000	18438	252.	500.	1700.
350	8285	11667	21511	343.	680.	2314.
400	9468	13333	24584	448.	889.	3022.
500	11835	16667	30730	700.	1389.	4721.
600	14202	20000	36876	1008.	2000.	6799.
700	16569	23333	43022	1372.	2722.	9254.
750	17753	25000	46095	1576.	3125.	10623.
800	18936	26667	49168	1793.	3556.	12087.
900	21303	30000	55314	2269.	4500.	15298.
1000	23670	33333	61460	2801.	5555.	18867.

FIG. 2-82 Values of current (at left) and the let-through energy (I^2t) withstand ratings for copper conductors as determined in accordance with three industry-recognized methods. Designers and installers should be aware that equipment grounding conductors selected in accordance with Sec. 250-95 and Table 250-95 of the NEC may *not* be adequate where large amounts of short-circuit current are available. In such applications, safety concerns for personnel and equipment dictate that the size of equipment grounding conductor be increased above the Code-recognized minimum.

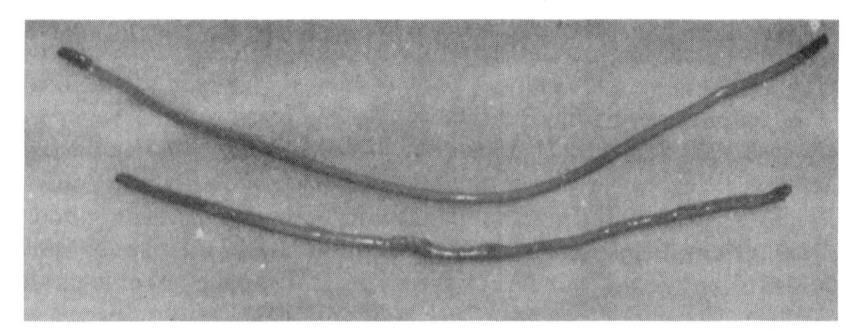

Here are two No. 12 AWG THHN copper conductors after being exposed to 40,000 A. The one No. 12 (at top) was protected in accordance with ICEA limits by a current-limiting overcurrent protective device. The other was not. As can be seen, in the first case, the conductor was not damaged because the current-limiting protective device opened the circuit before insulation temperature limits were exceeded. In the other conductor (at bottom), the operating characteristics of the overcurrent protective device resulted in a value of current flow that severely damaged the THHN (90°C rated) insulation.

method is adequate. But, for isolated ground applications, the ICEA insulation damage method might be more appropriate to ensure the desired isolation of the ground-return path even after a short circuit. The melt-point method would appear to be the least desirable, but could be construed as satisfying the wording of Sec. 250-51. Regardless of which method is used, certainly some evaluation of the grounding conductor's fault current-carrying capability must be performed where there are large values of available short-circuit current to assure compliance with the NEC and to ensure continuity of the ground-return path, which provides for automatic clearing of a faulted circuit.

How does one analyze the equipment grounding conductor for fault-carrying capability?

First, you will need to refer to the short-circuit analysis—which should have been performed to determine the minimum interrupting rating of the selected protective devices—and find the available fault current at that point in the system where the circuit originates. Next, using the protective device manufacturer's operating characteristics data, determine the amount of short-circuit current and the amount of let-through energy (I^2t) that the grounding conductor will be exposed to during a bolted fault. Then, refer to the data given in Fig. 2-82 to verify that the equipment grounding conductor, selected in accordance with Sec. 250-95 and Table 250-95, will be capable of carrying that value of current and withstanding the let-through energy. If not, then a larger conductor, that is, one capable of sustaining that value of current and withstanding the I^2t value, must be selected.

Although the problem of excessive fault current for the size of grounding conductor is more of a concern for feeders, branch circuits that originate in close proximity to the service equipment, such as at motor control centers located near the service equipment, should also be evaluated. Only through such an evaluation is it possible to ensure compliance with the NEC and ensure that the equipment grounding conductor will, in fact, be capable of facilitating operation of the circuit overcurrent protective device and provide automatic clearing of faulted conductors.

Bonding Jumpers

Any bonding jumper on the load side of the service should conform to the requirements for equipment grounding conductors. As a result, the bonding of conduits for a parallel-circuit makeup would have to comply with the NEC rule that requires equipment grounding conductors to be run in parallel "where conductors are run in parallel in multiple raceways." That rule would then be taken to require that bonding jumpers also be run in parallel for multiple-conduit circuits. In the case of the 800-A circuit run in two conduits, as described above, instead of a single No. 1/0 copper jumper from one bushing lug to the other bushing lug and then to the ground bus, it would be better to use a separate No. 1/0 copper jumper from each bushing to the ground bus, so the jumpers are run in parallel, as required for equipment grounding conductors. Figure 2-83 shows the two possible arrangements. Bonding jumpers on the load side of service equipment are sized and routed in the same way as equipment grounding conductors, because such bonding jumpers and equipment grounding conductors serve identical functions. And Sec. 250-95 requires the equipment grounding conductor for each of the conduits of a parallel circuit to be the full size determined from the circuit rating. Here, No. 1/0 copper for each conduit would be based on the 800-A rating of the feeder protective device. [*Note:* Although it is not recommended practice, NEC Sec. 250-79(d) does permit the above-described use of a single No. 1/0 jumper to bond both conduits to the equipment ground bus. In fact the NEC rule permits any number of metal conduits to be bonded to a switchboard or other equipment enclosure by a "single common continuous" jumper if it is sized from Table 250-95 for "the largest overcurrent device supplying" any of the circuits contained in all the conduits.]

Conductors in Raceway

Circuits of different voltages up to 600 V may occupy the same wiring enclosure (cabinet, box, housing), cable, or raceway, provided all the conductors are insulated for the maximum voltage of any circuit in the enclosure, cable, or raceway. For instance, motor power conductors and motor control conductors are permitted in the same conduit. In the past, there was a longstanding controversy about the use of control-circuit conductors in the same conduit with power leads to motors.

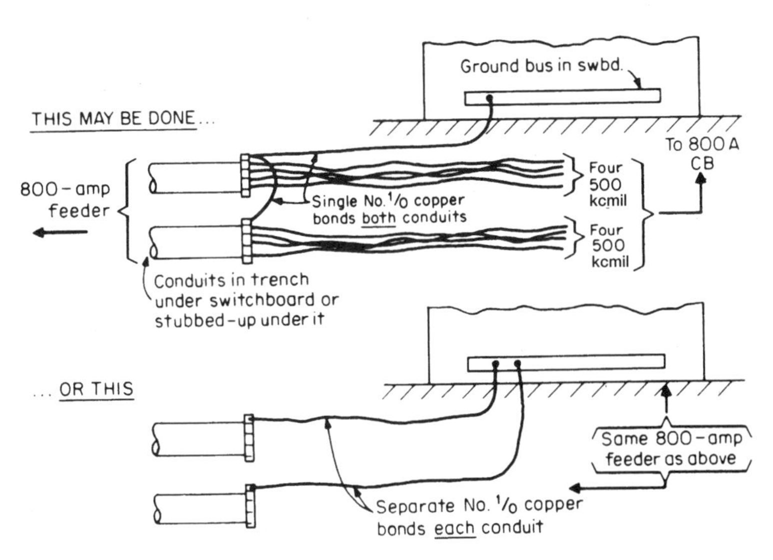

FIG. 2-83 Equipment bonding jumpers for multiple-conduit feeders provide better protection when run in parallel, as at bottom here.

Careful consideration should be given to sizing of the equipment grounding conductor where large amounts of short-circuit current are available. The equipment grounding conductors selected for the lighting branch circuits supplied above (arrow), which are fed from a panel in close proximity to the service equipment, should be evaluated to ensure the selected grounding conductors will be capable of withstanding the value of the fault current they are likely to carry.

As shown in Fig. 2-84, a common raceway may be used for both motor power conductors and control-circuit conductors only where two or more motors are required to be operated together to serve their load function. Many industrial and commercial installations contain machines, manufacturing operations, or processes in which a number of motors perform various parts or stages of a task. In such cases, either all the motors operate or none do; the placement of all control wires and power wires in the same raceway cannot produce a situation in which a fault in one motor circuit disables another circuit to a motor that might otherwise be kept operating.

But when a common raceway is used for power and control wires to supply separate, independent motors, a fault in one circuit could knock out others that do not have to be shut down when one goes out. With motor circuits so closely associated with vital equipment such as elevators, fans, and pumps in modern buildings, it is a matter of safety to separate such circuits and minimize the outage due to a fault in a single circuit. For safety's sake, do not "put all your eggs in one basket." But the objectionable loss of more than one motor on a single fault does not apply where all motors must be shut down when any one is stopped, as in multimotor machines and processes.

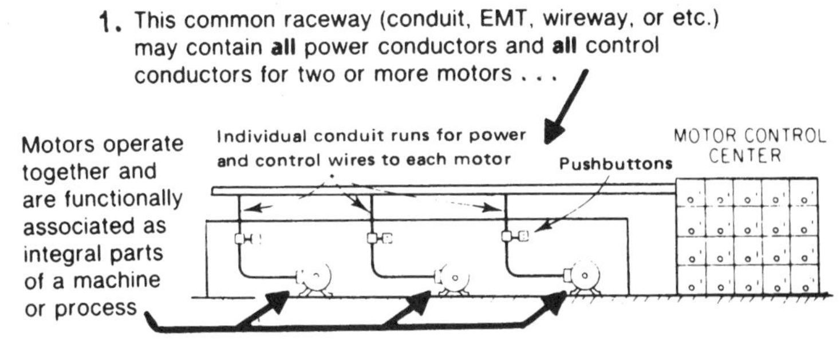

2 . . . **but**, the intent of this **Code** rule, along with that of Section 725-26 (b), permits a common raceway **only** where the power and control conductors are for a number of motors that operate integrally—such as a number of motors powering different stages or sections of a multi-motor process or production machine. Such usage complies with Section 725-26 (b) (last sentence), which permits power and control wires in the same raceway, cable, or other enclosure when the equipment powered is "functionally associated"—that is, the motors have to run together to perform their task.

FIG. 2-84 Mixing of power and control wires in a common raceway is limited.

For those cases where each of several motors serves a separate, independent load with no interconnection of control circuits and no mechanical interlocking of driven loads, the use of a separate raceway for each motor is required—but only when control wires are carried in the raceways, as shown in Fig. 2-85. For the three motors shown, it would be acceptable to run the power conductors for all the motors in a single raceway and all the control circuit wires in another raceway. Of course, deratings would have to be made, and there is the definite chance of losing more than one motor on a fault in only one of the circuits in either the power raceway or the control raceway.

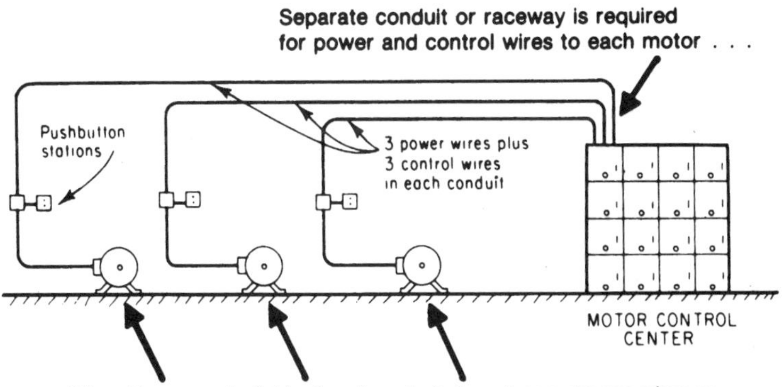

FIG. 2-85 Separate conduit must be used to carry power and control wires to independent motors.

Conduit Fill. For the makeup of all circuits, the NEC regulates the maximum number of conductors that may be pulled into rigid metal conduit, rigid nonmetallic conduit, intermediate metal conduit, electrical metallic tubing, flexible metal conduit, etc. The number of conductors permitted in a particular size of conduit or tubing is covered in NEC Chap. 9: Table 1 indicates the percentage of fill permitted, regardless of conductor type(s) used. Tables 4 and 5 permit calculation of conduit fill where different sizes and/or insulations are used in a common raceway; and App. C is used to determine fill where conductors are all the same size and insulation type. Tables 4 to 8 cover combinations of conductors of different sizes used for new work or rewiring. For all conductors, three or more to a conduit, the sum of the cross-sectional areas of the individual conductors must not exceed 40 percent of the interior cross-sectional area of the conduit or tubing, both for new work and for rewiring in existing conduit or tubing, as shown in Fig. 2-86. Note 4, which follows Table 1 in NEC Chap. 9, permits a 60 percent fill of conduit nipples not over 24 in long, without the need for derating of ampacities.

Tables C-1 through C-12 of Chap. 9 give the maximum allowable fill for conduit or tubing in which all conductors are the same size up to 2000 kcmil and for conduit sizes from ½ to 6 in. Notice that the tables in App. C are based on the type of raceway. That is, because raceways of the same size may have different internal dimensions, some raceways can hold more of a given size conductor. Therefore, the type of raceway must be considered.

Question What is the minimum size of EMT required for six No. 10 THHN wires?

Answer Table C-2 of App. C shows that six No. 10 THHN wires may be pulled into a ¾-in EMT.

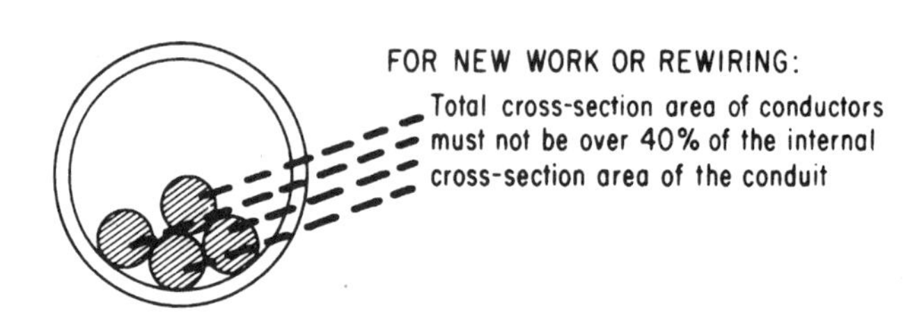

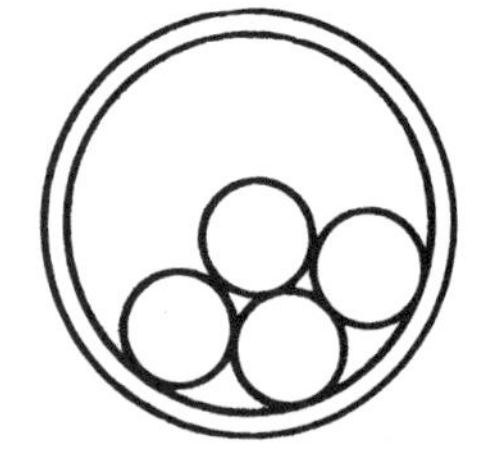

FIG. 2-86 For three or more conductors in a conduit, the sum of their cross-sectional areas must not exceed 40 percent of the conduit cross-sectional area.

Question What is the minimum rigid metal conduit size for use with four No. 6 RHH conductors with outer covering?

Answer Table C-8 of App. C shows that a 1¼-in minimum conduit size must be used for four to six No. 6 RHH conductors.

Question What is the minimum IMC size required for four 500-kcmil XHHW rigid metal conduit conductors?

Answer Table C-4 shows that 3-in IMC may contain four 500-kcmil XHHW (or THHN) conductors.

When not all the conductors in a conduit or tubing are the same size, the minimum required size of conduit or tubing must be calculated. Table 1 of Chap. 9 says that conduit containing three or more conductors of any type, for new work or rewiring, may be filled to 40 percent of the conduit cross-sectional area. Note 6 to this table refers to Tables 4, 5, and 5A of Chap. 9 for the dimensions of conductors, conduit, and tubing to be used in calculating conduit fill for combinations of conductors of different sizes.

Example. What is the minimum size of schedule 40 PVC rigid nonmetallic conduit required to enclose six No. 10 THHN, three No. 4 RHH (without outer covering), and two No. 12 TW conductors (Fig. 2-86A)?

Solution The cross-sectional areas of the conductors are found in Table 5 of Chap. 9:

No. 10 THHN: 0.0211 in^2

No. 4 RHH: 0.0973 in^2

No. 12 TW: 0.0181 in^2

(Note that RHH without outer covering has the same dimensions as THW.) The total area occupied by conductors is found as follows:

Six No. 10 THHN: 6 × 0.0211	0.1266 in^2
Three No. 4 RHH: 3 × 0.0973	0.2919 in^2
Two No. 12 TW: 2 × 0.0181	0.0362 in^2
Total	0.4547 in^2

In Table 4 of Chap. 9, the fifth column from the left gives 40 percent of the cross-sectional area of each of the conduit sizes listed in the first column at the left. The 40 per-

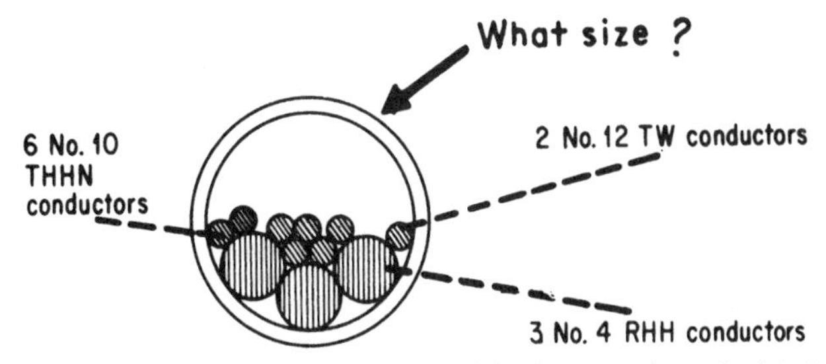

FIG. 2-86A Minimum permitted conduit size must be calculated when conductors of different sizes are used.

cent column shows that 0.333 in^2 is 40 percent fill for a 1-in conduit, and 0.581 in^2 is 40 percent fill for a 1¼-in schedule 40 PVC rigid nonmetallic conduit. Therefore, a 1-in conduit would be too small, and a 1¼-in schedule 40 PVC rigid nonmetallic conduit is the smallest that may be used for the 11 conductors.

Example What is the minimum size of IMC for four No. 4/0 TW and four No. 4/0 XHHW conductors?

Solution From Table 5, a No. 4/0 TW has a cross-sectional area of 0.3718 in^2. Four of these come to 4 × 0.3718, or 1.4872 in^2. From Table 5, four No. 4/0 XHHW have a cross-sectional area of 4 × 0.3197 or 1.2788 in^2. Then the total cross-sectional area is

$$1.4872 + 1.2788 = 2.7660 \text{ in}^2$$

The fifth column of Table 4 shows that 2½-in conduit would be too small. A 3-in IMC, with a 40 percent fill of 3.169 in^2, must be used.

Filling conduit to the NEC maximum is a minimal design practice; it is also frequently difficult or impossible from the practical standpoint of pulling the conductors into the conduit, owing to the usual twisting and bending of the conductors within the conduit. Bigger-than-minimum conduit should generally be used to provide some measure of spare capacity for load growth, and in many cases the conduit should be upsized considerably to allow for future installation of anticipated larger sizes of conductors.

In the sizing of conduit, neutral conductors are included in the total number of conductors because they occupy space. A completely separate consideration, however, is the relation of the neutral conductors to the number of *current-carrying* conductors in a conduit, which determines whether a derating factor must be applied:

Neutral conductors which carry only unbalanced current from phase conductors (as in the case of normally balanced 3-wire, single-phase or 4-wire, 3-phase circuits) are not counted in determining the current derating of conductors on the basis of the number in a conduit. Of course, a neutral conductor used with two phase legs of a 4-wire, 3-phase system to make up a 3-wire feeder is not a true neutral in the sense of carrying only current unbalance. Since the neutral carries the same current as the other two conductors under balanced load conditions, it must be counted as a phase conductor in derating for more than three conductors in a conduit.

Figure 2-87 shows four basic conditions of neutral loading relative to the need for counting the neutral conductor in derating a circuit to fluorescent or mercury ballasts:

Case 1. With balanced loads of equal power factor, there is no neutral current, and consequently no heating due to the neutral. For purposes of heat derating according to the NEC, this circuit produces the heating effect of three conductors.

Case 2. With two phases loaded and the third unloaded, the neutral carries the same current as the phases, but there is still the heating effect of only three conductors.

Case 3. With two phases fully loaded and the third phase partially loaded, the neutral carries the difference in current between the full phase value and the partial phase value, so again there is the heating effect of only three full-load phases.

Case 4. With a balanced load of fluorescent ballasts, third-harmonic generation causes a neutral current approximating the phase current, and there is the heating effect of four conductors. The neutral conductor should be counted with the phase conductors in determining conductor derating due to conduit occupancy.

Because the neutral of a 3-phase, 4-wire wye feeder to a load of fluorescent or mercury ballasts will carry even under balanced loading on the phases, such a neutral is not

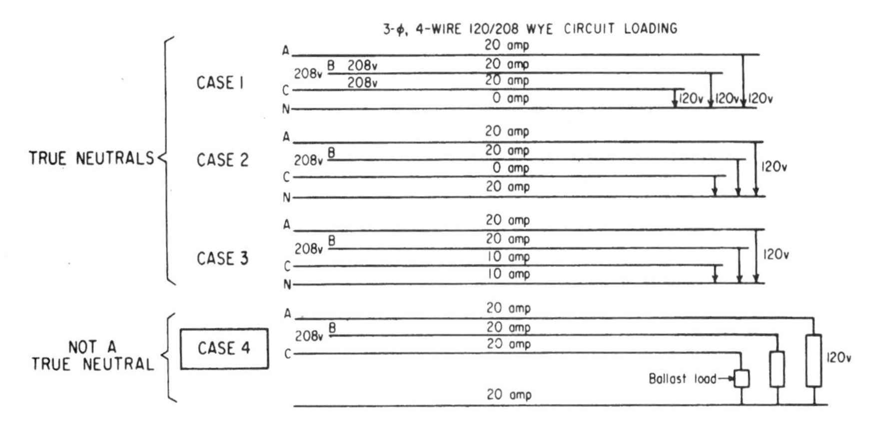

FIG. 2-87 All neutrals must be counted for conduit fill, but only "true" neutrals do not count for load-current derating.

a true non-current-carrying conductor and should be counted as a phase wire in determining the number of conductors to arrive at a derating factor for more than three conductors in a conduit. As a result, the conductors of a 3-phase, 4-wire feeder to a fluorescent load should have a load-current rating of only 80 percent of their normal ampacity from NEC Table 310-16.

Conductor Temperature Ratings

Important: As discussed in NEC Sec. 110-14(c), a general UL rule says that any electrical equipment rated over 100 A must be used with either 60°C conductors or 75°C conductors operating at the respective ampacities of those conductor insulation-temperature ratings unless the equipment is marked for use with conductors rated at higher temperatures. Equipment rated up to 100 A is generally limited to use with 60°C conductors or higher-temperature conductors operating at the 60°C ampacities for the particular conductor sizes, unless equipment is marked otherwise.

The UL temperature limitation is applied to the selection of feeder conductors in the steps that follow.

For the feeder in Fig. 2-88:

Step 1. Because the load on the feeder is continuous, the 100-A, 3-pole CB must have its load current limited to 80 A [80 percent of its rating, per NEC Sec. 220-10(b)]. This is sufficient for the 76-A load. (A CB or fused switch may be loaded continuously to 100 percent of the CB or fuse ampere rating only when the assembly is UL-listed for such use.)

Step 2. The CB load terminals are recognized by UL for use with 60°C conductors or higher-temperature conductors loaded not over the 60°C ampacity of the given size of conductor.

Step 3. The feeder phase conductors must have an ampacity of not less than 125 percent of the continuous load of 76 A, which means not less than 95 A, because that is also required by NEC Sec. 220-10(b). The feeder neutral conductor is not subject to this limitation because it is not connected to the terminals of a heat-producing device (such as

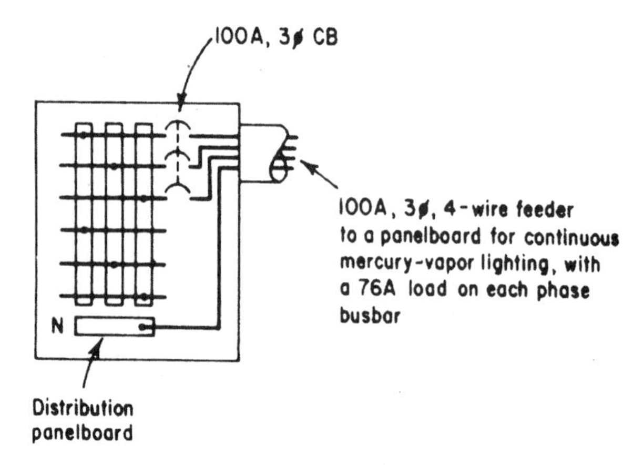

FIG. 2-88 Feeder conductors for equipment rated up to 100 A must not exceed their 60°C ampacity.

a CB or fusible switch). NEC Sec. 220-22 is the basic rule that covers the sizing of a feeder neutral. Section 220-10(b) covers the sizing of feeder phase conductors.

Step 4. If 60°C copper conductors are used for this feeder, reference must be made to the second column of Table 310-16. This feeder supplies electric-discharge lighting; therefore, Note 10(c) to the table requires that the feeder neutral be counted as a current-carrying conductor because of the harmonic currents present in the neutral. Then, because there are four current-carrying conductors in the conduit, the ampacity of each conductor must be derated to 80 percent of its value in the second column of Table 310-16. After the conductor is derated to 80 percent, it must have an ampacity of at least 95 A, as required by Sec. 220-10(b). From Table 310-16, a No. 1/0 TW conductor is rated at 125 A when only three conductors are used in a conduit. With four conductors in a conduit, the 125-A rating is reduced to 80 percent of that value (0.8 × 125 A), or 100 A, which is properly protected by the 100-A CB.

Note that this derating of ampacity of 80 percent, based on Note 8 of Tables 310-16 through 310-19, is in addition to the 80 percent load limitation (not a conductor derating) of Sec. 220-10(b). This is, in effect, a *double derating.* NEC Secs. 210-22(c) and 220-2(a) used to contain exceptions that seemed to make such a double derating unnecessary for branch circuits. *But* Sec. 220-10(b) does not contain an exception that eliminates the need for double derating of feeder conductors. It is certainly true that such a double derating provides a valuable and important amount of reserve capacity that is needed in feeder circuits to effectively minimize damage due to careless and/or "temporary" overloading during the operating life of the system.

The feeder circuit of four No. 1/0 TW conductors, rated at 100 A, would require a minimum of 2-in rigid metal conduit.

Note: A reduced size of neutral could be used, because Sec. 220-10(b) does not apply to the neutral.

Step 5. If 75°C conductors were used for this feeder circuit instead of 60°C conductors, the calculations would be different. For THW copper conductors, Table 310-16 shows that No. 1 conductors, rated at 130 A for not more than three current-carrying conductors in a conduit, would have an ampacity of 0.8 × 130 A, or 104 A, when four are used in the conduit and derated. Because 125 percent of 76 A is 95 A, the 104-A conductor ampacity would satisfy Sec. 220-10(b) as to the ampacity of the feeder conductors for a continuous load.

Although UL listing and testing of the CB is based on the use of 60°C conductors, the use of No. 1 75°C THW conductors is acceptable because the terminals of the breaker in this case would not be loaded to more than the ampere rating of a 60°C conductor of the same size. A No. 1 60°C TW conductor is rated at 110 A when not more than three current-carrying conductors are used in a conduit. When four conductors are in one conduit, the 60°C No. 1 wires are derated to 80 percent of 110 A, or 88 A. Because that value is greater than the load of 76 A on each CB terminal, the CB terminals are not loaded in excess of the 88-A allowable ampacity of 60°C No. 1 conductors, and the UL limitation is satisfied.

Four No. 1 THW conductors, rated at 104 A, would require a minimum of 1½-in conduit. Or, four No. 1 XHHW conductors could be used in 1½-in conduit.

CALCULATING MAXIMUM CONDUCTOR FILL FOR BOXES AND CONDUIT BODIES

Note that motor terminal housings are excluded from the rules on box conductor fill. And where any box or conduit body contains No. 4 or larger conductors, all the requirements of Sec. 370-28 on pull boxes must be satisfied. Refer to the discussion on "Minimum Dimensions for Larger Boxes," which follows.

Selection of any outlet or junction box for use in any electric circuit work must take into consideration the maximum number of wires permitted in the box by Sec. 370-16. Safe electrical practice demands that wires *not* be jammed into boxes because of the possibility of nicks or other damage to insulation, posing the threat of grounds or shorts.

As stated in part (a) of this section, Table 370-16(a) shows the maximum number of wires permitted in the *standard* metal boxes listed in that table. But that table applies only where all wires in a given box are of the same size, that is, all No. 14 or all No. 12, etc. Table 370-16(b) is provided for sizing a box where not all the wires in the box are the same size, by using so much cubic-inch space for each size of wire.

Table 370-16(a) includes the maximum number of No. 18 and No. 16 conductors that may be used in various sizes of boxes, and Table 370-16(b) gives the required box space for those sizes of conductors. Because of the extensive use of No. 18 and No. 16 wires for fixture wires and for control, signal, and communications circuits, these data are needed to ensure safe box fill for modern electrical systems.

Part (1) of Sec. 370-16(a) describes the detailed way of counting wires in a box and reducing the permitted number of wires shown in Table 370-16(a) where cable clamps, fittings, or devices like switches or receptacles take up box space.

Important details of the wire-counting procedure of part (1) are as follows:

1. From the wording, it is clear that no matter how many ground wires come into a box, whether they are ground wires in NM cable or ground wires run in metal or nonmetallic raceways, a deduction of only one conductor must be made from the number of wires shown in Table 370-16(a) (Fig. 2-89). Or, as will be shown in later examples, one or more ground wires in a box must be counted as a single wire of the size of the largest ground wire in the box. Any wire running unbroken through a box counts as one wire. Each wire coming into a splice device (crimp or twist-on type) is counted as one

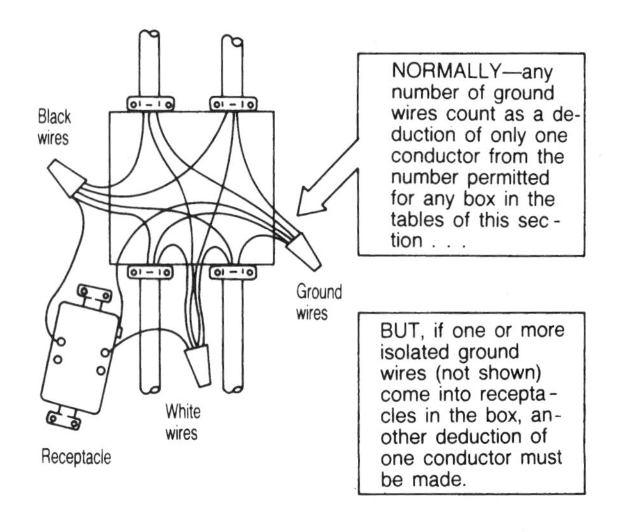

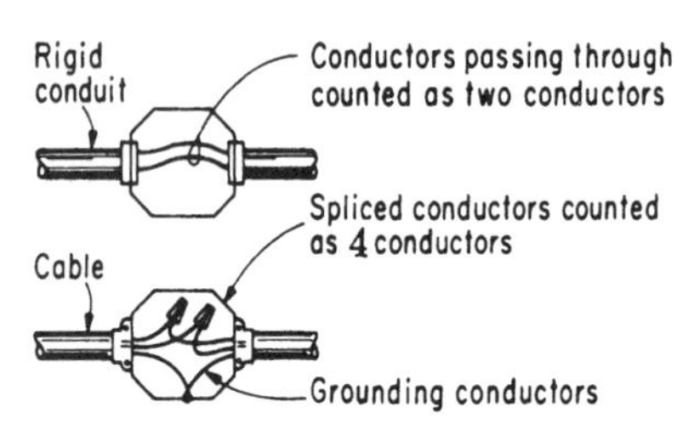

FIG. 2-89 Count all ground wires as *one* wire (or two wires if isolated-ground wires are also used) of the largest size of ground wire in the box.

wire. And each wire coming into the box and connecting to a wiring device terminal is *one* wire.

When a number of *isolated-ground* equipment grounding conductors for receptacles come into a box along with conventional equipment grounding wires, each type of equipment ground wire must be counted as one conductor for purposes of wire count when determining the maximum number of wires permitted in a box. When a number of isolated-ground receptacles are used in a box (as for computer wiring), all the isolated-ground conductors count as a deduction of 1 from the number of wires given in Table 370-16(a) as permitted for the particular size of box. *And then* another deduction of one conductor must be made for any other equipment grounding conductors (*not* isolated-ground wires).

2. Regarding the deduction of a wire from the Code-given number, Table 370-16(a), for fixture studs, cable clamps, and hickeys, does this apply to the above-mentioned items collectively regardless of number and combination, or does it apply to each item individually, such as clamps—minus one, studs—minus one, etc.?

The answer is that it is the intent of the second sentence in this paragraph to clarify that a deduction of 1 must be made from the number in the table for each *type* of device used in a box. A deduction of 1 must be made if the box contains cable clamps—whether one clamp or two clamps, a deduction of only 1 has to be made. A deduction of 1 must be made if the box contains a fixture stud. A deduction of 1 must be made if the box contains a hickey. Thus a box containing two clamps but no fixture studs or hickeys would have a deduction of 1 from the table number of wires for the clamps. If a box contained one clamp and one fixture stud, a deduction of 2 would be made because there are two *types* of devices in the box. Then, in addition to the deductions for clamps, hickeys, and/or studs, a deduction of two conductors must be made for each mounting strap that supports a receptacle, switch, or combination device. In the 1987 and previous NEC editions, a deduction of only *one* conductor had to be made for each wiring device mounting strap (or yoke) installed in the box.

3. Must unused cable clamps be removed from a box? And if clamps are not used at all in a box, must they be removed to permit removal of the one-wire reduction?

Unused cable clamps may be removed to gain space or fill in the box, or they may be left in the box if adequate space is available without the removal of the clamp or clamps. If one clamp is left, the one-wire deduction must be made. If no clamps are used at all in a box, such as where the cable is attached to the box by box connectors, the one-wire deduction is not made.

4. Is the short jumper installed between the grounding screw on a grounding-type receptacle and the box in which the receptacle is contained officially classified as a *bonding jumper?* And is this conductor counted when the box wire count is taken?

The jumper is classed as a bonding jumper. Section 250-74 uses the wording *bonding jumper* in the section pertaining to this subject. This conductor is not counted because it does not leave the box. The next-to-last sentence of Sec. 370-16(a)(1) covers that point.

The last sentence of Sec. 370-16(a)(1) requires that ganged boxes be treated as a single box of volume equal to the sum of the volumes of the sections that are connected to form the larger box. An example of wire counting and correct wire fill for ganged boxes is included in the following examples. *Note:* In the examples given here, the same rules apply to wires in boxes for any wiring method—conduit, EMT, BX, NM.

Examples of Box Wire Fill

The top example in Fig. 2-90 shows how deductions must be made from the maximum permitted wires in a box containing cable clamps and a fixture stud. The example at the bottom shows a nonmetallic-sheathed cable with three No. 14 copper conductors supplying a 15-A duplex receptacle (one ungrounded conductor, one grounded conductor, and one "bare" grounding conductor).

After supplying the receptacle, these conductors are extended to other outlets, and the conductor count is as follows:

Circuit conductors	4
Grounding conductors	1
For internal cable clamps	1
For receptacle	2
Total	8

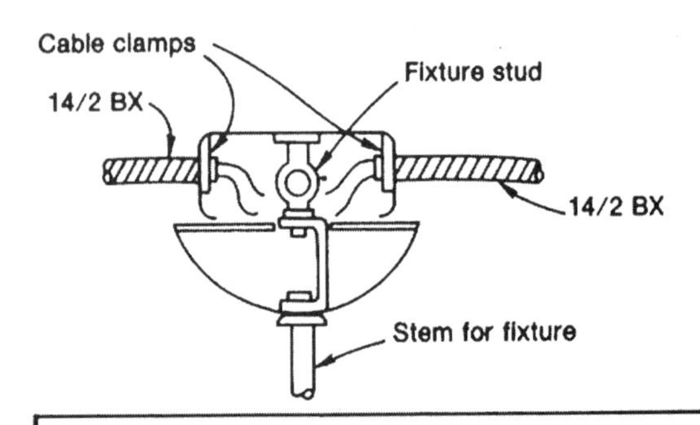

FOR A 4 × 1¼-IN. OCTAGONAL BOX:

From Table 370-16(a)	6 No. 14 wires
Minus one for the two cable clamps and minus one for the fixture stud	2 No. 14 wires
MAX. NUMBER PERMITTED	**4 No. 14 wires**

NOTE: If NM cable were used, another deduction of one for the two ground wires would make use of this box a violation.

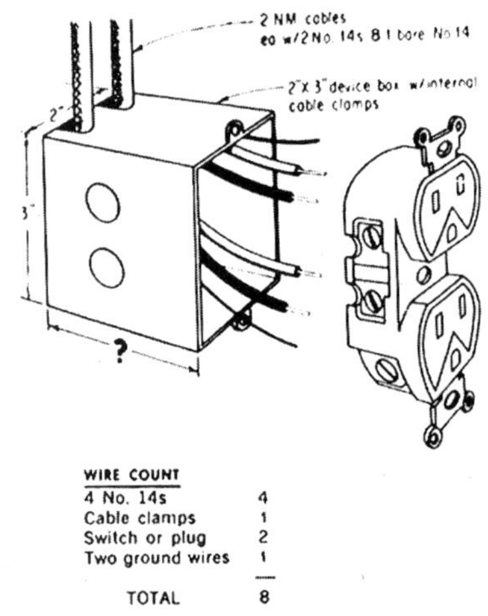

WIRE COUNT	
4 No. 14s	4
Cable clamps	1
Switch or plug	2
Two ground wires	1
TOTAL	8

Table 370-16 (a) shows that a 2" x 3" box which is suitable for use with 8 No. 14 wires must be 3½" deep.

FIG. 2-90 Correct wire count determines proper minimum size of outlet box.

The No. 14 conductor column of Table 370-16(a) indicates that a device box not less than 3 by 2 by ½ in is required. Where a square box with plaster ring is used, a minimum of 4- by 1¼-in size is required.

Table 370-16(a) includes the most popular types of metal "trade-size" boxes used with wires No. 14 to No. 6. Cubic-inch capacities are listed for each box shown in the table. According to paragraph (b), boxes other than those shown in Table 370-16(a) are required to be marked with the cubic-inch content so wire combinations can be readily computed.

Figure 2-91 shows another example with the counting data in the caption. The wire fill in this case violates the limit set by Sec. 370-16(a).

Figure 2-92 shows an example of wire-fill calculation for a number of ganged sections of sectional boxes. The photo shows a four-gang assembly of 3-in by 2-in by 3½-in box sections with six 14/2 NM cables, each with a No. 14 ground wire and one 14/3 NM cable with a No. 14 ground. The feed to the box is 14/3 cable (at right side), with its black wire supplying the receptacle which will be installed in the right-hand section. The red wire serves as feed to three combination devices—one in each of the other sections—each device consisting of two switches on a single strap. When finished, the four-gang box will contain a total of six switches and one duplex receptacle. Each of the 14/2 cables will feed a switched load. All the white neutrals are spliced together, and the seven bare No. 14 ground wires are spliced together, with one bare wire brought out to the receptacle ground terminal and one to the ground clip on the bottom of the left-hand section. The four-gang assembly is taken as a box of volume equal to 4 times the volume of one 3- by 2- by 3½-in box. From Table 370-16(a), that volume is 18 in^3 for each sectional box. Then for the four-gang assembly, the volume of the resultant box is 4×18, or 72 in^3. Then wire fill for the four-gang assembly may be 4 times that permitted for the basic single-gang box used in the assembly. Because a 3- by 2- by 3½-in box is shown in Table 370-16(a) to have a permitted fill of 9 No. 14 wires, the four-gang assembly may contain 4×9, or 36, No. 14 wires, with deductions made as required by Sec. 370-16(a)(1).

Deduct one wire for all the clamps; deduct one No. 14 for all the bare equipment ground wires; and deduct two No. 14s for each "strap containing one or more devices," which calls for a deduction of eight because there are four device "straps" (one of each of the three combination switches and one for the receptacle). The total deductions come to $1 + 1 + 8$, or 10.

Deducting 10 from 36 gives a permitted fill of 26 No. 14 insulated circuit wires. In the arrangement shown, there are six cables with two insulated wires and one with three insulated wires, for a total of 15 insulated No. 14 wires. Because that is well within the maximum permitted fill of 26 No. 14 wires, such an arrangement satisfies Sec. 370-16(a)(1).

The alternative method of counting wires and determining proper box size is as follows:

1. There are 15 No. 14 insulated circuit wires.
2. Add one wire for all the cable clamps.
3. Add one wire for all the No. 14 ground wires.
4. Add two wires for each of the four device straps.

The total of the wire count is $15 + 1 + 1 + 8$, or 25 No. 14 wires.

Then dividing that among the four box sections gives six-plus wires per section, which is taken as seven No. 14 wires per section. Referring to Table 370-16(a), note that a 3- by 2- by 2½-in box may contain six No. 14 wires. This calculation, therefore, estab-

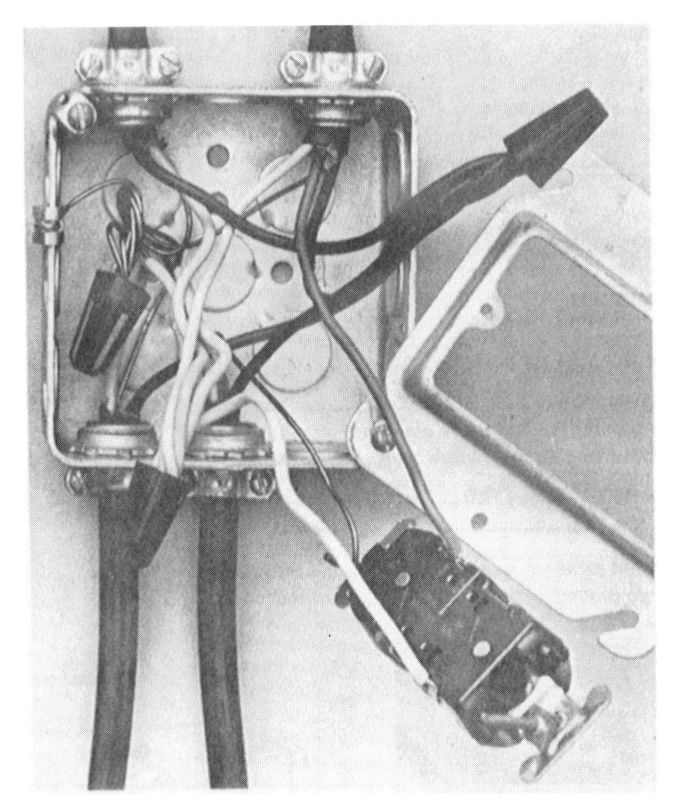

FIG. 2-91 *This is a code violation!* A 4- by 4- by $1\frac{1}{2}$-in square metal box, generally referred to as a "1900" box, has four NM cables coming into it. At upper right is a 14/3 cable with No. 14 ground. The other three cables are 14/2 NM, each with a No. 14 ground. The red wire of the 14/3 cable feeds the receptacle to be installed in the one-gang plaster ring. The black wire of the 14/3 feeds the black wires of the three 14/2 cables. All the whites are spliced together, with one brought out to the receptacle, as required by Sec. 300-13(b). All the ground wires are spliced together, with one brought out to the grounding terminal on the receptacle and one brought out to the ground clip on the left side of the box. The wire count is as follows: 9 No. 14 insulated wires, plus 1 for all the ground wires and 2 for the receptacle. That is a total of 12 No. 14s. Note that box connectors are used instead of clamps and there is, therefore, no addition of one conductor for clamps. But Table 370-16(a) shows that a 4- by $1\frac{1}{2}$-in square box may contain only 10 No. 14 wires.

lishes that the four-gang assembly could not be made up of 3- by 2- by $2\frac{1}{2}$-in boxes but would require 3- by 2- by $2\frac{3}{4}$-in boxes—with a permitted fill of seven No. 14 wires per section—to accommodate the seven No. 14 wires per section.

Although the Code wire-counting method in Sec. 370-16(a)(1) does not make reference to the counting method of Sec. 370-16(a)(2)—which applies where all the wires in a box are not the same size—that part (2) differs from the calculation made above. As

FIG. 2-92 Calculation of the proper minimum box size for the number of conductors used in ganged boxes must follow Sec. 370-16(a)(1), taking the assembly as a single box of the sum of the volumes of the ganged sections and filling it to the sum of the conductor count.

shown in Table 370-16(b), each No. 14 wire in a box must be allowed at least 2 in^3 of free space within the box. In the alternative calculation above, with a total of 25 No. 14 wires determined as the overall count, part (2) of Sec. 370-16(a) would require the box to have a minimum volume of 2×25, or 50, in^3. Each 3- by 2- by $2\frac{1}{2}$-in box has a volume of 12.5 [Table 370-16(a)]—for a total of 4×12.5, or 50, in^3 volume of the four-gang assembly. That volume satisfies the conductor volume and would permit use of 3- by 2- by $2\frac{1}{2}$-in boxes. But the rule of Sec. 370-16(a) would require at least 3- by 2- by $2\frac{3}{4}$-in boxes. Use of 3- by 2- by $3\frac{1}{2}$-in boxes would give more room and provide easier and safer installation.

When different sizes of wires are used in a box, part (2) of Sec. 370-16(a) requires that Table 370-16(b) be used in establishing adequate box size. Using the same method of counting conductors as described in Sec. 370-16(a)(1), the volume of cubic inches shown in Table 370-16(b) must be allowed for each wire depending upon its size. Where two or more ground wires of different sizes come into a box, they must all be counted as a single wire of the largest size used.

When deductions are made from the number of wires permitted in a box [Table 370-16(a)], as when devices, fixture studs, etc., are in the box, the deductions must "be based on the largest conductors entering the box" in any case where the conductors are of different sizes.

Figure 2-93 shows a calculation with different wire sizes in a box. When conduit or EMT is used, there are no internal box clamps and, therefore, no addition for clamps. In this example, the metal raceway is the equipment grounding conductor, so no addition has to be made for one or more ground wires. And the red wire is counted as one wire because it is run through the box without splice or tap. As shown in the wire count under

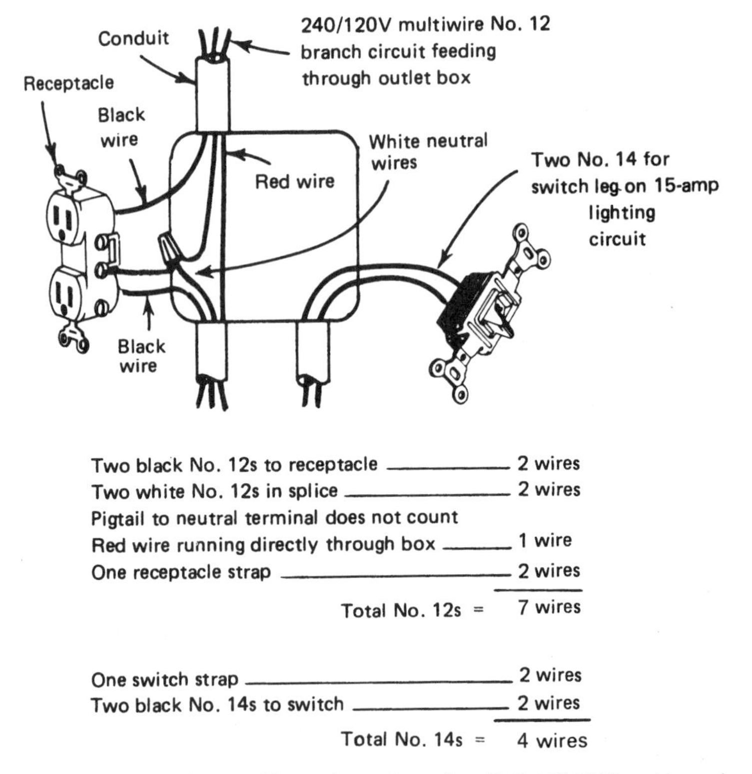

FIG. 2-93 When wires are different sizes, volumes from Table 370-16(b) must be used.

the sketch, the way to account for the space taken up by the wiring devices is to take each one as two wires of the same size as the largest wire coming into the box, that is, No. 12, as required in the end of the first sentence of part (a)(2). Note that the neutral pigtail required by Sec. 300-13(b) is excluded from the wire count, as it would be under Sec. 370-16(a)(1).

From Table 370-16(b) each No. 12 must be provided with 2.25 in^3—a total of 7 × 2.25, or 15.75, in^3 for the No. 12s. Then each No. 14 is taken at 2 in^3—a total of 4 × 2, or 8, in^3 for the No. 14s. Adding the two resultant volumes (15.75 + 8) gives a minimum required box volume of 23.75 in^3. From Table 370-16(a), a 4- by 4-in square box $2\frac{1}{8}$ in deep, with 30.3-in^3 interior volume, would satisfy this application.

For the many kinds of tricky control and power wire hookups so commonly encountered today, such as shown in Fig. 2-94, care must be taken to count all sizes of wires and make the proper volume provisions of Table 370-16(b).

FS and FD boxes—*Watch out!* Table 370-16(a) gives the maximum number of wires permitted in FS and FD boxes. But the last sentence of the first paragraph of Sec. 370-16(b) does indicate that FS and FD boxes may contain more wires if their internal volumes are marked and are greater than shown in Table 370-16(a).

Because the volumes in the table are minimums, most manufacturers continue to mark their products with the actual volume. This in many cases is considerably greater than the volumes shown in the table. The last sentence of Sec. 370-16(b) says that boxes

FIG. 2-94 Many boxes contain several sizes of wires, some running through, others spliced, and still others connected to wiring devices. Calculation of minimum acceptable box size must be carefully made. The combination switch and receptacle here is on a single mounting strap, which is taken as two wires of the size of wires connected to it.

that are marked to show a cubic-inch capacity greater than the minimums in the table may have conductor fill calculated in accordance with their actual volume, using the volume per conductor given in Table 370-16(b).

Part (b) of Sec. 370-16 covers boxes—metal and nonmetallic—that are not listed in Table 370-16(a) and conduit bodies. And the basic way of determining correct wire fill is to count wires in accordance with the intent of Sec. 370-16(a)(1) and then calculate required volume of the box or conduit body by totaling the volumes for the various wires from Table 370-16(b). The rules of part (b) can be broken down into two categories: boxes and conduit bodies.

1. *Boxes.* Part (b) covers wire fill for metal boxes, up to 100-in^3 volume, that are not listed in Table 370-16(a) and for nonmetallic outlet and junction boxes. Although Code rules have long regulated the maximum number of conductors permitted in metal wiring boxes [such as given in Table 370-16(a)], there was no regulation on the use of conductors in nonmetallic device boxes up to the 1978 NEC. Now Sec. 370-16(b) requires that *both* metal boxes not listed in Table 370-16(a) and nonmetallic boxes be durably and legibly marked by their manufacturer with their cubic-inch capacities to permit calculation of the maximum number of wires that the Code will permit in the box. Calculation of the conductor fill for these boxes will be based on the marked box volume and the method of counting conductors set forth in Sec. 370-16(a). The conductor volume will be taken at the values given in Table 370-16(b), and deductions of space as required for wiring devices or for clamps must be made in accordance with the rules of Sec. 370-16(a). This requirement for marking of both metal and nonmetallic boxes arises from the wording of Sec. 370-16(a), which refers to boxes other than those described in Table 370-16(a) and to nonmetallic boxes.

As shown in Fig. 2-95, a nonmetallic box for a switch has two 14/2 NM cables, each with a No. 14 ground. The wire count is four No. 14 insulated wires, plus two for the

FIG. 2-95 Every nonmetallic box must be "durably and legibly marked by the manufacturer" with its cubic-inch capacity to permit calculation of number of wires permitted in the box, using Table 370-16(b) and the additions of wire space required to satisfy Sec. 370-16(a)(1).

switch to be installed, and one for the two ground wires. That is a total of seven No. 14 wires. From Table 370-16(b), at least 2 in^3 of box volume must be allowed for each No. 14. This box must, therefore, be marked to show that it has a capacity of at least 7×2, or 14, in^3. [As shown, the ground wires are connected by a twist-on connector, with one end of the wire brought out to connect to a ground screw on the switch mounting yoke. Such a technique is required to provide grounding of a metal switchplate that is used on an outlet within reach of water faucets or other grounded objects. Refer to Secs. 250-42(e) and 410-56(d).]

2. *Conduit bodies.* Conduit bodies must be marked with their cubic-inch capacity, and conductor fill is determined on the basis of Table 370-16(b). Such conduit bodies may contain splices or taps. An example of such application is shown in Fig. 2-96. Each of the eight No. 12 wires that are "counted" as shown at bottom must be provided with at least 2.25 in^3, from Table 370-16(b). The T conduit body must, therefore, be marked to show a capacity of not less than 8×2.25, or 18, in^3.

For each No. 6 conductor used in the boxes or conduit bodies covered by Sec. 370-16(b), there must be at least 5 in^3 of box volume *and* a minimum space at least $1\frac{1}{2}$ in wide where any No. 6 is bent in a box or fitting.

Part (c) of Sec. 370-16 contains a number of provisions which must be carefully evaluated. Figure 2-97 shows the first rule. For instance, in that sketch, if a conduit body is connected to $\frac{1}{2}$-in rigid metal conduit, the conduit and the conduit body may contain seven No. 12 THHN wires, as indicated in Table C-8 in app. C; and the conduit body must have a csa at least equal to 2×0.3 in^2 (the csa of $\frac{1}{2}$-in conduit), or 0.6 in^2. That is really a matter for the fitting manufacturers to observe.

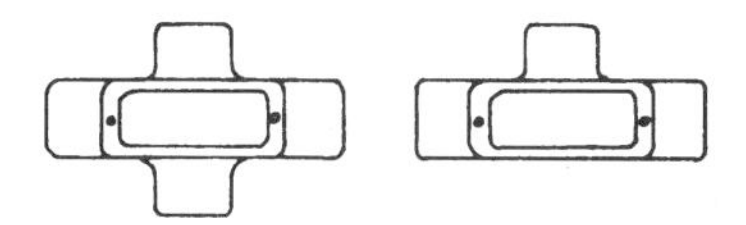

All such bodies must be durably and legibly marked by manufacturer with their cubic-inch capacities.

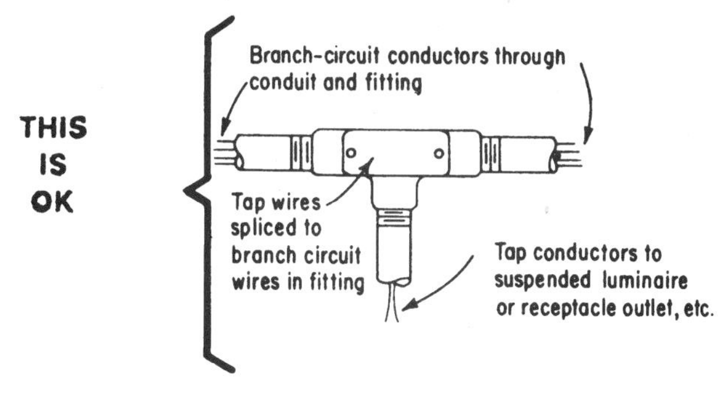

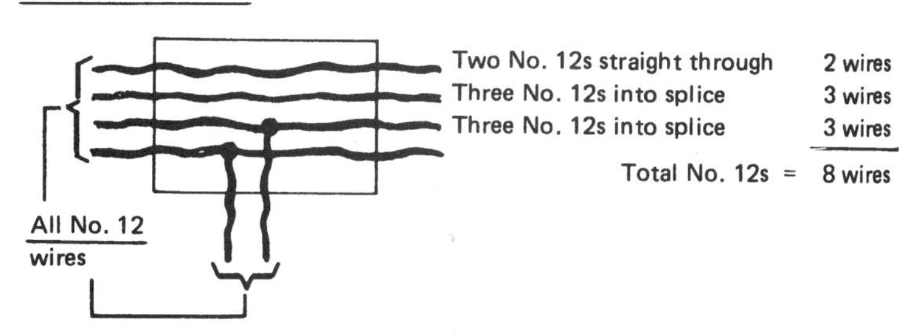

FIG. 2-96 Conduit bodies with more than *two* entries for conduit may contain splices or taps.

The second paragraph of part (c) covers the details shown in Fig. 2-98. The rule requires that where fittings are used as shown in the sketch, they must be supported in a rigid and secure manner. Because Sec. 370-23 establishes the correct methods for support of boxes and fittings, it must be observed, and that section refers to support by "conduits." Figure 2-99 shows typical applications of those conduit bodies for splicing.

MINIMUM DIMENSIONS FOR LARGER BOXES

As noted in Sec. 370-16, conduit bodies must be sized the same as pull boxes when they contain No. 4 or larger conductors.

For raceways containing conductors of No. 4 or larger size, the NEC specifies certain minimum dimensions for a pull or junction box installed in a raceway run. These rules also apply to pull and junction boxes in cable runs—but instead of using the cable diameter, the minimum trade size raceway required for the number and size of conductors in the cable must be used in the calculations. Basically there are two types of pulls:

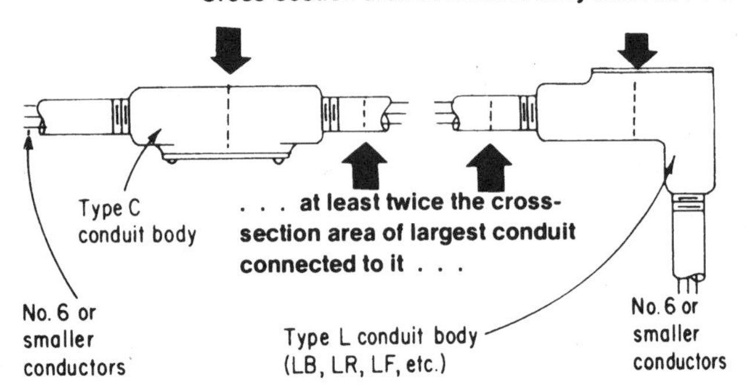

FIG. 2-97 For No. 6 and smaller conductors, conduit body must have a csa twice that of largest conduit.

straight pulls and angle pulls. Figure 2-100 covers straight pulls. Figure 2-101 covers angle pulls. In all the cases shown in those illustrations, the depth of the box only has to be sufficient to permit installation of the locknuts and bushings on the largest conduit. And the spacing between adjacent conduit entries is also determined by the diameters of locknuts and bushings, to provide proper installation. Depth is the dimension not shown in the sketches.

According to the rule of part (a)(2) in sizing a pull or junction box for an angle or "U" pull, if a box wall has more than one row of conduits, "each row shall be calculated separately and the single row that provides the maximum distance shall be used." Consider the following:

A pull box has two rows of conduits entering one side (or wall) of the box for a right-angle pull. What is the minimum required inside distance from the wall with the two rows of conduit entries to the opposite wall of the box?

Row 1: One $2\frac{1}{2}$-in and one 1-in conduit.

Row 2: One $\frac{1}{2}$-in, two $1\frac{1}{4}$-in, one $1\frac{1}{2}$-in, and two $\frac{3}{4}$-in conduits.

Interpretation of 1987 NEC Sec. 370-18(a)(2)

1. $6 \times$ diameter of "largest raceway" entering box wall:

$$6 \times 2\frac{1}{2} \text{ in} = 15 \text{ in}$$

2. Add "the maximum sum of diameters of all other raceway entries in any one row on the same wall of the box." Row 2 will give "the maximum sum": $\frac{1}{2} + (2 \times 1\frac{1}{4}) + 1\frac{1}{2} + (2 \times \frac{3}{4}) = 6$ in.

3. Add the two results: 15 in + 6 in = 21 in.

That is the minimum size box dimension to the wall opposite the wall where the conduits enter.

Although the basic rule still prohibits splicing and tapping in these fittings . . .

"C" conduit body

"L" conduit body

. . . Permission is **now** given to splice in such fittings, if Section 370-16(b) is satisfied; that is, if—

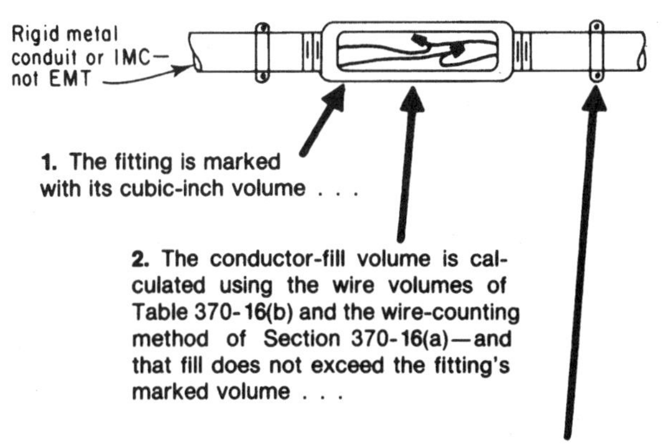

3. And the fitting is supported in "a rigid and secure manner"—such as by the "conduit," if the conduit is clamped on each side of the fitting as described in part (d) of Section 370-23.

FIG. 2-98 Splices may be made in "C" and "L" conduit bodies—if the conditions shown in this illustration are satisfied.

Calculating with Revised Rule of Present NEC

Calculate each row separately and take the box dimension from the row that gives the maximum distance.

Row 1: 6 × largest raceway ($2\frac{1}{2}$ in) + other entries (1 in) = 16 in

Row 2: 6 × the largest raceway ($1\frac{1}{2}$ in) + other entries [$\frac{1}{2}$ in + (2 × $1\frac{1}{4}$ in) + (2 × $\frac{3}{4}$ in)] = 9 in + $\frac{1}{2}$ in + 4 in = $13\frac{1}{2}$ in

Result: The minimum box dimension must be the 16-in dimension from run No. 1, which "provides the maximum distance" calculated.

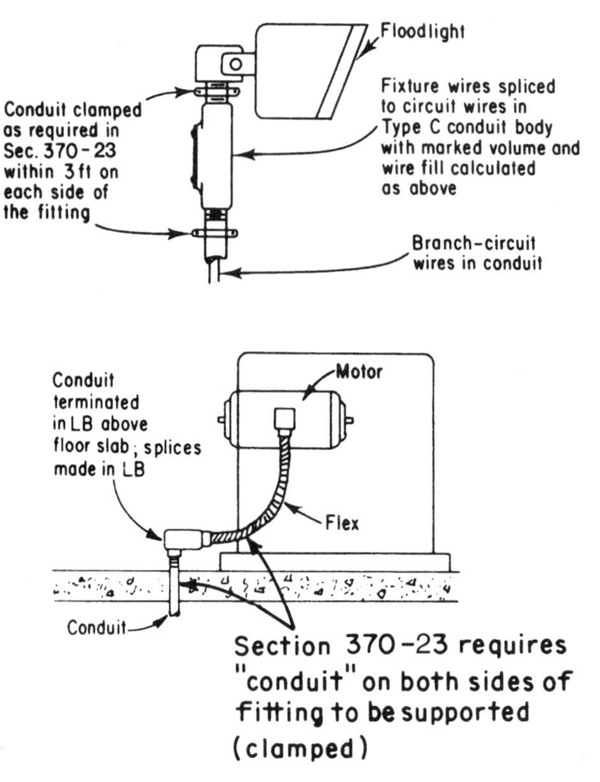

FIG. 2-99 Splicing in “C” or “L” conduit bodies is common practice.

Figure 2-102 shows a more complicated conduit and pull box arrangement, which requires more extensive calculation of the minimum permitted size. In this particular layout shown, the upper 3-in conduits running straight through the box represent a problem separate from the 2-in conduit angle pulls. In this case the 3-in conduit establishes the box length in excess of that required for the 2-in conduit. After computing the 3-in requirements, the box size was calculated for the angle pull involving the 2-in conduit.

Subparagraph (3) of Sec. 370-28(a) permits smaller pull or junction boxes where such boxes have been approved for and marked with the maximum number and size of conductors and the conduit fills are *less* than the maximum permitted in Table 1, Chap. 9. This rule provides guidelines for boxes which have been widely used for years, but which have been smaller than the sizes normally required in subparagraphs (1) and (2). These smaller pull boxes must be listed by UL under this rule.

There are many instances where an installation is made in which raceways and conductors are not matched so as to utilize maximum conduit fill as permitted by the Code. An example would be a 2-in conduit with six No. 4 THHN conductors. The Code would permit up to 16 conductors depending upon the type of insulation. It was felt that in such installations provisions should be made for the use of boxes or fittings which would not necessarily conform to the letter of the law as exemplified by the standards listed in sub-

EXAMPLES:

1.

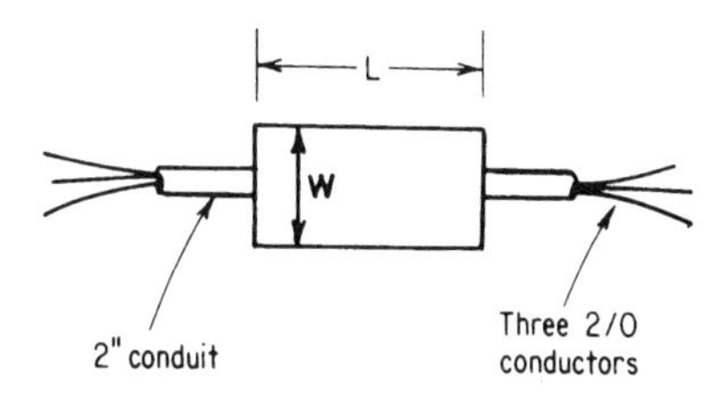

L = 8 × 2 in. = 16 in. minimum
W = Whatever width is necessary to provide proper installation of the conduit locknuts and bushings within the enclosure.

2.

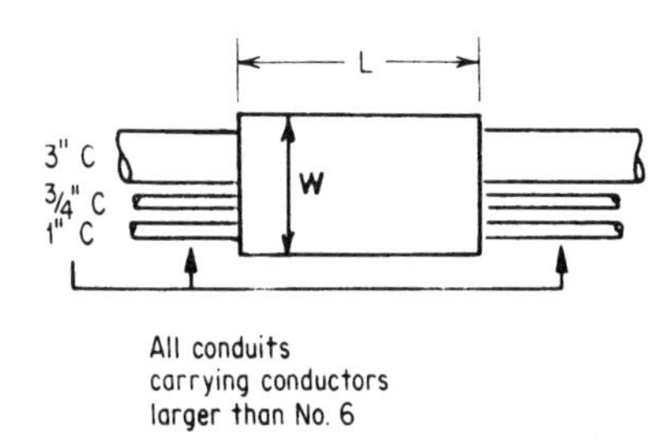

The 3-in. conduit is the largest.
Therefore—
L = 8 × 3 in. = 24 in. minimum
W = Width necessary for conduit locknuts and bushings.

FIG. 2-100 In straight pulls, the length of the box must be not less than 8 times the trade diameter of the largest raceway.

sections (1) and (2), but would compare favorably under test with a box sized as is required for the conductor and conduit.

A pull or junction box used with 2-in conduit and conductors No. 4 AWG or larger must be 16 in long if straight pulls are made and 12 in long if angle pulls are to be made. If we have a 2-in conduit and we are installing eight No. 4 RHH conductors, all pull or junction boxes would have to conform to these measurements. If, however, we are installing five No. 4 RHH conductors, a smaller box would be acceptable, provided it has been tested for and is marked with this number.

Figure 2-103 shows how the rules of Sec. 370-28(a) apply to conduit bodies. *Important:* The exception given in Sec. 370-28(a)(2) establishes the minimum dimen-

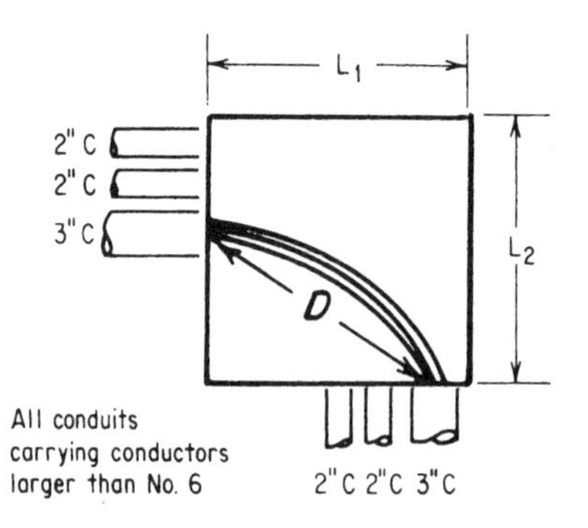

The 3-in. conduit is the largest.
Therefore—

$L_1 = 6 \times 3 \text{ in.} + (2 \text{ in.} + 2 \text{ in.}) = 22 \text{ in. min.}$

$L_2 = 6 \times 3 \text{ in.} + (2 \text{ in.} + 2 \text{ in.}) = 22 \text{ in. min.}$

$D = 6 \times 3 \text{ in.} = 18 \text{ in.}$, **minimum distance between raceway entries enclosing the same conductors**

FIG. 2-101 Box size must be calculated for angle pulls. For boxes in which the conductors are pulled at an angle or in a U, the distance between each raceway entry inside the box and the opposite wall of the box must not be less than 6 times the trade diameter of the largest raceway in a row. And the distance must be increased for additional raceway entries by the amount of the maximum sum of the diameters of all other raceway entries in the same row on the same wall of the box. The distance between raceway entries enclosing the same conductors must not be less than 6 times the trade diameter of the larger raceway.

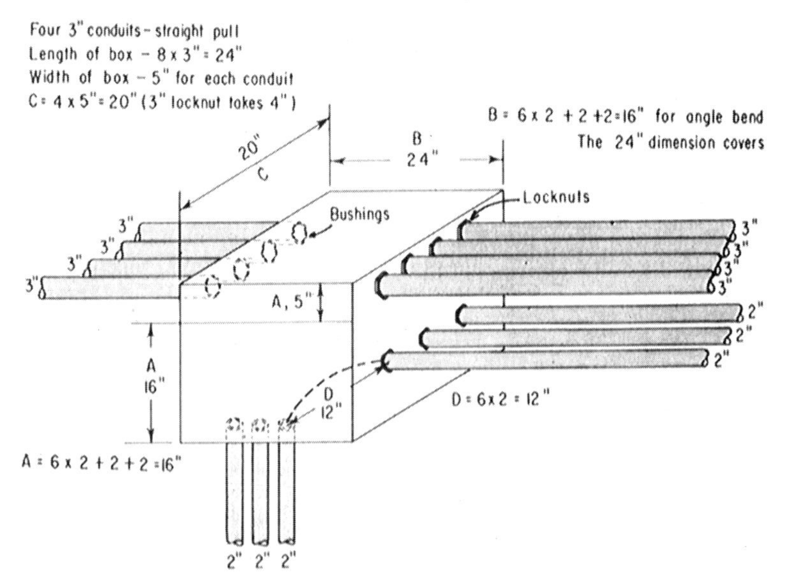

FIG. 2-102 A number of calculations are involved when angle and straight pulls are made in different directions and different planes.

STRAIGHT RUN

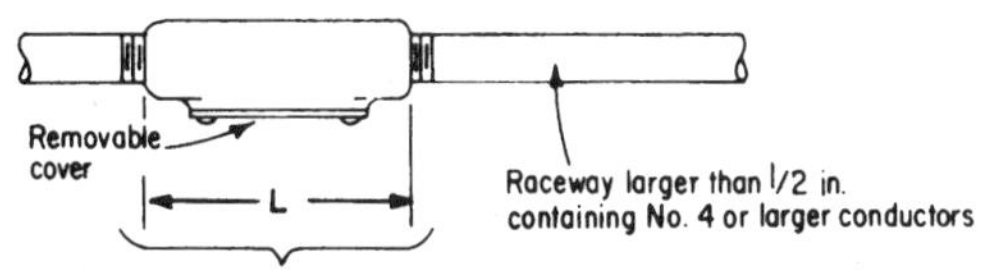

Type C conduit body must have length L equal to 8 times diameter of the raceway

Examples

If four No. 4 THHN are used in 1-in. conduit, conduit body must be at least 8 in. long.

If four 500 kcmil XHHW are used in 3-in. conduit, conduit body must be at least 24 in. long.

ANGLE RUN

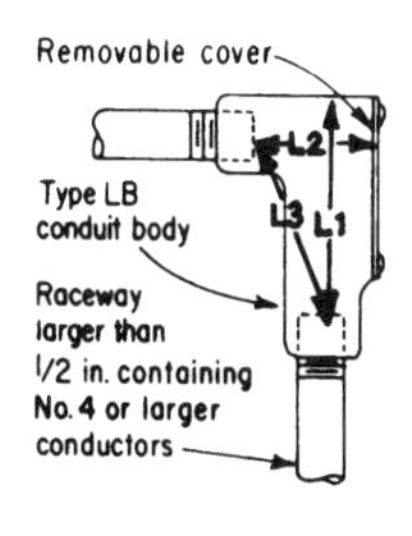

From Sec. 370-28(a)(2):

L1 = at least 6 times diameter of raceway (inside dimension)

L2 = at least equal to the distance given in Table 373-6(a) for the given size of conductor, as shown in the column for *one wire per terminal*

L3 = at least 6 times diameter of raceway

Examples

If four No. 4 THW conductors are used in 1¼-in. conduit, minimum dimensions would be calculated as follows:

L1 = 6 × 1¼ in. = 7.5 in.
L2 = 2 in., from Table 373-6(a) for one No. 4 conductor per terminal
L3 = 6 × 1¼ in. = 7.5 in.

If four 500 kcmil THW conductors are used in 3½-in. conduit, minimum dimensions would be:

L1 = 6 × 3½ in. = 21 in.
L2 = 6 in., from Table 373-6(a)
L3 = 6 × 3½ in. = 21 in.

FIG. 2-103 Conduit bodies must be sized as pull boxes under these conditions.

sion of L2 for angle runs, but this exception only applies to conduit bodies which have the removable cover opposite one of the entries, such as a type LB body. Types LR, LL, and LF do not qualify under that exception, and for such conduit bodies the dimension L2 would have to be at least equal to the dimension L1 (that is, 6 times raceway diameter).

Figure 2-104 shows the racking of cable required by part (b) of this section.

Figure 2-105 shows another consideration in sizing a pull box for angle conduit layouts. A pull box is to be installed to make a right-angle turn in a group of conduits consisting of two 3-in, two $2\frac{1}{2}$-in, and four 2-in conduits.

Subparagraph (2) of Sec. 370-28(a) gives two methods for computing the box dimensions, and both must be met.

First method:

$$
\begin{aligned}
6 \times 3 \text{ in} &= 18 \\
1 \times 3 \text{ in} &= 3 \\
2 \times 2\tfrac{1}{2} \text{ in} &= 5 \\
4 \times 2 \text{ in} &= \underline{8} \\
&= 34 \text{ in}
\end{aligned}
$$

Second method: Assuming that the conduits are to leave the box in the same order in which they enter, the arrangement is shown in Fig. 2-105 and the distance *A* between the ends of the two conduits must be not less than $6 \times 2 = 12$ in. It can be assumed that this measurement is to be made between the centers of the two conditions. By calculation, or by laying out the corner of the box, it is found that the distance *C* should be about $8\frac{1}{2}$ in.

The distance *B* should be not less than $30\frac{1}{2}$ in, approximately, as determined by applying practical data for the spacing between centers of conduits.

$$30\tfrac{1}{2} \text{ in} + 8\tfrac{1}{2} \text{ in} = 39 \text{ in}$$

In this case the box dimensions are governed by the second method. The largest dimension computed by either of the two methods is of course the one to be used. Of course, if conduit positions for conduits carrying the same cables are transposed, as in Fig. 2-101, then box size can be minimized.

The most practical method of determining the proper size of a pull box is to sketch the box layout with its contained conductors on a paper.

Section 370-28 applies particularly to the pull boxes commonly placed above distribution switchboards and which are often, and with good reason, termed *tangle boxes.* In such boxes, all conductors of each circuit should be cabled together by serving them with twine so as to form a self-supporting assembly that can be formed into shape, or the conductors should be supported in an orderly manner on racks, as required by part (b) of Sec. 370-28. The conductors should not rest directly on any metalwork inside the box, and insulating bushings should be provided wherever required by Sec. 373-6(c).

For example, the box illustrated in Fig. 2-105 could be approximately 5 in deep and accommodate one horizontal row of conduits. By making it twice as deep, two horizontal rows or twice the number of conduits could be installed.

Insulating racks are usually placed between conductor layers, and space must be allowed for them.

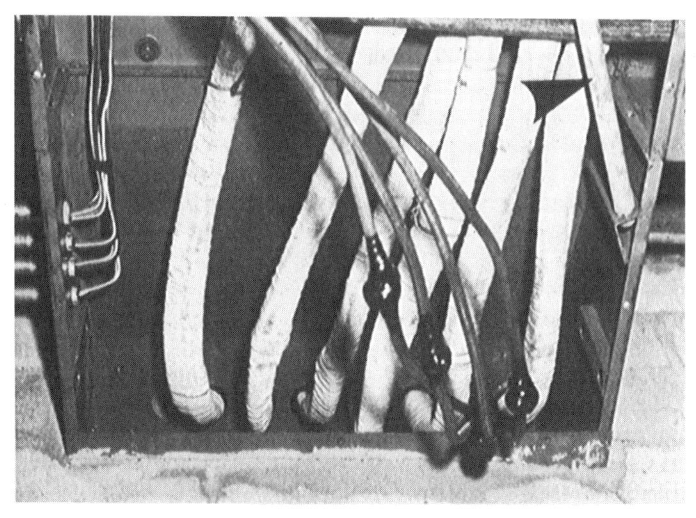

FIG. 2-104 If a pull box has *any* dimension over 6 ft, the conductors within it must be supported by suitable racking (arrow) or cabling, as shown here for arc-proofed bundles of feeder conductors, to keep the weight of the many conductors off the sheet-metal cover that attaches to the bottom of the box.

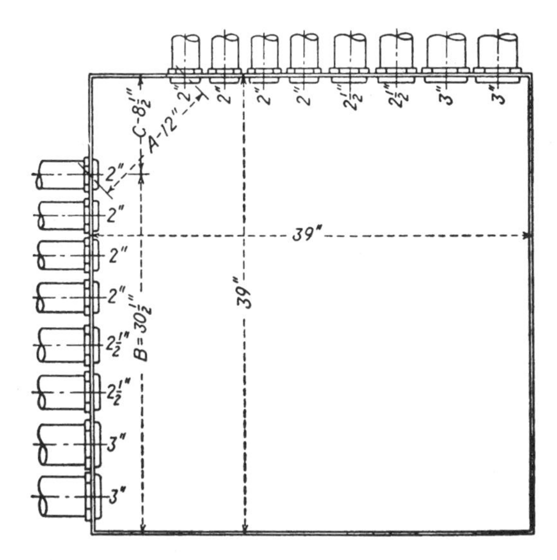

FIG. 2-105 Distance between conduits carrying same cables has great impact on overall box size.

SIZING PULL AND JUNCTION BOXES FOR CONDUCTORS RATED OVER 600 V

Figure 2-106 shows the rules on sizing of pull boxes for high-voltage circuits.

CALCULATING BENDING SPACE IN CABINETS AND CUTOUT BOXES

Deflection of Conductors

Parts (a) and (b) cover a basic Code rule that is referenced in a number of Code articles to ensure safety and effective conductor application by providing enough space to bend conductors within enclosures.

A basic concept of evaluating adequate space for bending conductors at terminals of equipment installed in cabinets is presented in this section. The matter of bending space for conductors at terminals is divided into two different configurations, as follows:

1. The conductor does not enter (or leave) the enclosure through the wall opposite its terminals. This would be any case where the conductor passes through a wall of the enclosure at right angles to the wall opposite the terminal lugs to which the conductor is connected or at the opposite end of the enclosure. In all such cases, the bend at the terminals is a single-angle bend (90° bend), and the conductor then passes out of the bending space. It is also called an *L bend,* as shown at the top of Fig. 2-107. For bends of that type, the distance from the terminal lugs to the wall opposite the lugs must conform to Table 373-6(a), which requires lesser distances than those of Table 373-6(b) because single bends are more easily made in conductors.

2. The conductor enters (or leaves) the enclosure through the wall opposite its terminals. This is a more difficult condition because the conductor must make an offset or double bend to go from the terminal and then align with the raceway or cable entrance. This is also called an *S* or a *Z bend* because of its configuration, as shown at the top right of Fig. 2-107. For such bends, Table 373-6(b) specifies a greater distance from the end of the lug to the opposite wall to accommodate the two 45° bends, which are made difficult by the short lateral space between lugs and the stiffness of conductors (especially with the plastic insulations in cold weather).

Table 373-6(b) provides increased bending space to accommodate use of factory-installed connectors that are not of the lay-in or removable type and to allow use of field-installed terminals that are not designated by the manufacturer as part of the equipment marking. Exception 1 to part (b)(1) is shown in the bottom sketch of Fig. 2-107.

Note: For providing Code-required bending space at terminals for enclosed switches or individually enclosed circuit breakers, refer to Sec. 380-18. For conductor bending space at panelboard terminals, refer to Sec. 384-35. In Fig. 2-107, the clearances shown are determined from Table 373-6(a) or Table 373-6(b), under the column for one wire per terminal. For multiple-conductor circuit makeups, the clearance at terminals and in side gutters has to be greater, as shown under two, three, four, etc., wires per terminal.

Exception 2 of part (b)(1) covers application of conductors entering or leaving a meter-socket enclosure, and was based on a study of 100- and 200-A meter sockets.

STRAIGHT PULLS

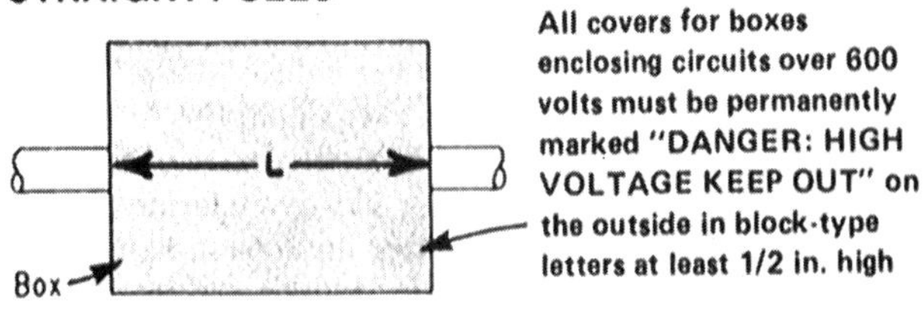

L - not less than 48 times the outside diameter, over sheath, of the largest shielded or lead-covered *conductor* or *cable* entering the box, *OR* not less than 32 times the outside diameter of the largest nonshielded conductor or cable.

NOTE: The box length must be 48 times the conductor or cable diameter, *not the conduit* diameter.

ANGLE PULLS

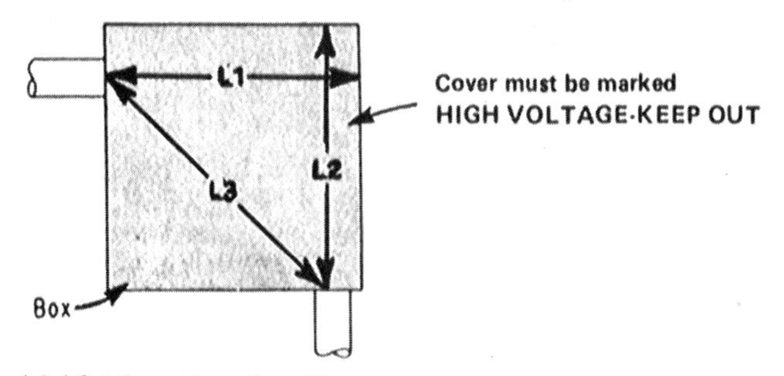

L1, L2, L3—not less than 36 times the outside diameter, over sheath, of the largest *conductor* or *cable*

FIG. 2-106 Minimum dimensions are set for high-voltage pull and junction boxes.

ENCLOSURES FOR SWITCHES AND/OR OVERCURRENT DEVICES CONTAINING SPLICES OR TAPS

Most enclosures for switches and/or overcurrent devices have been designed to accommodate only those conductors intended to be connected to terminals within such enclosures. And in designing such equipment it would be virtually impossible for manufacturers to anticipate various types of "foreign" circuits, feed-through circuits, or numerous splices or taps.

The rule here states that enclosures for switches, CBs, panelboards, or other operating equipment must not be used as junction boxes, troughs, or raceways for conductors

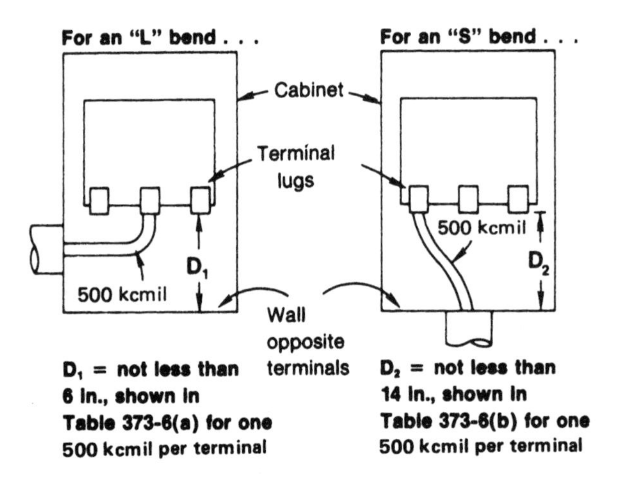

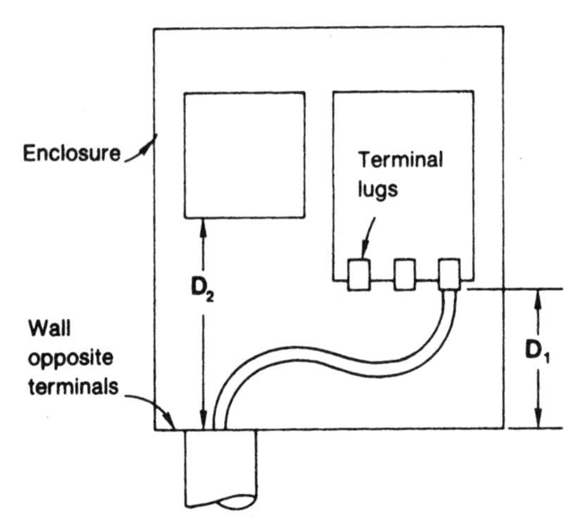

FIG. 2-107 These clearances are the minimum that must be observed.

feeding through or tapping off, unless designs suitable for the purpose are employed to provide adequate space. This rule affects installations in which a number of branch circuits or subfeeder circuits are to be tapped from feeder conductors in an auxiliary gutter, using fused switches to provide disconnect and overcurrent protection for the branch or subfeeder circuits. It also applies to feeder taps in panelboard cabinets.

In general, the most satisfactory way to connect various enclosures together is through the use of properly sized auxiliary gutters (Fig. 2-108) or junction boxes. Figure 2-109 shows a hookup of three motor disconnects, using a junction box to make the

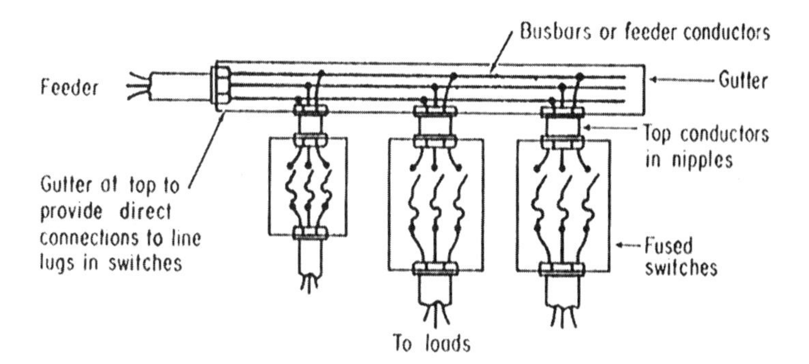

FIG. 2-108 Feeder taps in auxiliary gutter keep feeder cables and tap connectors out of switch enclosures.

feeder taps. Following this concept, enclosures for switches and/or overcurrent devices will not be overcrowded.

There are cases where large enclosures for switches and/or overcurrent devices will accommodate additional conductors, and this is generally where the 40 percent (conductor space) and 75 percent (splices or taps) at one cross section would apply. An example would be control circuits tapped off or extending through 200-A or larger fusible switches or CB enclosures. The csa within such enclosures is the *free gutter wiring space* intended for conductors.

The exception to this rule is shown in Fig. 2-110 and applied as follows:

> *Example:* If an enclosure has a gutter space of 3 by 3 in, the csa would be 9 in^2. Thus, the total conductor fill (use Table 5, Chap. 9) at any cross section (including conductors) could not exceed 6.75 in^2 (9×0.75).

In the case of large conductors, a splice other than a wire-to-wire "C" or "tube" splice would not be acceptable if the conductors at the cross section were near a 40 percent fill, because this would leave only a 35 percent space for the splice. Most splices for larger conductors with split-bolt connectors or similar types are usually twice the size of the conductors being spliced. Accordingly, where larger conductors are to be spliced within enclosures, the total conductor fill should not exceed *20 percent* to allow for any bulky splice at a cross section.

Figure 2-111 shows an example of feeder taps made in a panelboard side gutter where the cabinet is provided with adequate space for the large feeder conductors and for the bulk of the tap devices with their insulating tape wrap.

CONDUCTOR FILL IN METALLIC WIREWAYS

Wireways are sheet-metal troughs in which conductors are laid in place after the wireway has been installed as a complete system. Wireway is available in standard lengths of 1, 2, 3, 4, 5, and 10 ft, so that runs of any exact number of feet can be made up without cutting the duct. The cover may be a hinged or removable type. Unlike auxiliary gutters, wireways represent a type of wiring, because they are used to carry conductors between points located considerable distances apart.

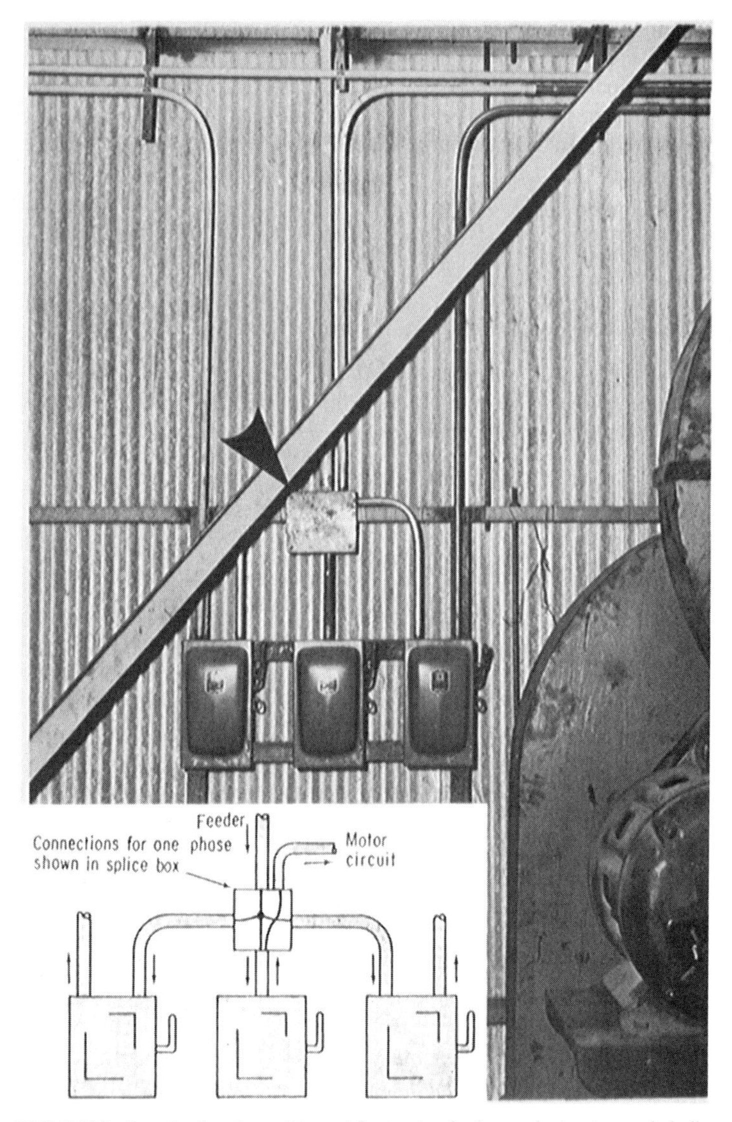

FIG. 2-109 Junction box (arrow) is used for tapping feeder conductors to supply individual motor branch circuits, as shown in inset diagram. (Sec. 373-8.)

The purpose of a wireway is to provide a flexible system of wiring in which the circuits can be changed to meet changing conditions, and one of its principal uses is for exposed work in industrial plants. Wireways are also used to carry control wires from the control board to remotely controlled stage switchboard equipment. A wireway is approved for any voltage not exceeding 600 V between conductors or 600 V to ground. An installation of wireway is shown in Fig. 2-112.

Wireways may contain up to 30 "current-carrying" conductors at any cross section (signal circuits and control conductors used for starting duty only between a motor and

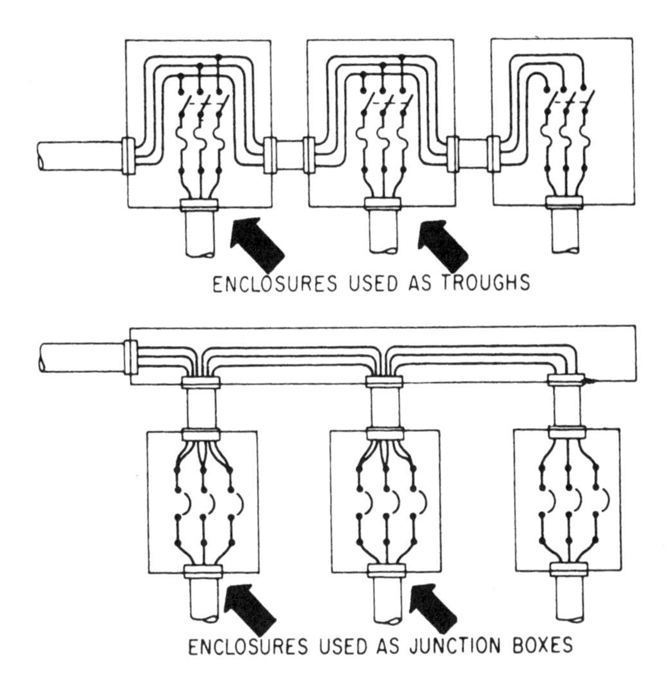

FIG. 2-110 These hookups are permitted where space in enclosure gutters satisfies the exception to basic rule.

its starter are not "current-carrying" conductors). The total cross-sectional area of the group of conductors must not be greater than 20 percent of the interior cross-sectional area of the wireway or gutter. And ampacity derating factors for more than three conductors do not apply to wireway as they do to wires in conduit. However, if the derating factors from Note 8 of NEC Tables 310-16 through 310-19 are used, there is no limit to the number of current-carrying wires permitted in a wireway or an auxiliary gutter. But, the sum of the cross-sectional areas of all contained conductors at any cross section of the wireway must not exceed 20 percent of the cross-sectional area of the wireway or auxiliary gutter. More than 30 conductors may be used under these conditions.

Exception 3 says that wireway used for circuit conductors for an elevator or escalator may be filled with any number of wires, occupying up to 50 percent of the interior cross section of the wireway, and no derating has to be made for fill.

The second sentence of the first paragraph has the effect of saying that any number of signal and/or motor control wires (even over 30) may be used in wireway provided the sum of their cross-sectional areas does not exceed 20 percent of wireway csa. And ampacity derating of those conductors is not required.

Figure 2-113 shows examples of wireway fill calculations. The example at the bottom shows a case where power and lighting wires (which *are* current-carrying wires) are mixed with signal wires. Because there are not more than 30 power and light wires, no derating of conductor ampacities is needed. If, say, 31 power and light wires were in the wireway, then the power and light conductors would be subject to derating. If all 49 conductors were signal and/or control wires, no derating would be required. But, in all cases, wireway fill must not be over 20 percent.

Deflected conductors in wireway must observe the rules on adequate enclosure space given in Sec. 373-6. This section is based on the following:

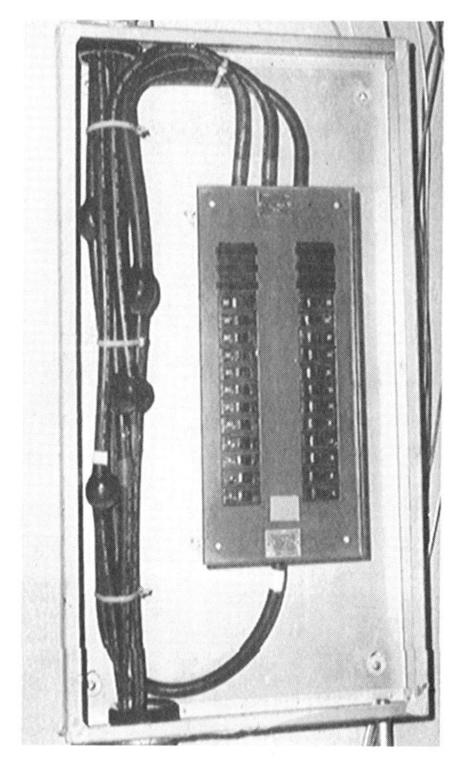

FIG. 2-111 The exception to Sec. 373-8 permits feeding through and tapping off in cabinets for panelboards on feeder risers, where the side gutter is specially oversized for the application.

Although wireways don't contain terminals or supplement spaces with terminals, pull boxes and conduit bodies don't either. This rule borrows language from both 374-9(d) and 370-28(a)(2), Exception, in an attempt to produce a consistent approach in the Code. Although in some cases the deflected conductors travel long distances in the wireway and are therefore easily inserted, in other cases the conductors are deflected again within inches of the first entry. The result is even more stress on the insulation than if they were entering a conduit body.

CALCULATING CONDUCTOR FILL FOR AUXILIARY GUTTERS

Auxiliary gutters are sheet-metal troughs in which conductors are laid in place after the gutter has been installed. Auxiliary gutters are used as parts of complete assemblies of apparatus such as switchboards, distribution centers, and control equipment, as shown in Fig. 2-114. But auxiliary gutters may not contain equipment even though they look like surface metal raceway (Art. 352), which may contain devices and equipment.

Auxiliary gutters are not intended to be a type of general raceway and are not permitted to extend more than 30 ft beyond the equipment which they supplement, except

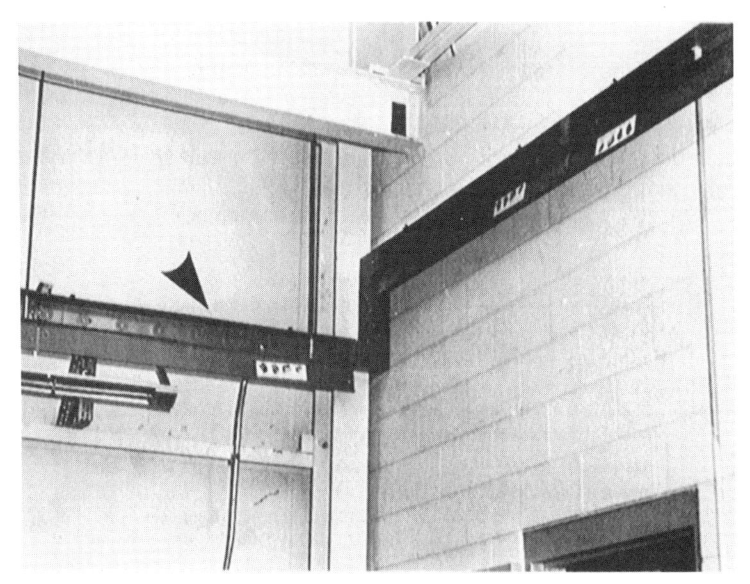

FIG. 2-112 Wireway in industrial plant, installed exposed, as required by Sec. 362-2, provides highly flexible wiring system that provides easy changes in the number, sizes, and routing of circuit conductors for machines and controls. Hinged covers swing down for ready access. Section 362-7 permits splicing and tapping in wireway. Section 362-11 covers use of conduit for taking circuits out of wireway. (Sec. 362-2.)

in elevator work. Where an extension beyond 30 ft is necessary, Art. 362 for wireways must be complied with. The label of Underwriters' Laboratories Inc. on each length of trough bears the legend "Wireways or Auxiliary Gutters," which indicates that they may be identical troughs but are distinguished one from the other only by their use.

The rules on permitted conductor fill for auxiliary gutters are basically the same as those for wireways. Refer to Sec. 362-5. Note that Exception 3 permits more than 30 general circuit wires; but where over 30 wires are installed, the correction factors specified in Note 8 to Tables 310-16 through 310-19 must be applied.

No limit is placed on the size of conductors that may be installed in an auxiliary gutter.

The csa of rubber-covered and thermoplastic-covered conductors given in Table 5, Chap. 9, must be used in computing the size of gutters required to contain a given combination of such conductors.

Figure 2-115 shows a typical gutter application where the conductor fill must be calculated to determine the acceptable csa of the gutter. There are several factors involved in sizing auxiliary gutters that often lead to selecting the wrong size. The two main factors are how conductors enter the gutter and the contained conductors at any cross section. The minimum required width of a gutter is determined by the csa occupied by the conductors and splices and the space necessary for bending conductors entering or leaving the gutter. The total csa occupied by the conductors at any cross section of the gutter must not be greater than 20 percent of the gutter interior csa at that point (Sec. 374-5). The total csa occupied by the mass of conductors and splices at any cross section of the gutter must not be greater than 75 percent of the gutter interior csa at that point [Sec. 374-8(a)].

Basic rule

1. Any number of current-carrying conductors up to a maximum of 30, without derating.

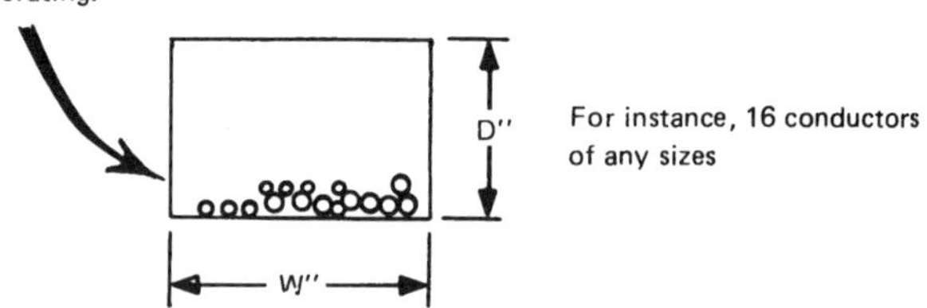

2. The sum of the cross-section areas of all the conductors (from table 5 in Chap. 9 of the *NEC*) must not be more than 20% X W'' X D''

Note: Signal and motor control wires are not considered to be current-carrying wires. Any number of such wires are permitted to fill up 20% of wireway cross-section area.

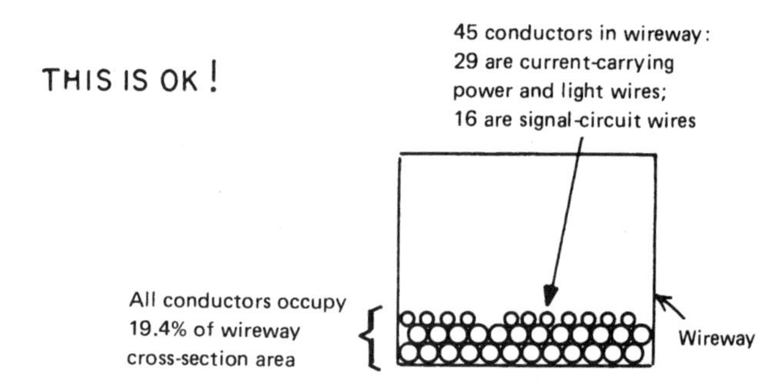

FIG. 2-113 Wireway fill and need for derating must be carefully evaluated.

In the gutter installation shown, assume that the staggering of the splices has been done to minimize the area taken up at any cross section—to keep the mass of splices from all adding up at the same cross section. The greatest conductor concentration is therefore either at section *x*, where there are three 300-kcmil and one 4/0 THW conductors, or at section *y*, where there are 3/0 THW conductors. To determine at which of these two cross sections the fill is greater, apply the appropriate csa's of THW conductors as given in Table 5, Chap. 9:

1. The total conductor csa at section x is 3×0.5581 in^2 plus 1×0.3904, or 2.0647 in^2.
2. The total conductor csa at section y is 8×0.3288 in^2, or 2.6304 in^2.

Section *y* is, therefore, the determining consideration. Because that fill of 2.6304 in^2 can at most be 20 percent of the gutter csa, the total gutter area must be at least 5 times this conductor fill area, or 13.152 in^2.

Assuming the gutter has a square cross section (all sides of equal width) and the sides have an integral number of inches, the nearest square value would be 16 in^2, indicating a 4- by 4-in gutter, and that would be suitable if the 300-kcmil conductors entered the

FIG. 2-114 Typical applications of auxiliary gutters provide the necessary space to make taps, splices, and other conductor connections involved where a number of switches or CBs are fed by a feeder (top) or for multiple-circuit routing, as at top of a motor control center (right) shown with a ground bus in gutter (arrow).

end of the gutter instead of the top. But because those conductors are deflected entering and leaving the gutter, the first two columns of Table 373-6(a) must also be applied to determine whether the width of 4 in affords sufficient space for bending the conductors. That consideration is required by Sec. 374-9(d). The worst condition (largest conductors) is where the supply conductors enter; therefore the 300-kcmil cable will determine the required space.

Table 373-6(a) shows that a circuit of one 300-kcmil conductor per phase leg (or wire per terminal) requires a bending space at least 5 in deep (in the direction of the entry of the 300-kcmil conductors), calling for a standard 6- by 6-in gutter for this application.

In Fig. 2-115, if the 300-kcmil conductors entered at the left-hand of the gutter instead of at the top, Sec. 374-9(d) would require Table 373-6(a) to be applied only to

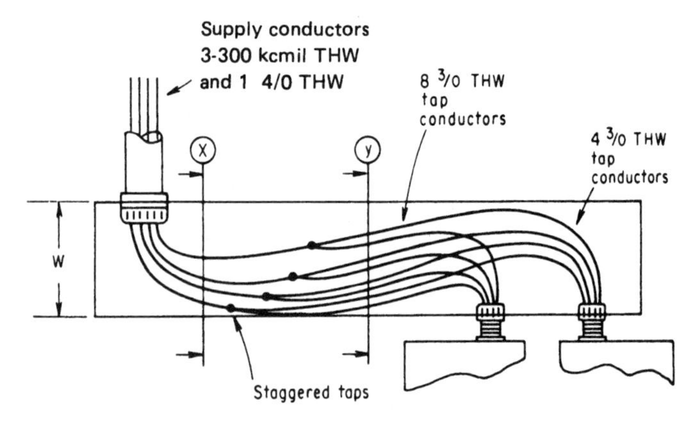

FIG. 2-115 Minimum acceptable gutter cross section and depth must be calculated.

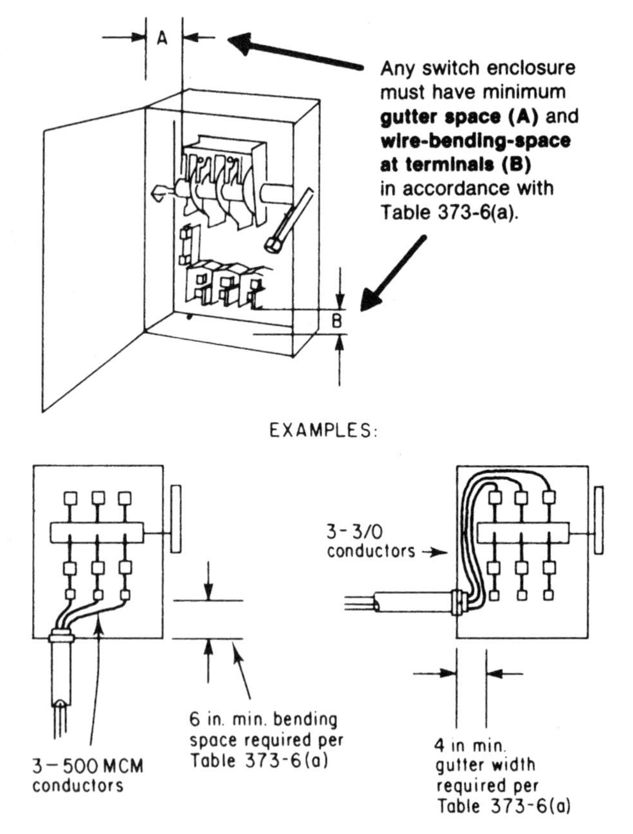

FIG. 2-116 Terminating and gutter space in switch enclosures must be measured.

the deflection of the No. 3/0 conductors. The table shows, under one wire per terminal, a minimum depth of 4 in is required. In that case, a 4- by 4-in gutter would satisfy.

SIZING ENCLOSURES FOR SWITCHES

Section 380-3 requires adequate wire-bending space at terminals and in side gutters of switch enclosures. In this section and in other sections applying to wiring space around other types of equipment, it is mandatory Code requirement that wire-bending space and side gutter wiring space conform to the requirements of Table 373-6(a) for side gutters and of Table 373-6(b) for wire-bending space at the line and load terminals, as described under Sec. 380-28. Those tables establish the minimum distance from wire terminals to enclosure surface or from the sides of equipment to enclosure side based on the size of conductors being used, as shown in Fig. 2-116.

This whole concern for adequate wiring space in all kinds of equipment enclosures reflects a repeated theme in many Code sections as well as in Art. 110 on general installation methods. One of the most commonly heard complaints from constructors and installers in the field concerns the inadequacy of wiring space at equipment terminals. Section 380-3 is designed to ensure sufficient space for the necessary conductors to run into and through switch enclosures.

CHAPTER 3
LIGHTING AND APPLIANCE BRANCH CIRCUITS

A *branch circuit* is any segment of a wiring system extending beyond the final automatic overcurrent protective device that is approved for use as branch-circuit protection and designated by the NEC as the branch-circuit protective device (Fig. 3-1). Branch circuits generally originate in panelboards, but individual branch circuits to motors commonly originate at either motor control centers or individual fused switches or circuit breakers tapped from busways or nippled out of auxiliary gutters.

In an electrical system, the branch circuits are the circuits of lowest capacity and current rating. They are the circuits to which load devices—lights, motors, etc.—are connected. Of course, it is possible (and not at all uncommon) to have branch circuits of very high rating, for instance, a branch circuit with conductors rated at 450 A to supply a 150-hp, 230-V, 3-phase motor. But the vast majority of branch circuits are rated not over 30 A.

Thermal cutouts or motor overload devices are not branch-circuit protection. Neither are fuses in luminaires or in plug connections, where they are used for ballast protection or individual fixture protection. Such "supplementary overcurrent protection" is the type of individual fixture or appliance protection that is connected on the load side of branch-circuit protection. NEC Secs. 424-22(c) to (e) cover the use of the supplementary overcurrent protection required for electric resistance duct heaters.

In its simplest form, a branch circuit consists of two wires which carry current at a particular voltage from protective device to utilization device. Although the branch circuit represents the last step in the transfer of power from the service or source of energy

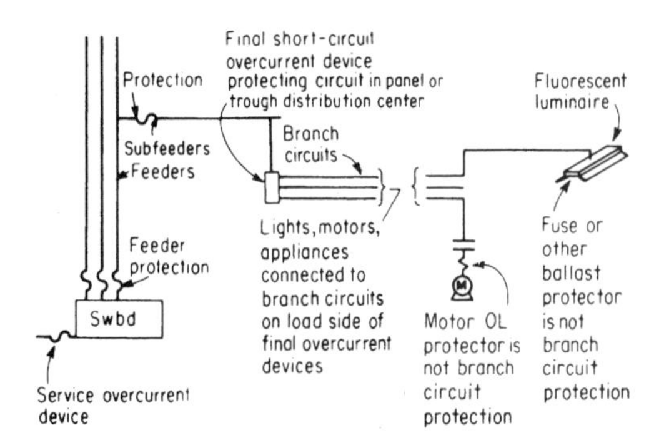

FIG. 3-1 Definition of a branch circuit.

to the utilization devices, it is the starting point for all design procedures. First, the loads are circuited. Then, the circuits are lumped on the feeders. Finally, the distribution system is connected to one or more sources of power.

Each and every branch circuit—whether for a power or lighting load in a commercial, industrial, or residential building—should be sized for its load, with spare capacity added where possible load growth is indicated, where necessary to reduce heating in continuously operating circuits, and/or for voltage stability. The design should also provide for the economical addition of circuits to handle future loads. Each circuit must provide the required power capacity at full utilization voltage at every outlet.

CONTINUOUS LOADS ON BRANCH CIRCUITS

Part (c) of Sec. 210-22 says that a branch circuit supplying a continuous current load must have a current rating (the rating of the circuit fuse or CB) and conductors that are not less than 125 percent of the continuous current load. The concept is the same as that described in Sec. 220-3(a), for branch-circuit continuous loading, and Sec. 220-10(b), which covers continuous loading on feeders and service conductors. The idea of the rule is to require the overcurrent (OC) device to be of a value that is equal to or greater than the value of continuous load times 1.25. And the conductors must have a table ampacity equal to or greater than 125 percent of the continuous load, *before* any derating. Where derating is required, the conductor's ampacity need *not* still be at least 125 percent of the continuous load, but must be properly protected in accordance with Sec. 240-3.

Section 210-22(c) says that (for loads other than motor loads) the rating of a branch circuit must be at least equal to 1.25 times the continuous load current when the load will constitute a continuous load, such as store lighting and similar loads that are on for periods of 3 h or more.

Because multioutlet branch circuits, such as lighting circuits, are rated in accordance with the rating or setting of the circuit overcurrent protective device, this rule has the effect of saying that the rating of the protective device must equal at least 1.25 times the continuous load current (Fig. 3-2). Where a circuit also supplies some noncontinuous load (not "on" for periods of 3 h or more) in addition to continuous load current, the branch-circuit protective device must have a rating not less than the noncontinuous load current plus 1.25 times the continuous load current.

For the branch-circuit conductors, Sec. 220-10(c) also requires additional capacity where supplying continuous loads. As indicated, the conductors must have a table ampacity (Tables 310-16 through 310-19) that is equal to or greater than the continuous load before application of any "adjustment" factors—for number of conductors (Note 8)—or "correction" factors—for elevated ambient temperature (ambient temperature correction factors at the bottom of the ampacity tables). Where adjustment or correction factors must be applied, the adjusted or corrected ampacity need only be properly protected in accordance with that ampacity provided by Sec. 240.3. The adjusted or corrected ampacity does *not* need to be equal to or greater than 125 percent of the continuous load.

Although the above-mentioned limitation applies only to loads other than motor loads, Sec. 384-16(c) says that "the total load on any overcurrent device located in a panelboard shall not exceed 80 percent of its rating where in normal operation the load will continue for 3 hours or more," which is the same rule stated reciprocally.

Note: In both of the above cases, neither the 125 percent of continuous load nor the 80 percent load limitation applies "where the assembly including the overcurrent device is approved for continuous duty at 100 percent of its rating."

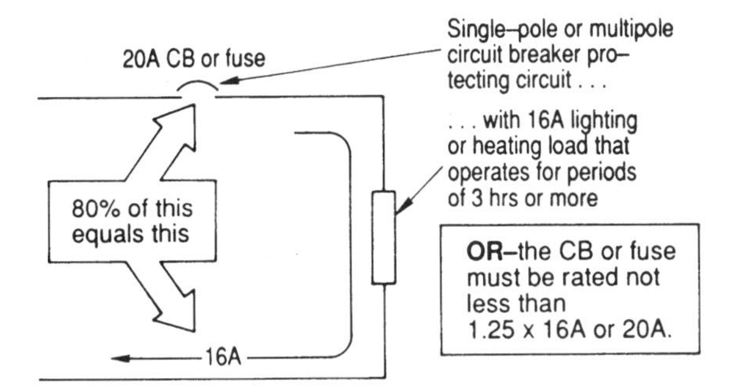

FIG. 3-2 Branch-circuit protective device must be rated not less than 125 percent of the continuous load current.

It is very important to understand that the UL and Code rules calling for load limitation to 80 percent of the rating of the protective device are based on the inability of the protective device itself to handle continuous load without overheating. And if a protective device has been designed, tested, and "listed" (such as by UL) for continuous operation at its full-load rating, then there is no requirement that the load current be limited to 80 percent of the breaker or fuse rating (or that the breaker or fuse rating be at least 125 percent of continuous load).

Section 384-16(c) requires any CB or fuse in a panelboard to have its load limited to 80 percent; and only one exception is made for such protective devices in a panel: A continuous load of 100 percent is permitted only when the protective device assembly (the CB unit or fuses in a switch) is approved for 100 percent continuous duty. (And there are no such devices rated less than 250 A at 250 V. UL has a hard-and-fast rule that any breaker not marked for 100 percent of continuous load must have its continuous load limited to 80 percent of its rating.) Based on Sec. 384-16(c) and the UL rules described, use of conductors that have had their ampacity derated because more than three are in a raceway (Note 8 to Tables 310-16 through 310-19) does not eliminate the requirement for an 80 percent limit on continuous load, but simply requires the additional capacity in the conductors *before* derating. After derating, the conductors need only be properly protected by the OC device.

CALCULATING CURRENT IN BRANCH CIRCUITS

1. Two-wire circuits at any power factor:

$$\text{Line current} = \frac{\text{voltamperes of connected load (or watts at unity PF)}}{\text{line voltage}}$$

$$\text{Line current } I = \frac{600\text{ W}}{120\text{ V}} = 5\text{ A}$$

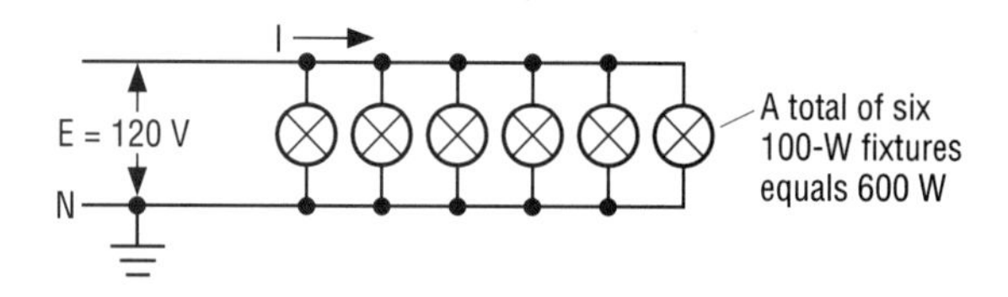

2. Three-wire circuits at any power factor:

Single-phase: Apply the same formula as for 2-wire branch circuit, considering each line to neutral separately. Use line-to-neutral voltage; result gives current in line conductors.

$$I_1 = \frac{1200 \text{ W}}{120 \text{ V}} = 10 \text{ A}$$

$$I_2 = \frac{720 \text{ W}}{120 \text{ V}} = 6 \text{ A}$$

$$I_N = I_1 - I_2 = 4 \text{ A}$$

I_1

120/240-V 3-wire, 1-ϕ circuit N

120 V

I_N

120 V

I_2

Total load of 1200 W

Total load of 720 W

3. Three-phase:

$$\text{Line current} = \frac{\text{voltamperes of balanced 3-phase load}}{\text{line voltage} \times 1.732}$$

$$\text{Line current in each phase leg} = \frac{11{,}940 \text{ VA}}{460 \text{ V} \times 1.732}$$

$$= \frac{11{,}940}{796} = 15 \text{ A}$$

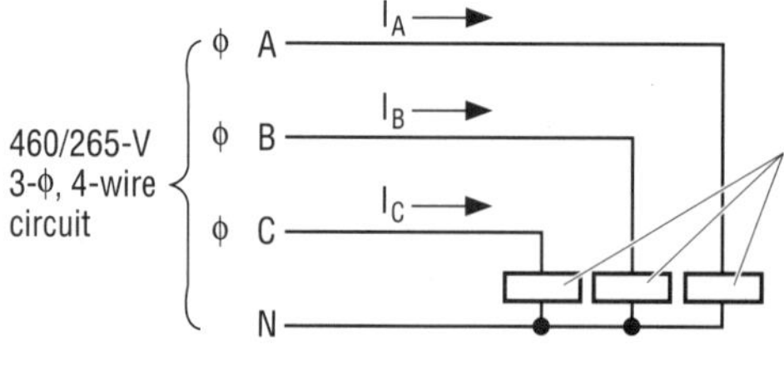

$I_A = I_B = I_C$, under balanced load

Each phase leg of this 3-phase, 4-wire circuit serves 1/3 of a balanced total fluorescent lighting load of 11,940 volt-amperes (determined by taking the volt-ampere rating of each ballast times the total number of ballasts in all of the fixtures)

A 3-wire, 240/120-V feeder supplies the panelboard shown in Example 1 below. Three of its branch circuits feed cooking appliances through receptacles. Circuits Nos. 1 and 2 are 2-wire, 120-V circuits; No. 3 is a 3-wire, 240/120 V circuit. The following examples illustrate the method of calculating branch-circuit and feeder currents under varying conditions of use.

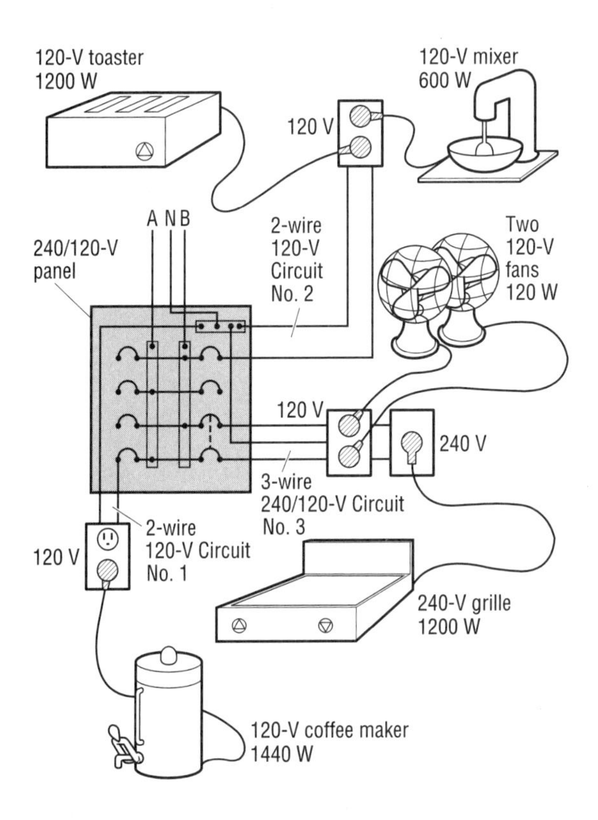

Example 1. Find the current in circuits 1 and 2. (Neglect voltage drop.)

Solution. Circuit No. 1 is connected across phase *A* and the neutral; thus $E = 120$ V, and

$$I = \frac{1440}{120} = 12 \text{ A}$$

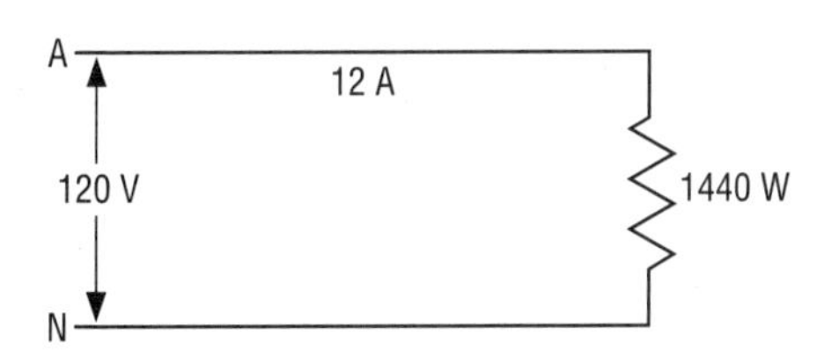

Circuit No. 2 is connected across phase B and the neutral; thus $E = 120$ V. There are two loads in parallel, and the total load is

$$600 + 1200 = 1800 \text{ W}$$

$$I = \frac{1800}{120} = 15 \text{ A}$$

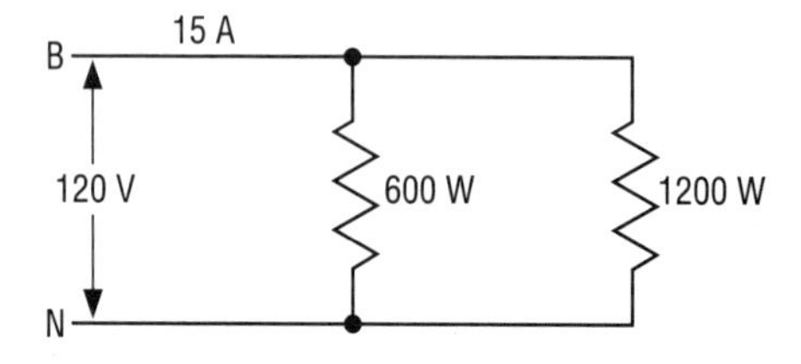

Example 2. What current will flow in feeder conductors A, B, and N if only circuits 1 and 2 are energized?

Solution. The only current flowing in phase A is that taken by circuit No. 1; therefore,

$$I_A = 12 \text{ A}$$

The only current flowing in phase B is that taken by circuit No. 2:

$$I_B = 15 \text{ A}$$

The neutral carries the unbalanced current:

$$I_N = 15 - 12 = 3 \text{ A}$$

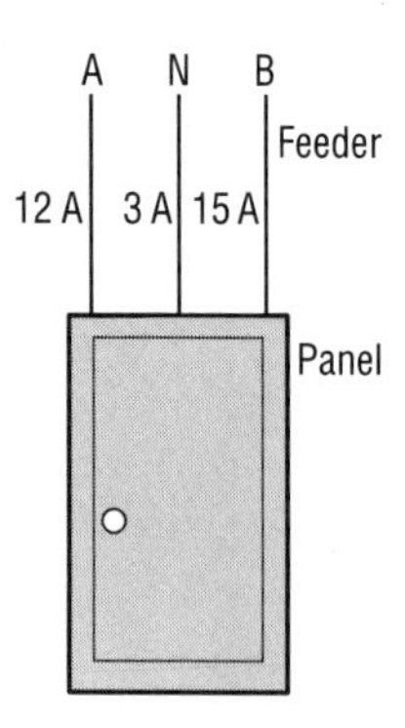

Example 3. If one of the fans on circuit No. 3 is turned off, what current will flow in each of the three circuit conductors?

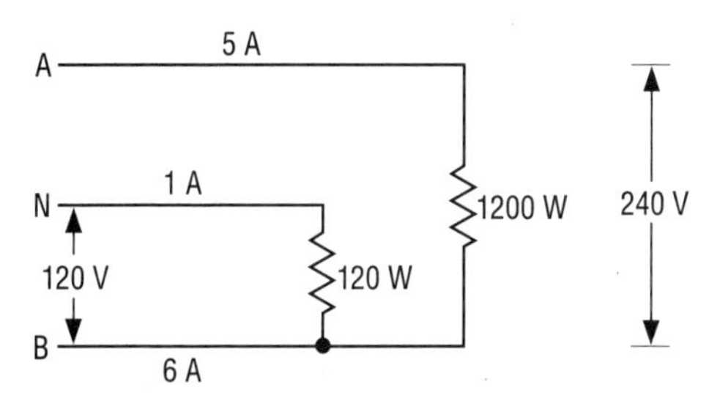

Solution. With one of the 120-V loads off, the neutral carries the current to the other 120-V load, so that

$$I_N = 1 \text{ A}$$

Phase *A* carries only the grille current

$$I_A = 5 \text{ A}$$

and phase *B* carries the sum of the other two currents:

$$I_B = 1 + 5 = 6 \text{ A}$$

Example 4. What is the current taken by circuit No. 3?

Solution. Each of the 120-V loads, since they are equal, will take

$$\frac{120 \text{ W}}{120 \text{ V}} \quad \text{or} \quad 1 \text{ A}$$

The 240-V, 1200-W load will take

$$\frac{1200 \text{ W}}{240 \text{ V}} \quad \text{or} \quad 5 \text{ A}$$

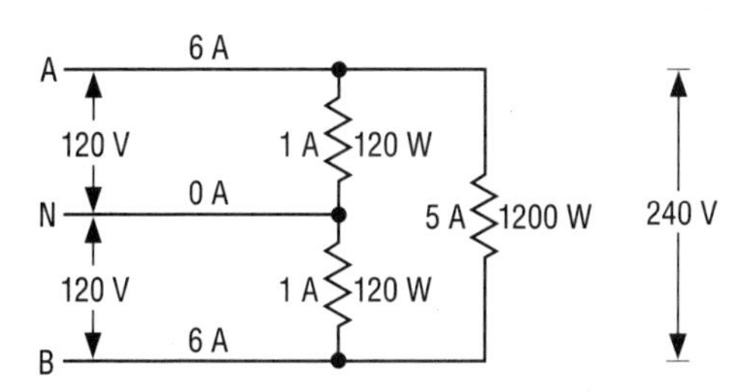

Thus each of the phase conductors of this circuit will take

$$5 + 1 = 6 \text{ A}$$

This is the total circuit current. Since the 120-V loads are equal, the neutral carries no current.

Example 5. What will be the current in each feeder conductor if all three circuits are energized and all appliances operating?

Solution

$$I_A = 12 + 6 = 18 \text{ A}$$

$$I_B = 15 + 6 = 21 \text{ A}$$

$$I_N = 21 - 18 = 3 \text{ A}$$

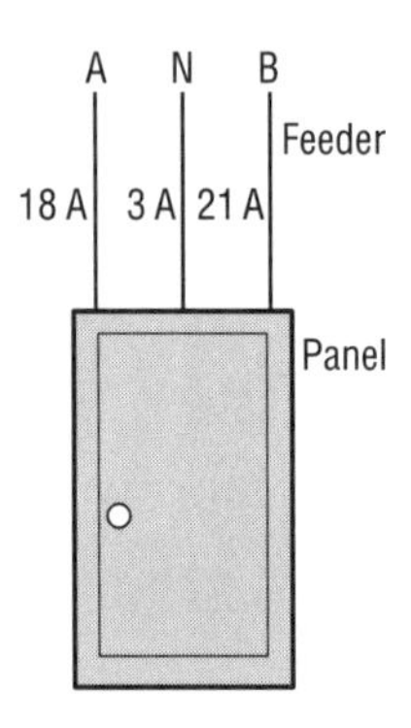

MINIMUM AMPACITY AND SIZE FOR BRANCH-CIRCUIT CONDUCTORS

In past NEC editions, the basic rule of Sec. 210-19 has said—and *still does* say—that the conductors of a branch circuit must have an ampacity that is not less than the maximum current load that the circuit will supply. Obviously, that is a simple and straightforward rule to ensure that the conductors are not operated overloaded (Fig. 3-3).

Additionally, the wording of this rule requires the circuit conductors to have an ampacity not less than "the rating of the branch circuit" for a multioutlet branch circuit that supplies receptacles for cord- and plug-connected loads. Section 210-3 clearly notes that the ampere rating of any circuit is set by the rating of the circuit OC device. Therefore, for multioutlet receptacle circuits, the conductor ampacity must be not less than the rating of the circuit protective device. The concept here is that plug receptacles provide for random, indeterminate loading or the circuit; and by matching conductor ampacity to the ampere rating of the circuit fuse or CB, overloading of the conductors can be avoided. But for multioutlet branch circuits that supply fixed outlets—such as lighting fixture outlets or hard-wired connections to electric heaters or other appliances—it is acceptable to have a condition in which the conductor ampacity is adequate for the load current but, where there is no standard rating of protective device that corresponds to the conductor ampacity, the circuit fuse or CB rating is the next-higher standard rating of protective device above the ampacity value of the conductor.

For multioutlet branch circuits (rated at 15, 20, 30, 40, or 50 A), the ampacities of conductors usually correspond to standard ratings of protective devices when there is

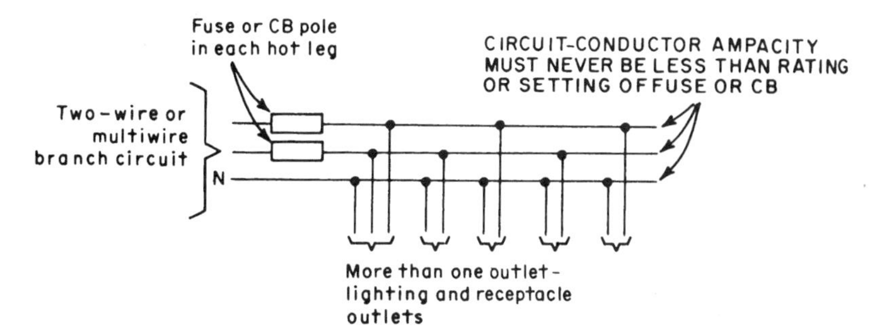

FIG. 3-3 This is the basic rule for any multioutlet branch circuit supplying one or more receptacles.

only one circuit in a cable or conduit. Standard rated protective devices of 15, 20, 30, 40, or 50 A can be readily applied to conductors that have corresponding ampacities from Tables 310-16 through 310-19 and their footnotes, that is, 15 A for No. 14, 20 A for No. 12, 30 A for No. 10, and 40 A for No. 8, with a 55-A rated No. 6 used for a 50-A circuit. But when circuits are combined in a single conduit so that more than three current-carrying conductors are involved, the ampacity derating factors of Note 8 to Table 310-16 often result in reduced ampacity values that do not correspond to standard fuse or CB ratings. It is to such cases that the rule of Sec. 210-19(a) may be applied.

For instance, assume that two 3-phase, 4-wire multioutlet circuits supplying fluorescent lighting are run in a single conduit. Two questions arise: (1) How much load current may be put on the conductors? (2) What is the maximum rating of overcurrent protection that may be used for each of the six hot legs?

The eight wires in the single conduit (six phases and two neutrals) must be taken as eight conductors when one is applying Note 8 of Table 310-16 because the neutrals to electric-discharge lighting carry harmonic currents and must be counted as current-carrying conductors [Note 10(c) of Table 310-16]. Note 8 then shows that the No. 14 wires must have their ampacity reduced to 70 percent (for 7 to 9 wires) of the 20-A ampacity given in Table 310-16 for No. 14 TW. With the eight No. 14 wires in the one conduit, then, each has an ampacity of 0.7×20, or 14 A. Because Sec. 210-19(c) requires circuit wires to have an ampacity at least equal to the rating of the circuit fuse or CB if the circuit is supplying receptacles, use of a 15-A fuse or 15-A circuit breaker would *not* be acceptable in such a case because the 14-A ampacity of each wire *is* less than "the rating of the branch circuit" (15 A). *But* because the circuits here are supplying fixed lighting outlets, as stated in the original assumption, Sec. 210-19(a) would accept the 15-A protection on wires with 14-A ampacity. In such a case, it is only necessary that the design load current on each phase not exceed 14 A if the load is noncontinuous. Or if the lighting load *is* continuous (operating steadily for 3 h or more), the load on each 15-A CB or fuse must not exceed 0.8×15 or 12 A [as required by Sec. 210-22(c) and Sec. 384-16(c)].

Note: If THHN-insulated No. 14s are used, the 70 percent derating is applied against the Table 310-16 value of 25 A. This means that the conductor ampacity for the 90°C THHN conductors is $25 \times 0.7 = 17.5$ A, which is "not less than the rating of the branch circuit," which is 15 A. Such an approach will satisfy Sec. 210-19 for multioutlet receptacles. Refer to the discussion of Sec. 210-20 and the discussion of ampacity and derating under Sec. 310-15.

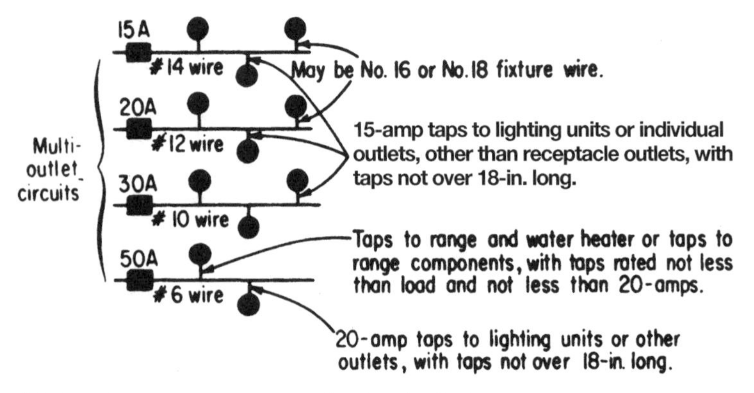

FIG. 3-4 Tap conductors may be smaller than circuit wires.

Section 210-19(c) sets No. 14 as the smallest size of general-purpose circuit conductors. But tap conductors of smaller sizes are permitted as explained in Exceptions 1 and 2 (Fig. 3-4). Number 14 wire, not longer than 18 in, may be used to supply an outlet unless the circuit is a 40- or 50-A branch circuit, in which event the minimum size of the tap conductor must be No. 12.

CALCULATING BRANCH CIRCUITS FOR RESIDENTIAL COOKING APPLIANCES

In part (b) of Sec. 210-19, the rule also calls for the same approach to sizing conductors for branch circuits to household electric ranges, wall-mounted ovens, counter-mounted cooking units, and other household cooking appliances (Fig. 3-5).

The maximum demand for a range of 12-kW rating or less is sized from NEC Table 220-19 as a load of 8 kW. And 8000 W divided by 230 V is approximately 35 A. Therefore, No. 8 conductors with an ampacity of 40 A may be used for the range branch circuit.

On modern ranges the heating elements of surface units are controlled by five heat-unit switches. The surface-unit heating elements will not draw current from the neutral unless the unit switch is in one of the low-heating positions. This is also true to a greater degree as far as the oven-heating elements are concerned, so that the maximum current in the neutral of the range circuit seldom exceeds 20 A. Because of that condition Exception 2 permits a smaller neutral than the ungrounded conductors, but not smaller than No. 10.

A reduced-size neutral for a branch circuit to a range, wall-mounted oven, or cooktop must have ampacity of not less than 70 percent of the circuit rating, which is determined by the current rating or setting of the branch-circuit protective device. This is a change from previous wording that required a reduced neutral to have an ampacity of at least 70 percent of "the ampacity of the ungrounded conductors." Under that wording, a 40-A circuit (rating of protective device) made up of No. 8 TW wires for the hot legs could use a No. 10 TW neutral—because its 30-A ampacity is at least 70 percent of the 40-A ampacity of a No. 8 TW hot leg ($0.7 \times 40 = 28$ A). But if No. 8 THHN (55-A

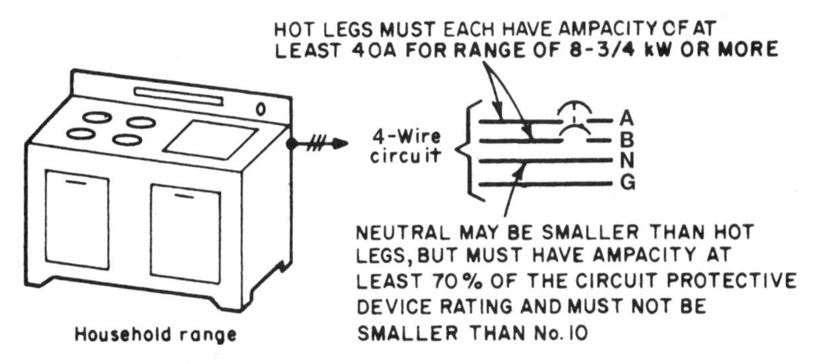

FIG. 3-5 Sizing circuit conductors for household electric range.

ampacity) is used for the hot legs with the same 40-A protected circuit, the neutral ampacity will have to be at least 70 percent of 55 A ($0.7 \times 55 = 38.5$ A) and a No. 10 TW (30 A) or a No. 10 THW (35 A) could not be used. The new wording bases the neutral size at 70 percent of the protective-device rating (0.7×40 A $= 28$ A), thereby permitting any of the No. 10 wires to be used, and does not penalize use of higher-temperature wires (THHN) for the hot legs.

Exception 1 permits taps from electric cooking circuits (Fig. 3-6). Because Exception 1 says that taps on a 50-A circuit must have an ampacity of at least 20 A, No. 14 conductors, which have an ampacity of 20 A in Table 310-16, may be used.

Exception 1 applies to a 50-A branch circuit run to a counter-mounted electric cooking unit and wall-mounted electric oven. The tap to each unit must be as short as possible and should be made in a junction box immediately adjacent to each unit. The words "no longer than necessary for servicing the appliance" mean that it should be necessary only to move the unit to one side in order for the splices in the junction box to be accessible.

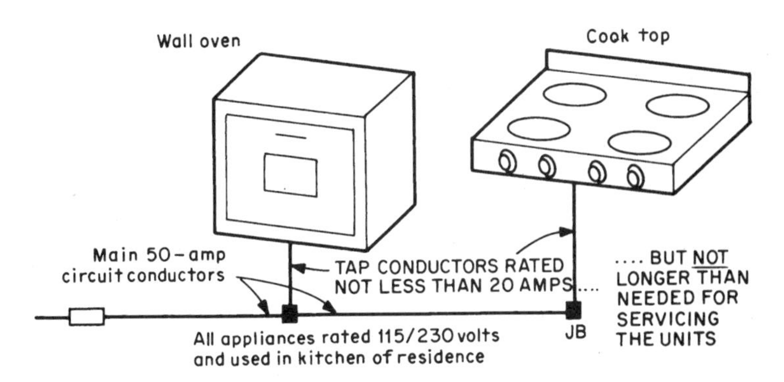

FIG. 3-6 Tap conductors may be smaller than wires of cooking circuit.

RATING OF BRANCH-CIRCUIT OVERCURRENT DEVICE

According to the basic Code rule of Sec. 210-20, the rating or setting of an overcurrent device in any branch circuit must not exceed the current-carrying capacity of the circuit conductor or may be the next-higher value of overcurrent device where conductor ampacity does not match the rating of a standard fuse or CB. Section 240-3 applies to the rating of overcurrent protection. Figure 3-7 shows the basic rules that apply to use of overcurrent protection for branch circuits. (Section 240-2 designates other Code articles that present data and regulations on overcurrent protection for branch circuits to specific types of equipment.)

Branch-circuit taps, as covered in Secs. 210-19 and 210-20, are considered protected by the branch-circuit overcurrent devices, even where the rating or setting of the protective device is greater than the ampere rating of the tap conductors, fixture wires, or cords.

When only three No. 12 TW or THW conductors of a branch circuit are in a conduit, each has a load-current rating of 20 A (see the footnote to Table 310-16, Art. 310) and must be protected by a fuse or CB rated not over 20 A. This satisfies Sec. 210-19, which requires branch-circuit conductors to have an ampacity not less than the rating of the branch circuit, and Sec. 210-3 notes that the rating of a branch circuit is established by the rating of the protective device. It also satisfies Sec. 210-20, which says

> Branch-circuit conductors... shall be protected by overcurrent protective devices having a rating or setting... *not exceeding that specified in Section 240-3* for conductors....

The basic rule of Section 240-3 says

> Conductors... shall be protected against overcurrent *in accordance with their ampacities* as specified in Tables 310-16 through 310-19 and all applicable notes to these tables.

That rule says that the conductors are required to be protected at a current value indicated by the table and its accompanying notes, such as Note 8, which reduces ampacities from the table values.

In Table 310-16, which applies to conductors in raceways and in cables and covers the majority of conductors used in electrical systems for power and light, the ampacities for sizes No. 14, No. 12, and No. 10 are particularly significant because copper conductors of those sizes are involved in the vast majority of branch circuits in modern electrical systems. Number 14 has an ampacity of 20, No. 12 has an ampacity of 25, and No. 10 has an ampacity of 30. The typical impact of that on circuit makeup and loading is as follows:

Number 12 TW or THW copper is shown to have an ampacity of 25; and based on the general UL requirements that equipment terminals be limited to use with conductors loaded not in excess of 60°C ampacities for wires up to No. 1 AWG, No. 12 THHN or XHHW copper conductors must also be treated as having a 25-A continuous rating (the ampacity of 60°C No. 12) and *not* 30 A, as shown in Table 310-16. *But* the footnote to Table 310-16 limits all No. 12 copper to not over 20-A load current by requiring that it be protected at not more than 20 A.

The ampacity of 25 A for No. 12 TW and THW copper wires interacts with Note 8 to Tables 310-16 to 310-19 where there are, say, six No. 12 TW current-carrying wires for the phase legs of two 3-phase, 4-wire branch circuits in one conduit supplying, say, receptacle loads. In such a case, the two neutrals of the branch circuits do not count in

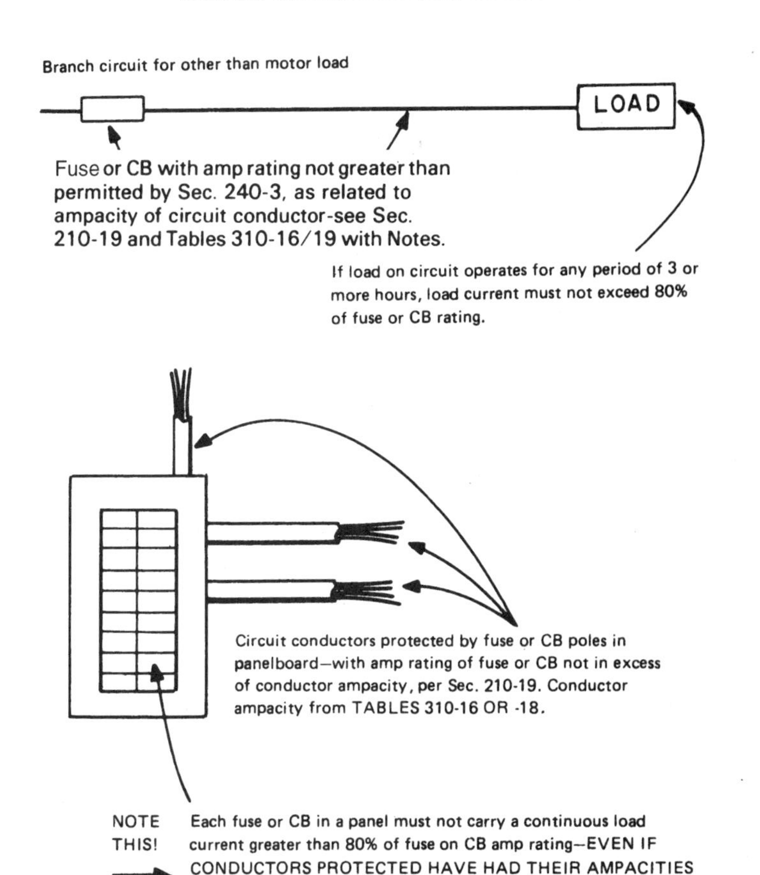

FIG. 3-7 Branch-circuit protection involves a number of rules.

applying Note 8, and only the six phase legs are counted to determine by how much all conductors must have their ampacities *derated* to the "Percent of Values in Tables," as stated at the top of the table in Note 8. In the case described here, that literally means that each No. 12 phase leg may be used at an ampacity of 0.8×25, or 20 A. And the footnote to Table 310-16 would require use of a fuse or CB rated not over 20 A to protect each No. 12 phase leg. Each No. 12 would be protected at the current value that represents the maximum I^2R heat input that the conductor insulation can withstand. In that example, the derated ampacity of the No. 12 conductors (20 A) is not in excess of "the rating of the branch circuit," that is, the 20-A rating of the fuse or CB protecting the circuit. Thus, Sec. 210-19(a) is completely and readily satisfied because the ampacity and protective device rating came out to the same value. Thus the circuits described could be used for supplying receptacles and/or fixed-load outlets. The only other possible qualification is that Sec. 384-16(c) would require the load current on each of the phase legs to be further limited to no more than 80 percent of the 20-A rating of the overcurrent device, that is, to 16 A, if the load current is "continuous" (operates steadily for 3 h or more), a condition not likely for receptacle-fed loads.

MAXIMUM LOADING FOR MULTIOUTLET BRANCH CIRCUITS

A single branch circuit to one outlet or load may serve any load and is unrestricted as to ampere rating. Circuits with more than one outlet are subject to NEC limitations on use as follows: (The word *appliance* stands for any type of utilization equipment.)

1. Branch circuits rated 15 and 20 A may serve lighting units and/or appliances. The rating of any one cord- and plug-connected appliance shall not exceed 80 percent of the branch-circuit rating. Appliances fastened in place may be connected to a circuit serving lighting units and/or plug-connected appliances provided the total rating of the fixed appliances fastened in place does not exceed 50 percent of the circuit rating (Fig. 3-8). *Example:* 50 percent of a 15-A branch circuit = 7.5 A. A room air-conditioning unit fastened in place, with a rating not in excess of 7.5 A, may be installed on a 15-A circuit having two or more outlets. Such units may not be installed on one of the small-appliance branch circuits required in Sec. 220-4(b).

However, modern design provides separate circuits for individual fixed appliances. In commercial and industrial buildings, separate circuits should be provided for lighting and separate circuits for receptacles.

2. Branch circuits rated 30 A may serve fixed lighting units (with heavy-duty-type lampholders) in other than dwelling units or appliances in any occupancy. Any individual cord- and plug-connected appliance which draws more than 24 A may not be connected to this type of circuit (Fig. 3-9).

Because an *individual* branch circuit—that is, a branch circuit supplying a single outlet or load—may be rated at any ampere value, it is important to note that the omission of recognition of a 25-A *multioutlet* branch circuit does not affect the full acceptability of a 25-A *individual* branch circuit supplying a single outlet. A typical application of such a circuit would be use of No. 10 TW aluminum conductors (rated at 25 A in Table 310-16), protected by 25-A fuses or circuit breakers, supplying, say, a 4500-W water heater at 230 V. The water heater is a load of 4500 ÷ 230, or 19.6 A, which is taken as a 20-A load. Then, because Sec. 422-14(b) designates water heaters as continuous loads (in tank capacities up to 120 gal), the 20-A load current multiplied by 125 percent equals 25 A and satisfies Sec. 422-4(a), Exception 2, on the required minimum branch-circuit rating. The 25-A rating of the circuit overcurrent device also satisfies Sec. 422-27(e),

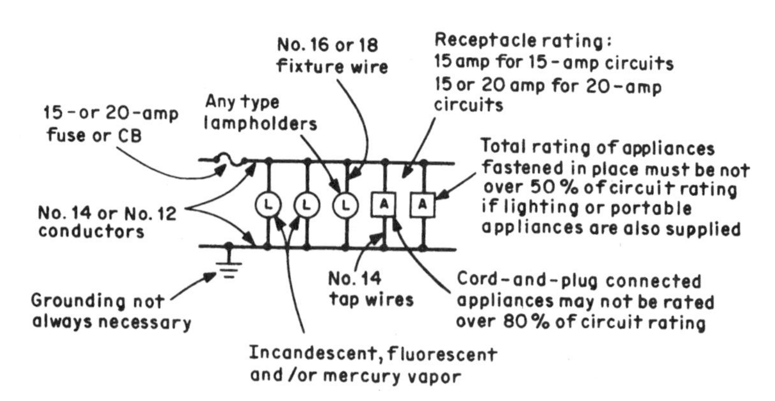

FIG. 3-8 General-purpose branch circuits—15 or 20 A.

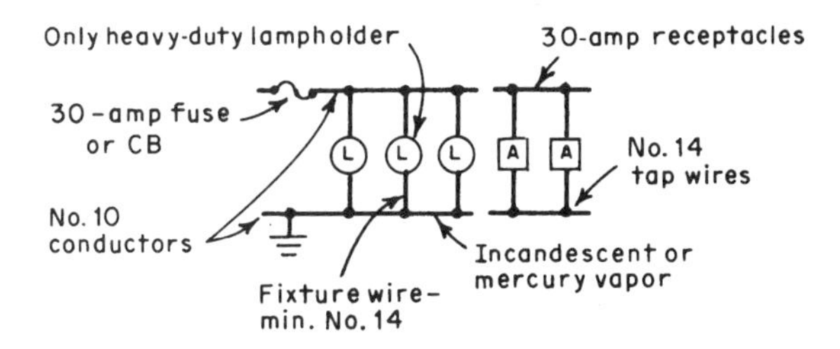

FIG. 3-9 Multioutlet 30-A circuits.

which says that the overcurrent protection must not exceed 150 percent of the ampere rating of the water heater.

Number 10 aluminum, with a 60°C ampacity of 25 A, may be used instead of No. 12 copper (rated 20 A). But the need for and the application possibilities of a 25-A *multioutlet* branch circuit have always been extremely limited. Such a circuit has never been considered suitable to supply lighting loads in dwelling units (where aluminum branch-circuit conductors have been primarily used). But for heavy-current appliances (16 to 20 A), realistic loading dictates use of an *individual* branch circuit, which *may* be rated at 25 A.

3. Branch circuits rated 40 and 50 A may serve fixed lighting units (with heavy-duty lampholders) or infrared heating units in other than dwelling units or cooking appliances in any occupancy (Fig. 3-10). Note that a 40- or 50-A circuit may be used to supply any kind of load equipment, such as a dryer or a water heater, where the circuit is an individual circuit to a single appliance. The conditions shown in that figure apply only where more than one outlet is supplied by the circuit. Figure 3-11 shows the combination of loads.

4. A multioutlet branch circuit rated over 50 A, as permitted by Sec. 210-3, is limited to use only for supplying industrial utilization equipment (machines, welders, etc.) and may *not* supply lighting outlets.

Except as permitted in Sec. 660-4 for portable, mobile, and transportable medical X-ray equipment, branch circuits having two or more outlets may supply only the loads specified in each of the above categories. Note that any other circuit is not permitted to have more than one outlet and would be an individual branch circuit.

Application of those rules—and other Code rules that refer to "dwelling unit"—must take into consideration the NEC definition for that phrase. A *dwelling unit* is defined as "one or more rooms" used "as a housekeeping unit" and must contain space or areas specifically dedicated to "eating, living, and sleeping" and must have "permanent provisions for cooking and sanitation." A one-family house is a dwelling unit. So is an apartment in an apartment house or a condominium unit. But a guest room in a hotel or motel or a dormitory room or unit is not a dwelling unit if it does not contain permanent provisions for cooking, which must mean a built-in range or counter-mounted cooking unit (with or without an oven).

Note that the requirement calling for heavy-duty-type lampholders for lighting units on 30-, 40-, and 50-A multioutlet branch circuits excludes the use of fluorescent lighting on these circuits because lampholders are not rated heavy-duty in accordance with Sec. 210-21(a) (Fig. 3-12). Mercury-vapor units with mogul lampholders may be used on these circuits provided tap conductor requirements are satisfied.

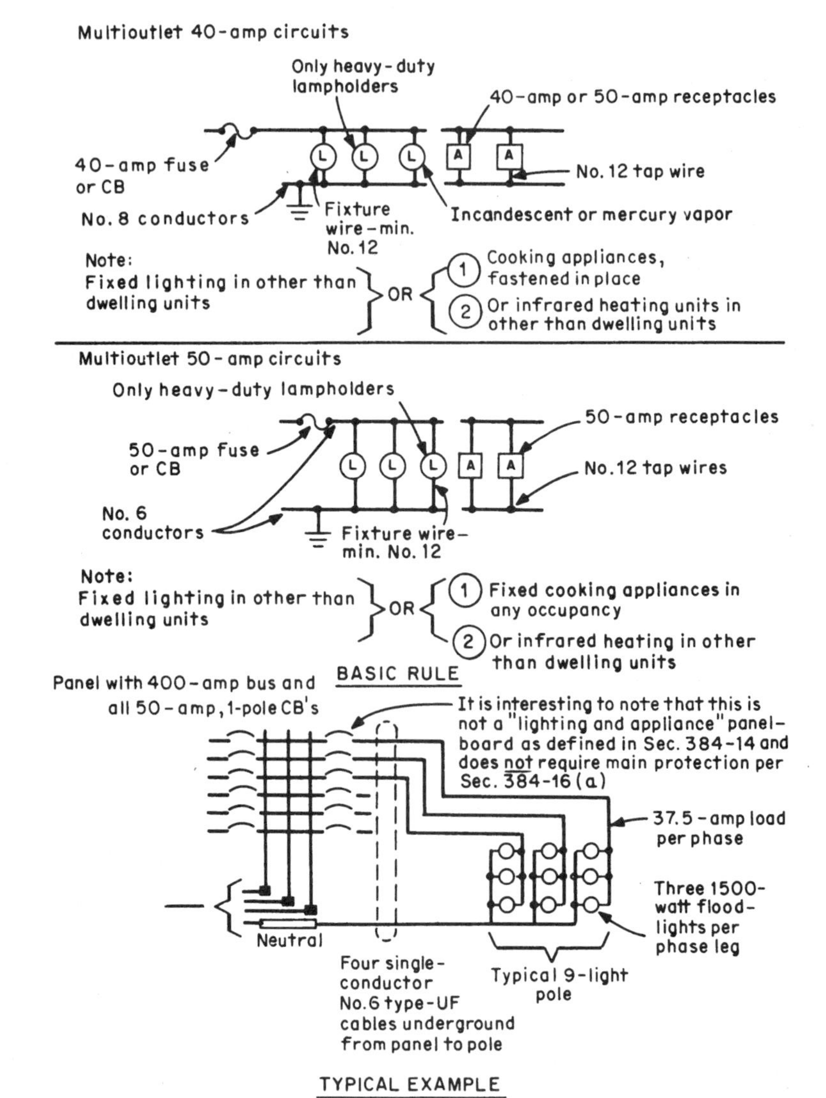

FIG. 3-10 Larger circuits.

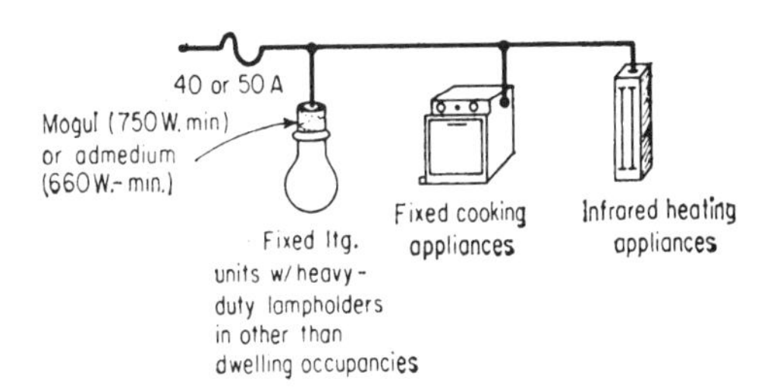

FIG. 3-11 Only specified loads may be used for multioutlet circuit.

As indicated, multioutlet branch circuits for lighting are limited to a maximum loading of 50 A. Individual branch circuits may supply any loads. Excepting motors, this means than an individual piece of equipment may be supplied by a branch circuit which has sufficient carrying capacity in its conductors, is protected against current in excess of the capacity of the conductors, and supplies only the single outlet for the load device.

Fixed outdoor electric snow-melting and deicing installations may be supplied by any of the above-described branch circuits. (See Sec. 426-4 in Art. 426, "Fixed Outdoor Electric De-Icing and Snow-Melting Equipment.")

Table 210-24 summarizes the requirements for the size of conductors where two or more outlets are supplied. The asterisk note also indicates that these ampacities are for copper conductors, where derating is not required. Where more than three conductors

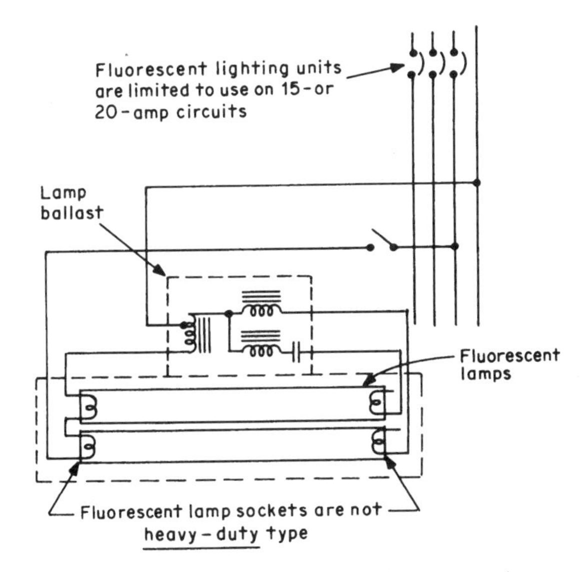

FIG. 3-12 Watch out for this limitation on fluorescent equipment.

are contained in a raceway or a cable, Note 8 to Tables 310-16 through 310-19 specifies the load-current derating factors to apply for the number of conductors involved. A 20-A branch circuit is required to have conductors which have an ampacity of 20 A and also must have the overcurrent protection rated 20 A where the branch circuit supplies two or more outlets. Refer to the detailed discussion of conductor ampacity and load-current limiting under Sec. 310-15, where Table 310-16 and its notes are explained.

As noted in the first sentence of Sec. 210-25, a branch circuit in a dwelling unit must not have outlets in any other dwelling unit. This rule is intended to prevent overloading of a circuit that could result from tenants in different dwelling units attempting to simultaneously connect heavy-current appliances to a single circuit. And as noted in the second sentence, common branch circuit(s) may be used to supply more than one dwelling unit in a two-family or multifamily dwelling (apartment house, condominiums, etc.) for the purposes of alarm, signal, communication, or similar safety or security functions. The section recognizes that apartment houses and similar multifamily dwelling units under single management do make use of *common* systems for the safety of all the tenants or occupants of the building (fire alarm, intrusion systems, communications systems, etc.). Such systems must be supplied from a common "house" panel or enclosure to and through individual dwelling units. This "common area" or "house" panel may *not* be supplied from any other panel. It must be fed from the service conductors and separately metered. That statement is related to the wording used, which mandates that such a panel *not* be supplied from any equipment that supplies a single dwelling. The same requirement is also extended to two-family dwellings.

Section 210-50(b) simply requires that wherever it is known that cord- and plug-connected equipment is going to be used, receptacle outlets must be installed. That is a general rule that applies to any electrical system in any type of occupancy or premise.

CALCULATING BRANCH-CIRCUIT LOAD

The task of calculating a branch-circuit load and then determining the size of circuit conductors required to feed that load is common to all electrical system calculations. Although it may seem to be a simple matter (and usually is), there are many conditions which make the problem confusing (and sometimes controversial) because of the NEC rules that must be observed.

The requirements for loading and sizing of branch circuits are covered in Art. 210 and in Sec. 220-3. In general, the following basic points must be considered.

- The ampacity of branch-circuit conductors must not be less than the maximum load to be served [Sec. 210-19(a)].
- The ampacity of branch-circuit conductors must generally not be less than the rating of the branch circuit. Section 210-19(a) requires that the conductors of a branch circuit that supplies any receptacle outlets have an ampacity not less than the rating of the branch circuit, which rating is determined by the rating or setting of the overcurrent device protecting the circuit.
- The rating of a branch circuit is established by the rating or setting of its OC protective device (Sec. 210-3).
- The normal, maximum, continuous ampacities of conductors in cables or raceways are given in Tables 310-16 to 310-19 for both copper and aluminum.

- These normal ampacities of conductors may have to be derated where there are more than three conductors in a cable or raceway (Note 8 to Tables 310-16 through 310-19).
- The current permitted to be carried by the branch-circuit protective device (fuse or CB) may have to be reduced if the load is continuous [Sec. 210-22(c) and Sec. 220-3(a)].

Computation of Branch Circuits

Section 210-20 says that the rating or setting of the branch-circuit overcurrent protective device is not to exceed that specified in Sec. 240-3 for conductors. The basic rule in Sec. 240-3 says that conductors shall be protected against overcurrent in accordance with their ampacities; but part (b) of Sec. 240-3 allows that where the ampacity of the conductor does not correspond with the standard ampere rating of a fuse or a circuit breaker, the next-higher standard device rating shall be permitted if this rating does not exceed 800 A and if the wire being protected is *not* part of a branch circuit supplying receptacles for plug-connected portable tools, appliances, or other plug-in loads. In selecting the size of the branch-circuit overcurrent device, the rule of *both* Secs. 210-19(a) and 210-20 must be satisfied, because *all* Code rules bearing on a particular detail must always be observed.

Section 210-19(a) does not permit *any* case where branch-circuit conductors supplying one or more receptacle outlets would have an ampacity of less than the ampere rating of the circuit protective device. Section 210-19(a) thereby correlates to Sec. 240-3(b) to prohibit using a protective device of "the next higher standard rating" on branch circuits to receptacles. However, a branch circuit that supplies only hard-wired outlets, such as lighting outlets or outlets to fixed electric heaters, may have its overcurrent protection selected as the next-higher standard rating of protective device above the ampacity of a conductor when the conductor ampacity does not correspond to a standard fuse or CB rating, as permitted by Sec. 240-3(b).

Section 220-3(a) must be evaluated against all those background data from Art. 210. Although determination of ampacities from Tables 310-16 through 310-19 yields the maximum allowable *continuous* current ratings of conductors, there are Code rules that limit the load that may be carried continuously (3 h or more) to no more than 80 percent of the rating of the circuit protective device. Section 210-22(c) says that the total load on a branch circuit must not be more than the sum of noncontinuous load *plus* 125 percent of the continuous load. Although the rating of the branch circuit is set by the ampere rating or setting of the overcurrent device protecting the circuit, the conductors of a branch circuit supplying any receptacles may not have ampacity (either normal or derated) less than the rating of the protective device [Sec. 210-19(a)]. As a repetition of the rule of Sec. 210-22(c), Sec. 220-3(a) requires a branch-circuit protective device to be rated not "less than the noncontinuous load plus 125 percent of the continuous load."

This wording is the reciprocal way of saying what Sec. 384-16(c) says—that the continuous load of a circuit must not exceed 80 percent of the rating of the branch-circuit protective device. (From Sec. 210-3, the *rating* of a branch circuit is determined by the "ampere" rating or setting of the overcurrent device.) The wording of both Sec. 210-22(c) and Sec. 220-3(a) states the need to limit heating effect. As shown in Fig. 3-13, although the circuit has a total load current of 20 A, the loading satisfies the wording that the continuous load shall not exceed 80 percent of the rating of the branch-circuit overcurrent protective device. ($0.8 \times 20 = 16$ A.) But according to the rule of Sec. 220-3(a), with 4 A of noncontinuous load, the above circuit could carry only that amount of continuous load which, when multiplied by 1.25, would equal 16 A. Then $16/1.25 = 12.8$ A. Thus the maximum continuous load that would be permitted in addition to the

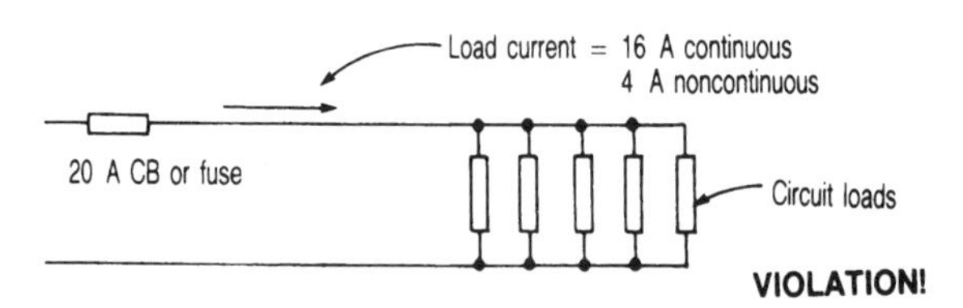

FIG. 3-13 Continuous load does not exceed 80 percent of the circuit rating (20 A); but the 20-A CB rating is *less* than "the noncontinuous load plus 125 percent of the continuous load" and violates Sec. 220-3(a).

noncontinuous load is 12.8 A. The branch-circuit rating (20 A) is "not less than the noncontinuous load (4 A) plus 125 percent of the continuous load" (12.5 × 12.8 A = 16 A).

[Section 220-10(b) also has the effect of limiting a continuous load to not more than 80 percent of the rating of any feeder CB or fuse protection that is not UL-listed for continuous loading to 100 percent of its rating.]

The continuous-current limitation, as set forth in those NEC sections, is not established because the conductors cannot carry 100 percent of their rated current continuously. The conductors still have the same ampacity—the same maximum allowable continuous-current rating. Likewise, a fused switch or circuit breaker can itself withstand the heat produced within it by 100 percent of its current rating. The 80 percent limit is set because of the following:

1. Conductors of any circuit must connect to the terminals of the fusible switch or circuit breaker that provides disconnect and protection for a branch circuit or feeder.
2. Current flow through a circuit produces heating in the fusible switch or circuit breaker as well as in the conductors.
3. The heat produced in the switch or CB does not generally harm the switch or CB itself, but that heat is readily conducted into the end lengths of conductors that are attached to the terminals.
4. Although the conductors can take the heat input from 100 percent of their own current rating, the extra heat conducted into the conductor from the switch or CB adds to the heat load on the conductors adjacent to the terminations.
5. For a continuous load, excessive heat will be produced in the conductor insulation if the conductor is already carrying its full rated current; and that can cause damage to the conductor insulation.

The effect of the rule of Sec. 220-3(a) is that any CB or fuse for a branch circuit supplying a total continuous load must have its load current limited to 80 percent, and only one exception is made for such protective devices—a continuous load of 100 percent of the fuse or CB rating is permitted *only* when the protective device assembly (fuse in switch or CB) is approved for 100 percent continuous duty. (And there are no such devices rated less than 225 A, so the exceptions referring to 100 percent rated devices do not apply to any branch circuits of less than 225 A. In addition, UL has a hard-and-fast rule that *any* breaker *not* marked for 100 percent continuous load must have its load limited to 80 percent of its rating.)

Code Table 220-3(b) lists certain occupancies (types of buildings) for which a minimum general lighting load is specified in voltamperes per square foot. In each type of building, there must be adequate branch-circuit capacity to handle the total load that is represented by the product of voltamperes per square foot and the square-foot area of the building. For instance, if one floor of an office building is 40,000 ft^2 in area, that floor must have a total branch-circuit capacity of 40,000 × 3½ VA/ft^2 [Code Table 220-3(b)] for general lighting. Note that the total load to be used in calculating the required circuit capacity must never be taken at less than the indicated voltamperes per square foot times square feet for those occupancies listed. Of course, if branch-circuit load for lighting is determined from a lighting layout of specific fixtures of known voltampere rating, the load value must meet the previous voltamperes-per-square-foot minimum; and if the load from a known lighting layout is greater, then the greater voltampere value must be taken as the required branch-circuit capacity.

Note that the bottom of Table 220-3(b) requires a minimum general lighting load of ½ VA/ft^2 to cover branch-circuit and feeder capacity for halls, corridors, closets, and all stairways.

As indicated in Sec. 220-3(b), when load is determined on a voltamperes-per-square-foot basis, open porches, garages, unfinished basements, and unused areas are not counted as part of the area. Area calculation is made using the *outside* dimensions of the "building, apartment, or other area involved."

When electric-discharge lighting (fluorescent or mercury-vapor lighting) is used on branch circuits, the presence of the inductive effect of the ballast or transformer creates a power factor consideration. Determination of the load in such cases must be based on the total of the voltampere rating of the units and not on the wattage of the lamps.

Based on extensive analysis of load densities for general lighting in office buildings, Table 220-3(b) requires a minimum unit load of only 3½ VA/ft^2—rather than the previous unit value of 5—for "office buildings" and for "banks."

A double-asterisk note at the bottom of the table requires the addition of another 1 VA/ft^2 to the 3½ value to cover the loading added by general-purpose receptacles in those cases where the actual number of receptacles is not known at the time feeder and branch-circuit capacities are being calculated. In such cases, a unit load of 4½ VA/ft^2 must be used, and the calculation based on that figure will yield minimum feed and branch-circuit capacity for both general lighting and all general-purpose receptacles that may later be installed.

Of course, where the actual number of general-purpose receptacles is known, the general lighting load is taken at 3½ VA/ft^2 for branch-circuit and feeder capacity, and each strap or yoke containing a single, duplex, or triplex receptacle is taken as a load of 180 VA to get the total required branch-circuit capacity, with the demand factors of Table 220-11 or Table 220-13 applied to get the minimum required feeder capacity for receptacle loads.

Part (c) covers rules on providing branch-circuit capacity for loads other than general lighting and designates specific amounts of load that must be allowed for each outlet. This rule establishes the minimum loads that must be allowed in computing the minimum required branch-circuit capacity for general-use receptacles and "outlets not used for general illumination." Item (3) requires that the actual voltampere rating of a recessed lighting fixture be taken as the amount of load that must be included in branch-circuit capacity. This permits local and/or decorative recessed lighting fixtures to be taken at their actual load value rather than having them be taken as "other outlets," which would require a load allowance of "180 voltamperes per outlet"—even if each such fixture were lamped at, say, 25 W. Or, in the case where a recessed fixture contained a 300-W lamp, allowance of only 180 VA would be inadequate.

Receptacle Outlets

Section 220-3(c)(7) calls for each single or each multiple receptacle *on one strap* to be taken as a load of "not less than 180 voltamperes"—in commercial, institutional, and industrial occupancies. The rule requires that every general-purpose, single or duplex or triplex convenience receptacle outlet in nonresidential occupancies be taken as a load of 180 VA, and that amount of circuit capacity must be provided for each such outlet (Fig. 3-14). Code intent is that each individual device strap—whether it holds one, two, or three receptacles—is a load of 180 VA. This rules makes clear that branch-circuit and feeder capacity must be provided for receptacles in nonresidential occupancies in accordance with loads calculated at 180 VA per receptacle strap.

If a 15-A, 115-V circuit is used to supply *only* receptacle outlets, then the maximum number of general-purpose receptacle outlets that may be fed by that circuit is

$$15 \text{ A} \times 115 \text{ V} \div 180 \text{ VA or 9 receptacle outlets}$$

For a 20-A, 115-V circuit, the maximum number of general-purpose receptacle outlets is

$$20 \text{ A} \times 115 \text{ V} \div 180 \text{ VA or 12 receptacle outlets}$$

See Fig. 3-15.

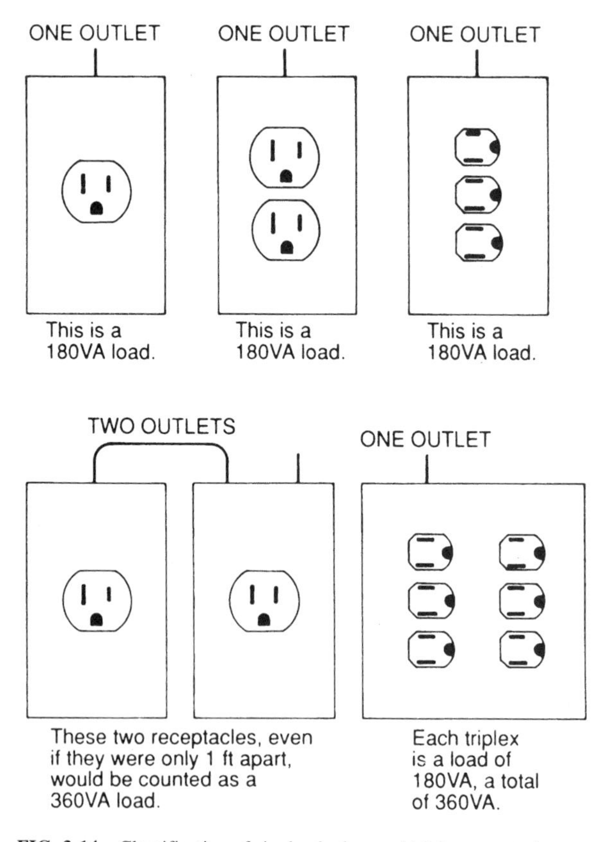

FIG. 3-14 Classification of single, duplex, and triplex receptacles.

15A, 115V CIRCUIT—Maximum of 9 receptacle outlets

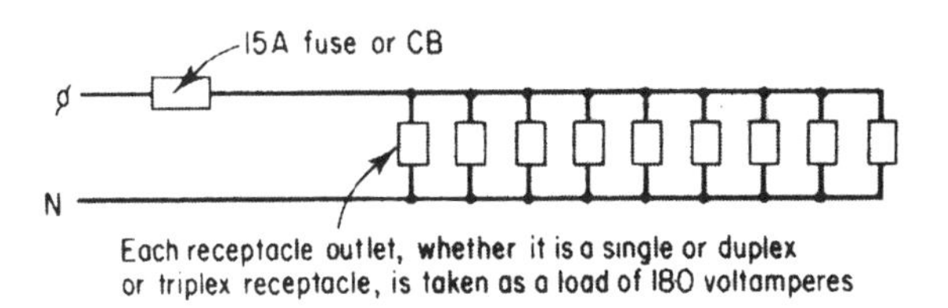

20A, 115V CIRCUIT—Maximum of 12 receptacle outlets

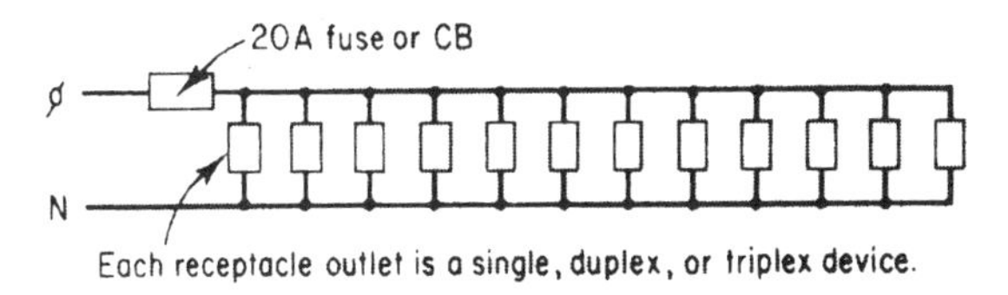

FIG. 3-15 Number of receptacles per circuit, nonresidential occupancy.

Note: In these calculations, the actual results work out to be 9.58 receptacles on a 15-A circuit and 12.77 on a 20-A circuit. Some inspectors round these values to the nearest integral numbers and permit 10 receptacle outlets on a 15-A circuit and 13 on a 20-A circuit.

Although manipulation of the Code gives the above-described data on maximum permitted number of receptacle outlets in commercial, industrial, institutional, and other nonresidential installations, there are no such limitations on the number of receptacle outlets on residential branch circuits. There are reasons for this approach.

In Sec. 210-50, the Code specifies where and when receptacle outlets are required on branch circuits. Note that there are no specific requirements for receptacle outlets in commercial, industrial, and institutional installations other than for store show windows in Sec. 210-62 and rooftop heating, refrigeration, and air-conditioning equipment in Sec. 210-63. There is the general rule that receptacles do have to be installed where flexible cords are used. In nonresidential buildings, if flexible cords are not used, there is no *requirement* for receptacle outlets. They have to be installed only where they are needed, and the number and spacing of receptacles are completely up to the designer. But because the Code takes the position that receptacles in nonresidential buildings only have to be installed where needed for connection of specific flexible cords and caps, it demands that where such receptacles are installed, each be taken as a load of 180 VA.

A different approach is used for receptacles in dwelling-type occupancies. The Code simply assumes that cord-connected appliances will always be used in all residential buildings and requires general-purpose receptacle outlets of the number and spacing indicated in Secs. 210-52 and 210-60. These rules cover one-family houses, apartments in multifamily houses, guest rooms in hotels and motels, living quarters in dormitories, etc. But because so many receptacle outlets are required in such occupancies and because use of plug-connected loads is intermittent and has great diversity of load values and operating cycles, the Code notes at the bottom of Table 220-3(b) that the loads connected to such receptacles are adequately served by the branch-circuit capacity required by Sec. 220-4, and no additional load calculations are required for such outlets.

In dwelling occupancies, it is necessary to first calculate the total "general lighting load" from Sec. 220-3(b) and Table 220-3(b) (at 3 VA/ft^2 or dwellings for 2 VA/ft^2 for hotels and motels, including apartment houses without provisions for cooking by tenants) and then provide the minimum required number and rating of 15-A and/or 20-A general-purpose branch circuits to handle that load, as covered in Sec. 220-4(a). As long as that basic circuit capacity is provided, any number of lighting outlets may be connected to any general-purpose branch circuit, up to the rating of the branch circuit if loads are known. The lighting outlets should be evenly distributed among all the circuits. Although residential lamp wattages cannot be anticipated, the Code method covers fairly heavy loading.

When the above Code rules on circuits and outlets for general lighting in dwelling units, guest rooms of hotels and motels, and similar occupancies are satisfied, general-purpose convenience receptacle outlets may be connected on circuits supplying lighting outlets; or receptacles only may be connected on one or more of the required branch circuits; or additional circuits (over and above those required by the Code) may be used to supply the receptacles. But no matter how general-purpose receptacle outlets are circuited, *any number* of general-purpose receptacle outlets may be connected on a residential branch circuit—with or without lighting outlets on the same circuit.

And when small-appliance branch circuits are provided in accordance with the requirements of Sec. 220-4(b), *any number* of small-appliance receptacle outlets may be connected on the 20-A small-appliance circuits—*but only* receptacle outlets may be connected to these circuits and only in the specified rooms.

Section 210-52(a) applies to spacing of receptacles connected on the 20-A small-appliance circuits, as well as spacing of general-purpose receptacle outlets. That section, therefore, establishes the *minimum* number of receptacles that must be installed for greater convenience of use.

Exception 1 to Sec. 220-3(c) requires branch-circuit capacity to be calculated for multioutlet assemblies (prewired surface metal raceway with plug outlets spaced along its length). Exception 1 says that each 1-ft length of such strip must be taken as a 180-VA load when the strip is used where a number of appliances are likely to be used simultaneously. For instance, in the case of industrial applications on assembly lines involving frequent, simultaneous use of plugged-in tools, the loading of 180 VA/ft must be used. (A loading of 180 VA for each 5-ft section may be used in commercial or institutional applications of multioutlet assemblies when use of plug-in tools or appliances is not heavy.) Figure 3-16 shows an example of such load calculation.

Exception 2 specifies that Table 210-19 be used to establish the load of household electric ranges. And Exception 3 permits branch-circuit capacity for the outlets required by Sec. 210-62 to be calculated as shown in Fig. 3-17—instead of using the load-per-outlet value from part (c).

The rule of Exception 4 states that no additional capacity is required for the outlets supplying telephone switching equipment. And as noted in Exception 5, in calculating the size of branch-circuit, feeder, or service conductors, a load of 5000 VA may be used for a household electric dryer when the actual dryer rating is not known. This is an exception to Sec. 220-3(c)(1), which specifies that the "ampere rating of appliance or load served" shall be taken as the branch-circuit load for an outlet for a specific appliance. And where more than one dryer is involved, the demand factors of Table 220-18 may be used.

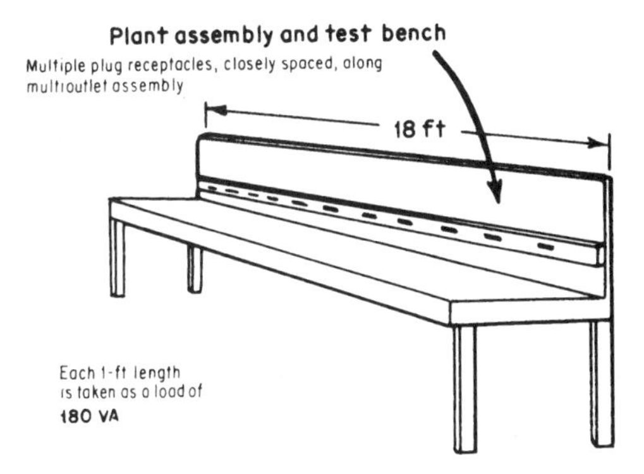

FIG. 3-16 Calculating required branch-circuit capacity for multioutlet assembly.

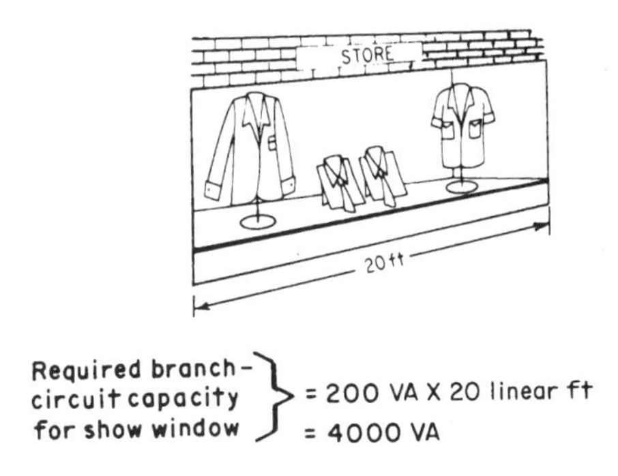

FIG. 3-17 Alternate method for calculating show window circuit capacity.

CALCULATING THE NUMBER OF BRANCH CIRCUITS

After following the rules of Sec. 220-3 to ensure that adequate branch-circuit capacity is available for the various types of load that might be connected to such circuits, Sec. 220-4(a) requires that the minimum required number of branch circuits be determined from the total computed load, as computed from Sec. 220-3, and from the load rating of the branch circuits used.

For example, a 15-A, 115-V, 2-wire branch circuit has a load rating of 15 A times 115 V, or 1725 VA. If the load is resistive, like incandescent lighting or electric heaters, that capacity is 1725 W. If the total load of lighting, say, computed from Sec. 220-3 were 3450 VA, then exactly two 15-A, 115-V, 2-wire branch circuits would be adequate to handle the load, provided that the load on the circuit was not a "continuous" load (one that operated steadily for 3 h or more). Because Sec. 220-3(a) requires that branch circuits supplying a continuous load be loaded to not more than 80 percent of the branch-circuit rating, if the above load of 3450 VA were a continuous load, it could not be supplied by *two* 15-A, 115-V circuits loaded to full capacity. A continuous load of 3450 VA could be fed by *three* 15-A, 115-V circuits—divided among the three circuits in such a way that no circuit has a load of over 15 A times 115 V times 80 percent, or 1380 VA. If 20-A, 115-V circuits are used, because each such circuit has a continuous load rating of 20 times 115 times 80 percent, or 1840 VA, the total load of 3450 VA could be divided between two 20-A, 115-V circuits. (A value of 120 V could be used instead of 115 V.)

Example. Given the required unit load of 3 VA/ft^2 for dwelling units [Table 220-3(b)], the Code minimum number of 20-A, 120-V branch circuits required to supply general lighting and general-purpose receptacles (not small-appliance receptacles in kitchen, dining room, etc.) in a 2200-ft^2 one-family house is three circuits. Each such 20-A circuit has a capacity of 2400 VA. The required total circuit capacity is 2200 times 3 VA/ft^2, or 6600 VA/ft^2. Then dividing 6600 by 2400 equals 2.75. Thus, at least three such circuits are needed.

Example. In Sec. 220-3(b), the NEC requires a minimum unit load of 3 VA/ft^2 for general lighting in a school, as shown in Table 220-3(b). For the school in this example, *minimum capacity for general lighting* would be

$$1500 \text{ ft}^2 \times 3 \text{ VA/ft}^2 \qquad \text{or} \qquad 4500 \text{ VA}$$

By using 115-V circuits, when the total load capacity of branch circuits for general lighting is known, it is a simple matter to determine how many lighting circuits are needed. By dividing the total load by 115 V (using 120 V would yield a lower current), the total current capacity of circuits is determined:

$$\frac{4500 \text{ VA}}{115 \text{ V}} = 39.1 \text{ A}$$

But because the circuits will be supplying continuous lighting loads (over 3 h), it is necessary to multiply that value by 1.25 in order to keep the load on any circuit to not more than 80 percent of the circuit rating. Then, using either 15- or 20-A, 2-wire, 115-V circuits (and dropping the fraction of an ampere) gives

$$\frac{1.25 \times 39 \text{ A}}{15 \text{ A}} = 3.25$$

which means four 15-A circuits, or

$$\frac{1.25 \times 39 \text{ A}}{20 \text{ A}} = 2.43$$

which means three 20-A circuits. And then each circuit must be loaded without exceeding the 80 percent maximum on any circuit.

Part (b) of Sec. 220-4 requires that two or more 20-A branch circuits be provided to supply all the receptacle outlets required by Sec. 210-52(b) in the kitchen, pantry, dining room, breakfast room, and any similar area of any dwelling unit—one-family houses, apartments, and motel and hotel suites with cooking facilities or serving pantries. That means two 2-wire, 20-A, 120-V circuits are equivalent to the 3-wire circuit and could be used. One 3-wire, 20-A, 240/120- or 208/120-V circuit could be provided to serve only receptacles for the small-appliance load in the kitchen, pantry, dining room, and breakfast room of any dwelling, except that such application will cause nuisance tripping of GFCI receptacles.

Part (c) of Sec. 220-4 requires that at least one 20-A branch circuit be provided for the one or more laundry receptacles installed, as required by Sec. 210-52(f), at the laundry location in a dwelling unit. Further, the last sentence of part (c) prohibits use of the laundry circuit for supplying outlets that are not for laundry equipment. And because laundry outlets are required by Sec. 210-50(c) to be within 6 ft of the intended location of the appliance, it would seem that any receptacle outlet more than 6 ft from laundry equipment could not be connected to the required 20-A laundry circuit (Fig. 3-18).

Part (d) of Sec. 220-4 makes clear that a feeder to a branch circuit panelboard and the main bus bars in the panelboard must have a minimum ampacity to serve the *calculated* total load of lighting, appliances, motors, and other loads supplied. And the amount of feeder and panel ampacity required for the general lighting load must not be less than the ampere value determined from the circuit voltage and the total voltamperes resulting from the minimum unit load from Table 220-3(b) (voltamperes per square foot) times the area of the occupancy supplied by the feeder—even if the actual connected load is less than the calculated load determined on the voltamperes-per-square-foot basis. (Of course, if the connected load is greater than that calculated on the voltamperes-per-square-foot basis, the greater value of load must be used in determining the number of branch circuits, the panelboard capacity, and the feeder capacity.)

It should be carefully noted that the first sentence of Sec. 220-4(d) states, "Where the load is computed on a voltamperes-per-square-foot (0.93 sq m) basis, the *wiring system* up to and including the branch-circuit panelboards(s) shall be provided to serve not less than the calculated load." Use of the phrase "wiring system up to and including" requires that a feeder have capacity for the total minimum branch-circuit load determined from square-foot area times the minimum unit load [voltamperes per square foot from Table 220-3(b)]. And the phrase clearly requires that amount of capacity to be allowed in every part of the distribution system supplying the load. The required capacity would, for instance, be required in a subfeeder to the panel, in the main feeder from which the subfeeder is tapped, and in the service conductors supplying the whole system.

Actually, reference to "wiring system" in the wording of Sec. 220-4(d) presents a requirement that goes beyond the heading "Branch Circuits Required" of Sec. 220-4 and, in fact, constitutes a requirement on *feeder* capacity that supplements the rule of the second sentence of Sec. 220-10(a). This requires a feeder to be sized to have enough capacity for "the computed load"—as determined by part A of this article (which means computed in accordance with Sec. 220-3).

A second part of Sec. 220-4(d) affects the required minimum number of branch circuits. Although the feeder and panelboard must have a minimum ampacity for the *calculated* load, it is only necessary to install the number of branch-circuit overcurrent devices and circuits required to handle the actual connected load in those cases where it is less than the calculated load. The last sentence of Sec. 220-4(d) is clearly an exception to the basic rule of the first sentence of Sec. 220-4(a), which says that "The minimum number of branch circuits *shall* be determined from the *total computed* load..."

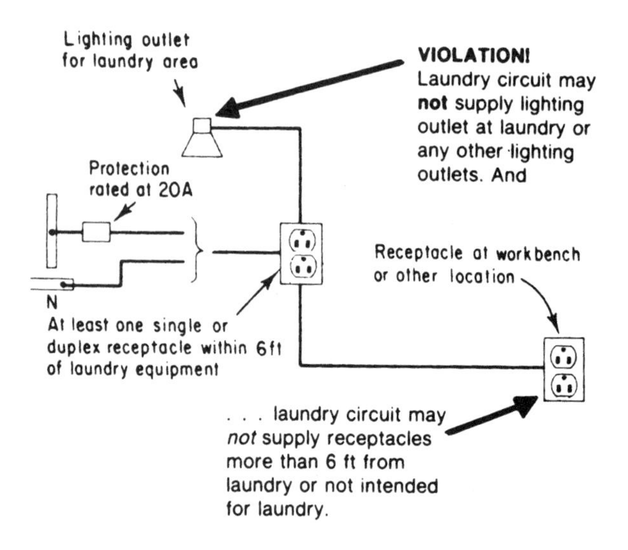

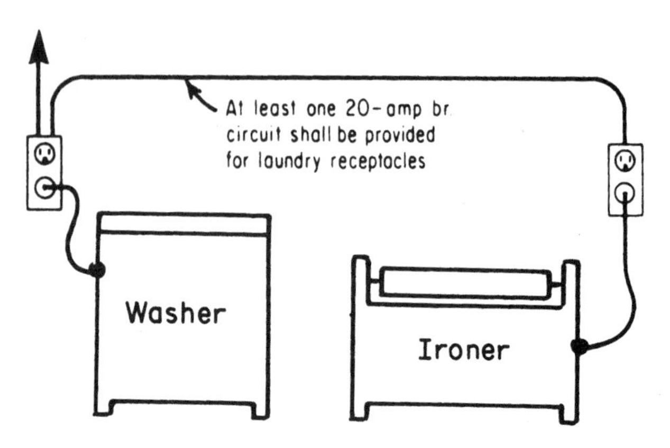

FIG. 3-18 *No* "other outlets" are permitted on a 20-A circuit required for laundry receptacle(s).

Instead of having to supply *that minimum* number of branch circuits, it is necessary to have only the number of branch circuits required for the actual total "connected load."

Example. For an office area of 200 × 200 ft, a 3-phase, 4-wire, 460/265-V feeder and branch-circuit panelboard must be selected to supply 277-V HID lighting that will operate continuously (3 h or more). The actual continuous connected load of all the lighting fixtures is 82 kVA. What is the minimum size of feeder conductors and panelboard rating that must be used to satisfy Sec. 220-4?

Solution

$$200 \text{ ft} \times 200 \text{ ft} = 40{,}000 \text{ ft}^2$$

$$40{,}000 \text{ ft}^2 \text{ @ minimum of } 3.5 \text{ VA/ft}^2 = 140{,}000 \text{ VA}$$

The minimum computed load for the feeder for the lighting is

$$140{,}000 \text{ VA} \div [(480)(1.732)] = 168 \text{ A per phase}$$

The actual connected lighting load for the area, calculated from the lighting design, is

$$92{,}000 \text{ VA} \div [(480)(1.732)] = 111 \text{ A per phase}$$

Sizing of the feeder and panelboard must be based on 168 A, *not* 111 A, to satisfy Sec. 220-4(d).

The next step is to correlate the rules of Sec. 220-4(a) and (d) with those of Sec. 220-10. Section 220-10(a) requires a feeder to be sized for the "computed load" as determined by part (A) [Sec. 220-3(b)]. The feeder to the continuous calculated load of 168 A must have an ampacity of at least 125 percent of that value before derating, if required; and the feeder protective device must be sized at 125 percent of the continuous load of 168 A, when using a CB or fused switch that is not UL-listed for continuous operation at 100 percent of rating, as required by Sec. 220-10(b).

$$168 \times 1.25 = 210 \text{ A} \qquad \text{[Sec. 220-10(b)]}$$

1. Assuming use of a non-100 percent rated protective device, the overcurrent device must be rated not less than 1.25 × 168 A, or 210 A, which calls for a standard 225-A circuit breaker or fuses (the standard rating above 210 A).
2. The conductors must have a 75°C ampacity (from Table 310-16) that is at least equal to the 210-A value.

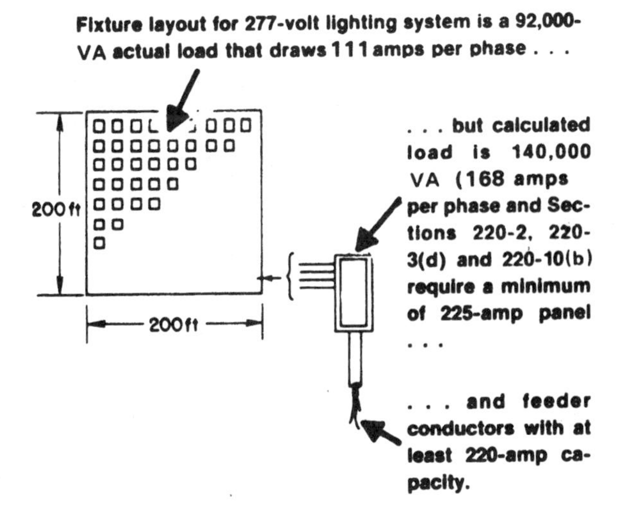

3. Using Table 310-16, the smallest size of feeder conductor that has a 75°C table ampacity that is equal to or greater than 210 A is a No. 4/0 copper with its 230-A table value.

4. If derating had been required, the derating could be applied against the 90°C table value for No. 4/0 copper, if it is insulated with one of the 90°C insulations. Then the conductors must simply be properly protected by the 225-A fuse or CB. In this case, the 230-A 75°C insulated No. 4/0s are properly protected.

Thus, all requirements of Sec. 220-10(b) and UL are satisfied.

Section 384-13 requires the panelboard here to have a rating not less than the minimum required capacity of the feeder conductors, which in this case means the panel must have a bus bar rating not less than 210 A. A 225-A panelboard (that is, the next standard rating of panelboard above the minimum calculated value of load current—210 A) is therefore required.

The number of branch-circuit protective devices required in the panel (the number of branch circuits) is based on the size of branch circuits used and their capacity related to connected load. If, say, all circuits are to be 20-A, 277-V phase-to-neutral, each pole may be loaded no more than 16 A because Sec. 220-3(a) requires the load to be limited to 80 percent of the 20-A protection rating. With 111 A of connected load per phase, a single-circuit load of 16 A calls for a minimum of 111 ÷ 16, or 8 poles per phase leg. Thus a 225-A panelboard with 24 breaker poles would satisfy the rule of Sec. 220-4(d).

CHAPTER 4
FEEDERS FOR POWER AND LIGHT

CALCULATING FEEDER LOAD

The key to accurate determination of required feeder conductor capacity in amperes is effective calculation of the total load to be supplied by the feeder. Feeders and subfeeders are sized to provide sufficient power to the circuits they supply. For the given circuit voltage, they must be capable of carrying the amount of current required by the load, plus any current which may be required in the future. The size of a feeder depends upon known load, future load, and voltage drop.

The minimum load capacity which must be provided in any feeder or subfeeder can be determined by considering NEC requirements on feeder load. As presented in Sec. 220-10, these rules establish the minimum load capacity to be provided for all types of loads.

Part (a) of Sec. 220-10 requires feeder conductors to have ampacity at least equal to the sum of loads on the feeder, as determined from Sec. 220-3. Then part (b) rules on the rating of any feeder protective device and conductor sizing where supplying continuous or continuous and noncontinuous loads.

If an overcurrent protective device for feeder conductors is not UL-listed for continuous operation at 100 percent of its rating, the load on the device must not exceed the noncontinuous load plus 125 percent of the continuous load. The first paragraph of part (b) applies to feeder overcurrent devices—circuit breakers and fuses in switch assemblies—and requires that the rating of any such protective device generally never be less than the amount of noncontinuous load of the circuit (that amount of current that will not be flowing for 3 h or longer) plus 125 percent of the amount of load current that will be continuous (flowing steadily for 3 h or longer) (Fig. 4-1).

For any given load to be supplied by a feeder, after the minimum rating of the overcurrent device is determined from the above calculation (noncontinuous plus 125 percent of continuous), then a suitable size of feeder conductor must be selected. For each ungrounded leg of the feeder (the so-called phase legs of the circuit), the conductor ampacity must be at least equal to the amount of noncontinuous current plus 125 percent of the amount of continuous current, in the 75°C column of the NEC tables of ampacity (Tables 310-16 through 310-19) and before derating (the accompanying notes).

The rule of Sec. 220-10(b), for feeders, is the same as that given in Sec. 210-22(c) and Sec. 220-3(a). Feeder and service conductors must have a Table 310-16 ampacity in the 75°C column that is equal to or greater than the noncontinuous load plus 125 percent of the continuous load. This is the same requirement for the OC device protecting the feeder conductor. And the feeder conductors must satisfy all requirements related to

THE RULE

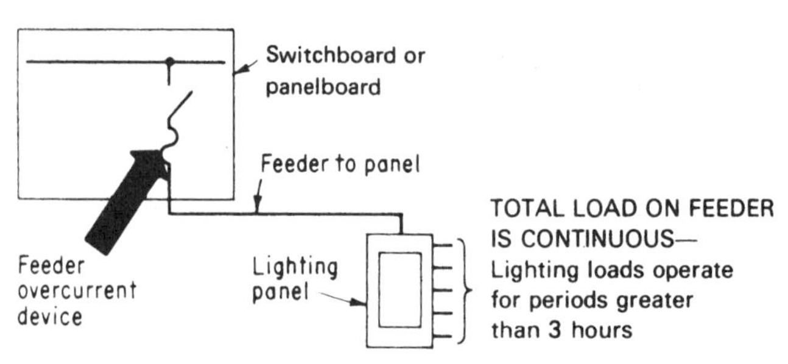

FEEDER OVERCURRENT DEVICE **must be rated not less than 125% of the continuous load *and* the feeder conductors must be sized so they have a 75° C table ampacity before derating, so that they are properly protected by the rating of the feeder CB or fuses, as required by Sec. 240-3 after derating. Another way of saying that is "the continuous load must not exceed 80% of the rating of the protection."**

EXAMPLE: For this feeder, with conductors rated at 420A in **Table 340-16**, the maximum continuous load permitted for a conventional fused switch is 80% of the 400-A fuse rating [400X0.8 = 320A].

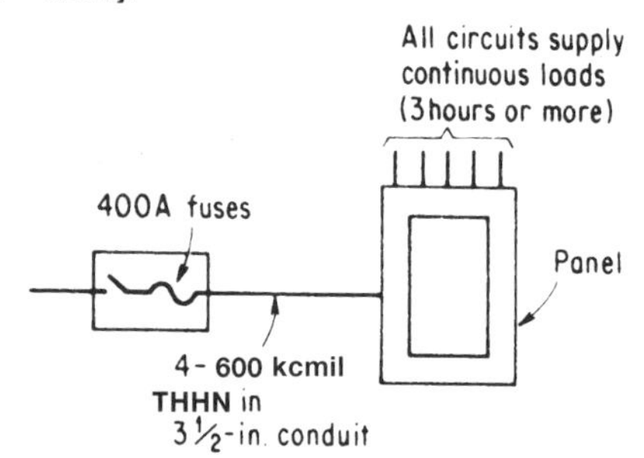

Rating of fuses must be at least equal to 125% times the continuous load and the 400-A rating is proper protection for conductors with an ampacity of 380 A, **after derating.**

FIG. 4-1 Feeders must generally be loaded to no more than 80 percent for a continuous load.

derating—number of conductors and/or ambient temperature. Then they need to be properly protected by the OC device.

Note that if this were electric-discharge lighting, the 80 percent derating of the 600-kcmil THHN conductors would give them a final ampacity of 380 A (475 × 0.8). If THW or other 75°C insulation were used, the required derating would give them an ampacity of 336 A [420 A (from the 75°C column in Table 310-16) × 0.8 = 336 A]. This does *not* satisfy the rule of Sec. 240-3(b) because the next standard rating of protective device above 336 A is 350 A. The use of the 400-A device would be a violation with 75°C insulated 600-kcmil conductors supplying this discharge lighting load.

Note that the conductor size increase described above applies only to the ungrounded or phase conductors because they are the ones that must be properly protected by the rating of the protective device. A neutral or grounded conductor of a feeder does not have to be increased; its size must simply have ampacity sufficient for the neutral load as determined from Sec. 220-22.

The exception for Sec. 220-10(b) notes that a circuit breaker or fused switch that is UL-listed for continuous operation at 100 percent of its rating may be loaded right up to a current equal to the device rating. Feeder underground conductors must be selected to have ampacity equal to the noncontinuous load plus the continuous load. The neutral conductor is sized in accordance with Sec. 220-22, which permits reduction of neutral size for feeders loaded over 200 A that do not supply electric-discharge lighting, data-processing equipment, or similar loads that generate high levels of harmonic currents in the neutral.

Fuses for Feeder Protection

The rating of a fuse is taken as 100 percent of rated nameplate current when enclosed by a switch or panel housing. But, because of the heat generated by many fuses, the maximum continuous load permitted on a fused switch is restricted by a number of NEMA, UL, and NEC rules to 80 percent of the rating of the fuses. Limitation of circuit-load current to *no more than* 80 percent of the current rating of fuses in equipment is designed to protect the switch or other piece of equipment from the heat produced in the fuse element—and to protect attached circuit wires from excessive heating close to the terminals. The fuse can actually carry 100 percent of its current rating continuously without damage to itself, but its heat is conducted into the adjacent wiring and switch components.

NEMA standards require that a fused, enclosed switch be marked, as part of the electrical rating, "Continuous Load Current Not to Exceed 80 Percent of the Rating of Fuses Employed in Other Than Motor Circuits." That derating compensates for the extra heat produced by continuous operation. Motor circuits are excluded from that rule, but a motor circuit is required by the NEC to have conductors rated at least 125 percent of the motor full-load current, which in effect limits the load current to 80 percent of the conductor ampacity and limits the load on the fuses rated to protect those conductors. But the UL *Electrical Construction Materials Directory* does recognize fused bolted-pressure switches and high-pressure butt-contact switches for use at 100 percent of their rating on circuits with available fault currents of 100,000, 150,000, or 200,000 rms symmetrical A—as marked (Fig. 4-2). (See "Fused Power Circuit Devices" in that UL publication.)

Manual and electrically operated switches designed to be used with class L current-limiting fuses rated 601 to 4000 A, 600 V ac are listed by UL as "fused power circuit devices." This category covers bolted-pressure-contact switches and high-pressure, butt-type-contact switches suitable for use as feeder devices or service switches if marked "Suitable for Use as Service Equipment." Such devices "have been investigated for use at *100 percent of their rating* on circuits having available fault currents of 100,000, 150,000, or 200,000 rms symmetrical amperes" as marked.

CB for Feeder Protection

The nominal or theoretical continuous-current rating of a CB generally is taken to be the same as its trip setting—the value of current at which the breaker will open, either instantaneously or after some intentional time delay. But as described above for fuses, the real continuous-current rating of a CB—the value of current that it can safely and

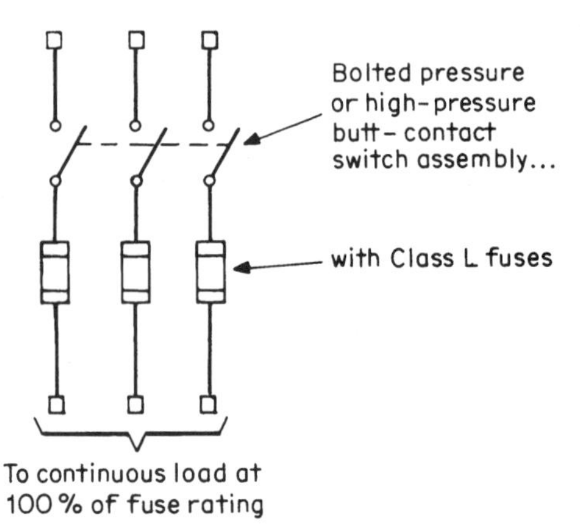

FIG. 4-2 Some fused switches may be used at 100 percent rating for continuous load.

properly carry for periods of 3 h or more—frequently is reduced to 80 percent of the nameplate value by codes and standards' rules.

The UL *Electrical Construction Materials Directory* contains a clear, simple rule in the instructions under "Circuit Breakers, Molded-Case." It says:

> Unless otherwise marked, circuit breakers should not be loaded to exceed 80 percent of their current rating, where in normal operations the load will continue for three or more hours.

A load that continuous for 3 h or more is a *continuous* load. If a breaker is marked for *continuous* operation, it may be loaded to 100 percent of its rating and operate continuously.

There are some CBs available for continuous operation at 100 percent of their current rating, but they must be used in the mounting and enclosure arrangements established by UL for 100 percent rating. Molded-case CBs of the 100 percent continuous type are made in ratings from 250 A at 250 V up. Information on use of 100 percent rated breakers is given on the nameplates.

Figure 4-3 shows two examples of CB nameplate data for two types of UL-listed 2000-A, molded-case CBs that are specifically tested and listed for continuous operation at 100 percent of their 2000-A rating—*but* only under the conditions described on the nameplate. These two typical nameplates clearly indicate that ventilation may or may not be required. Because most switchboards have fairly large interior volumes, the "minimum enclosure" dimensions shown on these nameplates (45 by 38 by 20 in) usually are readily achieved. *But* special UL tests must be performed if these dimensions are *not* met. Where bus bar extensions and lugs are connected to the CB within the switchboard, the caution about copper conductors does not apply, and aluminum conductors may be used.

If the ventilation pattern of a switchboard does not meet the ventilation pattern and the required enclosure size specified on the nameplate, the CB must be applied at 80 percent rating. Switchboard manufacturers have UL tests conducted with a CB installed in a specific enclosure, and the enclosure may receive a listing for 100 percent rated operation even though the ventilation pattern or overall enclosure size may not meet the

EXAMPLE 1

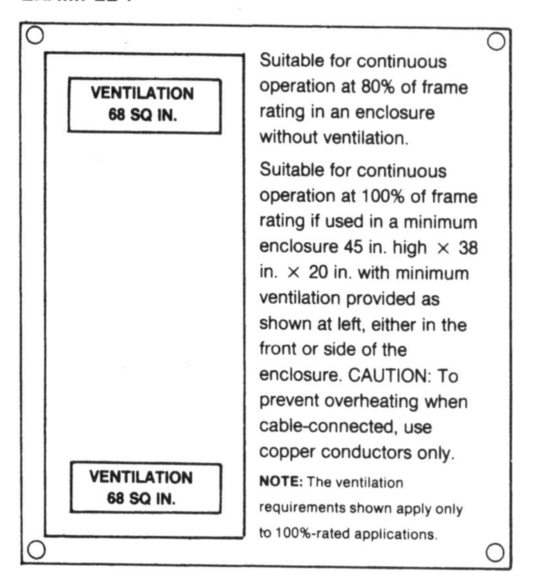

EXAMPLE 2

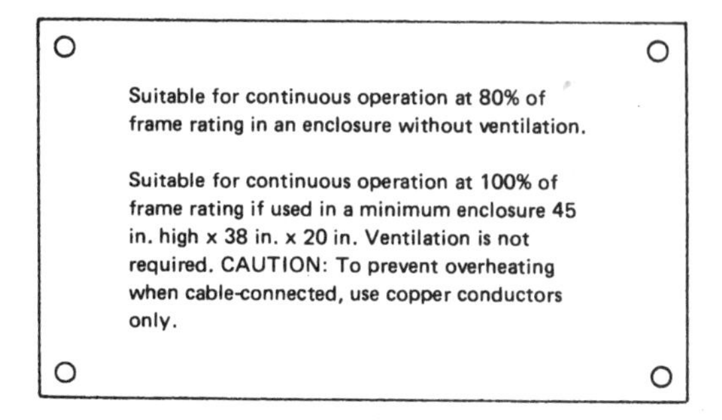

FIG. 4-3 Nameplates from CBs rated for 100 percent continuous loading.

specifications. In cases where the breaker nameplate specifications are not met by the switchboard, the customer has to request a letter from the manufacturer certifying that a 100 percent rated listing has been received. Otherwise, the breaker must be applied at 80 percent.

To realize savings with devices listed by UL at 100 percent of their continuous-current rating, use must be made of a CB manufacturer's data sheet to determine the types and ampere ratings of breakers available that are 100 percent rated, along with the frame sizes, approved enclosure sizes, and ventilation patterns required by UL, if any.

It is essential to check the instructions given in the UL listing to determine *if* and under what conditions a CB (or a fuse in a switch) is rated for continuous operation at 100 percent of its current rating.

FEEDERS SUPPLYING GENERAL LIGHTING

For general illumination, a feeder must have capacity to carry the total load of lighting branch circuits determined as part of the lighting design and not less than a minimum branch-circuit load determined on a voltamperes-per-square-foot basis from the table in Sec. 220-3(b).

Demand factor permits sizing of a feeder according to the amount of load which operates simultaneously.

Demand factor is the ratio of the maximum amount of load that will be operating at any one time on a feeder to the total connected load on the feeder under consideration. This factor is frequently less than 1. The sum of the connected loads supplied by a feeder is multiplied by the demand factor to determine the load which the feeder must be sized to serve. This load is termed the *maximum demand* of the feeder:

$$\text{Maximum demand load} = \text{connected load} \times \text{demand factor}$$

Tables of demand and diversity factors have been developed from experience with various types of load concentrations and various layouts of feeders and subfeeders supplying such loads. Table 220-11 of the NEC presents common demand factors for feeders to general lighting loads in various types of buildings (Fig. 4-4).

The demand factors given in Table 220-11 may be applied to the total branch-circuit load to get required feeder capacity for lighting (but must not be used in calculating branch-circuit capacity). Note that a feeder may have capacity of less than 100 percent of the total branch-circuit load for only the types of buildings designated in Table 220-11, that is, for dwelling units, hospitals, hotels, motels, and storage warehouses. In all other types of occupancies, it is assumed that *all* general lighting will be operating at the same time, and each feeder in those occupancies must have capacity (ampacity) for 100 percent of the voltamperes of branch-circuit load of general lighting that the feeder supplies.

Example. If a warehouse feeder fed a total branch-circuit load of 20,000 VA of general lighting, the minimum capacity in that feeder to supply that load must be equal to

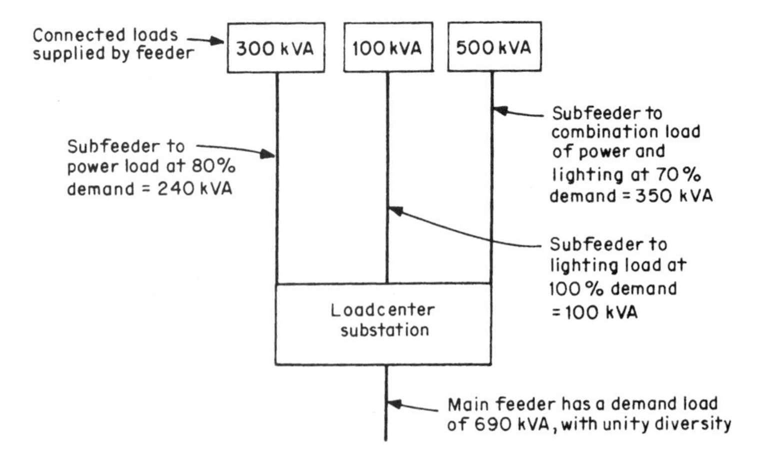

FIG. 4-4 How demand factors are applied to connected loads.

12,500 VA plus 50 percent times (20,000−12,500) VA. That works out to be 12,500 plus 0.5 × 7500 = 16,250 VA.

But the note to Table 220-11 warns against using any value less than 100 percent of branch-circuit load for sizing any feeder that supplies load that will be energized at the same time.

SHOW-WINDOW LIGHTING

If show-window lighting is supplied by a feeder, capacity must be included in the feeder to handle 200 VA per linear foot of show-window length. Because that is the same loading as given in Sec. 220-3(c), Exception 3, it works out to be a 100 percent demand for the entire branch-circuit load of show-window lighting.

RECEPTACLE LOADS—NONDWELLING UNITS

This rule permits two possible approaches to determining the required feeder ampacity to supply receptacle loads in "other than dwelling units," where a load of 180 VA of feeder capacity must be provided for all general-purpose 15- and 20-A receptacle outlets. (In dwelling units and in guest rooms of hotels and motels, no feeder capacity is required for 15- or 20-A general-purpose receptacle outlets. Such load is considered sufficiently covered by the load capacity provided for general lighting.) But in other than dwelling units, where a load of 180 VA of feeder capacity must be provided for all general-purpose 15- and 20-A receptacle outlets, a *demand factor* may be applied to the total calculated receptacle load as follows. Wording of this rule makes clear that *either* Table 220-11 or Table 220-13 may be used to apply demand factors to the total load of 180-VA receptacle loads when calculating required ampacity of a feeder supplying receptacle loads connected on branch circuits.

In other than dwelling units, the branch-circuit load for receptacle outlets, for which 180 VA was allowed per outlet, may be added to the general lighting load and may be reduced by the demand factors in Table 220-11. That is the basic rule of Sec. 220-13 and, in effect, requires any feeder to have capacity for the total number of receptacles it feeds and requires that capacity to be equal to 180 VA (per single or multiple receptacles) times the total number of receptacles (straps)—with a reduction from 100 percent of that value permitted only for the occupancies listed in Table 220-11.

Because the demand factor of Table 220-11 is shown as 100 percent for "all other" types of occupancies, the basic rule of Sec. 220-13 as it appeared prior to the 1978 NEC required a feeder to have ampacity for a load equal to 180 VA times the number of general-purpose receptacle outlets that the feeder supplied. That is no longer required. Recognizing that there is great diversity in use of receptacles in office buildings, stores, schools, and all the other occupancies that come under "all others" in Table 220-11, Sec. 220-13 contains a table to permit reduction of feeder capacity for receptacle loads on feeders. Those demand factors apply to *any* "nondwelling" occupancy.

The amount of feeder capacity for a typical case in which a feeder, say, supplies panelboards that serve a total of 500 receptacles is shown in Fig. 4-5.

Although the calculation of Fig. 4-5 cannot always be taken as realistically related to usage of receptacles, it is realistic relief from the 100 percent demand factor, which presumed that all receptacles were supplying 180-VA loads simultaneously.

Take the total number of general-purpose receptacle outlets fed by a given feeder. . .

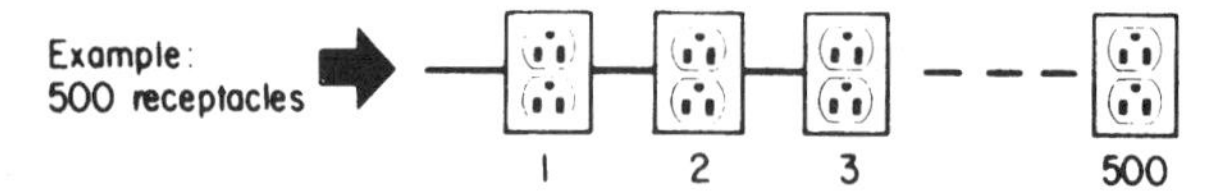

. . . multiply the total by 180 voltamperes [required load of Section 220-2(c) (5) for each receptacle]. . .

500 × 180 VA = 90,000 VA

Then apply the demand factors from Table 220-13:

First 10 kVA or less @ 100% demand = 10,000 VA
Remainder over 10 kVA @ 50% demand
= (90,000 − 10,000) × 50%
= 80,000 × 0.5 = 40,000 VA
Minimum demand-load total = 50,000 VA

Therefore, the feeder must have a capacity of 50 kVA for the total receptacle load. Required minimum ampacity for that load is then determined from the voltage and phase-makeup (single- or 3-phase) of the feeder.

FIG. 4-5 Table 220-13 permits demand factor in calculating feeder demand load for general-purpose receptacles.

MOTORS

Any feeder that supplies a motor load or a combination load (motors plus lighting and/or other electrical loads) must satisfy the indicated NEC sections of Art. 430. Feeder capacity for motor loads is usually taken at 125 percent of the full-load current rating of the largest motor supplied, *plus* the sum of the full-load currents of the other motors supplied. (For further information on motor circuits, see Chap. 6.)

FIXED ELECTRIC SPACE HEATING

Capacity required in a feeder to supply fixed electric space heating equipment is determined on the basis of a load equal to the total connected load of heaters on all branch circuits served from the feeder. Under conditions of intermittent operation or where not all units can operate at the same time, permission may be granted for use of less than a 100 percent demand factor in sizing the feeder. Sections 220-30, 220-31, and 220-32 permit alternate calculations of electric heat load for feeders or service-entrance conductors (which constitute a service feeder) in dwelling units. But reduction of the feeder capacity to less than 100 percent of connected load must be authorized by the local electrical inspector. Feeder load current for heating must not be less than the rating of the largest heating branch circuit fed.

SMALL-APPLIANCE AND LAUNDRY LOADS—DWELLING UNIT

For a feeder or service conductors in a single-family dwelling, in an individual apartment of a multifamily dwelling with provisions for cooking by tenants, or in a hotel or motel suite with cooking facilities or a serving pantry, at least 1500 VA of load must be provided for each 2-wire, 20-A small-appliance circuit (to handle the small-appliance load in kitchen, pantry, and dining areas). The total small-appliance load determined in this way may be added to the general lighting load, and the resulting total load may be reduced by the demand factors given in Table 220-11.

A feeder load of at least 1500 VA must be added for each 2-wire, 20-A laundry circuit installed, as required by Sec. 220-4(c). And that load may also be added to the general lighting load and subjected to the demand factors in Table 220-11.

APPLIANCE LOAD—DWELLING UNIT(S)

For fixed appliances (fastened in place) other than ranges, clothes dryers, air-conditioning equipment, and space-heating equipment, feeder capacity in dwelling occupancies must be provided for the sum of these loads; but if there are at least four such fixed appliances, the total load of four or more such appliances may be reduced by a demand factor of 75 percent (NEC Sec. 220-17). Wording of this rule makes clear that a "fixed appliance" is one that is "fastened in place."

As an example of application of this Code provision, consider the following calculation of feeder capacity for fixed appliances in a single-family house. The calculation is made to determine how much capacity must be provided in the service-entrance conductors (the service feeder):

Water heater	2500 W	230 V =	11.0 A
Kitchen disposal	½ hp	115 V = 6.5 A + 25% =	8.1 A
Furnace motor	¼ hp	115 V =	4.6 A
Attic fan	¼ hp	115 V = 4.6 A	0.0 A
Water pump	½ hp	230 V =	3.7 A
Load in amperes on each ungrounded leg of feeder =			27.4 A

To comply with Sec. 430-24, 25 percent is added to the full-load current of the ½-hp, 115-V motor because it is the highest-rated motor in the group. Since it is assumed that the load on the 115/230-V feeder will be balanced and each of the ¼-hp motors will be connected to different ungrounded conductors, only one is counted in the above calculation. Except for the 115-V motors, all the other appliance loads are connected to both ungrounded conductors and are automatically balanced. Since there are four or more fixed appliances in addition to a range, clothes dryer, etc., a demand factor of 75 percent may be applied to the total load of these appliances. Seventy-five percent of 27.4 is 20.5 A, which is the current to be added to that computed for the lighting and other loads, to determine the total current to be carried by the ungrounded (outside) service-entrance conductors.

The above demand factor may be applied to similar loads in two-family or multifamily dwellings.

ELECTRIC CLOTHES DRYERS—DWELLING UNIT(S)

This rule prescribes a *minimum* demand of 5 kVa for 120/240-V electric clothes dryers in determining branch-circuit and feeder sizes. Note that this rule applies only to "household" electric clothes dryers, and not to commercial applications. This rule is helpful because the ratings of electric clothes dryers are not usually known in the planning stages when feeder calculations must be determined (Fig. 4-6).

When sizing a feeder for one or more electric clothes dryers, a load of 5000 VA or the nameplate rating, whichever is larger, shall be included for each dryer—subject to the demand factors of Table 220-18 when the feeder supplies a number of clothes dryers, as in an apartment house.

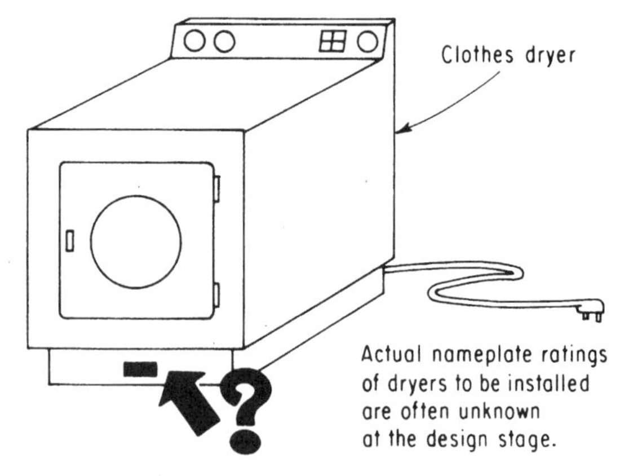

FIG. 4-6 Feeder load of 5 kVA per dryer must be provided if actual load is not known.

ELECTRIC RANGES AND OTHER COOKING APPLIANCES—DWELLING UNIT(S)

Feeder capacity must be allowed for household electric cooking appliances rated over $1\frac{3}{4}$ kW, in accordance with Table 220-19 of the Code. Feeder demand loads for a number of cooking appliances on a feeder may be obtained from Table 220-19.

Note 4 to Table 220-19 permits sizing of a branch circuit to supply a single electric range, a wall-mounted oven, or a counter-mounted cooking unit in accordance with that table. That table is also used in sizing a feeder (or service conductors) that supplies one or more electric ranges or cooking units. Note that Sec. 220-19 and Table 220-19 apply only to such cooking appliances in a dwelling unit and do not cover commercial or institutional applications (except for Note 5, which applies to cooking appliances over $1\frac{3}{4}$ kW and used in instructional programs).

Figure 4-7 shows a typical NEC calculation of the minimum demand load to be used in sizing the branch circuit to the range. The same value of demand load is used in sizing a feeder (or service conductors) from which the range circuit is fed. Calculation is as follows:

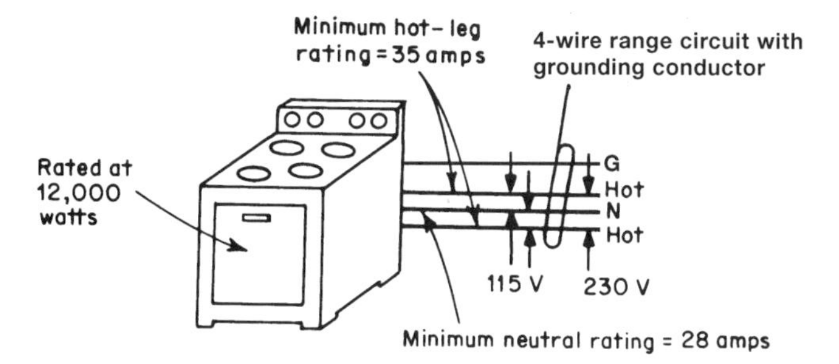

FIG. 4-7 Minimum ampere rating of branch-circuit conductors for a 12-kW range.

A branch circuit for the 12-kW range is selected in accordance with Note 4 of Table 220-19, which says that the branch-circuit load for a range may be selected from the table itself. Under the heading "Number of Appliances," read across from 1. The maximum demand to be used in sizing the range circuit for a 12-kW range is shown under the heading "Maximum Demand" to be not less than 8 kW. The minimum rating of the range-circuit ungrounded conductors will be

$$\frac{8000\text{ W}}{230\text{ V}} = 34.78 \text{ or } 35\text{ A}$$

NEC Table 310-16 shows that the minimum size of copper conductors that may be used is No. 8 (TW—40 A, THW—45 A, XHHW or THHN—50 A). Number 8 is also designated in Sec. 210-19(b) as the minimum size of conductor for any range rated $8\frac{3}{4}$ kW or more.

The overload protection for this circuit of No. 8 TW conductors would be *40-A fuses or a 40-A circuit breaker.* If THW, THHN, or XHHW wires are used for the circuit, they must be taken as having an ampacity of not more than 40 A and protected at that value. That requirement follows from the UL rule that conductors up to No. 1 AWG size must be used at the 60°C ampacity for the size of conductor, regardless of the actual temperature rating of the insulation, which may be 75 or 90°C. The ampacity used must be that of TW wire of the given size.

Although the two hot legs of the 230/115-V, 3-wire circuit must be not smaller than No. 8, Exception 2 to Sec. 210-19(b) permits the neutral conductor to be smaller, but it specifies that it must have an ampacity not less than 70 percent of the rating of the branch-circuit CB or fuse and may never be smaller than No. 10.

For the range circuit in this example, the neutral may be rated

$$70\% \times 40\text{ A (rating of branch-circuit protection)} = 28\text{ A}$$

This calls for a No. 10 neutral.

Figure 4-8 shows a more involved calculation for a range rated over 12 kW. Figure 4-9 shows two units that total 12 kW and are taken at a demand load of 8 kW, as if they were a single range. Figure 4-10 shows another calculation for separate cooking units on one circuit. And a feeder that would be used to supply any of the cooking installations shown in Figs. 4-7 through 4-10 would have to include capacity equal to the demand load used in sizing the branch circuit.

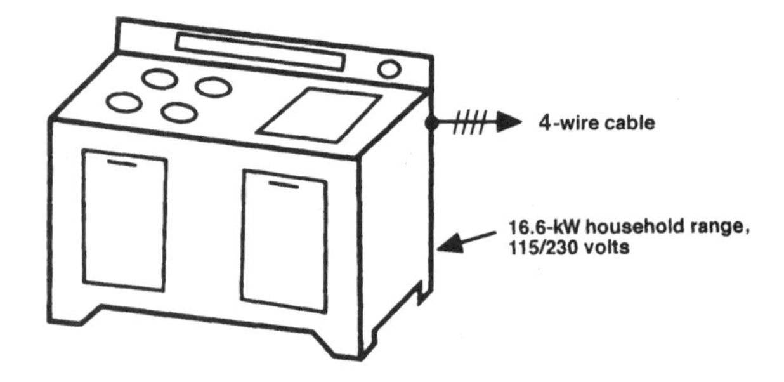

Refer to *NE Code* Table 220-19.

1. Column A applies to ranges rated not over 12 kW, but this range is rated 16.6 kW.
2. Note 1, below the Table, tells how to use the Table for ranges over 12 kW and up to 27 kW. For such ranges, the maximum demand in Column A must be increased by 5% for each additional kW of rating (or major fraction) above 12 kW.
3. This 16.6-kW range exceeds 12 kW by 4.6 kW.
4. 5% of the demand in Column A for a single range is 400 watts (8000 watts x 0.05).
5. The maximum demand for this 16.6-kW range must be increased above 8 kW by 2000 watts:

   400 watts (5% of Column A) X 5 (4 kW + 1 for the remaining 0.6 kW)
6. The required branch circuit must be sized, therefore, for a total demand load of

   8000 watts + 2000 watts = 10,000 watts
7. Required size of branch circuit—

$$\text{amp rating} = \frac{10{,}000\ \text{w}}{230\ \text{v}} = 43\ \text{amps}$$

USING 60C CONDUCTORS, AS REQUIRED BY UL, THE UNGROUNDED BRANCH CIRCUIT CONDUCTORS WOULD CONSIST OF NO. 6 TW CONDUCTORS.

FIG. 4-8 Sizing a branch circuit for a household range over 12 kW.

A feeder supplying more than one range (rated not over 12 kW) must have ampacity sufficient for the maximum demand load given in Table 220-19 for the number of ranges fed. For instance, a feeder to 10 such ranges would have to have ampacity for a load of 25 kW.

Other Calculations on Electric Cooking Appliances

The following "roundup" points out step-by-step methods of wiring the various types of household electric cooking equipment (range, counter-mounted cooking units, and wall-mounted ovens) according to the NEC.

Tap Conductors. Section 210-19(b), Exception 2, gives permission to reduce the size of the neutral conductor of a 3-wire range branch circuit with ground to 70 percent of the rating of the CB or fuses protecting the branch circuit. However, this rule does not apply to smaller taps connected to a 50-A circuit—where the smaller taps (none less than 20-A ratings) must all be the same size. Further, it does not apply when individual branch circuits supply each wall- or counter-mounted cooking unit and all circuit conductors are of the same size and less than No. 10.

Section 210-19(b), Exception 1, permits tap conductors, rated not less than 20 A, to be connected to 50-A branch circuits that supply ranges, wall-mounted ovens, and counter-mounted cooking units. These taps cannot be any longer than necessary for servicing. Figure 4-11 illustrates the application of this rule.

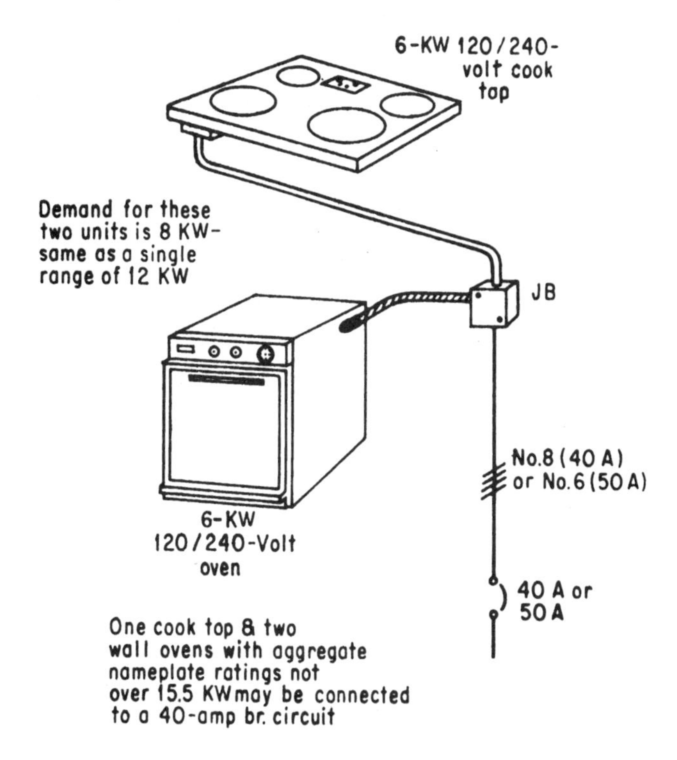

REMEMBER! All new branch circuits to ranges, cooktops, dryers, etc., and junction/outlet boxes in the circuit must be grounded with an equipment grounding conductor that is run with the circuit conductor. (Sec. 250-60)

FIG. 4-9 Two units treated as a single-range load.

In Sec. 210-19(b), Exception 1, the wording "no longer than necessary for servicing" encourages the location of circuit junction boxes as close as possible to each cooking and oven unit connected to 50-A circuits. A number of counter-mounted cooking units have integral supply leads about 36 in long, and some ovens come with supply conduit and wire in lengths of 48 to 54 in. Therefore, a box should be installed close enough to connect these leads.

Feeder and Circuit Calculations. Section 220-19 permits the use of Table 220-19 for calculating the feeder load for ranges and other cooking appliances that are individually rated more than $1\frac{3}{4}$ kW.

Note 4 of the table reads: "The branch-circuit load for one wall-mounted oven or one counter-mounted cooking unit shall be the nameplate rating of the appliance." Figure 4-12 shows a separate branch circuit to each cooking unit, as permitted.

Common sense dictates that there is no difference in demand factor between a single range of 12 kW and a wall-mounted oven and surface-mounted cooking unit totaling 12 kW. This is explained in the last sentence of Note 4 of Table 220-19. The mere division of a complete range into two or more units does not change the demand factor.

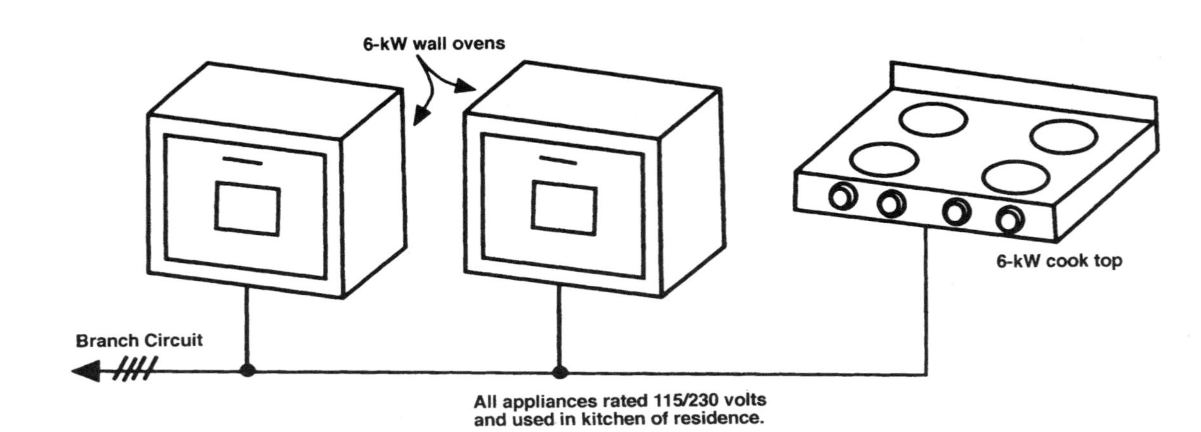

1. Note 4 of Table 220-19 says that the branch-circuit load for a counter-mounted cooking unit and not more than two wall-mounted ovens, all supplied from a single branch circuit and located in the same room, shall be computed by adding the nameplate ratings of the individual appliances and treating this total as a single range.

2. Therefore, the three appliances shown may be considered to be a single range of 18-kW rating (6 kW + 6 kW + 6 kW).

3. From Note 1 of Table 220-19, such a range exceeds 12 kW by 6 kW and the 8-kW demand of Column A must be increased by 400 watts (5% of 8000 watts) for each of the 6 additional kilowatts above 12 kW.

4. Thus, the branch-circuit demand load is—

8000 WATTS + (6 x 400 WATTS) = 10,400 WATTS

A 50-AMP CIRCUIT IS REQUIRED.

MINIMUM SIZE OF EQUIPMENT GROUNDING CONDUCTION IS NO. 10 (TABLE 250 = 95).

FIG. 4-10 Determining branch-circuit load for separate cooking appliances on a single circuit.

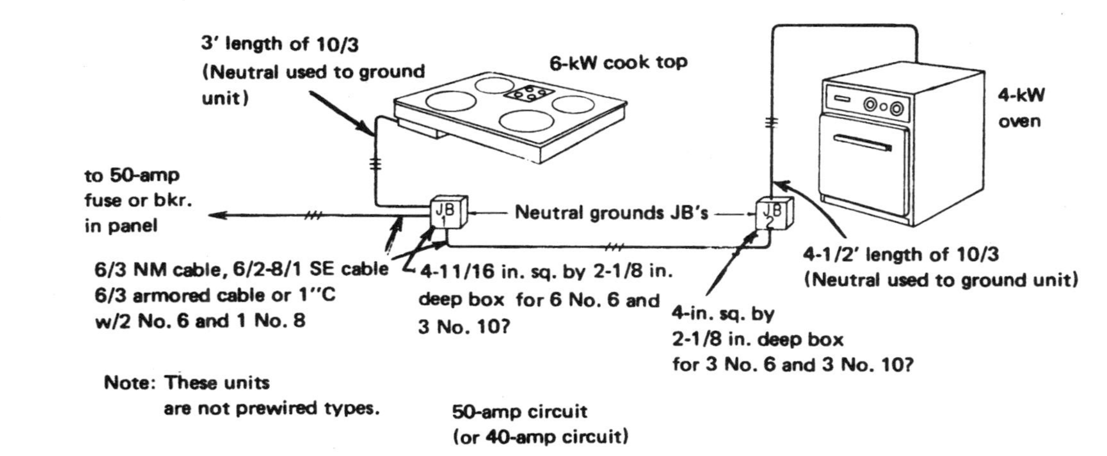

NEC rules permit a 50-amp circuit to supply cook tops and ovens. Typical arrangement shows such a circuit. Junction box sizes are computed from Table 370-16(a) and Table 370-16(b) for No. 6 conductor combinations. Taps to each unit are No. 10 to permit the use of the neutral as an equipment ground. Using the neutral to ground the junction boxes is permitted by Sec. 250-60.

FIG. 4-11 One branch circuit to cooking units.

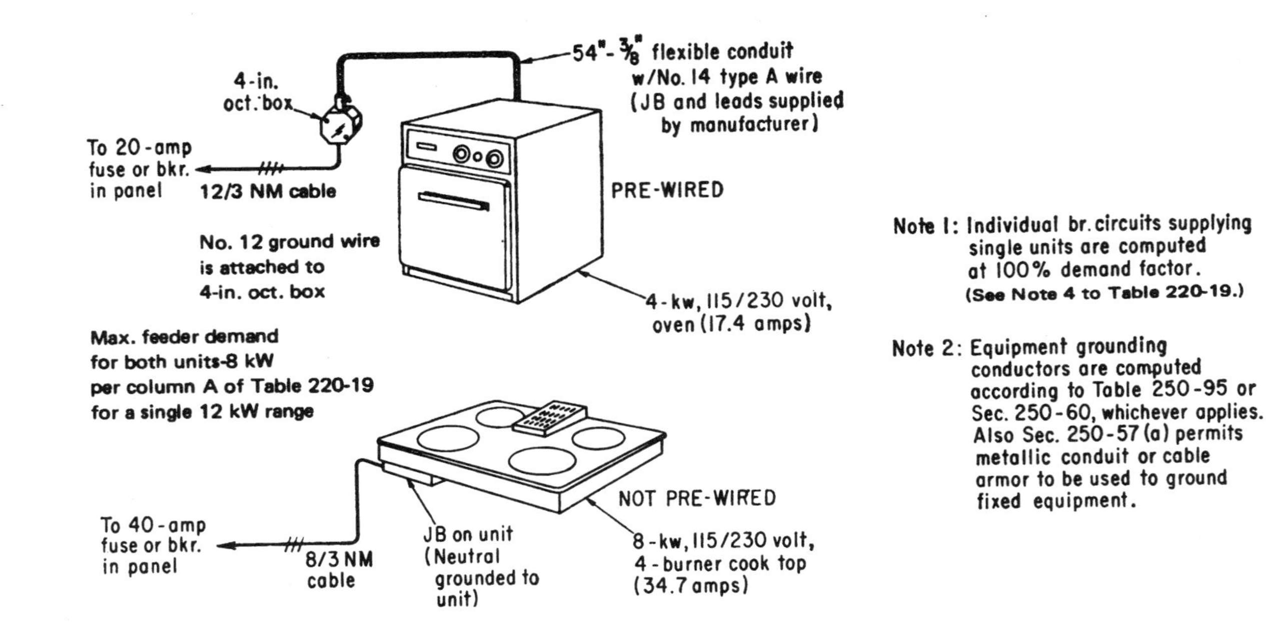

FIG. 4-12 Separate branch circuit to cooking units.

Therefore, the most direct and accurate method of computing the branch-circuit and feeder calculations for wall-mounted ovens and surface-mounted cooking units within each occupancy is to total the kilowatt ratings of these appliances and treat this total kilowatt rating as a single range of the same rating. For example, a particular dwelling has an 8-kW, four-burner, surface-mounted cooking unit and a 4-kW wall-mounted oven. This is a total of 12 kW, and the maximum permissible demand given in column A of Table 220-19 for a single 12-kW range is 8 kW.

Similarly, it follows that if the ratings of a two-burner, counter-mounted cooking unit and a wall-mounted oven are each 3.5 kW, the total of the two is 7 kW—the same total as a small 7-kW range. Because the 7-kW load is less than $8\frac{3}{4}$ kW, Note 3 of Table 220-19 permits column C of Table 220-19 to be used in lieu of column A. The demand load is 5.6 kW (7 kW $\times$ 0.80). Range or total cooking and oven unit ratings less than $8\frac{3}{4}$ kW are more likely to be found in small apartment units of multifamily dwellings than in single-family dwellings.

Because the demand loads in column A of Table 220-19 apply to ranges not exceeding 12 kW, they also apply to wall-mounted ovens and counter-mounted cooking units within each individual occupancy by totaling their aggregate nameplate kilowatt ratings. Then if the total rating exceeds 12 kW, Note 1 to the table should be used as if the units were a single range of equal rating. For example, assume that the total rating of a counter-mounted cooking unit and two wall-mounted ovens is 16 kW in a dwelling unit. The maximum demand for a single 12-kW range is given as 8 kW in column A. Note 1 requires that the maximum demand in column A be increased 5 percent for each additional kilowatt or major fraction thereof that exceeds 12 kW. In this case 16 kW exceeds 12 kW by 4 kW. Therefore, 5 percent times 4 equals 20 percent, and 20 percent of 8 kW is 1.6 kW. The maximum feeder and branch-circuit demand is then 9.6 kW (8 kW + 1.6 kW). A 9600-W load will draw over 40 A at 230 V, thereby requiring a circuit rated over 40 A.

For the range or cooking unit demand factors in a multifamily dwelling, say a 12-unit apartment building, a specific calculation must be made, as follows:

1. Each apartment has a 6-kW counter-mounted cooking unit and a 4-kW wall-mounted oven. And each apartment is served by a separate feeder from a main switchboard. The maximum cooking demand in each apartment feeder is computed in the same manner as previously described for single-family dwellings. Since the total rating of cooking and oven units in each apartment is 10 kW (6 kW + 4 kW), column A of Table 220-19 for one appliance should apply. Thus, the maximum cooking demand load on each feeder is 8 kW.
2. In figuring the size of the main service feeder, column A should be used for 12 appliances. Thus, the demand is 27 kW.

As an alternate calculation, assume that each of the 12 apartments has a 4-kW counter-mounted cooking unit and 4-kW wall-mounted oven. This totals 8 kW per apartment. In this case column C of Table 220-19 can be used to determine the cooking load in each separate feeder. By applying column C on the basis of a single 8-kW range, the maximum demand is 6.4 kW (8 kW $\times$ 0.80). Therefore, 6.4 kW is the cooking load to be included in the calculation of each feeder. Notice that this is 1.6 kW less than that of the previous example, where cooking and oven units totaling 10 kW had a demand load of 8 kW. And this is logical, because smaller units should produce a smaller total kilowatt demand.

On the other hand, it is better to use column A instead of column C for computing the main service feeder capacity for twelve 8-kW cooking loads. The reason for this is that column C is inaccurate where more than eight 8-kW ranges (or combinations) and more than twelve 7-kW ranges (or combinations) are to be used. In these instances, calculations made on the basis of column C result in a demand load greater than that of column A for the same number of ranges. As an example, twelve 8-kW ranges have a demand load of 30.72 kW (12×8 kW $\times 0.32$) in applying column C, but only a demand load of 27 kW in column A. And in column A the 27 kW is based on twelve 12-kW ranges. This discrepancy dictates use of column C only on the limited basis previously outlined.

Branch-Circuit Wiring. Where individual branch circuits supply each counter-mounted cooking unit and wall-mounted oven, there appears to be no particular problem. Figure 4-12 gives the details for wiring units on individual branch circuits.

Figure 4-11 shows an example of how typical counter-mounted cooking units and wall-mounted ovens are connected to a 50-A branch circuit.

Several manufacturers of cooking units provide an attached flexible-metal conduit with supply leads and a floating 4-in octagonal box as part of each unit. These units are commonly called *prewired types.* With this arrangement, an electrician does not have to make any supply connections in the appliance. Where such units are connected to a 50-A circuit, the 4-in octagonal box is removed, and the flexible circuit is connected to a large circuit junction box, which contains the No. 6 circuit conductors.

On the other hand, some manufacturers do not furnish supply leads with their cooking units. As a result, the electrical contractor must supply the tap conductors to these units from the 50-A circuit junction box. See Fig. 4-11. In this case, connections must be made in the appliance as well as in the junction box.

Figure 4-13 shows a single branch circuit supply the same units shown in Fig. 4-11.

40-A Circuits. The NEC does not recognize a 40-A circuit for two or more outlets, as noted in Sec. 210-23(c). Because a No. 8 (40-A) circuit can supply a single range rated not over 15.4 kW, it can also supply counter- and wall-mounted units not exceeding the same total of 15.4 kW. The rating of 15.4 kW is determined as the maximum rating of equipment that may be supplied by a 40-A branch circuit, which has a capacity of 9200 W (40 A $\times$ 230 V). From Note 1 to Table 220-19, a 15.4-kW load requires a demand capacity equal to $8000 \text{ W} + [(15.4 - 12) \times 0.05 \times 8000] = 8000 \text{ W} + (3 \times 0.05 \times 8000) = 8000 + 1200 = 9200$ W.

Figure 4-14 shows an arrangement of a No. 8 (40-A) branch circuit supplying one 7.5-kW cooking unit and one 4-kW oven. Or individual branch circuits may be run to the units.

KITCHEN EQUIPMENT—OTHER THAN DWELLING UNIT(S)

Commercial electric cooking loads must comply with Sec. 220-20 and its table of feeder demand factors for *commercial* electric cooking equipment, including dishwasher booster heaters, water heaters, and other kitchen equipment. Space-heating, ventilating, and/or air-conditioning equipment is excluded from the phrase "other kitchen equipment."

At one time, the Code did not recognize demand factors for such equipment. Code Table 220-20 is the result of extensive research on the part of electric utilities. The demand factors given in Table 220-20 may be applied to *all* equipment (except the excluded heating, ventilating, and air-conditioning loads) that *either* is thermostatically

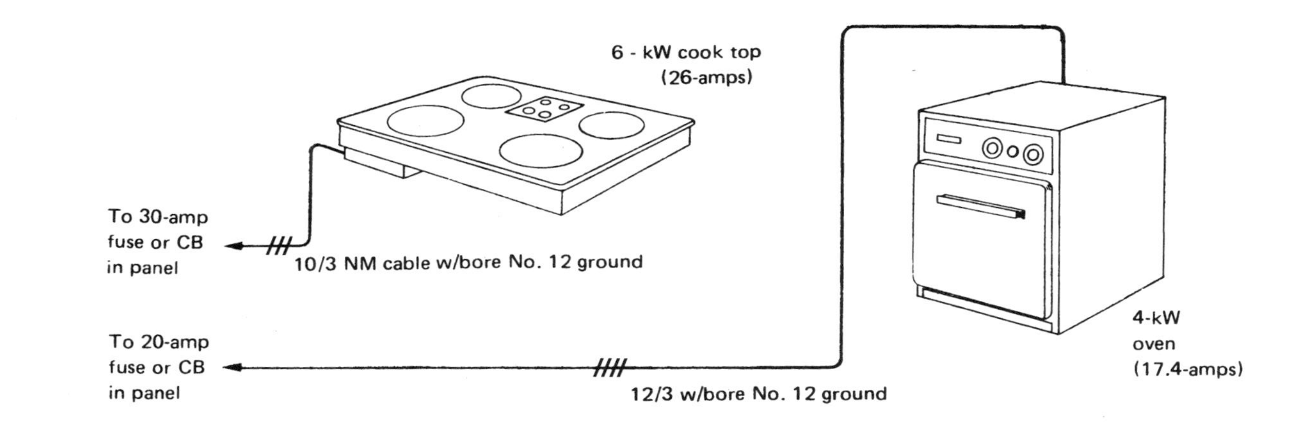

Individual branch circuits supply the same units that appear in Fig. 220-18. With this arrangement, smaller branch circuits supply each unit with no JBs required. Although two additional fuse or CB poles are required in a panelboard, overall labor/material costs are less than the 50-amp circuit shown in Fig. 220-18. However, one disadvantage to individual circuits is that smaller size circuits will not handle larger units, which may be installed at a later date.

FIG. 4-13 Separate circuits have advantages.

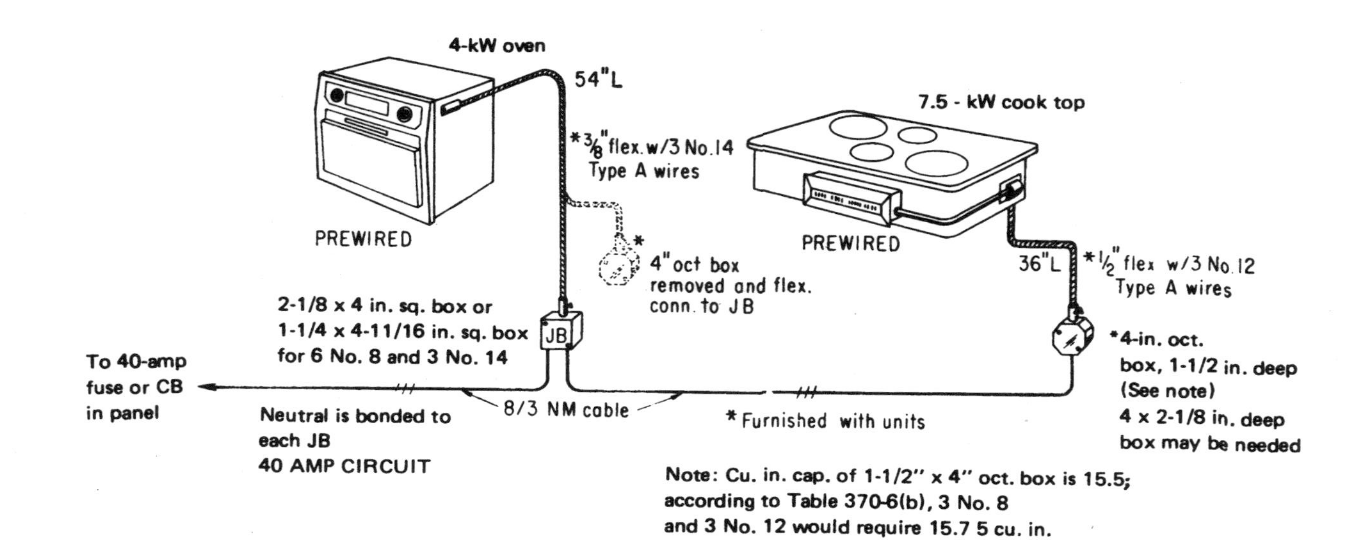

The *NEC* permits 40-amp circuits in lieu of 50-amp circuits where the aggregate nameplate rating of cook tops and ovens is less than 15.5 kW. Most ranges are less than 15.5 kW and so are most combinations of cook tops and ovens.

FIG. 4-14 A single 40-A circuit may supply units.

controlled *or* is used only on an intermittent basis. Continuously operating loads, such as infrared heat lamps used for food warming, are taken at 100 percent demand and not counted in the "number of units" that are subject to the demand factors of Table 220-20.

The rule says that the minimum load to be used in sizing a feeder to commercial-kitchen equipment must not be less than the sum of the largest two kitchen equipment loads. If the feeder load—determined by using Table 220-20 for the total number of appliances that are controlled or intermittent and then adding the sum of load ratings of continuous loads like heat lamps—is less than the sum of load ratings of the two largest load units, then the minimum feeder load must be taken as the sum of the two largest load units.

Example. Find the minimum demand load to be used in sizing a feeder supplying a 20-kW quick-recovery water heater, a 5-kW fryer, and four continuously operating 250-W food-warmer infrared lamps with a 208Y/120-V, 3-phase, 4-wire supply.

Solution. Although the water heater, the fryer, and the four lamps are a total of 1 + 1 + 4 = 6 unit loads, the 250-W lamps may not be counted in using Table 220-20 because they are continuous loads. For the water heater and the fryer, Table 220-20 indicates that a 100 percent demand must be used where the "number of units of equipment" is 2. The feeder minimum load must then be taken as

Water heater @100%	20 kW
Fryer @100%	5 kW
Four 250-W lamps @100%	1 kW + 25 kW = 26 kW

Then the feeder must be sized for a minimum current load of

$$\frac{26 \times 1000}{208 \times 1.732} = 72 \text{ A}$$

The two largest equipment loads are the water heater and the fryer

$$20 \text{ kW} + 5 \text{ kW} = 25 \text{ kW}$$

and they draw

$$\frac{25 \times 1000}{208 \times 1.732} = 69 \text{ A}$$

Therefore, the 72-A demand load calculated from Table 220-20 satisfies the last sentence of the rule because that value is "not less than" the sum of the largest two kitchen equipment loads. The feeder must be sized to have at least 72 A of capacity (a minimum of No. 3 TW, THW, THHN, RHH, etc.).

Figure 4-15 shows another example of reduced sizing for a feeder to kitchen appliances.

NONCOINCIDENT LOADS (SEC. 220-21)

When dissimilar loads (such as space heating and air cooling in a building) are supplied by the same feeder, the smaller of the two loads may be omitted from the total capacity required for the feeder if it is unlikely that the two loads will operate at the same time.

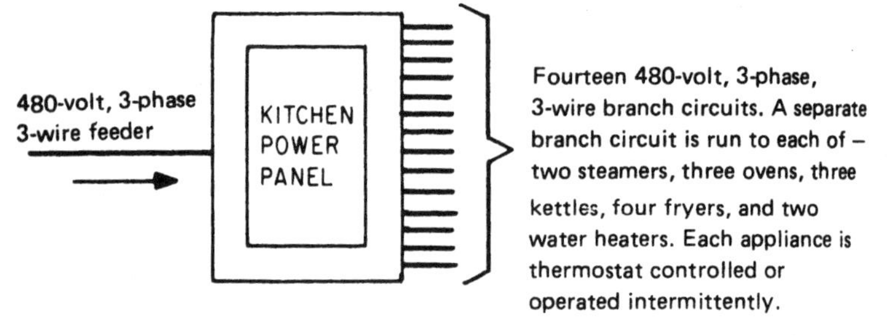

Kitchen panel supplies fourteen 480-volt, 3-phase, 3-wire branch circuits.

A separate branch circuit is run to each of—

Two steamers,
Three ovens,
Three kettles,
Four fryers, and
Two water heaters

The 14 appliances make up a total connected load of 303.3 kVA

QUESTION:

Is a full-capacity feeder (303.3 kVA/480 X 1.73 = 366 amps) required here? Or can a demand factor be applied?

ANSWER:

Although it is possible that all of the appliances might operate simultaneously, it is not expected that they will all be operating at full connected load. Table 220-20 of the *NE Code* does permit use of a demand factor on a feeder for commercial electric cooking equipment (including dishwasher, booster heaters, water heaters and other kitchen equipment). As shown in the Table, for six or more units, a demand factor of 65% can be applied to the feeder sizing:

366 amps × 0.65 = 238 amps

The feeder must have a least that much capacity, and that amp rating must be at least equal to or greater than the sum of the amp ratings of the two largest load appliances served. Capacity must be included in the building service entrance conductors for this load

FIG. 4-15 Demand factor for commercial-kitchen feeder.

FEEDER NEUTRAL LOAD

Section 220-22 covers requirements for sizing the neutral conductor in a feeder, that is, determining the required ampere rating of the neutral conductor. The basic rule of this sections says that minimum required ampacity of a neutral conductor must be at least equal to the "feeder neutral load," which is the "maximum unbalance" of the feeder load.

"The maximum unbalanced load shall be the maximum net computed load between the neutral and any one ungrounded conductor. . . ." In a 3-wire, 120/240-V, single-phase feeder, the neutral must have a current-carrying capacity at least equal to the current drawn by the total 120-V load connected between the more heavily loaded hot leg and the neutral. As shown in Fig. 4-16, under unbalanced conditions, with one hot leg fully loaded to 60 A and the other leg open, the neutral will carry 60 A and must have the same rating as the loaded hot leg. Thus No. 6 THW hot legs require No. 6 THW neutral (copper).

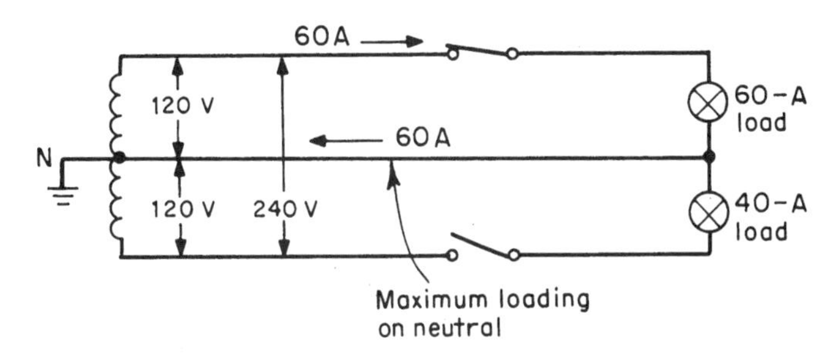

FIG. 4-16 Neutral must be sized the same as hot leg with heavier load.

Note that straight 240-V loads, connected between the two hot legs, do not place any load on the neutral. As a result, the neutral conductor of such a feeder must be sized to make up a 2-wire, 120-V circuit with the more heavily loaded hot leg. Actually, the 120-V circuit loads on such a feeder would be considered as balanced on both sides of the neutral. The neutral, then, would be the same size as each of the hot legs if only 120-V loads were supplied by the feeder. If 240-V loads also were supplied, the hot legs would be sized for the total load; but the neutral would be sized for only the total 120-V load connected between one hot leg and the neutral, as shown in Fig. 4-17.

But there are qualifications on the basic rule of Sec. 220-22:

1. When a feeder supplies household electric ranges, wall-mounted ovens, counter-mounted cooking units, and/or electric dryers, the neutral conductor may be smaller than the hot conductors but must have a carrying capacity at least equal to 70 percent of the current capacity required in the ungrounded conductors to handle the load (that is, 70 percent of the load on the ungrounded conductors). Table 220-19 gives the demand loads to be used in sizing feeders which supply electric ranges and other cooking appliances. Table 220-18 gives demand factors for sizing the ungrounded circuit conductors for feeders to electric dryers. The 70 percent demand factor may be applied to the minimum required size of a feeder phase (or hot) leg in order to determine the minimum permitted size of neutral, as shown in Fig. 4-18.

2. For feeders of three or more conductors—3-wire, dc; 3-wire, single-phase; and 4-wire, 3-phase—a further demand factor of 70 percent may be applied to that portion of the unbalanced load in excess of 200 A. That is, in a feeder supplying only 120-V loads evenly divided between each ungrounded conductor and the neutral, the neutral conductor must be the same size as each ungrounded conductor up to 200-A capacity, but may be reduced from the size of the ungrounded conductors for loads above 200 A by adding to the 200 A only 70 percent of the amount of load current above 200 A in computing the size of the neutral. Note that this 70 percent demand factor is applicable to the unbalanced load in excess of 200 A and not simply to the total load, which in many cases may include 240-V loads on 120/240-V, 3-wire, single-phase feeders or 3-phase loads or phase-to-phase connected loads on 3-phase feeders. Figure 4-19 shows an example of neutral reduction as permitted by Sec. 220-22.

Watch out! The size of a feeder neutral conductor may *not* be based on less than the current load on the feeder phase legs when the load consists of electric-discharge lighting, data-processing equipment, or similar nonlinear loads. The foregoing reduction of the neutral to 200 A plus 70 percent of the current over 200 A does not apply when all

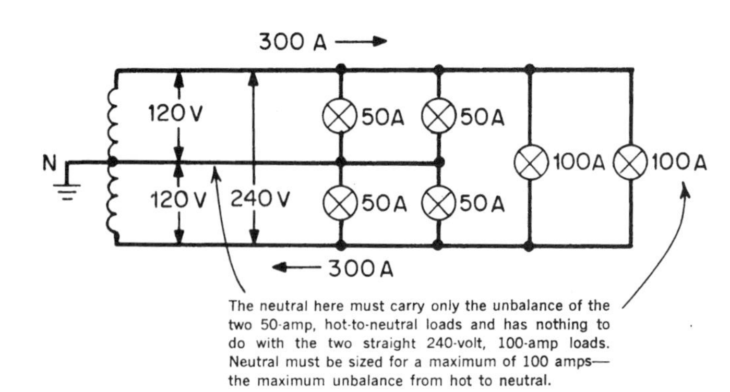

FIG. 4-17 Neutral sizing is not related to phase-to-phase loads.

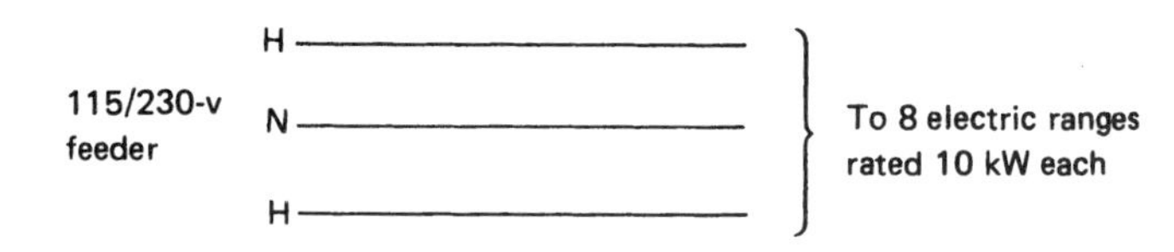

From Table 220-19—DEMAND LOAD for 8 10-kW ranges = 23 kW

LOAD ON EACH UNGROUNDED LEG $= \frac{23{,}000 \text{ W}}{230 \text{ V}} =$ 100 amps (e. g., No. 1TW)

Required minimum neutral capacity = 70% X 100 amps = 70 amps (e.g., No. 4TW)

FIG. 4-18 Sizing the neutral of a feeder to electric ranges.

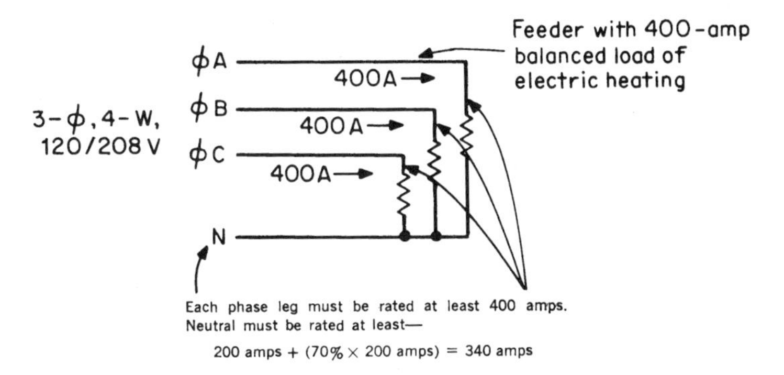

FIG. 4-19 Neutral may be smaller than hot-leg conductors on feeders over 200 A.

or most of the load on the feeder consists of electric-discharge lighting, electronic data-processing equipment, and similar electromagnetic or solid-state equipment. In a feeder supplying ballasts for electric-discharge lamps or computer equipment, there must not be a reduction of the neutral capacity for that part of the load which consists of discharge light sources, such as fluorescent mercury-vapor or other HID lamps. For feeders supplying only electric-discharge lighting or computers, the neutral conductor must be the same size as the phase conductors, no matter how big the total load may be (Fig. 4-20). Full sizing of the neutral of such feeders is required because, in a balanced circuit supplying ballasts or computer loads, neutral current approximating the phase current is produced by third (and other odd-order) harmonics developed by the ballasts. For large electric-discharge lighting or computer loads, this factor affects sizing of neutrals all the way back to the service. It also affects the rating of conductors in conduit because such a feeder circuit consists of *four* current-carrying wires, which requires application of an 80 percent reduction factor. [See Note 8 and Note 10(c) of "Notes to Tables 310-16 through 310-19" in the NEC.]

In the case of a feeder supplying, say, 200 A of fluorescent lighting and 200 A of incandescent, there can be no reduction of the neutral below the required 400-A capacity of the phase legs, because the 200 A of fluorescent lighting load cannot be used in any way to take advantage of the 70 percent demand factor on that part of the load in excess of 200 A.

Note that the Code wording in Sec. 220-22 permits reduction in the size of the neutral when electric-discharge lighting and/or computers are used, if the feeder supplying the electric-discharge lighting load over 200 A happens to be a 120/240-V, 3-wire, single-phase feeder. In such a feeder, the third-harmonic currents in the hot legs are 180° out of phase with each other and, therefore, would not be additive in the neutral as they are in a 3-phase, 4-wire circuit. In the 3-phase, 4-wire circuit, the third-harmonic components of the phase currents are in phase with each other and add in the neutral instead of canceling. Figure 4-21 shows a 120/240-V circuit.

Figure 4-22 shows a number of circuit conditions involving the rules on sizing a feeder neutral.

Remember that these examples are intended to provide an explanation of the method used for sizing the *neutral* conductor in a circuit in accordance with Sec. 220-22. Sizing

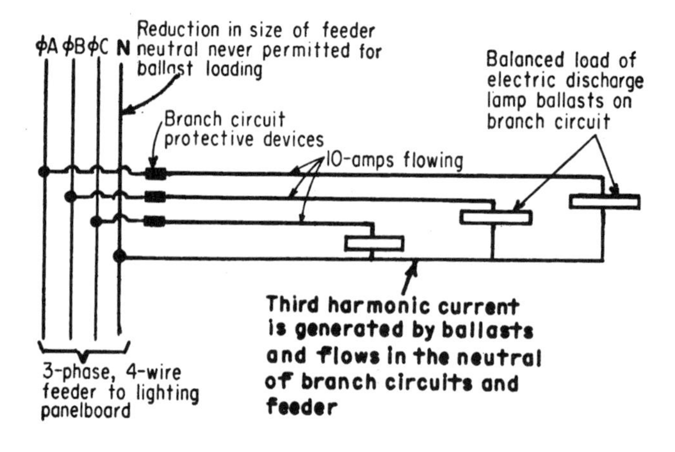

EXAMPLE:

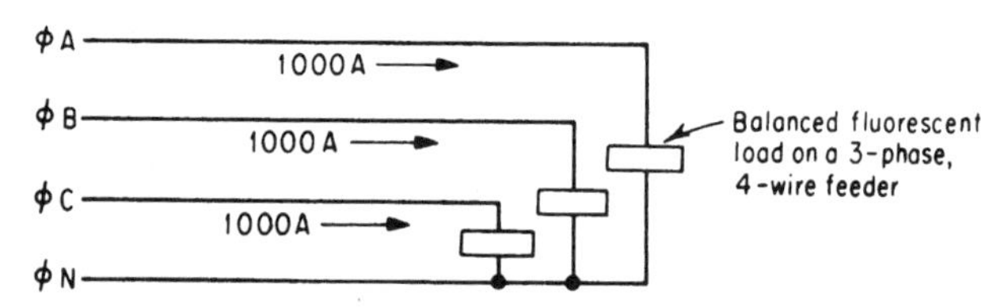

There must be no reduction in amp rating of this neutral.
It must have 1000-amp rating.

FIG. 4-20 Full-size neutral for feeders to ballast loads, computers, or other phase-to-neutral nonlinear loads.

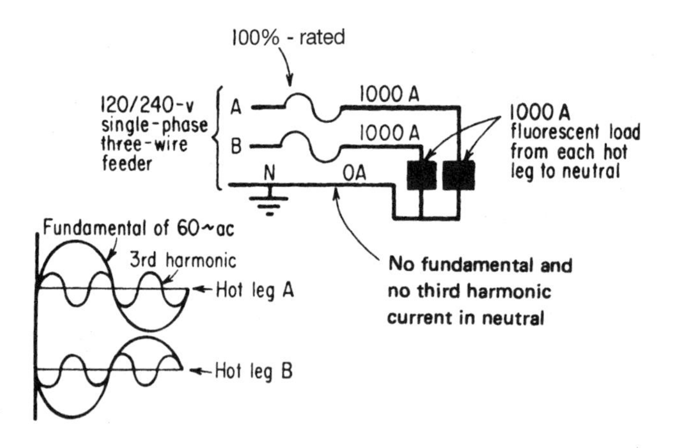

As shown, both the fundamental and harmonic currents are 180° out of phase and cancel in the neutral. Under balanced conditions, the neutral current is zero. And the literal wording of Sec. 220-22 says there can be a reduction in neutral capacity when fluorescent lighting is supplied from a single-phase,3-wire system. As a result, the 70% factor for current over 200 amps should be applied as it would be for incandescent loading. Neutral here must be rated for 760 (200 x [800 x .7]) amps.

FIG. 4-21 Reduction of neutral capacity with zero neutral current.

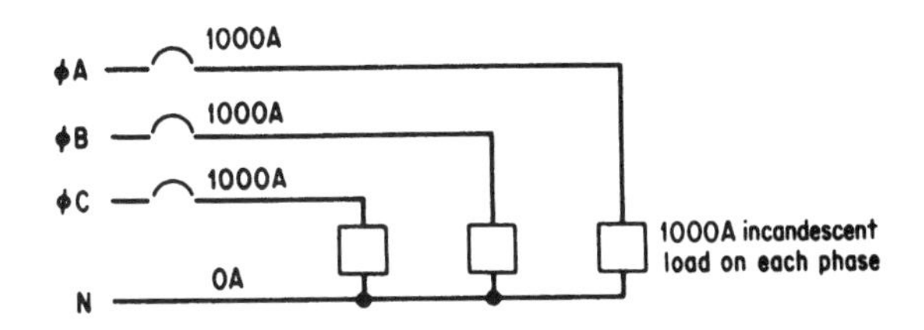

1. Incandescent lighting only
Serving an incandescent load, each phase conductor must be rated for 1000 amps. But neutral only has to be rated for 200 amps plus (70% x 800 amps) or 200 + 560 = 760 amps.

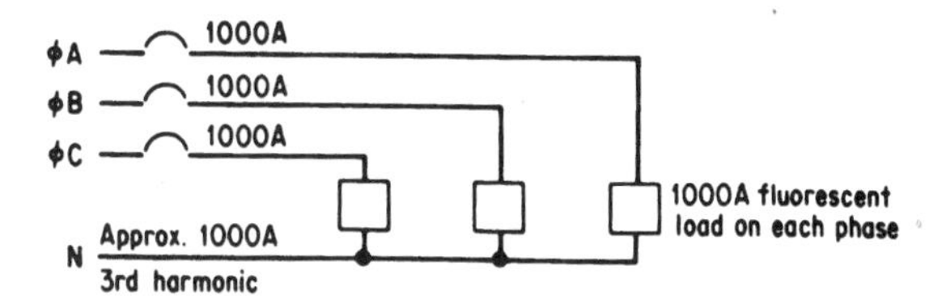

2. Electric discharge lighting only
Because load is electric discharge lighting, there can be no reduction in the size of the neutral. Neutral must be rated for 1000 amps, because the third harmonic currents of the phase legs add together in the neutral. This applies also when the load is mercury-vapor or other metallic-vapor lighting.

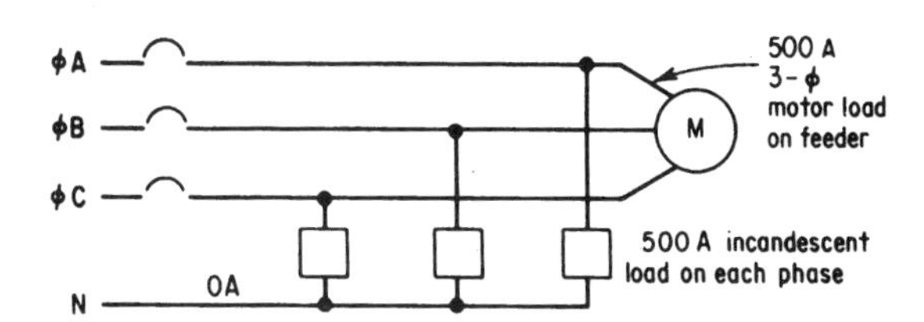

3. Incandescent plus motor load
Although 1000 amps flow on each phase leg, only 500 amps is related to the neutral. Neutral, then, is sized for 200 amps plus (70% x 300 amps) or 200 + 210 = 410 amps. The amount of current taken for 3-phase motors cannot be "unbalanced load" and no capacity has to be provided for this in the neutral.

FIG. 4-22 Sizing the feeder neutral for different conditions of loading.

NOTE: All OC devices are 100%-rated (ie., UL-listed for continuous loading at the devices current rating. Where non-100%-rated devices are used the OC device and the conductor size must be not less than the sum of noncontinuous current plus the continuous current times 1.25. [(See Sec. 210-22(c), 220-10(b).] Motor conductors sized in accordance with Art. 430.

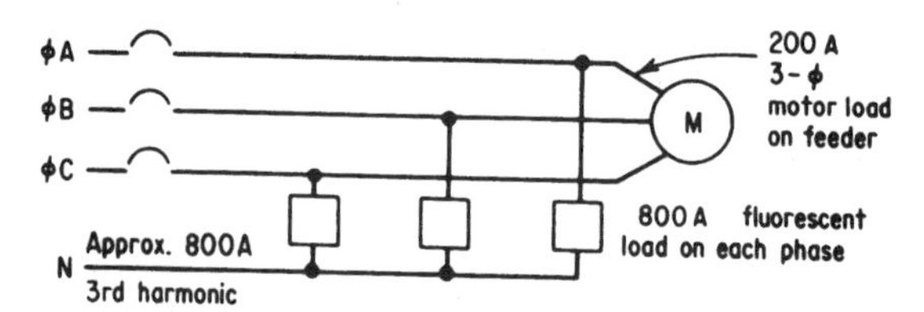

4. Electric discharge lighting plus motor load

Here, again, the only possible load that could flow on the neutral is the 800 amps flowing over each phase to the fluorescent lighting. But because it is fluorescent lighting there can be no reduction of neutral capacity below the 800-amp value carried on each phase. The 70% factor for that current above 200 amps DOES NOT APPLY in such cases.

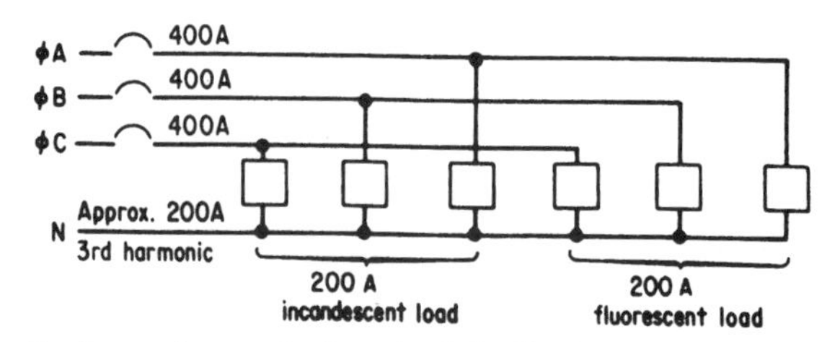

5. Incandescent plus electric discharge lighting

Each phase leg carries a total of 400 amps to supply the incandescent load plus the fluorescent load. But because there can be no reduction of neutral capacity for the fluorescent and because the incandescent load is not over 200 amps, the neutral must be sized for the maximum possible unbalance, which is 400 amps.

FIG. 4-22 (*Continued*) Sizing the feeder neutral for different conditions of loading.

of the circuit protective device and the phase conductors *must* satisfy Sec. 210-22(c), 220-3(a), or 220-10(b) depending on the type of circuit (that is, branch circuit or feeder and service conductors).

OPTIONAL CALCULATION—DWELLING UNIT

Section 220-30 sets forth an optional method of calculating service demand load for a residence. This method may be used instead of the standard method as follows:

- Only for a one-family residence or an apartment in a multifamily dwelling, or other "dwelling unit"
- Served by a 120/240-V or 120/208-V, 3-wire, 100-A or larger service or feeder
- Where the total load of the dwelling unit is supplied by one set of service-entrance or feeder conductors

This method recognizes the greater diversity attainable in large-capacity installations. It therefore permits a smaller size of service-entrance conductors for such installations than would be permitted by using the load calculations of Sec. 220-10 through Sec. 220-22.

In making this calculation, the heating load or the air-conditioning load may be disregarded as a "noncoincident load," where it is unlikely that two dissimilar loads (such as heating and air conditioning) will be operated simultaneously. In previous NEC editions, 100 percent of the air-conditioning load was compared with 100 percent of the total connected load of four or more separately controlled electric space-heating units, and the smaller of the two loads was not counted in the calculation. In the present NEC, 100 percent of the air-conditioning load is compared with only *40 percent* of the total connected load of four or more electric space heaters, or *65 percent* for less than four separately controlled units, and the lower value is omitted from the calculation.

For instance, if a dwelling unit had 3 kW of air-conditioning load and 6 kW of electric heaters, the electric heating load of 6 kW was formerly used in the calculation because it is greater than the 3-kW air-conditioning load—even though the heating load was then subjected to the diversity factor of 40 percent in the calculation itself. *Now,* with these same loads, the 3-kW air-conditioning load is greater than 40 percent of the 6-kW heating load ($0.40 \times 6 = 2.4$ kW), and the calculation will be made using the 3-kW air-conditioning load at 100 percent demand, and the electric heating load will be disregarded.

Example. A typical application of the data and table of Sec. 220-30, in calculating the minimum required size of service conductors, is as follows:

A 1500-ft^2 house (excluding unoccupied basement, unfinished attic, and open porches) contains the following specific electric appliances:

12-kW range

2.5-kW water heater

1.2-kW dishwasher

9 kW of electric heat (in five rooms)

5-kW clothes dryer

6-A, 230-V ac unit

When the optional method is used, if a house has air conditioning as well as electric heating, there is recognition in Sec. 220-21 that if "it is unlikely that two dissimilar loads will be in use simultaneously," it is permissible to omit the smaller of the two in calculating required capacity in feeder or in service-entrance conductors. In Sec. 220-30, that concept is spelled out in the table to require adding only the larger of either the total air-conditioning load *or* 40 percent of the connected load of *four or more* separately controlled electric space-heating units (65 percent for less than four). For the residence considered here, these loads would be as follows:

$$\text{Air conditioning} = 6\text{ A} \times 230\text{ V} = 1.38\text{ kVA}$$

$$40\%\text{ of heating (five separate units)} = 9\text{ kW} \times 0.4 = 3.6\text{ kW (3600 VA)}$$

Because 3.6 kW is greater than 1.38 kVA, it is permissible to omit the air-conditioning load and provide a capacity of 3.6 kW in the service or feeder conductors to cover *both* the heating and air-conditioning loads.

The "other loads" must be totaled up in accordance with Sec. 220-30:

	Voltamperes
1. 1500 VA for each of two small-appliance circuits (2-wire, 20-A) required by Sec. 220-4(b)(1)	3,000
Laundry branch circuit (3-wire, 20-A)	1,500
2. 3 VA/ft² of floor area for general lighting and general-use receptacles (3 × 1500 ft²)	4,500
3. Nameplate rating of fixed appliances:	
Range	12,000
Water heater	2,500
Dishwasher	1,200
Clothes dryer	5,000
Total	29,700

In reference to Table 220-30, load categories 1, 2, 3, and 4 are not applicable here: "Air-conditioning" has already been excluded as a load because 40 percent of the heating load is greater. There is no "central" electric space heating; there is no electric thermal storage heating; and there are *not* less than four separately controlled electric space-heating units.

The total load of 29,700 VA, as summed up above, includes "all other load," as referred to in Code Table 220-30. *Then*

1. Take 40% of the 9000-W heating load	3,600
2. Take 10 kVA of "all other load" at 100% demand	10,000
3. Take the "remainder of other load" at 40% demand factor:	
(29,700 − 10,000) × 40% = 19,700 × 0.4	7,880
Total demand	21,480

Using 230- and 115-V values of voltage rather than 240- and 120-V values, ampacities may then be calculated. At 230 V, single phase, the *ampacity of each service hot leg would then have to be*

$$\frac{21{,}480\ \text{W}}{230\ \text{V}} = 89.5 \text{ or } 90\ \text{A}$$

But Sec. 230-42(b)(2) requires a minimum conductor rating when demand load is 10 kW (10 kVA) or more:

$$\text{Minimum service conductor required} = 100\ \text{A}$$

Then the neutral service-entrance conductor is calculated in accordance with Sec. 220-22, based on Sec. 220-10. All 230-V loads have no relation to required neutral capacity. The water heater and electric space-heating units operate at 230 V, 2-wire and have no neutrals. By considering only those loads served by a circuit with a neutral conductor and determining their maximum unbalance, the minimum required size of neutral conductor can be determined.

When a 3-wire, 230/115-V circuit serves a total load that is balanced from each hot leg to neutral—that is, one-half of the total load is connected from one hot leg to neutral, and the other half of total load from the other hot leg to neutral—the condition of maximum unbalance occurs when all the load fed by one hot leg is operating and all the load fed by the other hot leg is off. Under that condition, the neutral current and hot-leg current are equal to half the total load watts divided by 115 V (half the volts between hot legs). But that current is exactly the same as the current that results from dividing the *total* load (connected hot leg to hot leg) by 230 V (which is twice the voltage from hot leg to neutral). Because of this relationship, it is easy to determine neutral-current load by simply calculating hot-leg current load—total load from hot leg to hot leg divided by 230 V.

In the example here, the neutral-current load is determined from the following steps that sum up the components of the neutral load:

	Voltamperes
1. Take 1500 ft² at 3 VA/ft² [Table 220-3(b)]	4500
2. Add three small-appliance circuits (two kitchen, one laundry) at 1500 VA each (Sec. 220-16)	4500
Total lighting and small-appliance load	9000
3. Take 3000 VA of that value at 100% demand factor (Secs. 220-11 and 220-16; Table 220-11)	3000
4. Take the balance of the load (9000−3000) at 35% demand factor: 6000 VA × 0.35	2100
Total of 3 and 4	5100

Assuming an even balance of this load on the two hot legs, the neutral load under maximum unbalance will be the same as the total load (5100 VA) divided by 230 V (Fig. 4-23):

$$\frac{5100 \text{ VA}}{230 \text{ V}} = 22.17 \text{ A}$$

And the neutral unbalanced current for the range load can be taken as equal to the 8000-W range demand load multiplied by the 70 percent demand factor permitted by Sec. 220-22 and then divided by 230 V (Fig. 4-24):

$$\frac{8000 \times 0.7}{230} = \frac{5600}{230} = 24.34 \text{ A}$$

Then the neutral-current load that is added by the 115-A, 1200-W dishwasher must be added (Fig. 4-25):

$$\frac{1200 \text{ W}}{115 \text{ V}} = 10.43 \text{ A}$$

The clothes dryer contributes neutral load due to the 115-V motor, its controls, and a light. As allowed in Sec. 220-22, the neutral load of the dryer may be taken at 70 percent of the load on the ungrounded hot legs. Then (5000 VA × 0.7) ÷ 230 V = 15.2 A.

The minimum required neutral capacity is, therefore,

$$
\begin{array}{l}
22.17\ \text{A} \\
24.34\ \text{A} \\
10.43\ \text{A} \\
\underline{15.2\ \ \text{A}} \\
72\ \ \ \ \text{A} \qquad \text{(Fig. 4-25)}
\end{array}
$$

From Code Table 310-16, the neutral minimum for 72 A would be:

No. 3 copper TW, THW, THHN, or XHHW
No. 2 aluminum TW, THW, THHN, or XHHW

And the 75 or 90°C conductors must be used at the ampacity of 60°C conductors, as required by UL instructions in UL's *Electrical Construction Materials Directory* and by Sec. 110-14(c) of the NEC.

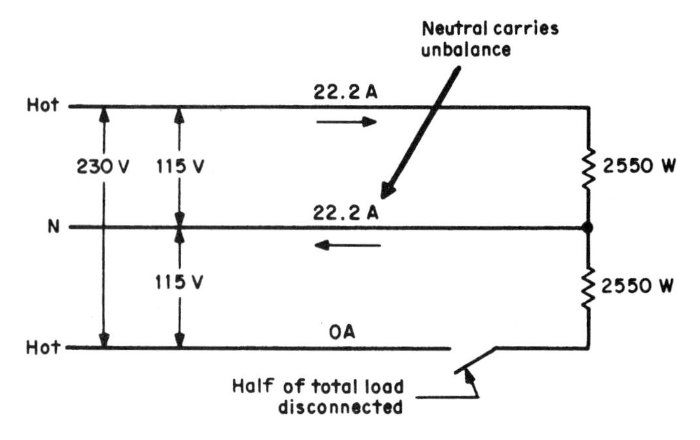

FIG. 4-23 Neutral current for lighting and receptacles.

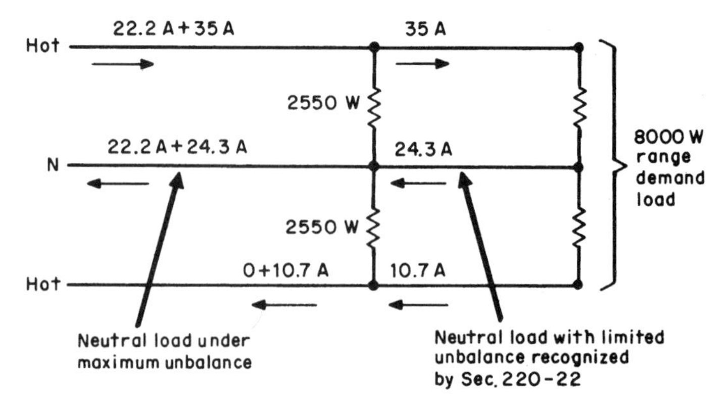

FIG. 4-24 Neutral for lighting, receptacles, and range.

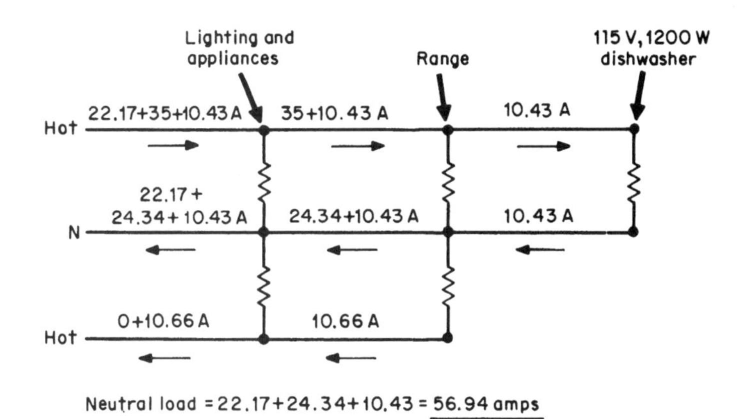

FIG. 4-25 Neutral current for all but dryer.

Note: The above calculation of the minimum required capacity of the neutral conductor differs from the calculation and results shown in Example 2(a) in Chap. 9 in the NEC book. In the book, the 1200-W dishwasher load is added as a 230-V load to the range load and general lighting and receptacle load. To include a 115-V load as a 230-V load (and then divide the total by 230 V, as shown) does not accurately represent the neutral load that the 115-V, 1200-W dishwasher will produce. In fact, it yields exactly half the neutral load that the dishwasher represents. The optional calculation method of Sec. 220-30 does indicate in part (3) that fixed appliances should be added at nameplate load and does not differentiate between 115-V devices and 230-V devices. It simply totals all load and then applies the 100 percent and 40 percent demand factors as indicated. That method clearly is based on well-founded data about diversity of loads and is aimed at determining a reasonable size of the service hot legs. But calculation of the feeder neutral in accordance with Sec. 220-22 is aimed at determining the *maximum unbalanced current* to which the service neutral might be subjected.

Although the difference is small between the NEC book value of 64 A (or 67 A if 15,400 V is divided by 230 V) and the value of 72 A determined here, precise calculation should be made to ensure real adequacy in conductor ampacities. The difference actually changes the required minimum size of neutral conductor from No. 4 up to No. 3 for copper. A load like a dishwasher, which draws current for a considerable time and is not just a few-minute device like a toaster, should be factored into the calculation with an eye toward adequate capacity of conductors.

OPTIONAL CALCULATION FOR ADDITIONAL LOADS IN EXISTING DWELLING UNIT

Section 220-31 covers an optional calculation for additional loads in an existing dwelling unit that contains a 120/240- or 208/120-V, 3-wire service of any current rating. The method of calculation is similar to that in Table 220-30.

The purpose of this section is to permit the *maximum load* to be applied to an *existing* service without the necessity of increasing the size of the service. The calculations are based on numerous load surveys and tests made by local utilities throughout the country. This optional method would seem to be particularly advantageous when smaller loads such as window air conditioners or bathroom heaters are to be installed in a dwelling with, say, an existing 60-A service, as follows:

If there is an existing electric range, say, 12 kW (and no electric water heater), it would not be possible to add any load of substantial rating. Based on the formula 13,800 VA (230 V $\times$ 60 A) = 8000 + 0.4(X − 8000), the total "gross load" that can be connected to an existing 115/230-V, 60-A service would be X = 22,500 VA. Actually, it can be greater if a value of 240 V is used (240 V $\times$ 60 A = 14,400 VA).

Example. Thus, an existing 1000-ft² dwelling with a 12-kW electric range, two 20-A appliance circuits, a 750-W furnace circuit, and a 60-A service would have a gross load of:

	Voltamperes
1000 ft² $\times$ 3 VA/ft²	3,000
Two 20-A appliance circuits @1500 VA each	3,000
One electric range	12,000
Furnace circuit	750
Gross voltamperes	18,750

Since the *maximum* permitted gross load is 22,500 VA, an appliance not exceeding *3750 VA* could be added to this existing 60-A service. However, the tabulation at the end of this section lists air-conditioning equipment, central space heating, and less than four separately controlled space-heating units at 100 percent demands; and if the appliance to be added is one of these, then it would be limited to *1500 VA:*

From the 18,750-VA gross load we have 8000 VA at 100 percent demand + [10,750 VA (18,750 − 8000) $\times$ 0.40] = 12,300 VA. Then 13,800 VA (60 A $\times$ 230 V) − 12,300 VA = 1500 VA for an appliance listed at *100 percent demand.*

OPTIONAL CALCULATION—TWO DWELLING UNITS

Section 220-33 provides an optional calculation for sizing a feeder to "two dwelling units." It notes that if calculation of such a feeder according to the basic long method given in part B of Art. 220 exceeds the minimum load ampacity permitted by Sec. 220-32 for three identical dwelling units, then the *lesser* of the two loads may be used. This rule was added to eliminate the obvious illogic of requiring a greater feeder ampacity for two dwelling units than for three units of the same load makeup. Now optional calculations provide for a feeder to one dwelling unit, two dwelling units, or three or more dwelling units.

OPTIONAL METHOD—SCHOOLS

The optional calculation for feeders and service-entrance conductors for a school makes clear that feeders "within the building or structure" must be calculated in accordance with the standard long calculation procedure established by part B of Art. 220. *But* the ampacity of any individual feeder does not have to be greater than the minimum required ampacity for the whole building, regardless of the calculation results from part B.

The last sentence in this section excludes portable classroom buildings from the optional calculation method to prevent the possibility that the demand factors of Table 220-34 would result in a feeder or service of lower ampacity than the connected load. Such portable classrooms have air-conditioning loads that are not adequately covered by using a watts-per-square-foot calculation with the small area of such classrooms.

OPTIONAL CALCULATIONS FOR ADDITIONAL LOADS TO EXISTING INSTALLATIONS

Because of the universal practice of adding loads to feeders and services in all kinds of existing premises, this calculation procedure is given in the Code. To determine how much more load may be added to a feeder or set of service-entrance conductors, at least one year's accumulation of measured maximum-demand data must be available. Then the required spare capacity may be calculated in accordance with Sec. 220-35 as follows:

$$\text{Additional load capacity} = \text{ampacity of feeder or service conductors} - [(1.25 \times \text{existing demand kVA} \times 1000) \div \text{circuit voltage}]$$

where circuit voltage is the phase-to-phase value for single-phase circuits and 1.732 times the phase-to-phase value for 3-phase circuits.

A third required condition is that the feeder or service conductors be protected against overcurrent, in accordance with applicable Code rules on such protection.

THE TAP RULES

The basic rule of Sec. 240-21(a) is shown in Fig. 4-26. Note that the permission given in what used to be Exception 1 in the 1990 and previous editions of the Code has been dropped. However, inspectors and common sense indicate such application *is* still acceptable.

Although the basic Code requirement dictates the use of an overcurrent device at the point at which a conductor receives its supply, the other subsections present "exceptions" to this rule in the case of taps to feeders. That is, to meet the practical demands of field application, certain lengths of unprotected conductors may be used to tap energy from protected feeder conductors (Fig. 4-26).

These "exceptions" to the rule for protecting conductors at their points of supply are made in the case of 10-, 25-, and 100-ft taps from a feeder, as described in Secs. 240-21(b), (c), (d), (e), and (m). Application of the tap exceptions should be made carefully to effectively minimize any sacrifice in safety. The tap exceptions are permitted without overcurrent protective devices at the point of supply.

Section 240-21(b) says that unprotected taps not over 10 ft long (Fig. 4-27) may be made from feeders or transformer secondaries provided:

1. The smaller conductors have a current rating that is not less than the combined computer loads of the circuits supplied by the tap conductors and have ampacity of—

Not less than the rating of the "device" supplied by the tap conductors, or
Not less than the rating of the overcurrent device (fuses or CB) that might be installed at the termination of the tap conductors.

Important limitation: For any 10-ft unprotected feeder tap installed in the field, the rule limits its connection to a feeder that has protection rated *not* more than 1000 percent of (10 times) the ampacity of the tap conductor. Under the rule, unprotected No. 14 tap conductors are not permitted to tap a feeder any larger than 1000 percent of the 20-A ampacity of No. 14 copper conductors—which would limit such a tap for use with a maximum feeder protective device of not over 10 × 20 A, or 200 A.

2. The tap does not extend beyond the switchboard, panelboard, or control device which it supplies.
3. The tap conductors are enclosed in conduit, EMT, metal gutter, or other approved raceway when not a part of the switchboard or panelboard.

Section 240-21(b) specifically recognizes that a 10-ft tap may be made from a transformer secondary in the same way it has always been permitted from a feeder. In either case, the tap conductors must not be over 10 ft long and must have ampacity not less than the ampere rating of the switchboard, panelboard, or control device—or the tap

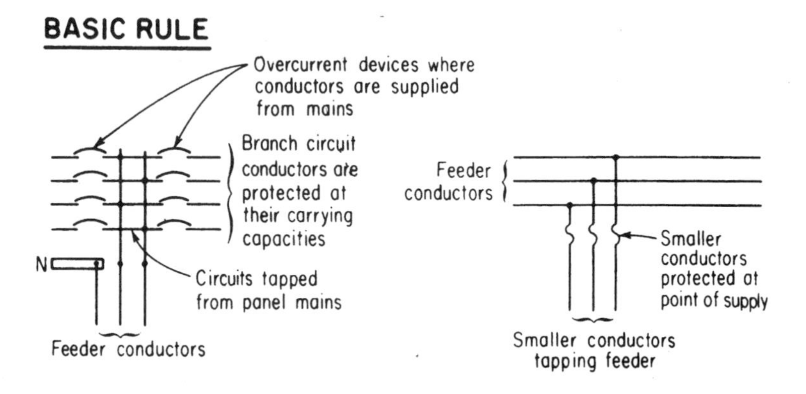

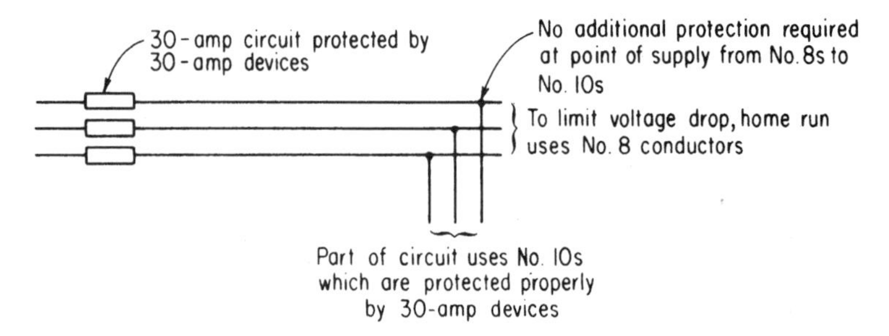

FIG. 4-26 Conductors must be protected at their supply ends.

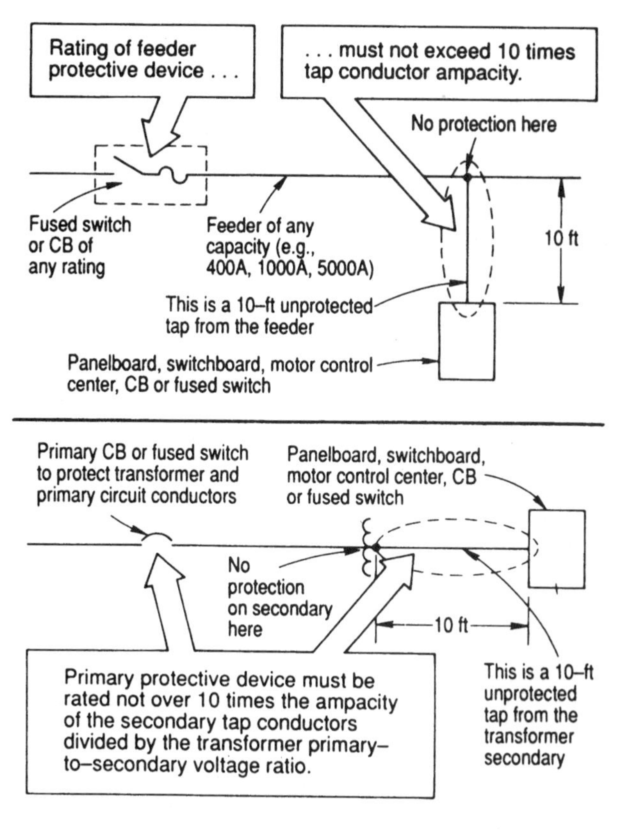

FIG. 4-27 Ten-ft taps may be made from a feeder or a transformer secondary.

conductors may be terminated in an overcurrent protective device rated not more than the ampacity of the tap conductors. In the case of an unprotected tap from a transformer secondary, the ampacity of the 10-ft tap conductors would have to be related through the transformer voltage ratio to the size of the transformer primary protective device, which in such a case would be "the overcurrent device on the line side of the tap conductors."

Taps not over 25 ft long [Sec. 240-21(c)] (Fig. 4-28) may be made from feeders provided:

1. The smaller conductors have a current rating at least one-third that of the feeder overcurrent device rating or of the conductors from which they are tapped.

2. The tap conductors are suitably protected from physical damage or enclosed in a raceway. This will allow for the use of a cable or raceways.
3. The tap is terminated in a single CB or set of fuses which will limit the load on the tap to the ampacity of the tap conductors.

Examples of Taps

Figure 4-29 shows use of a 10-ft feeder tap to supply a single motor branch circuit. The conduit feeder may be a horizontal run or a vertical run, such as a riser. If the tap conductors are of such size that they have a current rating at least one-third that of the feeder conductors (or protection rating) from which they are tapped, they could be run a distance of 25 ft without protection at the point of tap-off from the feeder because they would comply with the rules of Sec. 240-21(c) which permit a 25-ft tap if the conductors terminate in a single protective device rated not more than the conductor ampacity. Although Sec. 364-12 generally requires that any busway used as a feeder have overcurrent protection on the busway for any subfeeder or branch circuit tapped from the busway, Exception 1 allows the use of a cable-tap box on the busway without overcurrent protection (as shown in the conduit installation of Fig. 4-29). Refer to Secs. 240-24 and 364-12.

A common application of the 10-ft tap exception is the supply of panelboards from conduit feeders or busways, as shown in Fig. 4-30. The case shows an interesting requirement that arises from Sec. 384-16, which requires that lighting and appliance panelboards be protected on their supply side by overcurrent protection rated not more than the rating of the panelboard bus bars. The 100-A panel protection would have to be in the panel or just ahead of it. It could not be at the junction box on the conduit because that would make it not readily accessible and therefore a violation of Sec. 240-24. With a conduit feeder, a fused-switch or CB main in the panelboard could be rated up to the 100-A main rating.

As covered in Sec. 240-21(d), transformer taps may be made where the total length of unprotected primary plus secondary conductors does *not* exceed 25 ft. If the primary conductors *are* protected in accordance with their ampacity, the secondary conductors may be any length up to 25 ft. For transformer applications, typical tap considerations are shown in Fig. 4-31.

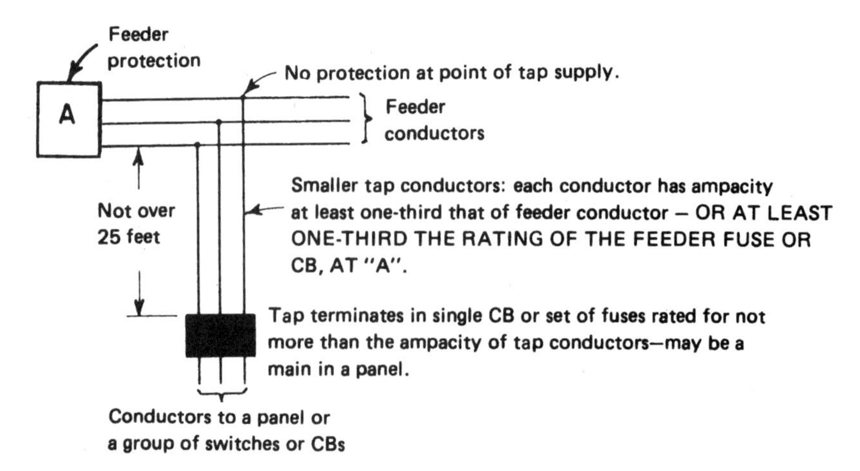

FIG. 4-28 Sizing feeder taps not over 25 ft long.

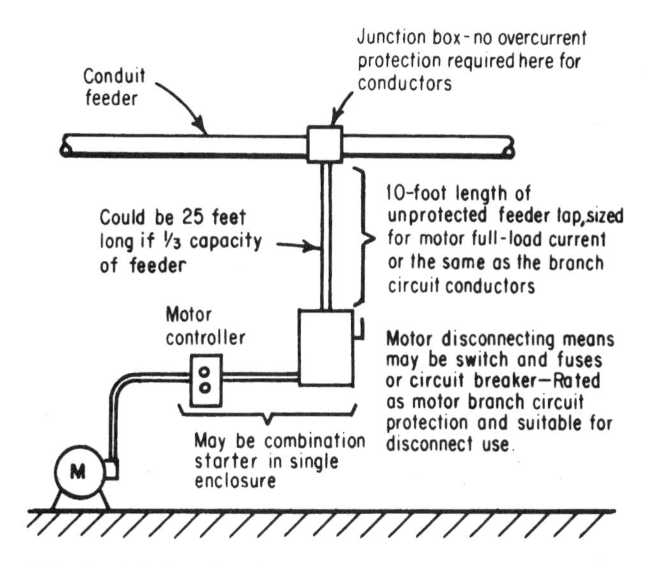

FIG. 4-29 A 10-ft tap for a single motor circuit.

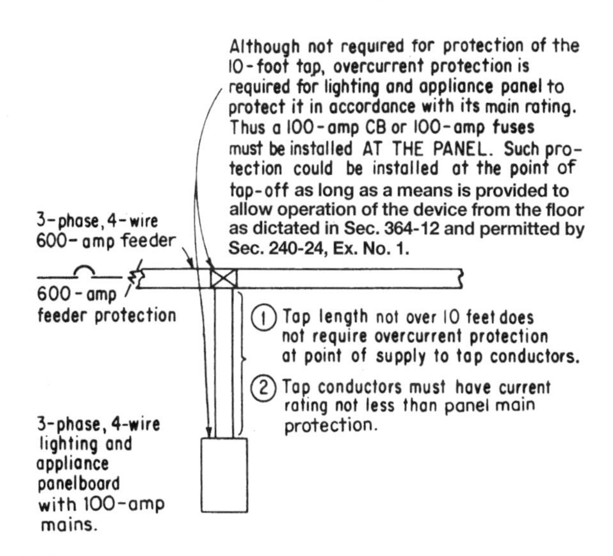

FIG. 4-30 A 10-ft tap to lighting panel with unprotected conductors.

25-FT TAP

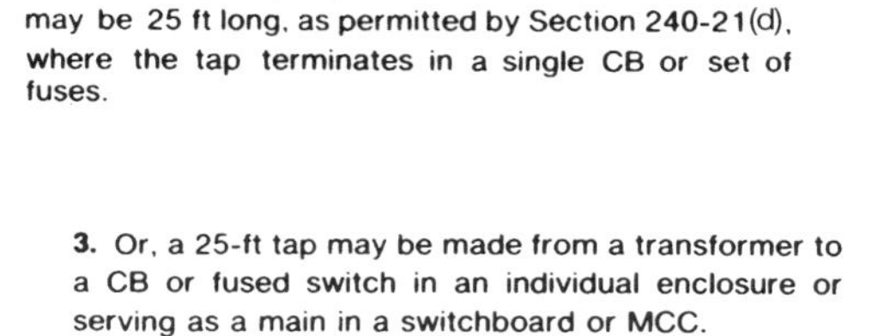

1. If transformer secondary feeds the panel **having a main CB or fused switch,** then . . .

2. . . . secondary tap conductors from transformer may be 25 ft long, as permitted by Section 240-21(d), where the tap terminates in a single CB or set of fuses.

3. Or, a 25-ft tap may be made from a transformer to a CB or fused switch in an individual enclosure or serving as a main in a switchboard or MCC.

NOTE: From a single transformer secondary of adequate capacity, more than one set of 10-ft tap conductors may be run to more than one panel or other distribution equipment.

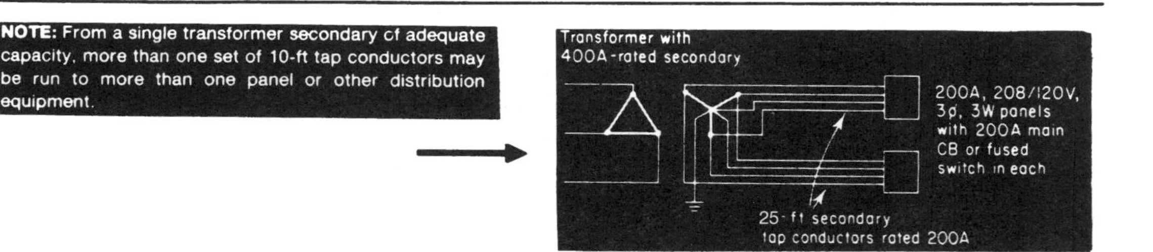

FIG. 4-31 Taps from transformer secondaries.

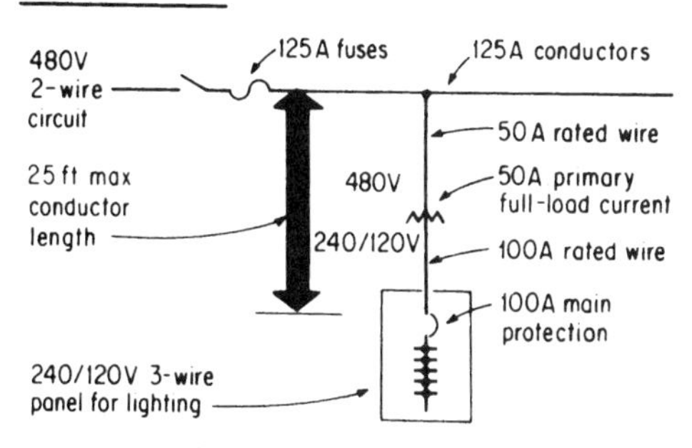

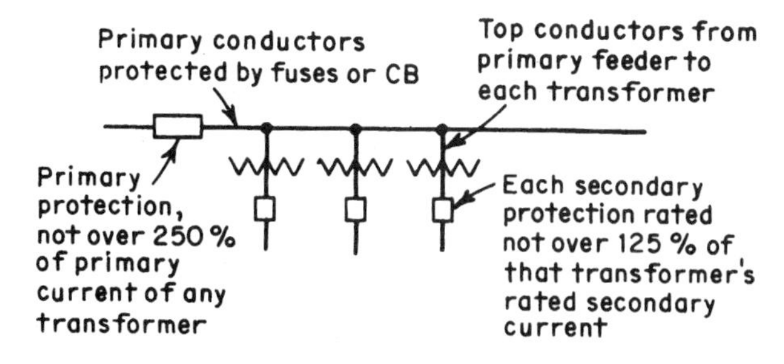

FIG. 4-32 Feeder tap of primary plus secondary not over 25 ft long.

Figure 4-32 shows application of Sec. 240-21(d) in conjunction with the rule of Sec. 450-3(b)(2) covering transformer protection. As shown in example 1 in Fig. 4-32, the 100-A main protection in the panel is sufficient protection for the transformer and the primary and secondary conductors when these conditions are met:

1. Tap conductors have ampacity at least one-third that of the 125-A feeder OC protective device.
2. Secondary conductors are rated at least one-third the ampacity of the 125-A feeder OC device, based on the primary-to-secondary transformer ratio.
3. Total tap is not over 25 ft, primary plus secondary.
4. All conductors are suitably protected from physical damage.
5. Secondary conductors terminate in the 100-A main protection that limits secondary load to the ampacity of the secondary conductor and simultaneously provides the protection required by the lighting panel and is not rated over 125 percent of transformer secondary current.
6. Primary feeder protection is not over 250 percent of transformer rated primary current, as recognized by Sec. 450-3(b)(2), and the 100-A main breaker in the panel is satisfactory as the required "overcurrent device on the secondary side rated or set at not more than 125 percent of the rated secondary current of the transformer."

In example 2 of Fig. 4-32, each set of tap conductors from the primary feeder to each transformer may be the same size as primary feeder conductors *or* may be smaller than primary conductors if sized in accordance with Sec. 240-21(d), which permits a 25-ft tap from a primary feeder to be made up of both primary and secondary tap conductors. The 25-ft tap may have any part of its length on the primary or secondary but must not be longer than 25 ft and must terminate in a single CB or set of fuses.

Figure 4-33 shows another example of Sec. 240-21(d) and Sec. 450-3(b)(2). Because the primary wires tapped to each transformer from the main 100-A feeder are also rated 100 A and are therefore protected by the 100-A feeder protection, all the primary circuit to each transformer is excluded from the allowable 25-ft tap to the secondary main protective device. The 100-A protection in each panel is not over 125 percent of the rated transformer secondary current. It therefore provides the transformer protection required by Sec. 450-3(b)(2). The same device also protects each panel at its main bus bar rating of 100 A.

Figure 4-34 compares the two different 25-ft tap techniques covered by Sec. 240-21(c) and (d).

As shown in Fig. 4-35, Sec. 240-21(i) gives permission for unprotected taps to be made from generator terminals to the first overcurrent device the generator supplies, such as in the fusible switch or circuit breakers used for control and protection of the circuit that the generator supplies. As the rule is worded, no maximum length is specified for the generator tap conductors. But because the tap conductors terminate in a single circuit breaker or set of fuses rated or set for the tap conductor ampacity, tap conductors up to 25 ft long would comply with the basic concept given in Sec. 240-21(c) for 25-ft taps. And Sec. 445-5, which is referenced, requires the tap conductors to have an ampacity of at least 115 percent of the generator nameplate current rating.

Section 240-21(e) is another departure from the rule that conductors must be provided with overcurrent protection at their supply ends, where they receive current from

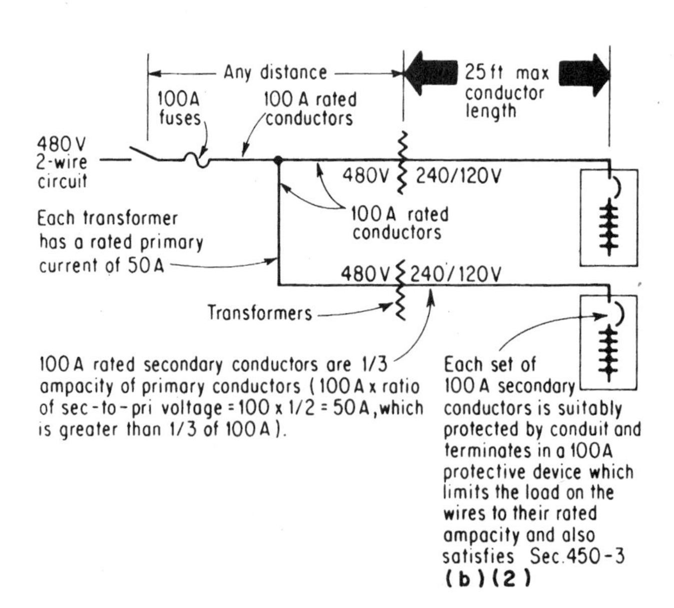

FIG. 4-33 Sizing a 25-ft tap and transformer protection.

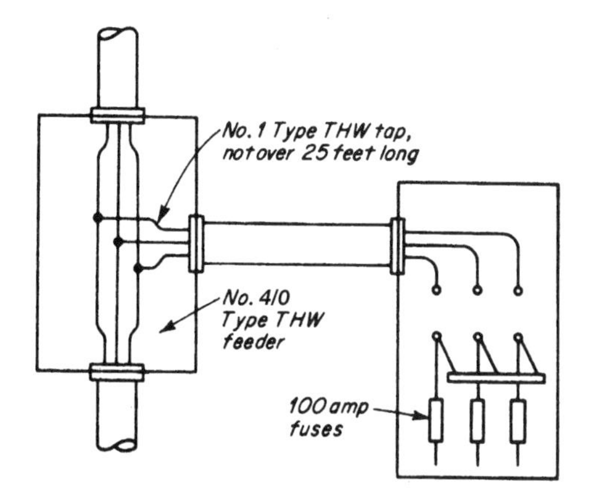

25 - ft tap — 240-21(1c)

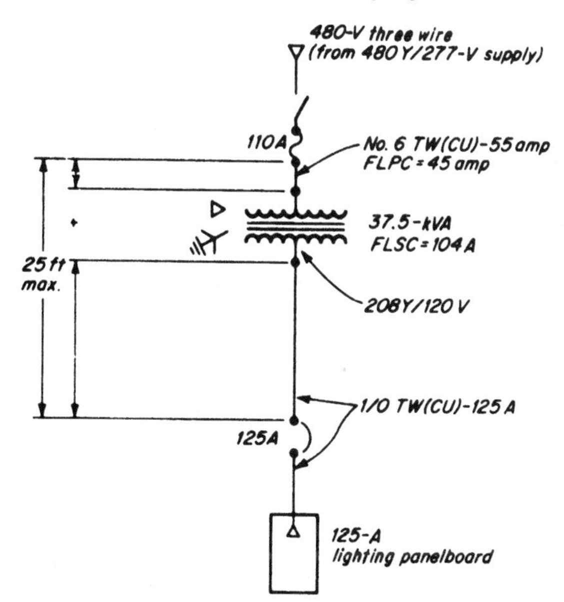

25 - ft tap — 240-21(d)

FIG. 4-34 Examples show the difference between the two types of 25-ft taps.

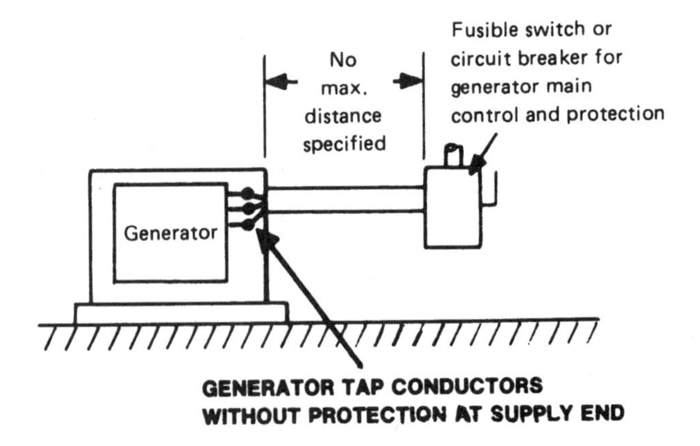

FIG. 4-35 Unprotected tap may be made from a generator's output terminals to the first overcurrent device.

other larger conductors or from a transformer. Section 240-21(e) permits a longer length than the 10-ft unprotected tap of Sec. 240-21(b) and the 25-ft tap of Sec. 240-21(c). Under specified conditions that are similar to the requirements of the 25-ft-tap exception, an unprotected tap up to 100 ft in length may be used in "high-bay manufacturing buildings" that are over 35 ft high *at the walls*—but only "where conditions of maintenance and supervision assure that only qualified persons will service the system." Obviously, that last phrase can lead to some very subjective and individualistic determinations by the authorities enforcing the Code. And the phrase "35 ft high at the walls" means that this rule cannot be applied where the height is over 35 ft at the peak of a triangular or curved roof section but less than 35 ft at the walls.

The 100-ft-tap exception must meet specific conditions:

1. From the point at which the tap is made to a larger feeder, the tap must not have more than 25 ft of its length run horizontally, and the sum of horizontal run and vertical run must not exceed 100 ft. Figure 4-36 shows some of the almost limitless configurations of tap layout that would fall within the dimension limitations.
2. The tap conductors must have an ampacity equal to at least one-third of the rating of the overcurrent device protecting the larger feeder conductors from which the tap is made.
3. The tap conductors must terminate in a circuit breaker or fused switch, where the rating of overcurrent protection is not greater than the tap conductor ampacity.
4. The tap conductors must be protected from physical damage and must be installed in metal or nonmetallic raceway.
5. There must be no splices in the total length of each of the conductors of the tap.
6. The tap conductors must not be smaller than No. 6 copper or No. 4 aluminum.
7. The tap conductors must not pass through walls, floors, or ceilings.
8. The point at which the tap conductors connect to the feeder conductors must be at least 30 ft above the floor of the building.

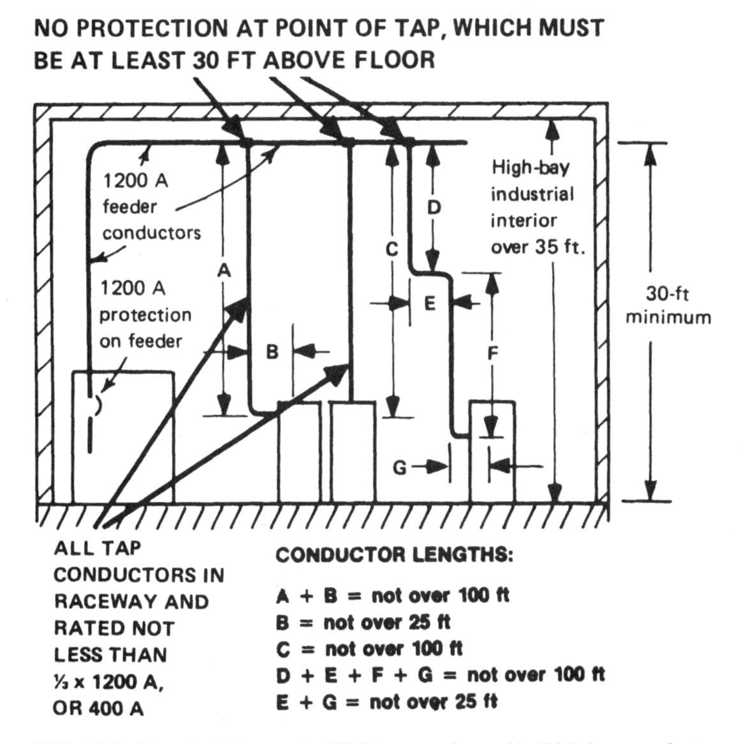

FIG. 4-36 Unprotected taps up to 100 ft long may be used in "high-bay manufacturing buildings."

As shown in Fig. 4-36, the tap conductors from a feeder protected at 1200 A are rated at not less than one-third the protection rating, or 400 A. Although 500-kcmil THW copper is rated at 380 A, that value does not satisfy the minimum requirement for 400 A. But if 500-kcmil THHN or XHHW copper, with an ampacity of 430 A, were used for the tap conductors, the rule would be satisfied. However, in such a case, those conductors would have to be used as if their ampacity were 380 A for the purpose of load calculation because of Sec. 110-14(c) of the Code, the general UL rule of 75°C conductor terminations for connecting to equipment rated over 100 A, such as the panelboard, switch, motor-control center, or other equipment fed by the taps. And the conductors for the main feeder being tapped could be rated less than the 1200 A shown in the sketch if the 1200-A protection on the feeder was selected in accordance with Sec. 430-62 or Sec. 430-63 for supplying a motor load or motor and lighting load. In such cases, the overcurrent protection may be rated considerably higher than the feeder conductor ampacity. But the tap conductors must have ampacity at least equal to one-third *the feeder protection rating.*

Section 240-21(j) applies exclusively to industrial electrical systems. Conductors up to 25 ft long may be tapped from a transformer secondary without overcurrent protection at their supply end and without need for a single circuit breaker or set of fuses at their load end. Normally, a transformer secondary tap up to 25 ft long must comply with the rules of Sec. 240-21(d), which call for such a transformer secondary tap to be made with conductors that require no overcurrent protection at their supply end but are required to terminate at their load end in a single CB or single set of fuses with a setting or rating not over the conductor ampacity. However, Sec. 240-21(j) permits a 25-ft tap from a transformer secondary without termination in a single main overcurrent device, *but* it limits the application to "industrial installations." The tap conductor ampacity must be at least equal to the transformer's secondary current rating and must be at least equal to the sum of the ratings of overcurrent devices supplied by the tap conductors. The conductors could come into main lugs only of a power panel if the conductor ampacity is at least equal to the sum of the ratings of the protective devices supplied by the bus bars in the panel. Or the tap could be made to an auxiliary gutter from which a number of individually enclosed circuit breakers or fused switches are fed from the tap—provided that the ampacity of the tap conductors is at least equal to the sum of the ratings of protective devices supplied.

An example of the application of Sec. 240-21(j) is shown at the top of Fig. 4-37. If that panel contains eight 100-A circuit breakers (or eight switches fused at 100 A), then

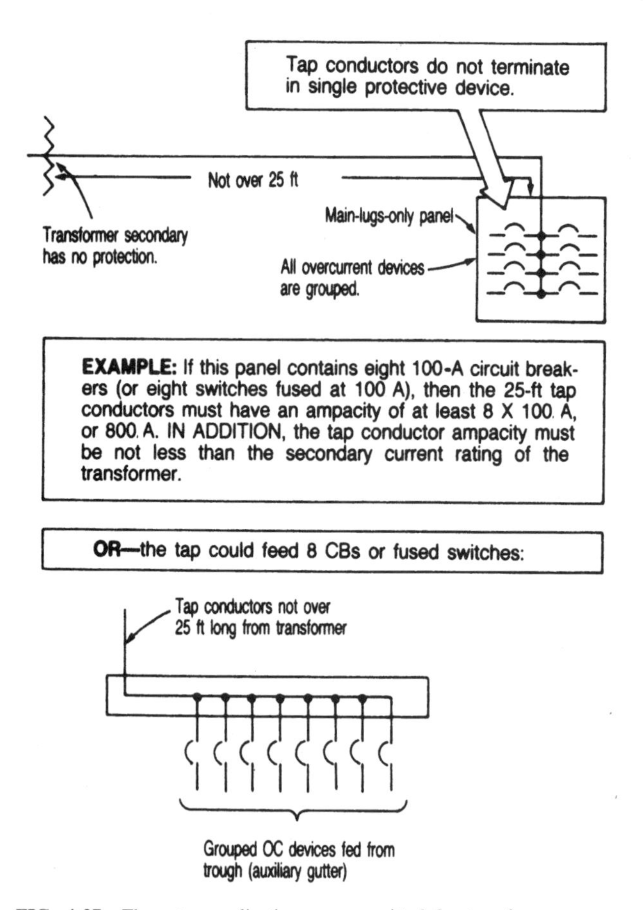

FIG. 4-37 These tap applications are permitted for transformer secondaries only in "industrial" electrical systems.

the 25-ft tap conductors must have an ampacity of at least 8×100 A, or 800 A. In addition, the tap conductor ampacity must be not less than the secondary current rating of the transformer. The layout of a similar application at an auxiliary gutter is shown at the bottom of Fig. 4-37.

FEEDER SIZING VERSUS PANELBOARD RATING

The sizing of feeder conductors for continuous loading (such as lighting in schools, office buildings, stores, and other occupancies where lighting is on all day) raises problems in panelboard sizing. NEC Sec. 384-13 requires that any panelboard have bus bar ampacity at least equal to the NEC minimum feeder conductor capacity required by Art. 220. (See Fig. 4-38.) In effect, the clear, straightforward rule of Sec. 384-13 calls for a panelboard

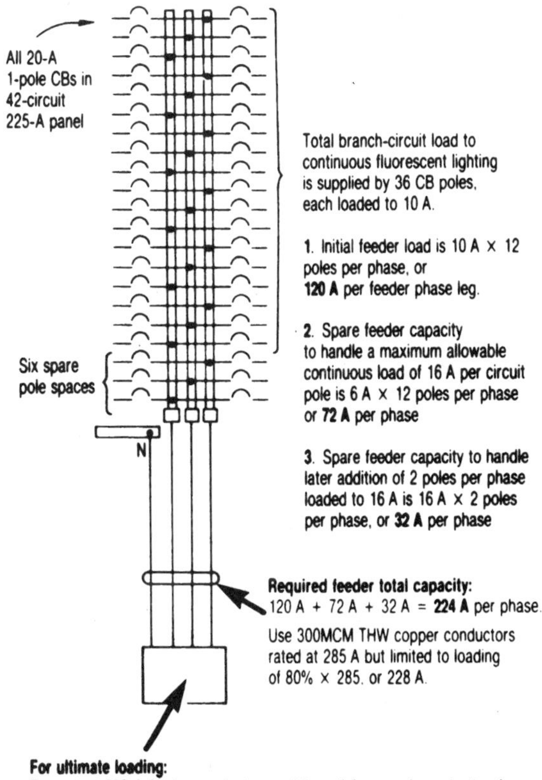

FIG. 4-38 For real spare capacity, the feeder and all circuit components must be sized to carry the ultimate load.

to have a bus bar ampacity that is not less than the minimum ampacity required for the feeder conductors that supply the sum of the branch-circuit loads fed from the panel. When the minimum required feeder ampacity is calculated for the total of branch-circuit loads in accordance with Sec. 220-10, the panelboard supplied by that feeder must have bus bars with an ampacity rating at least equal to the minimum feeder ampacity. But here is where a controversy starts for feeders with continuous (3-h or more) loading:

1. Section 220-10(b) sets two different possible minimum values for feeder ampacity, depending upon the type of overcurrent device protecting the feeder.

2. If the circuit breaker or fusible switch protecting the feeder conductors is not "listed" (such as by UL) for continuous operation at 100 percent of its current rating, the CB or fuses in the switch must be rated at least at 125 percent of the continuous-load current; the feeder conductors must also have ampacity not less than 125 percent of the continuous-load current. If the feeder protective device is listed for 100 percent continuous loading, then the minimum required feeder conductor ampacity is simply the feeder load current.

3. As a result of the interaction of Secs. 384-13 and 220-10(b), the minimum required panel bus bar rating depends on the type of protective device (100 percent rated or non-100 percent rated) used in the feeder that supplies the panel.

4. A continuous minimum feeder load current of 176 A must have feeder conductors (and therefore a panelboard bus bar rating) of not less than 125 percent of 176 A, or 220 A, if the CB or fusible switch to the panel is not listed for 100 percent continuous loading. Only a small percentage of feeder protective devices are so listed.

5. As shown in Fig. 4-39, the impact of these rules on panelboard bus bar rating is to produce illogical conditions, because panelboard sizing is not related to the reason behind the 125 percent multiplier in Sec. 220-10(b). A circuit breaker or fusible-switch assembly that is not listed for 100 percent continuous loading would generate excessive heat within its own parts if it were operated at its full-current rating continuously, and such heat would damage conductor insulation. For that reason, the 125 percent multiplier of Sec. 220-10(b) establishes the same limitation as the general UL rule that (unless marked otherwise) a circuit breaker must not be loaded continuously to more than 80 percent of its rating (which is the reciprocal of having a rating of at least 125 percent of the continuous-load current). Circuit breakers and fusible switches that are listed for 100 percent continuous loading are constructed and/or ventilated to prevent the generation of excessive heat.

6. Although the letter of the NEC rules requires the panel bus bar ratings shown in Fig. 4-39, there is no safety compromise in sizing the bus bars at no more than 100 percent of continuous load, because application of the panelboard is independent of the conditions that dictate the 125 percent multiplier for feeder protection and feeder conductors. For instance, sizing of the panel bus bars at 125 percent of 176 A calls for a panel of not less than 1.25 × 176 A, or 220-A, bus bar capacity, dictating the use of a 225-A panel. But safety and the rule of NEC Sec. 220-3(d) would be fully served by using a panel with bus bars rated for 100 percent of the minimum "computed" load of 176 A (although a 225-A panel is the next-larger standard size above 176 A). The use of a panel rated at 125 A (on the basis that the actual load is only 116 A) would be contrary to the requirement of Sec. 220-3(d) calling for the "wiring system" (including the panelboard) to be sized as a minimum on a watts-per-square-foot basis.

Important: In applications such as described above, it is extremely important to distinguish between a "lighting and appliance" panelboard (as defined in Sec. 384-14) and other panelboards. Because a lighting and appliance panelboard requires main overcur-

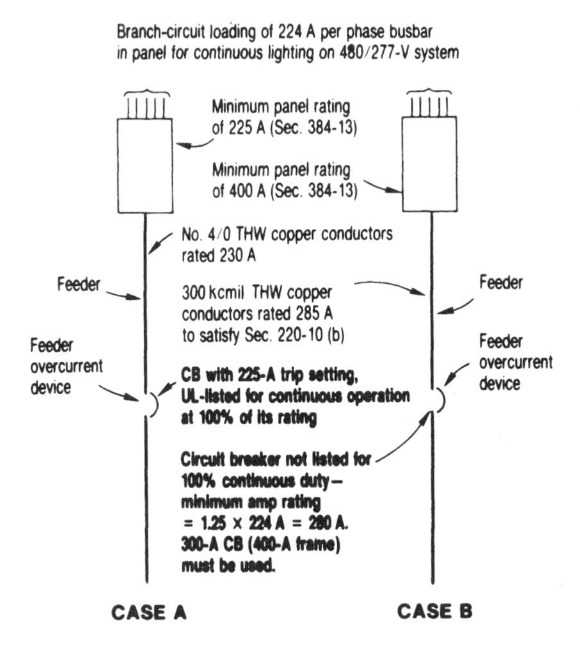

FIG. 4-39 Sizing of feeder conductors and overcurrent protection affects the sizing of panelboards.

rent protection (on the supply side of the panel) rated not over the panel bus bar rating, when the rating of the feeder overcurrent device is increased to 125 percent of the continuous feeder load, a panel for which the feeder overcurrent device serves as main protection must have a bus bar rating not lower than the current rating of the feeder protection, as shown in case B of Fig. 4-39. The feeder CB may not be rated less than 300 A, and that value of protection would not satisfy as "mains protection" for a 225-A panel at the load end of the feeder. The next standard size of panelboard above 225 A is 400 A, and that would be properly protected by the 300-A feeder CB (or fuses). Of course, a 225-A panel with its own 100 percent rated main 225-A CB or fused switch could be used in case B, instead of the 400-A panel. This is a typical example of the need to interrelate and satisfy all NEC rules bearing on each design situation.

Figure 4-40 shows how the rules of Secs. 220-10(b) and 384-13 combine to dictate the minimum bus bar rating of a branch-circuit or feeder panel that is not subject to the mains protection requirement for a lighting and appliance panel. The load on the panel is exactly the same in cases A and B. But because the feeder protection in case A is not rated for 100 percent continuous loading, the sizes of the feeder protection, feeder conductors, and panel must be increased because of the continuous load. In case B, the 125 percent multiplier is not required. Because the panel in case A does not require a main protective device at its rating, there should be no problem in having an 800-A panel there instead of the 1200-A (or 1000-A) panel—even though the 900-A feeder protection exceeds the 800-A panel bus bar rating. The size of the panel should not be required to be increased to 125 percent of the continuous load simply because of the unrelated heat characteristics of a remotely located feeder protective device. However, the wording of Sec. 384-13 *does* dictate such upsizing of the panel bus bar rating.

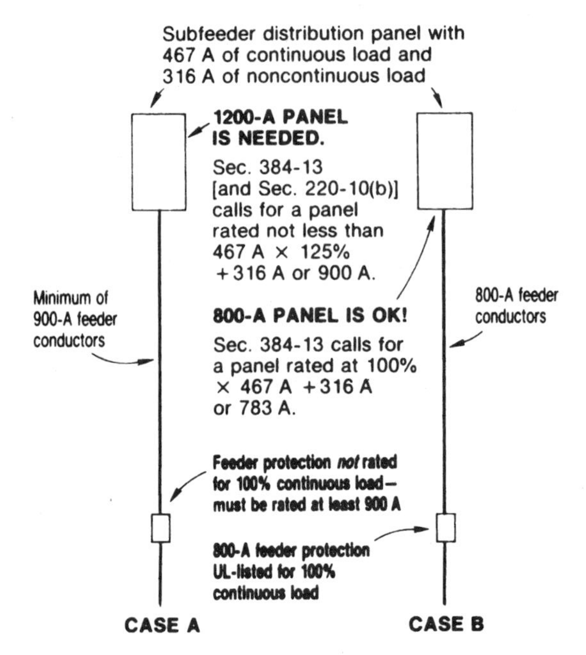

FIG. 4-40 NEC rules produce puzzling conflicts in sizing panels.

LIGHTING AND APPLIANCE BRANCH-CIRCUIT PANELBOARD

Even though a panelboard may be used largely for other than lighting purposes, it is to be judged under the requirements of Sec. 384-14 for lighting and appliance branch-circuit panelboards if it conforms to the specific conditions stated in the definition of Sec. 384-14.

Watch out for this definition! There are many panel makeups that supply no lighting and appear to be power panels or distribution panels, yet are technically lighting and appliance panels in accordance with the above definition and must have protection for the bus bars. Figure 4-41 shows an example of how it is determined whether a panelboard *is* or *is not* "a lighting and appliance branch-circuit panelboard." The determination is important because it indicates whether main protection is required for any particular panel, to satisfy Sec. 384-16.

Figure 4-42 shows panels that do not need main protection because they are not lighting and appliance panels, which are the only types of panels required by Sec. 384-16 to have main protection. Just as it is strange to identify a panel that supplies no lighting as a lighting panel, as in Fig. 4-41, it is also strange that some panels that supply *only* lighting, as in Fig. 4-42, are technically *not* lighting panels. Because of the definition of Sec. 384-14, if the protective devices in a panel are *all* rated over 30 A *or* if there are *no* neutral connections provided in the panel, then the panel is not a lighting and appliance panel and it does not require main overcurrent protection.

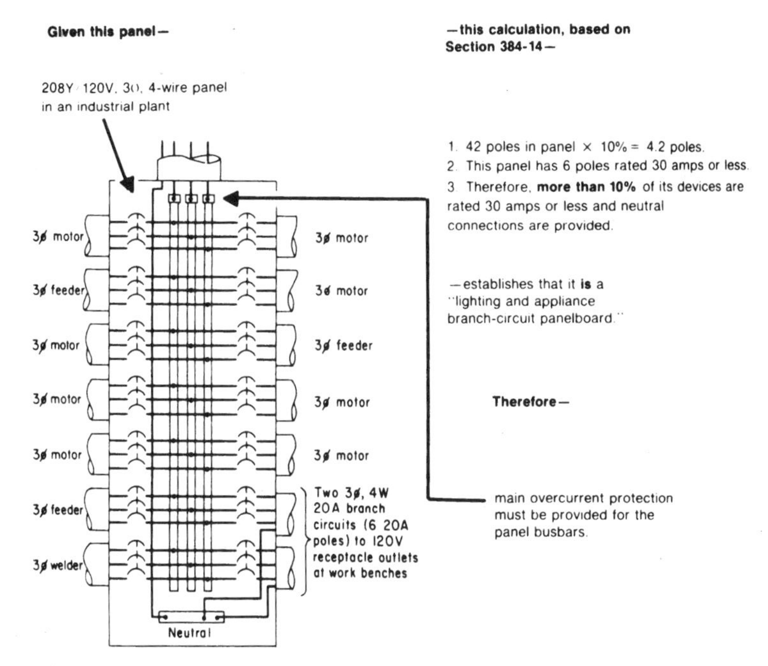

FIG. 4-41 Definition of a "lighting and appliance" panelboard hinges on a specific calculation.

PANELBOARD OVERCURRENT PROTECTION

Rules in this section of the NEC concern the protection of "lighting and appliance branch-circuit panelboards." In general, lighting and appliance branch-circuit panels must be individually protected on the supply side by not more than two main CBs or two sets of fuses having a combined rating not greater than that of the panelboard, as shown in Fig. 4-43. Individual protection is not required when a lighting and appliance branch-circuit panelboard is connected to a feeder which has overcurrent protection not greater than that of the panelboard (case 1 in Fig. 4-44), as noted in Exception 1.

Because of the wording of the definition in Sec. 384-14, it is vitally important to evaluate a panel carefully to determine if main protection is required.

Where a number of panels are tapped from a single feeder protected at a current rating higher than that of the bus bars in any of the panels, the main protection may be installed as a separate device just ahead of the panel or as a device within the panel feeding the bus bars (case 2 and case 3 in Fig. 4-44). The main protection would normally be a CB or fused switch of the number of poles corresponding to the number of bus bars in the panel.

Figure 4-45 shows other variations on the same protection requirements. As shown in the sketch with 400-A panels, it is often more economical to order the panels with bus bar capacity higher than required for the load on the panels so the panel bus rating matches the feeder protection, thereby eliminating the need for panel main protection.

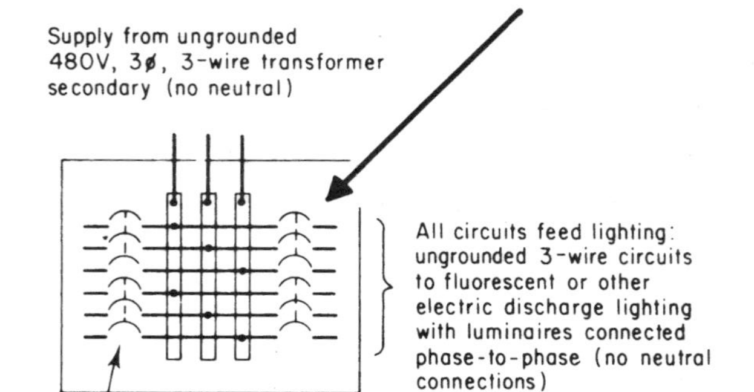

FIG. 4-42 Some panels that supply only lighting are technically not "lighting and appliance" panels.

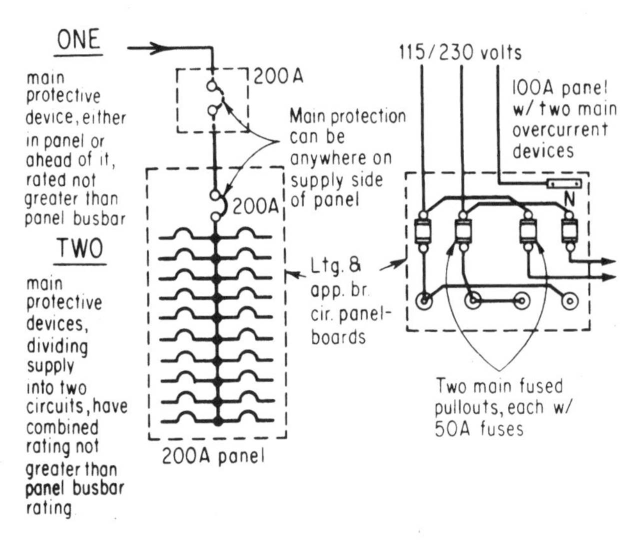

FIG. 4-43 "Main protection" may consist of one or two CBs or sets of fuses.

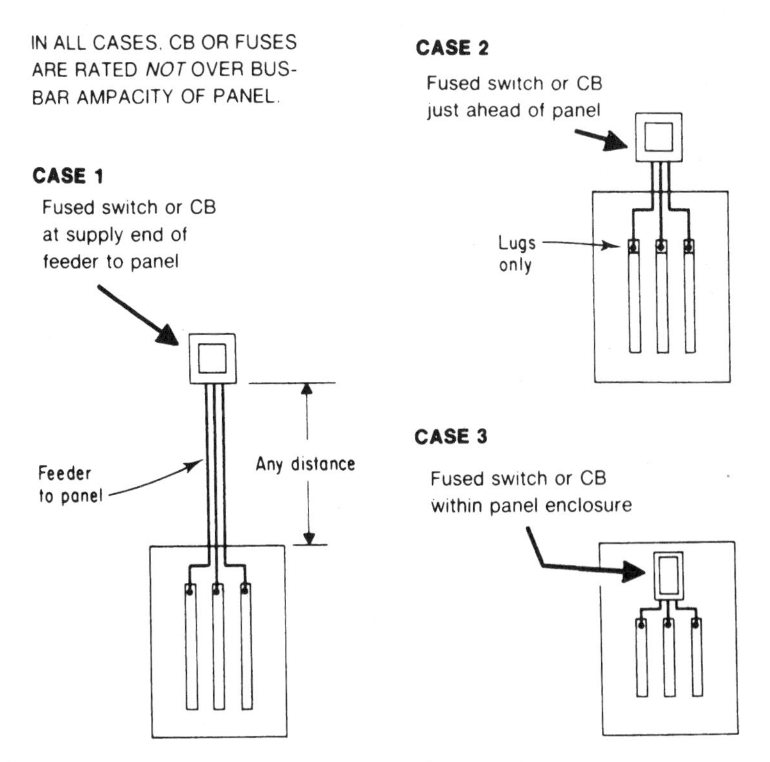

FIG. 4-44 Main panel protection may be at any of these locations.

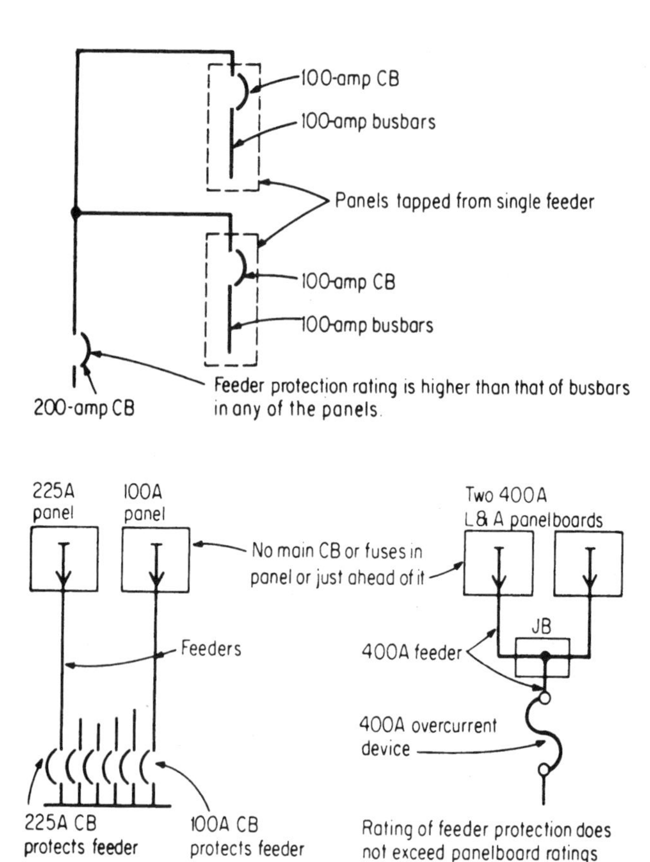

FIG. 4-45 Panel protection may be provided in a variety of ways.

Although part (a) of this section does spell out those general requirements for main protection of lighting and appliance branch-circuit panelboards, Exception 2 of that section notes that a panelboard used as residential service equipment for an "existing" installation may have up to *six* main protective devices [as permitted by Sec. 230-90(a), Exception 3]. Such usage is limited to "individual residential occupancy," such as a private house or an apartment in multifamily dwellings where the panel *is* truly service equipment and is not a subpanel fed from service equipment in the basement or from a meter bank on the load side of a building's service.

But note that the phrase "for existing installations" makes it a violation to use a service panelboard (load center) with more than two main service disconnect and protection devices for a residential occupancy in a new building. See Fig. 4-46. That applies to one-family houses and to "dwelling units" in apartment houses, condominiums, and the like. Use of, say, a split-bus panelboard with four or six main protective devices (such as 2-pole CBs) in the bus bar section that is fed by the service-entrance conductors is limited to use only for a service panel in "an individual residential occupancy" in an existing building, such as service modernization. See Fig. 4-47.

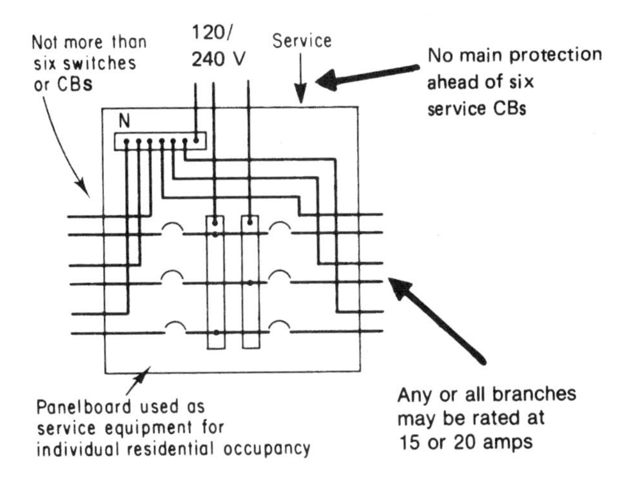

FIG. 4-46 Exception 2 eliminates need for main in residential service panel, *but only* where such a panel is installed as a replacement in an existing installation.

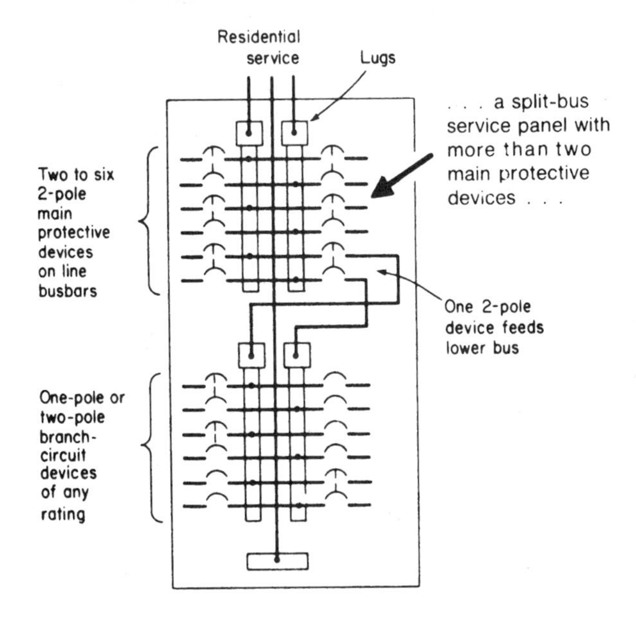

FIG. 4-47 Use of a split-bus load center with more than two main overcurrent devices is permitted for residential service equipment *only* in "existing installations."

FIG. 4-48 Any panelboard containing snap switches rated 30 A or less must have main or feeder protection rated not over 200 A.

Exception 2 brings the rule on overcurrent protection for residential service panels into agreement with the basic rule of Sec. 384-16(a) as it applies to all lighting and appliance panelboards in new construction.

The rules of Sec. 384-16(a) have been altered in past editions of the NEC as follows:

1. In the 1975 NEC, Exception 2 of Sec. 384-16(a) permitted up to six main protective devices in a residential service panel for *either* new work or additions. But the rule prohibited use of any 15- or 20-A CBs or fuses without main overcurrent protection ahead of them. Under that rule, a split-bus panel with up to six main CBs or sets of fuses could be used for a residential service, but any 15- or 20-A protective devices had to be installed in the panel bus section fed by the service-entrance conductors. The rule prohibiting 15- and 20-A protective devices without a main ahead of them ruled out use of small residential panels that had only six single- or 2-pole devices rated 15 or 20 A, with no main ahead. Such panels had been used for a long time for service to one-family houses that required only two small-appliance circuits and a few lighting circuits where the dwelling unit had no electric water heating, space heating, or major appliances.

2. In the 1978 NEC, Exception 2 was changed to permit use of 15- or 20-A protective devices without need for a main ahead of them, such as in the main bus bar section of a split-bus panel or in small six-circuit panels for service in not-all-electric houses or apartments.

3. Now, in the present NEC, for any new job, any panelboard—for service or otherwise, in any type of occupancy—may have no more than two main protective devices if the panel is a lighting and appliance panel (as determined from Sec. 384-14). Residential service panels do qualify as lighting and appliance panelboards.

Panelboards used for service equipment are required by NEC Sec. 230-70 to be marked as "suitable for use as service equipment," and panelboards are so marked.

The rule of part (b) of this section is covered in Fig. 4-48. Any panel, a lighting panel or a power panel, which contains snap switches (and CBs are not snap switches) rated 30 A or less must have overcurrent protection and not in excess of 200 A. Panels which are not lighting and appliance panels and do not contain snap switches rated 30 A or less do not have to be equipped with main protection and may be tapped from any size feeder. Figure 4-49 shows these two examples of overcurrent protection requirements for panelboards.

Part (c) applies to *any* overcurrent device in a panel and is a similar rule to those of Sec. 210-22(c) or Sec. 220-3(a), as shown in Fig. 4-50. The only exception to the rule is for overcurrent device assemblies that are "approved" (which means UL-listed) for continuous loading at 100 percent of their current rating. Refer to Sec. 220-10(b) on feeder protection.

Part (d) of this section applies to a panelboard fed from a transformer. The rule requires that overcurrent protection for such a panel, as required in (a) and (b) of the same section, be located on the secondary side of the transformer. An exception is made for a panel fed by a 2-wire, single-phase transformer secondary. Such a panel may be protected by a primary device. This concept of prohibiting use of panel protection on the primary side of a transformer feeding the panel is consistent with the rules covered under Sec. 240-3(i). Refer to the discussion there.

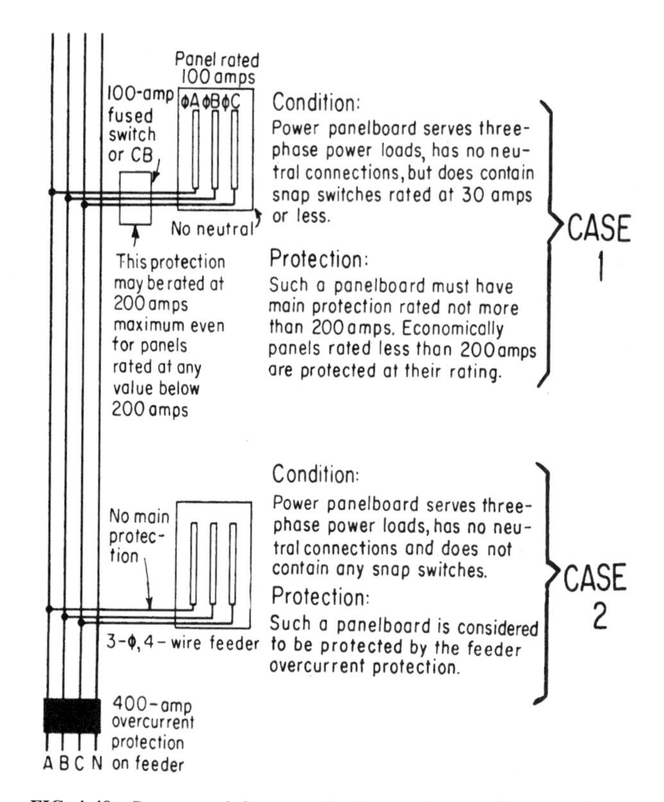

FIG. 4-49 Power panels have very limited requirements for protection.

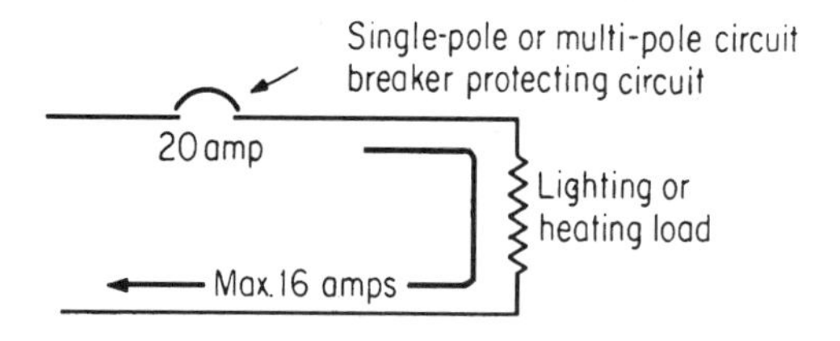

Total load on circuit must not exceed 80% of circuit rating when load is continuous (operates for 3 hours or more) - such as store lighting. CB must not be loaded over 80% of its rating.

FIG. 4-50 This applies for all fuses *and* breakers in a panelboard.

CHAPTER 5
SERVICES

The service for any building consists of the conductors and equipment used to deliver electric energy from the utility supply lines (or from an on-site generator) to the distribution system of the building (or other premises being supplied). Service may be brought to a building either overhead or underground—from a utility pole line or from an underground transfer vault. Figure 5-1 shows the various component parts of services.

The word *service* includes all the materials and equipment involved with the transfer of electric power from the utility distribution line to the electrical wiring system of the premises being supplied. Although service layouts vary widely, depending upon the voltage and ampere rating, the type of premises being served, and the type of equipment selected to do the job, every service generally consists of *service drop* conductors (for overload service from a utility pole line) or *service lateral* conductors (for an underground service from either an overhead or underground utility system), along with metering equipment, some type of switch or circuit-breaker control, overcurrent protection, and related enclosures and hardware. A typical layout of "service" for a one-family house breaks down as shown in Fig. 5-1.

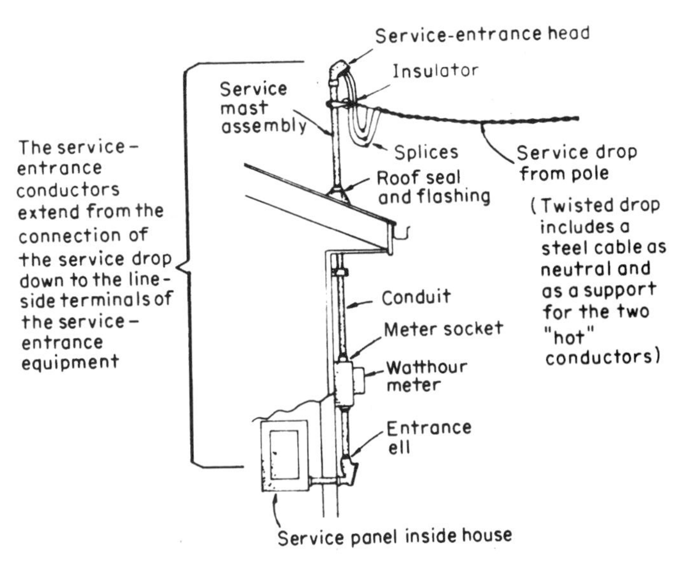

FIG. 5-1 Typical service arrangement for a residential building.

That part of the electrical system which directly connects to the utility supply line is referred to as the *service entrance.* Depending upon the type of utility line serving the building, there are two basic types of service entrances: overhead and underground.

The overhead service has been the most commonly used type. In a typical example, the utility supply line is run on wood poles along the street property line or back-lot line of the building, and a cable connection is made overhead from the utility line to a bracket installed high up on the building (Fig. 5-2). This line of wood poles also carries the telephone lines, and the poles are generally called *telephone* poles.

The aerial cable that runs from the overhead utility lines to the bracket on the outside wall of the building is called the *service drop.* This cable is installed by the utility line worker. At the bracket which terminates the service drop, conductors are then spliced to the drop cable conductors to carry power down to the electric meter and into the building.

In the underground service, the conductors that run from the utility line to the building are carried underground. Such an underground run to a building may be tapped from either an overhead utility pole line or an underground utility distribution system. Figure 5-3 compares overhead with underground services. Although underground utility services tapped from a pole line at the property line have been used for many years to eliminate the unsightliness of overhead wires coming to a building, the use of underground service tapped from an underground utility system has only recently started to gain widespread usage in residential areas. This latter technique is called *URD,* for underground residential distribution.

As noted above, when a building is supplied by an overhead drop, an installation of conductors must be made on the outside of the building to pick up power from the drop conductors and carry it into the meter enclosure and service-entrance equipment (switch, CB, panelboard, or switchboard) for the building. On underground services, the supply conductors are also brought into the meter enclosure on the building and then run into the service equipment installed, usually, within the building. *Service conductors* is a general term that covers all the conductors used to connect the utility supply circuit or transformer to the service equipment of the premises served. This term includes *service drop* conductors, *service lateral* (underground service) conductors, and *service-entrance* conductors. In an overhead distribution system, the service conductors begin at the line pole where connection is made. If a primary line is extended to transformers installed outdoors on private property, the service conductors to the building proper begin at the secondary terminals of the transformers. Where the supply is from an underground distribution system, the service conductors begin at the point of connection to

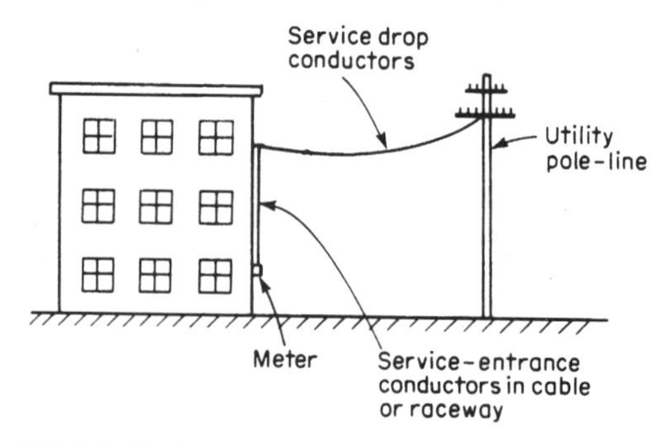

FIG. 5-2 Basic overhead service.

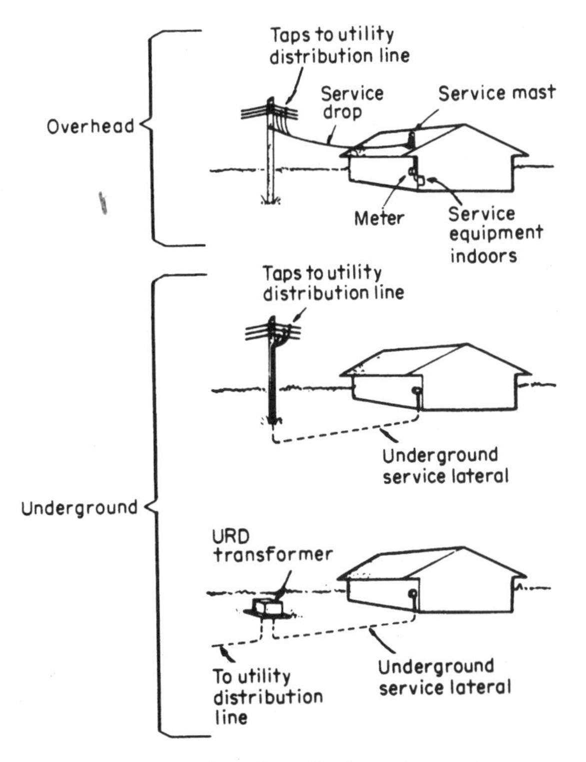

FIG. 5-3 Service may be made overhead *or* underground.

the underground street mains. In every case the service conductors terminate at the service equipment.

NUMBER OF SERVICES

Part (b) of Section 230-2 introduces a requirement that applies to any installation where more than one service is permitted by the Code to supply one building. It requires a "permanent plaque or directory" to be mounted or placed "at each service drop or lateral or at each service-equipment location" to advise personnel that there are other services to the premises and to tell where such other services are and what building parts they supply.

For any building, the service consists of the conductors and equipment used to deliver electric energy from the utility supply lines to the interior distribution system. Service may be made to a building either overhead or underground, from a utility pole line or from an underground transformer vault.

The first sentence of Sec. 230-2 requires that a building or structure be supplied by "only one service." Because *service* is defined in Art. 100 as "the *conductors* and equipment for delivering energy from the electricity supply system to the wiring system of the premises supplied," use of one "service" corresponds to use of one "service drop" or one

FIG. 5-4 One set of service drop conductors supplies building from utility line (coming from upper left) and two sets of SE conductors are tapped through separate metering CTs. (Sec. 230-2.)

"service lateral." Thus, the basic rule of this section requires that a building or other structure be fed by only one service drop (overhead service) or by only one set of service lateral conductors (underground service). As shown in Fig. 5-4, a building with only one service drop to it satisfies the basic rule even when more than one set of service-entrance (SE) conductors are tapped from the single drop (or from a single lateral circuit).

The seven exceptions to that basic rule cover cases where two or more service drops or laterals may supply a single building or structure.

Exceptions 1 and 2 permit a separate drop or lateral for supply to a fire pump and/or to emergency electrical systems, such as emergency lighting or exit lights.

Exception 2 to the basic rule that a building "shall be supplied by only one service" recognizes use of an additional power supply to a building from any "parallel power production systems." This would permit a building to be fed by a solar photovoltaic, wind, or other electric power source—in addition to a utility service—just as an emergency or standby power source is also permitted (Fig. 5-5).

Exception 3 recognizes another situation in which more than one service (that is, more than one service drop or lateral) may be used. By "special permission" of the inspection authority, more than one service may be used for a multitenant building when there is no single space that would make service equipment available to all tenants.

Two or more services to one building are permitted when the total demand load of all the feeders is more than 2000 A, up to 600 V, where a single-phase service needs more than one drop, or by special permission (Fig. 5-6). Exception 4 relates capacity to permitted services. Where requirements exceed 2000 A, two or more sets of service conductors may be installed. Below this value, special permission is required to install more than one set. The term *capacity requirements* appears to apply to the total calculated load for sizing service-entrance conductors and service equipment for a given installation.

Exception 5 requires special permission to install more than one service to buildings of large area. Examples of large-area buildings are high-rise buildings, shopping centers, and major industrial plants. In granting special permission the authority having

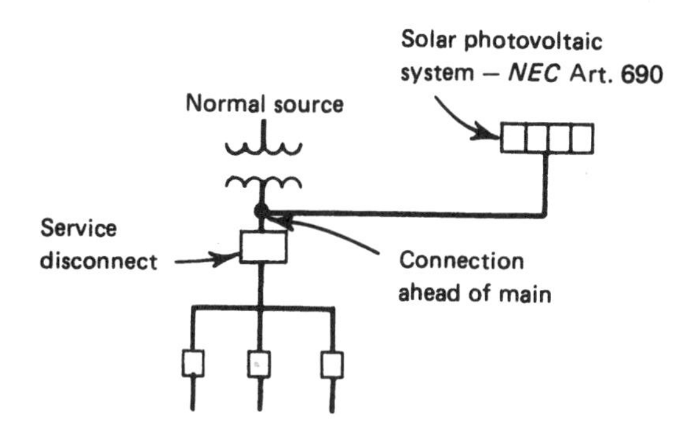

FIG. 5-5 Electric power generated by a solar voltaic assembly or by a wind-driven generator may be used as a source of power in "parallel" with the normal service.

jurisdiction must examine the availability of utility supplies for a given building, load concentrations within the building, and the ability of the utility to supply more than one service. Any of the special-permission clauses in the exceptions in Sec. 230-2 require close cooperation and consultation between the authority having jurisdiction and the serving utility.

Cases of separate light and power services to a single building and separate services to water heaters for purposes of different rate schedules are also exceptions to the general rule of single service. And if a single building is so large that one service cannot handle the load, special permission can be given for additional services. Exception 6 is illustrated at the bottom of Fig. 5-6.

Exception 7 to the basic rule requiring that any "building or other structure" be supplied by "only one service" adds an important qualification of that rule as it applies only to Sec. 230-40, Exception 2, covering service-entrance layouts where two to six service disconnects are to be fed from one drop or lateral and are installed in separate individual enclosures at one location, with each disconnect supplying a separate load. As described in Sec. 230-40, Exception 2, such a service equipment layout may have a separate set of service-entrance conductors run to *"each or several"* of the two to six enclosures. Section 230-2, Exception 7, notes that where a separate set of underground conductors of size 1/0 or larger is run to each or several of the two to six service disconnects, the several sets of underground conductors are considered to be one service (that is, one service lateral) even though they are run as separate circuits, that is, connected at their supply end (at the transformer on the pole or in the pad-mount enclosure or vault) *but not* connected at their load ends. The several sets of conductors are taken to be "one service" in the meaning of Sec. 230-2, although they actually function as separate circuits (Fig. 5-7).

Although Sec. 230-40, Exception 2, applies to "service-entrance conductors" and service equipment layouts fed by *either* a "service drop" (overhead service) or a "service lateral" (underground service), Exception 7 is addressed specifically and only to service "lateral" conductors (as indicated by the word *underground*) because of the need for clarification based on the Code definitions of *service drop; service lateral; service-entrance conductors, overhead system;* and *service-entrance conductors, underground system.* (Refer to these definitions in the Code book to clearly understand the intent of Exception 7 and its relation to Sec. 230-40, Exception 2.)

Ex. No. 4

1. . . .when the total demand load of all feeders is greater than 2000 amps (up to 600 volts), or

2. . . .when the load demand of a single-phase installation is higher than the utility's normal maximum for a single service, or

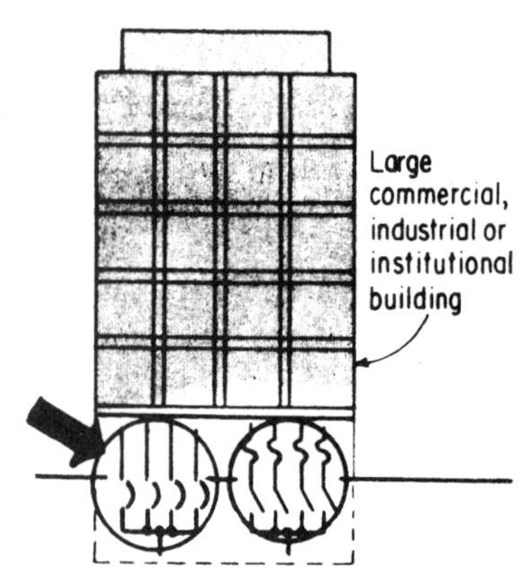

3. . . .when special permission is obtained from the inspection authority.

NOTE: "Two or more services" means two or more service drops or service laterals—not sets of service-entrance conductors tapped from one drop or lateral.

Ex. No. 6

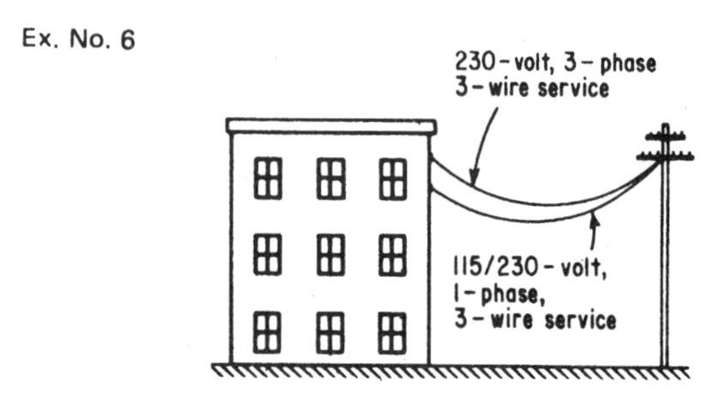

FIG. 5-6 Exceptions to Sec. 230-2 permit two or more services under certain conditions.

The 1975 *NE Code* had this limitation on service laterals (and this is still acceptable)—

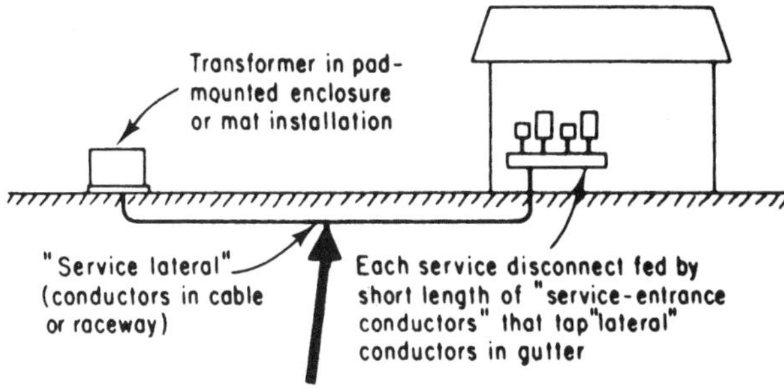

"Service lateral" conductors are not "service entrance" conductors and were, therefore, not applicable to the subdivision permission of Section 230-40, Exception No. 2. The requirement of Section 230-2 for one set of service lateral conductors demanded one circuit of single-conductor or parallel-conductor makeup.

Now, Exception No. 7 considers this type of hookup to be one set of service lateral conductors —

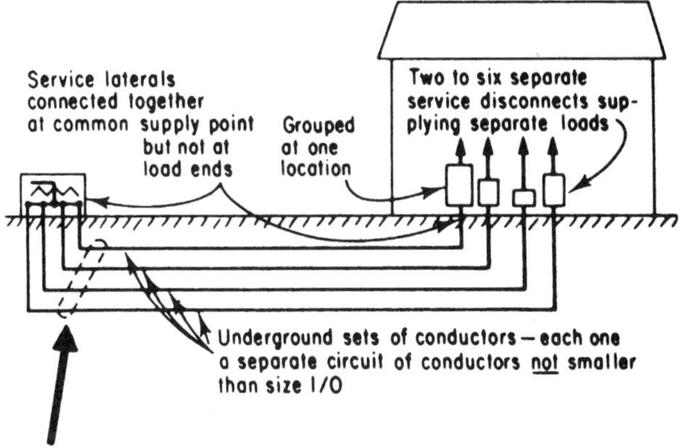

This is **one** service lateral, in the meaning of the basic rule of Section 230-2.

FIG. 5-7 "One" service lateral may be made up of several circuits.

The matter involves these separate but related considerations:

1. Because a "service lateral" may (and usually does) run directly from a transformer on a pole or in a pad-mount enclosure to gutter taps where short tap conductors feed the terminals of the service disconnects, most layouts of that type literally do not have any "service-entrance conductors" that would be subject to the application permitted by Sec. 230-40, Exception 2—other than the short lengths of tap conductors in the gutter or box where splices are made to the lateral conductors.

2. Because Sec. 230-40, Exception 2, refers only to sets of "service-entrance conductors" as being acceptable for individual supply circuits tapped from *one* drop or lateral to feed the separate service disconnects, that rule clearly does not apply to service lateral conductors which by definition are not service-entrance conductors. So there is no permission in Sec. 230-40, Exception 2, to split up service lateral capacity. And the basic rule of Sec. 230-2 has the clear, direct requirement that a building or structure be supplied through only *one* lateral for any underground service. That is, either a service lateral must be a single circuit of one set of conductors; or if circuit capacity requires multiple conductors per phase leg, the lateral must be made up of sets of conductors in parallel—connected together at *both* the supply and load ends—in order to constitute a single circuit (that is, one lateral).

3. Exception 7 permits "laterals" to be subdivided into separate, nonparallel sets of conductors in the way that Sec. 230-40, Exception 2, permits such use for "service-entrance conductors"—*but only* for conductors size 1/0 and larger and *only* where each separate set of lateral conductors (each separate lateral circuit) supplies *one* or *several* of the two to six service disconnects.

Exception 7 recognizes the importance of subdividing the total service capacity among a number of sets of smaller conductors rather than a single parallel circuit (that is, a number of sets of conductors connected together at *both* their *supply and load* ends). The single parallel circuit would have much lower impedance and would, therefore, require a higher short-circuit interrupting rating in the service equipment. The higher impedance of each separate set of lateral conductors (not connected at their load ends) would limit short-circuit current and reduce short-circuit duty at the service equipment, permitting lower IC (interrupting capacity) rated equipment and reducing the destructive capability of any faults at the service equipment.

NUMBER OF SERVICE-ENTRANCE CONDUCTOR SETS

As a logical follow-up to the basic rule of Sec. 230-2, which requires that a single building or structure be supplied "by only one service" (that is, only one service drop or lateral), this rule calls for only one set of SE conductors to be supplied by each service drop or lateral that is permitted for a building. Exception 1 covers a multiple-occupancy building (a two-family or multifamily building, a multitenant office building, or a store building, etc.). In such cases, a set of SE conductors for each occupancy or for groups of occupancies is permitted to be tapped from a single drop or lateral (Fig. 5-8).

When a multiple-occupancy building has a separate set of SE conductors run to each occupancy, in order to comply with Sec. 230-70(a), the conductors should either be run on the outside of the building to each occupancy or, if run inside the building, be encased in 2 in of concrete or masonry in accordance with Sec. 230-6. In either case the service equipment should be located "nearest to the entrance of the conductors inside the building," and each occupant would have "access to his disconnecting means."

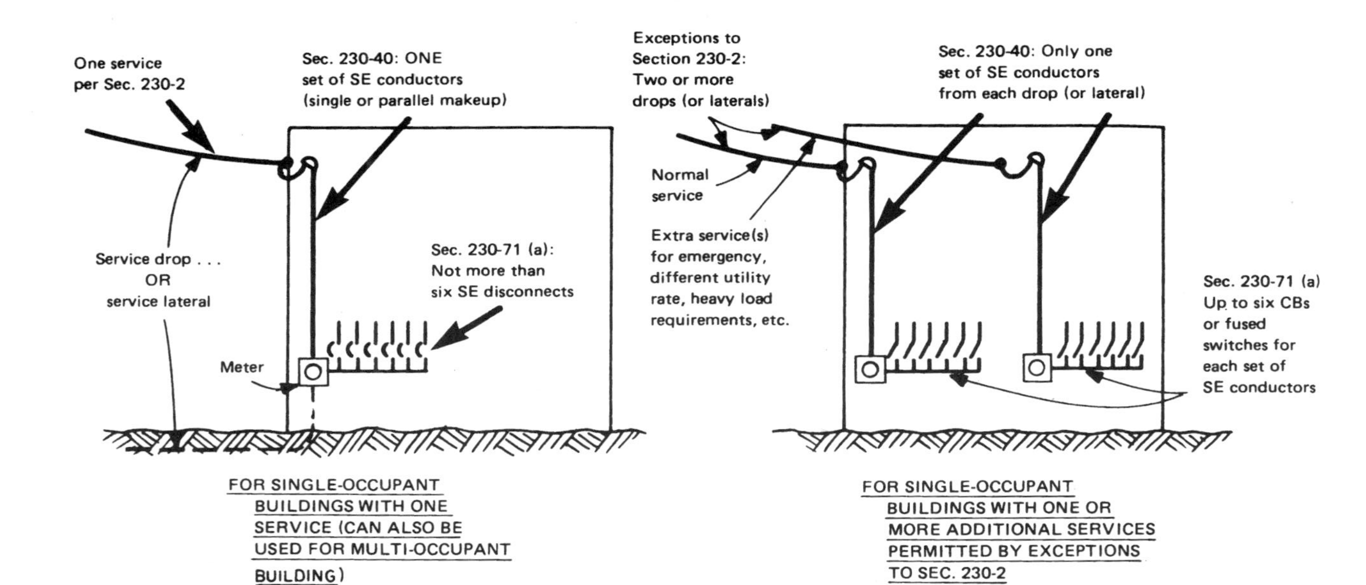

FIG. 5-8 Service layouts must simultaneously satisfy Secs. 230-2, 230-40, and 230-71 and all other NEC rules that are applicable.

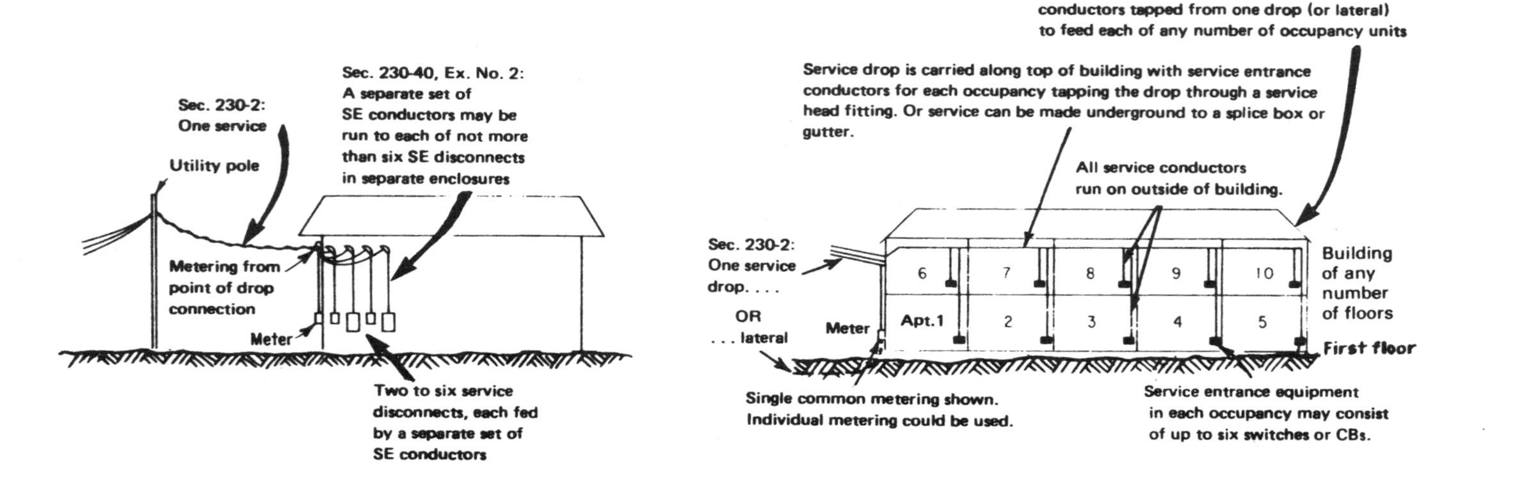

FIG. 5-8 (*Continued*) Service layouts must simultaneously satisfy Secs. 230-2, 230-40, and 230-71 and all other NEC rules that are applicable.

Any desired number of sets of service-entrance conductors may be tapped from the service drop or lateral, or two or more subsets of service-entrance conductors may be tapped from a single set of main service conductors, as shown for the multiple-occupancy building in Fig. 5-8.

Exception 2 permits two to six disconnecting means to be supplied from a single service drop or lateral where each disconnect supplies a separate load (Fig. 5-9). Exception 2 recognizes the use of, say, six 400-A sets of service-entrance conductors to a single-occupancy or multiple-occupancy building in lieu of a single main 2500-A service. It recognizes the use of up to six subdivided loads extending from a single drop or lateral in a *single-occupancy* as well as multiple-occupancy building. Where single metering is required, doughnut-type CTs could be installed at the service drop.

The real importance of this rule is to eliminate the need for "paralleling" conductors of large-capacity services, as widely required by inspection authorities to satisfy previous editions of the NEC (Fig. 5-9). This same approach could be used in subdividing services into smaller load blocks to avoid the use of the equipment ground-fault circuit protection required by Sec. 230-95.

This rule can also facilitate expansion of an existing service. Where less than six sets of service-entrance conductors were used initially, one or more additional sets can be installed subsequently without completely replacing the original service. Of course, metering considerations will affect the layout. *But* the two to six disconnects (circuit breakers or fused switches) must be installed close together at one location, and not spread out in a building.

SIZE AND RATING

Sizing of service-entrance conductors involves the same type of step-by-step procedure as set forth for sizing feeders covered in Art. 220. A set of service-entrance conductors is sized just as if it were a feeder. In general, the service-entrance conductors must have a minimum ampacity—current-carrying capacity—selected in accordance with the ampacity tables and rules of Sec. 310-15, sufficient to handle the total lighting and power load as calculated in accordance with Art. 220. Where the Code gives demand factors to use or allows the use of acceptable demand factors based on sound engineering determination of less than 100 percent demand requirement, the lighting and power loads may be modified.

According to the exception of Sec. 230-42(a), the maximum usable ampacity of busways used as service-entrance conductors must be taken to be the ampere value for which the busway has been listed or labeled. This is an exception to the basic rule that requires the ampacity of service-entrance conductors to be "determined from Section 310-15"—which does not give ampacities of busways.

From the analysis and calculations given in the feeder-circuit section, a total power and lighting load can be developed to use in sizing service-entrance conductors. Of course, where separate power and lighting services are used, the sizing procedure should be divided into two separate procedures.

When a total load has been established for the service-entrance conductors, the required current-carrying capacity is easily determined by dividing the total load in kilovoltamperes (or kilowatts with proper correction for power factor of the load) by the voltage of the service.

From the required current rating of conductors, the required size of conductors is determined. Sizing of the service neutral is the same as for feeders. Although suitably insulated conductors must be used for the phase conductors of service-entrance feeders,

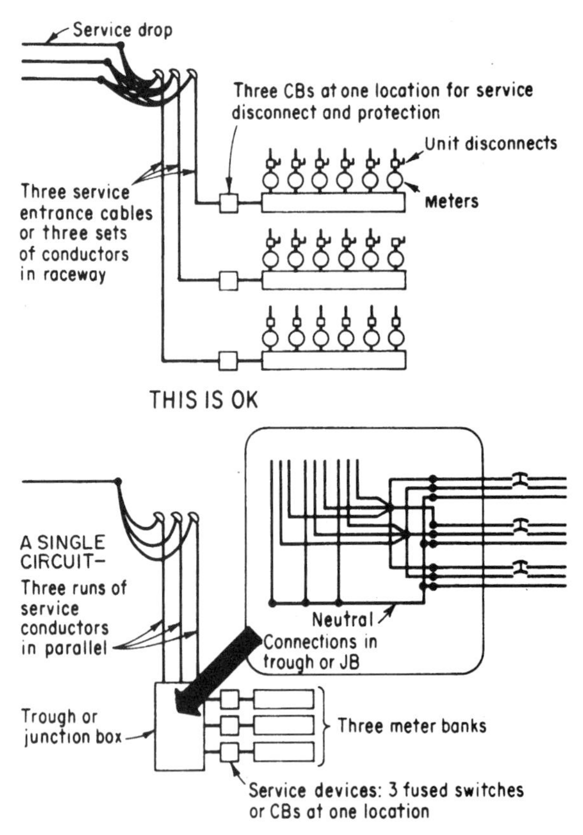

THIS WAS COMMONLY REQUIRED TO SATISFY PREVIOUS *NE CODE* BUT IS NOT NOW NECESSARY

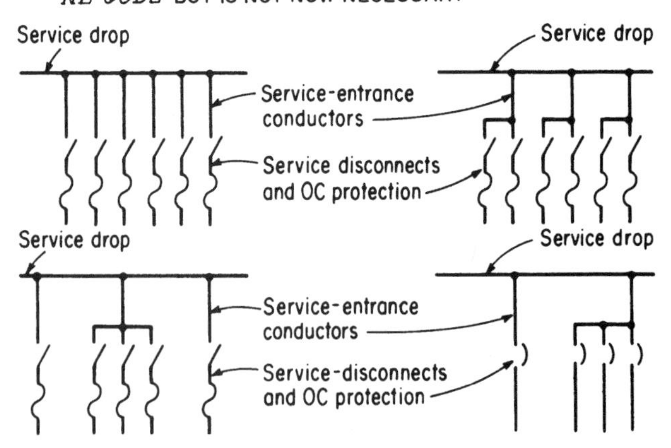

From two to six separate sets of service-entrance conductors may be supplied by a single service drop for either single- or multiple-occupancy buildings. Disconnects can be of same or different ratings, and each set of service-entrance conductors can be installed using any approved wiring method.

FIG. 5-9 Tapping sets of service-entrance conductors from one drop (or lateral).

the NEC does permit use of bare grounded conductors (such as neutrals) under the conditions covered in Secs. 230-30 and 230-41.

An extremely important element of service design is consideration of faults. Service busway and other service conductor arrangements must be sized and designed to ensure safe application with the service disconnect and protection. That is, service conductors must be capable of withstanding the let-through thermal and magnetic stresses on a fault.

After calculating the required circuits for all the loads in the electrical system, the next step is to determine the minimum required size of service-entrance conductors to supply the entire connected load. The NEC procedure for sizing SE conductors is the same as that for sizing feeder conductors for the entire load, as set forth in Sec. 220-10. Basically, the service "feeder" capacity must be not less than the sum of the loads on the branch circuits for the different applications.

The *general lighting load* is subject to demand factors from Table 220-11, which takes into account the fact that simultaneous operation of all branch-circuit loads, or even a large part of them, is highly unlikely. Thus, feeder capacity does not have to equal the connected load. The other provisions of Art. 220 are then factored in.

Part (b) of Sec. 230-42 makes a 100-A service conductor ampacity a mandatory minimum if the system supplied is a one-family dwelling with more than five 2-wire branch circuits (or the equivalent of that for multiwire circuits) or if a one-family dwelling has an initial computed load of 10,000 W. Now that three 20-A small-appliance branch circuits are required in a single-family dwelling, the average new home will need a 100-A, 3-wire service, because even without *electric* cooking, heating, drying, or water-heating appliances, more than five 2-wire branch circuits will be installed.

Just as the sizing of a feeder involves adding the continuous and noncontinuous current loads for lighting, heating, motors, and/or other power equipment and then applying any applicable demand factors, the service-entrance conductors are sized for the entire connected load in a building or other premises. The following simple example shows the procedure.

Example. Assume a 120,000-ft^2 store building has the following loads:

252 kVA of continuous lighting

73-kVA demand load for receptacles

116 kVA of electric heating

346 kVA of motor loads

38 kVA of appliances

The general approach to the sizing of service-entrance conductors might be as follows:

Step 1. Although the conductors must have sufficient capacity for 252 kVA of continuous lighting, the NEC requires a minimum feeder capacity of 3 W/ft^2 for general lighting, for a total of $120{,}000 \times 3 = 360{,}000$ W. Assuming the use of high-power-factor lighting equipment, which can be taken as 360 kVA, NEC Table 220-11 requires it be taken at 100 percent demand.

Step 2. The 73-kVA demand load for receptacles was calculated from NEC Sec. 220-13.

Step 3. The 116 kVA of electric space heating is a 100 percent demand figure, based on connected load and control provisions.

Step 4. The 346-kVA motor load is the sum of 125 percent of the load of the largest motor in the building and the demand loads of all the other motors (in accordance with NEC Secs. 430-24, 430-26, and 430-33, second paragraph).

Step 5. The 38 kVA of appliance load is the total load of all appliances that could be operating simultaneously.

The service-entrance conductors would thus seem to be required to have sufficient capacity for a total of

$$360 + 73 + 116 + 346 + 38 = 933 \text{ kVA}$$

On a 480Y/277-V system, the minimum required current capacity is most effectively calculated using a voltage value of 460 V, which is a multiple of 115 and 230 V. Using the lower value of voltage yields a higher value of current, with greater assurance of adequacy in the conductor size. (Where calculations are to be performed for a test—journeyman, master electrician, contractor, inspector, etc.—*remember,* Sec. 220-2 requires the use of "nominal voltages." For the purpose of testing, use these voltage values.) This gives

$$\frac{933{,}000}{460 \times 1.732} = 1171 \text{ A}$$

This result indicates that service-entrance conductors with 1171-A usable capacity would be adequate for the initial demand load.

But immediately the critical need to add even minimal spare capacity would dictate the use of conductors rated for at least 1200 continuous A (which is the lowest standard NEC rating of overcurrent protection above 1171 A). And it is always necessary to focus on "usable" ampacity, because NEC Sec. 220-10(b) requires "feeder" conductors (and, therefore, service-entrance feeder conductors) to have ampacity at least equal to "the noncontinuous load plus 125 percent of the continuous load" if the service protective device is not a circuit breaker listed for continuous loading at 100 percent of its rating or if the device is not a switch-and-fuse assembly listed for 100 percent continuous load. And Note 8 to NEC Table 310-16, if applicable here, requires ampacity derating for continuous or noncontinuous loads where more than three current-carrying conductors are installed in a raceway or in a cable or are directly buried in the earth.

In this example, where service conductors are being sized for a store, the minimum calculated lighting load of 360 kVA must be taken at 125 percent of its value because it is continuous (operating steadily for over 3 h). And any parts of the other loads that might operate continuously must also be taken at 125 percent of their values. Ventilation fans or air conditioning that would run for 3 h or more would be included. It might be determined, for instance, that 20 percent of the nonlighting load is continuous. Then of the 573 kVA of nonlighting load, 114.6 kVA (0.2 × 573 kVA) would be taken at 125 percent and the balance at 100 percent.

Still another, typical factor that would have to be considered is the effect of "noncoincident" loads on the required service conductor capacity. If it can be determined that none of the 116-kVA heating load will operate at the same time as the air-conditioning load, then either the heating load or the air-conditioning load, whichever is smaller, may be omitted from the total demand load, because capacity available for the larger of the two loads will more than satisfy the current draw of the smaller load. Of course, all conditions must be accounted for, such as simultaneous operation of the air conditioning and electric-resistance reheat coils in air-conditioning ducts.

If the air-conditioning load is greater than the 116-kVA electric space-heating load in the store building, then the nonheating part of the nonlighting load is 73 kVA + 346 kVA + 38 kVA, or a total of 457 kVA, of which, say, 20 percent is continuous. Then the sizing of the service conductors proceeds as follows:

Lighting load:	360 kVA × 1.25	450 kVA
Nonlighting load:	457 kVA × 0.2 × 1.25	114 kVA
Nonlighting load:	457 kVA × 0.8 × 1.00	366 kVA
Total demand		930 kVA

To handle that demand load, the conductor ampacity has to be

$$\frac{930{,}000}{460 \times 1.732} = 1167 \text{ A}$$

which indicates that service conductors with an ampacity of 1167 A would be adequate for the initial design load; but, again, there would be no spare capacity for load growth. And note that 1167 A of ampacity from Table 310-16 would be adequate only if the makeup of the circuit did not require derating because more than three current-carrying conductors were installed in a single raceway or in a cable, or directly buried.

If the 1167-A circuit is made up of conductors installed in conduit, multiple conductors will be used for the phase legs and the neutral, with three phases and a neutral in each of the multiple conduits. If the major part of the load on a service feeder is electric-discharge lighting, such as fluorescent, metal-halide, or high-pressure sodium, then the neutral conductor of the feeder must be counted as a current-carrying conductor. In such a case, the three-phase conductors and the neutral conductor have to be taken as four current-carrying conductors in each of the conduits making up the multiple 1167-A circuit, requiring the application of the rule of Note 8 to NEC Table 310-16, which calls for an ampacity derating factor of 80 percent to be applied to all the conductors in the service conduits.

If that 80 percent ampacity reduction factor must be applied to the conductors making up the phase legs for the above 1167-A service feeder, then the conductor selected from Table 310-16 to be used in multiple for each phase leg must have an ampacity value such that its derated ampacity, multiplied by the number of conductors making up the phase leg, equals the required ampacity of 1167 A. For instance, if the service is made up of four conduits, with each one containing a neutral conductor plus three 500-kcmil THW copper conductors for phases *A, B,* and *C* in that conduit, the derating factor of 80 percent has to be applied to the ampacity of 380 A shown in Table 310-16 for a 500-kcmil THW copper conductor. With each service feeder phase leg made up of four 500-kcmil THW copper conductors (one in each of the four conduits), the ampacity of each phase leg is

$$4 \text{ (conductors)} \times 380 \text{ A (per conductor)} \times 0.8 \text{ (derating factor)} = 1216 \text{ A}$$

This derated ampacity of 1216 A is the basic ampacity of the conductors under those conditions of application. It fully satisfies the previous calculated requirement for an ampacity of at least 1167 A in the service feeder. In addition, protection of that feeder by a 1200-A single main circuit breaker or 1200-A fuses in a 1200-A switch also satisfies the rule of Sec. 220-10(b) that calls for overcurrent protection to have a rating not less than the value of 1167 A.

On a 480Y/277-V, 3-phase, 4-wire service to the example store building, a neutral conductor will be brought in to provide for 277-V lighting and other 277-V loads, so that each set of conductors of the parallel circuit will consist of four conductors: three phase legs and a neutral. But that neutral will be providing for current flow to a load that is only 252 kVA, out of a much larger total. Although the electric-discharge lighting load will be carrying significant third-harmonic current, even under balanced phase-loading conditions, the lighting is not a major portion of the load and the neutral is not required by Note 10(c) of NEC Table 310-16 to be counted as a current-carrying conductor. As a consequence, if the circuit is made up of multiple sets of four conductors in each conduit, Note 8 of Table 310-16 does not require derating of the ampacity of each of the four conductors, and conductors for the above-calculated 1167 A may be selected directly from Table 310-16.

The value of 1167-A ampacity for the service conductor is, of course, based solely on providing adequate rating for the initial load. But spare capacity must be added. Although economics always constrains the decision to provide spare capacity for unforeseen future loads, experience consistently confirms that such provision is not only prudent but also economical in the long run. Loads start getting added the day the premises are occupied, and they keep piling up with the passage of time. Many designers argue that spare capacity must always be at least 25 percent of the connected demand load in commercial and institutional buildings and at least 10 percent in industrial buildings, where loads are better accounted for and future changes are more readily anticipated.

In the store building, the original connected load is 825 kVA (252 + 73 + 116 + 346 + 38); adding a full 25 percent would give 825 × 1.25 = 1031 kVA. Dividing 1,031,000 VA by 460 × 1.732 yields an ampacity of 1294 A in the service conductors. The 1294-A value is the minimum load rating for conductors whose ampacity is required by Sec. 220-10(b) to be equal to 125 percent of the load current for continuous operation. That calls for a minimum ampacity of 1.25 × 1294 A = 1617 A. And even the minimum value of 1167 A for service conductors to the initial load is based on a spare capacity of 108 kVA (360 kVA of calculated general lighting minus the actual connected load of 252 kVA).

Service conductors with an ampacity of 1600 A constitute an economically realistic basis of effective design for present and future loads, to ensure the full capacity of 1294 A, with 1600 A being the next-larger standard rating of protective device above 1200 A. But it also becomes important to make sure that the conductors selected actually have that ampacity.

Assuming use of aluminum service-entrance conductors in conduit, each phase leg of the circuit could be made up of four 700-kcmil THHN conductors, each with an ampacity of 420 A (Table 310-16), for a total of 1680 A per phase leg, with each of four conduits containing a phase *A* conductor, a phase *B* conductor, a phase *C* conductor, and a neutral conductor (which would be sized differently than the phase legs and would be smaller). Because electric-discharge lighting does not constitute the *major* portion of the load on the service conductors (only 252 kVA out of the total), Note 10(c) to Table 310-16 does not require that the neutral be counted as a current-carrying conductor in each conduit, and derating for conduit fill (Note 8) is not required. Dividing the 1294-A design load among the four conductors per phase leg of the service feeder shows that each 700-kcmil THHN aluminum conductor is loaded to 1294 ÷ 4 = 324 A. Because that value is less than the 375-A ampacity of 700-kcmil THW (75°C) aluminum conductors, the THHN (90°C) conductors are not being used in excess of the 75°C ampacity of 700-kcmil aluminum conductors. They would, therefore, satisfy the UL (and NEC) requirement that conductor connections to equipment terminals (the one or more service CBs or switches) not operate above the ampacities of 75°C conductors for equipment rated over 100 A (or above 60°C ampacities up to a 100-A equipment rating) unless the equipment is marked otherwise. To satisfy that UL rule, the maximum load must not exceed 4 × 375 A, or 1500 A.

As will be discussed later, NEC Sec. 220-10(b) is also a factor in the selection of the one or more circuit breakers or fused switches that are required for disconnect and over-

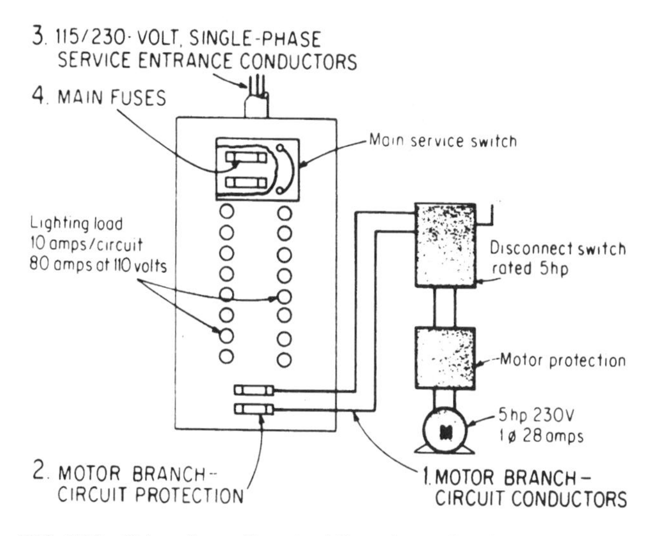

FIG. 5-10 Sizing of a small service follows the numbered steps to obtain the required ampacity and protection.

current protection of service-entrance conductors. Section 220-10(b) requires a rating of 125 percent of the load current for conductor ampacities and overcurrent protection, applied to that part of the load current that is continuous (operates steadily for 3 h or longer). But the exception to that rule permits continuous operation with conductors and overcurrent protection rated at 100 percent (instead of 125 percent) *if* the CB or fused switch is UL-listed for full-load continuous operation. In the foregoing analysis, the use of service conductors rated at 1200 A and protected by a CB or fused switch rated for continuous loading to 100 percent of its rating could be considered an effective and economical alternative in supplying spare capacity.

Although the foregoing might seem to be an involved and almost laborious evaluation of required service conductor ampacity, there is really no alternative to that type of procedure if all the related factors are to be effectively included. Any other approach quickly becomes a guess and includes the risk of being inadequate, dangerous, and costly.

Figure 5-10 shows a service and distribution panel for a small building, where sizing of the service conductors and overcurrent protection would follow a typical feeder-sizing approach:

1. Size of motor branch-circuit conductors: 125 percent of 28 A = 35 A. This requires No. 8 TW, THW, RHH, or THHN copper, or No. 6 TW, THW, RHH, or THHN aluminum. Conductors are sized on the basis of the ampacity of 60°C conductor insulation.
2. Size of motor branch-circuit fuses (using one-time, non-time-delay fuses): 300 percent of 28 A = 84 A. This requires a maximum fuse size of 90 A. Smaller fuses, such as the time-delay type, may be used.
3. Size of service-entrance conductors: 125 percent × [28 A + 80 A (lighting load)] = 115 A. Conductors for the hot legs must be No. 1/0 TW or No. 2 THW, RHH, or THHN copper. (Note that if the lighting load is continuous, the SE conductors must be sized for 125 percent of 28 A + 125 percent of 80 A.)

4. Size of main fuses: 90 A (from item 2 above) + 80 A = 170 A. This requires a maximum fuse size of 175 A, in a 200-A switch. Again, smaller fuses may and should be used where possible to improve the overload protection on the circuit conductors.

Residential Service-Entrance Conductors

The sizing of service-entrance conductors for a residential building—an apartment house, condominium, or two- or four-family house—follows the feeder-sizing procedures that apply to other types of buildings. And that same general procedure may be followed in sizing service-entrance conductors for a one-family house or an apartment in a multifamily dwelling.

Example. A house has two floors with outside dimensions 30 by 25 ft. For both floors, that gives a total living area of 30 ft × 25 ft × 2 = 1500 ft². The house has oil-fired heating and hot water and only one large electric appliance, a 12-kW electric range. What is the minimum required size of circuit conductors?

Solution. Using the standard NEC method for feeder sizing (as covered in NEC Secs. 220-10 through 220-21), the loads are totaled as follows:

General lighting: 1500 ft² × 3 W/ft²	4500 W
Two kitchen-appliance circuits at 1500 W	3000 W
One laundry circuit	1500 W
Total	9000 W

Then the demand factors are applied:

300 W at 100%	3000 W
6000 (9000 − 3000) W at 35%	2100 W
Basic service entrance load	5100 W

The feeder demand load for the 12-kW range must be added to 5100 W. As stated in NEC Sec. 220-19, the range feeder demand load is selected from Table 220-19. In this case, it is 8 kW (column A, one appliance):

Basic load	5,100 W
Range feeder capacity	8,000 W
Total feeder load	13,100 W

The minimum required ampacity for the ungrounded service-entrance conductors of a 230/115-V, 3-wire, single-phase service is readily found by dividing the total feeder load by 230 V:

$$\frac{13{,}000 \text{ W}}{230 \text{ V}} = 56.96 \text{ or } 57 \text{ A}$$

Although a load current of that rating could be readily supplied by No. 4 copper TW, THW, THHN, or XHHW conductors, an important provision of NEC Sec. 230-42(b)(2) comes into play here. Where the initial computed total feeder load is 10 kW or more (and it is 13.1 kW here) for a single-family dwelling, the ungrounded conductors of a service feeder must be rated at least 100 A; that is, the minimum capacity of the service entrance hot legs is 100 A.

That would call for a minimum of one of the following (using 60°C ampacity):

- No. 1 TW copper conductors (110 A)
- No. 1 THW copper conductors (130 A)
- No. 1 THHN copper conductors (150 A)
- No. 1 XHHW copper conductors (150 A)
- No. 1/0 TW, THW, THHN, or XHHW aluminum conductors (100 to 125 A)

The above example represents the NEC minimum sizing of service-entrance conductors in accordance with the steps indicated in Secs. 220-10 through 220-21. The calculation accounts only for the initial loads and includes no capacity for load growth. A simplified calculation procedure for residential services can be based on the same NEC capacity requirements but can include a measure of extra capacity to cover the virtually inevitable growth in connected loads. Such a design procedure is as follows:

1. Calculate the general lighting load and the general-purpose receptacle outlet load for those outlets served by 20-A, 115-V, general-purpose circuits. Although each circuit has a full capacity of $20 \times 115 = 2300$ W, the circuit will be considered as being rated at only 2000 W for purposes of this calculation, thereby allowing some spare capacity.

For each 500 ft^2 of floor area in the house (excluding porches, garages, unused spaces, unfinished areas, etc.), allow one circuit. Allow an extra circuit for any part of 500 ft^2 left over.

Multiply the required number of circuits by 200 W to get the total load in watts. (The same result can be obtained by multiplying the total floor area in square feet by 4 W/ft^2.)

2. Find the total circuit capacity—in watts—to be allowed for the appliance load in the kitchen, dining room, pantry, laundry, and utility area that is to be served by 20-A, 115-V small-appliance circuits. This total capacity can be obtained by multiplying the number of such circuits by 2000 W. If the exact number of such circuits is not known, a 6000-W load can be assumed (two kitchen-appliance circuits and one laundry circuit).

3. Take 3000 W of the sum of the amounts of steps 1 and 2 at 100 percent demand.

4. Add to this figure 35 percent (demand) of the remainder of the sum of the amounts of steps 1 and 2. The result is the capacity that must be provided in the service-entrance conductors to supply the general lighting and receptacle loads.

5. Add 8000 W for an electric range (rated not over 12 kW). (Refer to NEC Table 220-19.) If the electric cooking appliances consist of a built-in oven and rangetop, the NEC must be consulted to get the proper demand load. The NEC allows the nameplate ratings of a counter-mounted cooking unit and not more than two wall-mounted ovens to be added and treated as a single range of that total rating.

6. Add the wattage ratings of all appliances to be served by individual circuits not previously accounted for in the calculation. A demand factor of 75 percent may be applied to the sum of the nameplate ratings of these appliances—excluding cooking units, clothes dryers, air-conditioning units, and any electric space-heating load.

If both electric heating and air conditioning are to be used in the house, the rating (in watts) of only the larger of these two connected loads need be used in this total, since the two loads will not be used simultaneously.

7. Add the following: the general lighting and general-purpose receptacle load (from step 4), the electric-range demand load (from step 5), and the appliance, air-conditioning, and heating loads (from step 6).

8. Divide the total number of watts obtained in step 7 by 230 V (for 230/115-V, 3-wire, single-phase service) to get the required ampere rating of the service conductors.

Optional Service-Entrance Calculation

The NEC provides an optional method for sizing residential service-entrance conductors in Sec. 220-30. That optional calculation may be used only for a one-family residence or an apartment in a multifamily dwelling, or other "dwelling unit," that is served by a 115/230- or 120/208-V, 3-wire, 100-A or larger service, where the total load is supplied by one set of service-entrance conductors.

The method recognizes the greater diversity attainable in large-capacity installations. It therefore permits a smaller size of service-entrance conductors for such installations than would be permitted by using the load calculations of Secs. 220-10 through 220-21.

A typical example of the optional calculation of Sec. 220-30, based on Table 220-30 (shown in Fig. 5-11), follows.

Load (in kW or kVA)	Demand Factor Percent
Largest of [see Section 220-30(c)]	
Air conditioning and cooling, including heat pump compressors	100
Central electric space heating including integral supplemental heating in heat pumps	65
Less than four separately controlled electric space heating units.	65
Plus:	
First 10 kW of all other load	100
Remainder of other load	40

The loads identified in Table 220-30 as "other load" and as "remainder of other load" shall include the following:

(1) 1500 watts for each 2-wire, 20-ampere small appliance branch circuit and each laundry branch circuit specified in Section 220-16.

(2) 3 watts per square foot (0.093 sq m) for general lighting and general-use receptacles.

(3) The nameplate rating of all fixed appliances, ranges, wall-mounted ovens, counter-mounted cooking units, and including four or more separately controlled space heating units.

(4) The nameplate ampere or kVA rating of all motors and of all low-power-factor loads.

FIG. 5-11 Optional calculation for residential service conductors is based on this table.

Example. A 1500-ft^2 house (excluding unoccupied basement, unfinished attic, and open porches) contains the following specific electric appliances:

12-kW range

2.5-kW water heater

1.2-kW dishwasher

9 kW of electric heat (in five rooms)

5-kW clothes dryer

6-A, 230-V air-conditioning unit

There is recognition in Sec. 220-21 that if "it is unlikely that two dissimilar loads will be in use simultaneously," then it is permissible to omit the smaller of the two in calculating the required capacity of feeder or service-entrance conductors. In Sec. 220-30, that concept is spelled out to require adding only the larger of either an air-conditioning load or the connected load of four or more separately controlled electric space-heating units. For the residence considered here, these loads are as follows:

Air conditioning: 6 A $\times$ 230 V = 1.38 kW

Heating (five separate units): 9.00 kW

It is permissible, therefore, to disregard the air-conditioning load in using the optional method.

The heating and other loads (except air conditioning) must be totaled in accordance with Sec. 220-30:

1. 1500 W for each of two small-appliance circuits (2-wire, 20-A) required by Sec. 220-3(b)(1): 3000 W
2. Laundry branch circuit (3-wire, 20-A): 1500 W
3. 3 W/ft^2 of floor area for general lighting and general-use receptacles (3 $\times$ 1500 ft^2): 4500 W
4. Nameplate rating of fixed appliances:

   Range: 12,000 W

   "Four or more" separately controlled heating units: 9000 W

   Water heater: 2500 W

   Dishwasher: 1200 W

   Clothes dryer: 5000 W
5. Total: 38,700 W

The first four load categories given in Table 220-30 are not applicable here. Air conditioning has already been excluded as a load because the heating load is greater. There is no electric thermal storage or other heating system where connected load is expected to be continuous; there is no "central" electric space heating; and there are not less than four separately controlled electric space-heating units.

The total load of 38,700 W, as summed above, is classified as "all other load," as referred to on line 6 of NEC Table 220-30. The last two references in Table 220-30 constitute all the "optional" calculations in this particular example:

1. Take 10 kW of "all other load" at 100 percent demand to obtain 10,000 W.
2. Take the "remainder of other load" at 40 percent demand factor to obtain 0.4 (38,700 − 10,000) = 11,480 W.
3. Add the results of steps 1 and 2 to obtain 10,000 W + 11,480 W = 21,480 W.

At 230 V, single phase, the minimum ampacity of each service hot leg has to be

$$\frac{21{,}480\text{ W}}{230\text{ V}} = 93.39 \text{ or } 93\text{ A}$$

But Sec. 230-41(b)(2) requires a minimum conductor rating when the demand load is 10 kW or more. Thus,

$$\text{Minimum service conductor rating} = 100\text{ A}$$

The neutral service-entrance conductor size is calculated in accordance with Sec. 220-22, based on Sec. 220-10. The 230-V loads have no relation to the required neutral capacity. The water heater, clothes dryer, and electric space-heating units operate at 230 V, 2-wire, and have no neutrals. By considering only those loads served by a circuit with a neutral conductor and determining the maximum unbalance, the minimum required size of neutral conductor can be determined.

When a 3-wire, 230/115-V circuit serves a total load that is balanced from each hot leg to neutral—that is, half the total load is connected from one hot leg to neutral, and the other half of the total load from the other hot leg to neutral—the condition of maximum unbalance occurs when all the load fed by one hot leg is operating and all the load fed by the other hot leg is deenergized. Under that condition, the neutral current and hot-leg current are equal to half the total load wattage divided by 115 V (half the voltage between hot legs). But that current is exactly the same as the current that results from dividing the total load (connected hot leg to hot leg) by 230 V (which is twice the voltage from hot leg to neutral). Because of this relationship, it is easy to determine the neutral-current load by simply calculating the hot-leg current load—the total load from hot leg to hot leg, divided by 230 V.

In this example, calculation of the neutral-current load begins with the following steps, in which the components of the neutral load are summed:

1. Multiply 1500 ft^2 by 3 W/ft^2 to obtain 4500 W.
2. Add three small-appliance circuits (two kitchen, one laundry) at 1500 W each to obtain 4500 W.
3. Add the results of steps 1 and 2 to obtain the total lighting and small-appliance load as 9000 W.
4. Take 3000 W of that value at 100 percent demand factor to obtain 3000 W.
5. Take the balance of the load at 35 percent demand factor to obtain 0.35 (9000 − 3000) = 2100 W.
6. Add the results of steps 4 and 5 to obtain 5100 W.

The remainder of the calculation is performed with reference to Fig. 5-12.

Assuming an even balance of this 5100-W load on the two hot legs, under maximum unbalance the corresponding neutral current will be the same as the total load (5100 W) divided by 230 V:

$$\frac{5100\text{ W}}{230\text{ V}} = 22.17\text{ A}$$

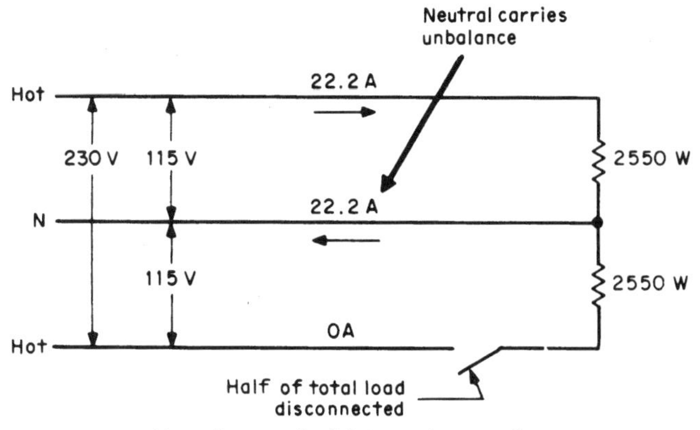

Neutral current for lighting and receptacles.

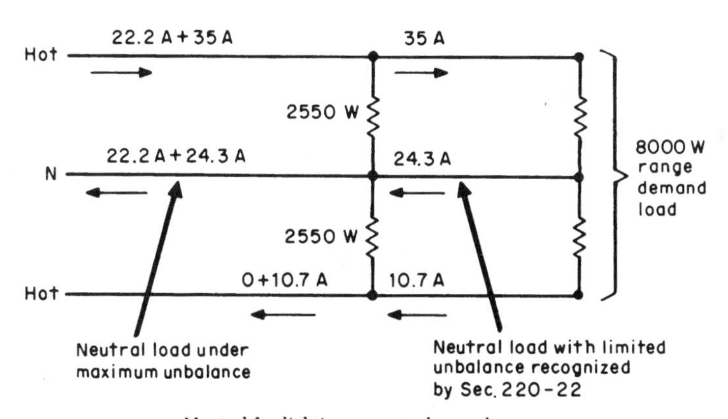

Neutral for lighting, receptacles, and range.

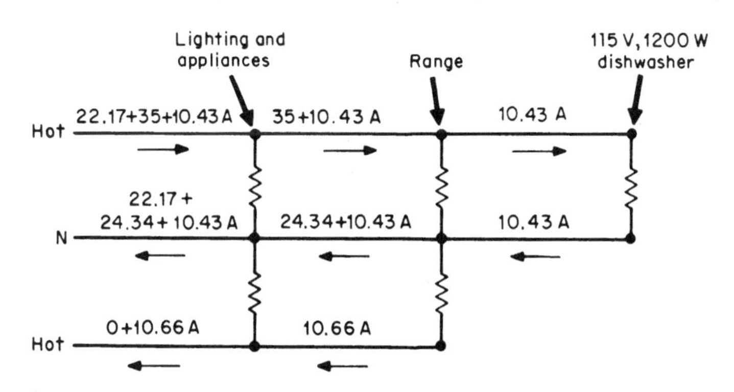

Neutral current for total load.

FIG. 5-12 These loads determine the required neutral-conductor ampacity.

The neutral unbalanced current for the range load can be taken as equal to the 8000-W range demand load multiplied by the 70 percent demand factor permitted by Sec. 220-22 and then divided by 230 V:

$$\frac{8000 \times 0.7}{230} = \frac{5600}{230} = 24.34 \text{ A}$$

The neutral-current load that is added by the 115-V, 1200-W dishwasher must also be included:

$$\frac{1200 \text{ W}}{115 \text{ V}} = 10.43 \text{ A}$$

The minimum required neutral capacity is, therefore, the sum 22.17 A + 24.34 A + 10.43 A = 56.94 A, or, after rounding, 57 A. From NEC Table 310-16, the minimum neutral conductors are

Copper: No. 4 TW, THW, THHN, or XHHW

Aluminum: No. 3 TW, THW, THHN, or XHHW

And the 75 or 90°C conductors must be used at the ampacity of 60°C conductors, as required by the UL *Electrical Construction Materials Directory* and Sec. 110-14(c) of the NEC.

Note: The above calculation of the minimum required capacity of the neutral conductor differs from the calculation and results shown in Example 2(a) in NEC Chap. 9. There, the 1200-W dishwasher load is added as a 230-V load to the range load and general lighting and receptacle load. To include this 115-V load as a 230-V load (and then to divide the total by 230 V, as shown) does not accurately represent the neutral load that the 115-V, 1200-W dishwasher will produce. In fact, it yields exactly half the neutral load that the dishwasher represents. The optional calculation method of Sec. 220-30 does indicate, in part (3), that fixed appliances are added at nameplate load, and it does not differentiate between 115-V devices and 230-V devices. It simply totals all loads and then applies the 100 percent and 40 percent demand factors as indicated. That method clearly is based on well-founded data about load diversity and is aimed at obtaining a reasonable size of service hot legs. But calculation of the feeder neutral in accordance with Sec. 220-22 is aimed at determining the maximum unbalanced current to which the service neutral might be subjected.

Although there is only a small difference between the NEC value of 51.7 A and the value of 56.9 A determined here, precise calculation should be made to ensure the adequacy of conductor ampacities. The difference of 5.2 A actually changes the required minimum size of neutral conductor from No. 6 up to No. 4 for copper, and from No. 4 up to No. 3 for aluminum. A load like a dishwasher, which draws current for a considerable time and is not used for just a few minutes like a toaster, should be factored into the calculation with an eye toward ensuring adequate capacity for conductors.

Another example of the application of Sec. 220-30 to the sizing of residential service conductors shows the very heavy loading that is permitted on a 100-A service.

Example. The following loads apply to a 1500-ft^2 house with all-electric utilization:

- 1500 W for laundry receptacle circuit
- 1500 W for each of two (minimum of two) required kitchen-appliance circuits: 3000 W

- 1500 ft² at 3 W/ft² for general lighting and receptacles: 4500 W
- 14 kW of electric space heating from more than four separately controlled units: 14,000 W
- 12-kW electric range: 12,000 W
- 3-kW water heater: 3000 W
- 5-kW clothes dryer: 4500 W
- 3-kW load of unit air conditioners (because this load is smaller than the space-heating load and will not be operated simultaneously with it, no load need be added): 0 W
- Total: 42,500 W

Solution

First 10 kW at 100 percent	10,000 W
Remainder at 40 percent: 32,500 W × 0.4	13,000 W
Total demand load	23,000 W

Size of service = 23,000 W ÷ 230 V = 100-A service

Under certain load conditions, the calculated required service capacity may be substantially less than 100 A. In such cases, however, 100 A is the minimum size service which can be used.

RATING OF DISCONNECT

Aside from the limited conditions covered in parts (a) and (b), Sec. 230-79 requires that service equipment (in general) have a rating not less than 60 A, applicable to both fusible and CB equipment. Part (c) requires 100-A minimum rating of a single switch or CB used in the service disconnect for any "one-family dwelling" with an *initial* load of 10 kVA or more or where the *initial* installation contains more than *five 2-wire* branch circuits. Note that the rule applies to one-family houses only, because of the definition of "one-family dwelling" as given in Art. 100. It does not apply to apartments or similar dwelling units that are in two-family or multifamily dwellings.

If the demand on a total connected load, as calculated from Sec. 220-10 through Sec. 220-21, is 10 kVA or more, a 100-A service disconnect as well as 100-A rated service-entrance conductors [Sec. 230-42(b)(2)] must be used. Any one-family house with an electric range rated 8¾ kW must always have a 100-A rated disconnect (or service equipment) because such a range is a demand load of 8 kW and the two required 20-A kitchen appliance circuits come to a demand load of 3000 W [Sec. 220-16(a)] at 100 percent from Table 220-11—and the 8 kW plus 3 kW exceeds the 10-kW (or 10-kVA) level at which a minimum 100-A rated service is required.

If a 100-A service is used, the demand load may be as high as 23 kVA. By using the optional service calculations of Table 220-30, a 23-kVA demand load is obtained from a connected load of 42.5 kVA. This shows the effect of diversity on large-capacity installations.

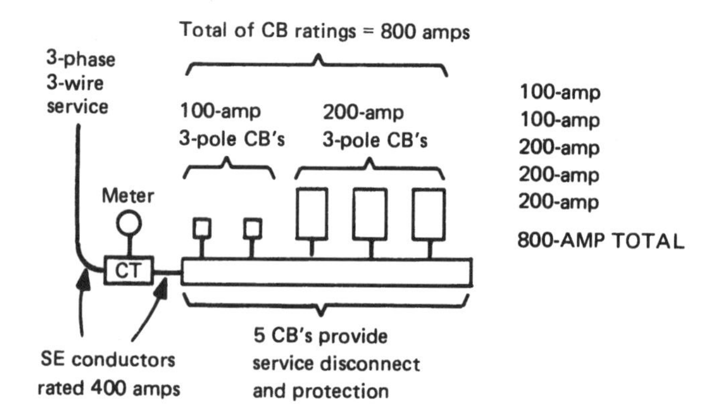

From Art. 220 (Secs. 220-10 through 220-21), calculation of demand load indicated that a single disconnect for this service must be rated at least 400 amps. The rating of multiple disconnects must total at least that value.

FIG. 5-13 Multiple disconnects must have their sum of ratings at least equal to the minimum rating of a single disconnect.

COMBINED RATING OF DISCONNECTS

Figure 5-13 shows an application of the rule of Sec. 230-80, based on determining what rating of a single disconnect would be required *if* a single disconnect were used instead of multiple ones. Note that the sum of ratings above 400 A does comply with the rule of this section and with Exception 3 of Sec. 230-90(a) even though the 400-A service-entrance conductors could be heavily overloaded. Exception 3 exempts this type of layout from the need to protect the conductors at their rated ampacity, as required in the basic rule of Sec. 230-90. The Code assumes that the 400-A rating of the service-entrance conductors was carefully calculated from Art. 220 to be adequate for the maximum sum of the demand loads fed by the five disconnects shown in the layout.

OVERCURRENT PROTECTION

The intent in paragraph (a) of Sec. 230-90 is to ensure that the overcurrent protection required in the service-entrance equipment protects the service-entrance conductors from "overload." It is obvious that these overcurrent devices cannot provide "fault" protection for the service-entrance conductors if the fault occurs in the service-entrance conductors, but can protect them from overload where so selected as to have proper rating. Conductors on the load side of the service equipment are considered as feeders or branch circuits and are required by the Code to be protected as described in Arts. 210, 215, and 240.

Each ungrounded service-entrance conductor must be protected by an overcurrent device in series with the conductor (Fig. 5-14). The overcurrent device must have a rating or setting not higher than the allowable current capacity of the conductor, with the exceptions noted.

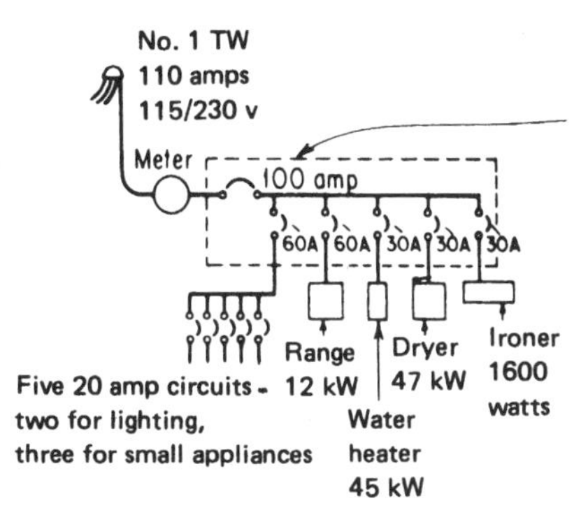

Service equipment with one overall 100-amp main disconnect and fuse. Current through service conductors limited to 100 amperes. Without a main disconnect and overcurrent device, current is not limited and current over 110 amps could flow. Sum of protective devices is 210 amps per hot leg.

NOTE: Service-entrance conductors must be selected with adequate ampacity for the calculated service demand load, from Secs. 220-10 through 220-21.

FIG. 5-14 Single main service protection must not exceed conductor ampacity (or may be next-higher rated device above conductor ampacity).

Exception 1

If the service supplies one motor in addition to other load (such as lighting and heating), the overcurrent device may be rated or set in accordance with the required protection for a branch circuit supplying the one motor (Sec. 430-52) plus the other load, as shown in Fig. 5-15. Use of 175-A fuses where the calculation calls for 170-A conforms to Exception 2 of Sec. 230-90—next-higher standard rating of fuse (Sec. 240-6). For motor branch circuits and feeders, Arts. 220 and 430 permit the use of overcurrent devices having ratings or settings higher than the capacities of the conductors. Article 230 makes similar provisions for services where the service supplies a motor load or a combination load of both motors and other loads.

If the service supplies two or more motors as well as other load, then the overcurrent protection must be rated in accordance with the required protection for a feeder supplying several motors plus the other load (Sec. 430-63). Or if the service supplies only a multimotor load (with no other load fed), then Sec. 430-62 sets the maximum permitted rating of overcurrent protection.

Exception 3

Not more than six CBs or six sets of fuses may serve as overcurrent protection for the service-entrance conductors even though the sum of the ratings of the overcurrent devices is in excess of the ampacity of the service conductors supplying the devices, as illustrated in Fig. 5-16. The grouping of single-pole CBs as multipole devices, as permitted for disconnect means, may also apply to overcurrent protection. And a set of fuses is all the fuses required to protect the ungrounded service-entrance conductors.

This exception ties into Sec. 230-80. Service conductors are sized for the *total* maximum demand load—applying permitted demand factors from Table 220-11. Then each of the two to six feeders fed by the SE conductors is also sized from Art. 220 based on the load fed by each feeder. When those feeders are given overcurrent protection in accordance with their ampacities, frequently the sum of those overcurrent devices is greater than the ampacity of the SE conductors which were sized by applying the applicable demand fac-

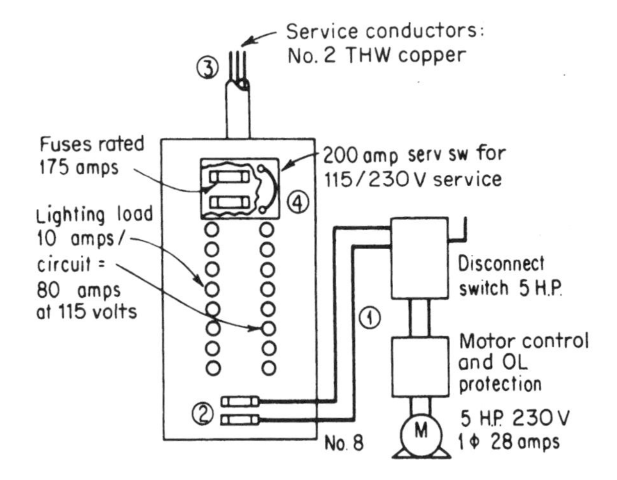

1. Size of motor branch circuit conductors: 125% x 28 amps equals 35 amps. This requires No. 8's.

2. Size of motor branch circuit fuses: 300% x 28 amps equals 84 amps. This requires maximum fuse size of 90 amps. Smaller fuses, such as time-delay type, may be used.

3. Size of service entrance conductors must be adequate for a load of 125% x 28 amps plus 80 amps (lighting load) or 115 amps.

4. Size of main fuses: 90 amps (from 2 above) plus 80 amps equals 170 amps. This requires maximum fuse size of 175 amps. Again, smaller fuses may and should be used where possible to improve the overload protection on the circuit conductors.

FIG. 5-15 Service protection for lighting plus motor load.

tors to the total connected load of all the feeders. Exception 3 recognizes that possibility as acceptable even though it departs from the rule in the first sentence of Sec. 230-90(a). The assumption is that if calculation of demand load for the SE conductors is correctly made, there will be no overloading of those conductors because the diversity of feeder loads (some loads on, some off) will be adequate to limit load on the SE conductors.

Assume that the load of a building computed in accordance with Art. 220 is 255 A. Under Sec. 240-3(b), 300-A fuses or a 300-A CB may be considered as the proper-size overcurrent protection for service conductors rated between 255 and 300 A if a single service disconnect is used.

If the load is separated in such a manner that six 70-A CBs could be used instead of a single service disconnect means, total rating of the CBs will be greater than the ampacity of the service-entrance conductors. And that will be acceptable.

Section 240-3(a) is shown in Fig. 5-17 and is intended to prevent opening of the fire-pump circuit on any overload up to and including stalling or even seizing of the pump

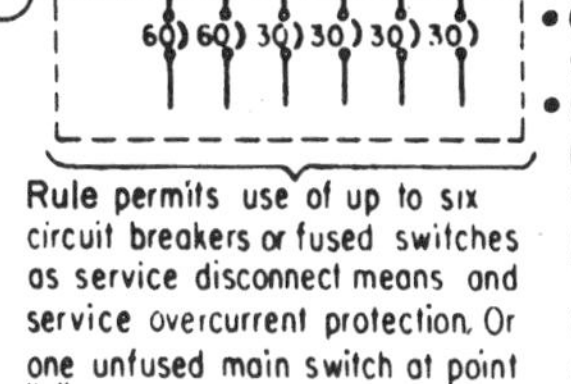

FIG. 5-16 With six subdivisions of protection, conductors could be overloaded.

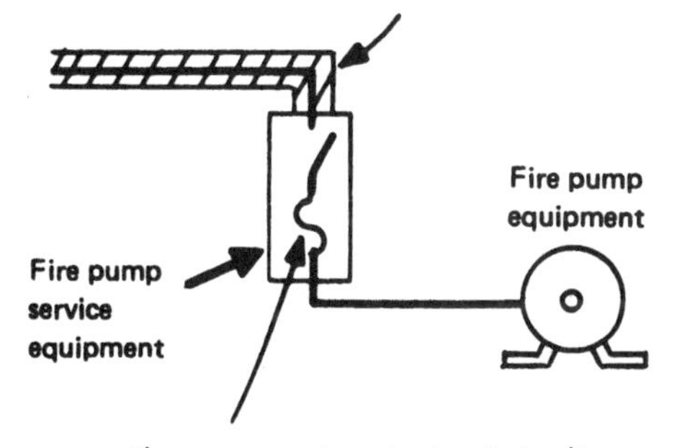

FIG. 5-17

motor. Because the conductors are "outside the building," operating overload is no hazard; and under fire conditions, the pump must have no prohibition on its operation. It is better to lose the motor than attempt to protect it against overload when it is needed.

SERVICE GROUND-FAULT PROTECTION

Fuses and CBs, applied as described in the previous section on overcurrent protection, are sized to protect conductors in accordance with their current-carrying capacities. The function of a fuse or CB is to open the circuit if current exceeds the rating of the protective device. This excessive current might be caused by operating overload, by a ground fault, or by a short circuit. Thus, a 1000-A fuse will blow if current in excess of that value flows over the circuit. It will blow early on heavy overcurrent and later on low overcurrents. But it will blow, and the circuit and equipment will be protected against the damage of the overcurrent. But there is another type of fault condition which is very common in grounded systems and will not be cleared by conventional overcurrent devices. That is the phase-to-ground fault (usually arcing) which has a current value less than the rating of the overcurrent device.

On any high-capacity feeder, a line-to-ground fault (that is, a fault from a phase conductor to a conduit, to a junction box, or to some other metallic equipment enclosure) can, and frequently does, draw current of a value less than the rating or setting of the circuit protective device. For instance, a 500-A ground fault on a 2000-A protective device which has only a 1200-A load will not be cleared by the device. If such a fault is a *bolted* line-to-ground fault, a highly unlikely fault, there will be a certain amount of heat generated by the I^2R effect of the current; but this will usually not be dangerous, and such fault current will merely register as additional operating load, with wasted energy (wattage) in the system. But bolted phase-to-ground faults are very rare. The usual phase-to-ground fault exists as an intermittent or arcing fault, and an arcing fault of the same current rating as the essentially harmless bolted fault can be fantastically destructive because of the intense heat of the arc.

Of course, any ground-fault current (bolted or arcing) above the rating or setting of the circuit protective device will normally be cleared by the device. In such cases, bolted-fault currents will be eliminated. But even where the protective device eventually operates, in the case of a heavy ground-fault current which adds to the normal circuit load current to produce a total current in excess of the rating of the normal circuit protective device (fuse or CB), the time delay of the device may be minutes or even hours—more than enough time for the arcing-fault current to burn out conduit and enclosures, acting just like a torch and even propagating flame to create a fire hazard.

In spite of the growth of effective and skilled application of conventional overcurrent protective devices, the problem of ground faults continues to persist and even grows with expanding electrical usage. In the interests of safety, definitive engineering design must account for protection against such faults. Phase overcurrent protective devices are normally limited in their effectiveness because (1) they must have a time delay and a setting somewhat higher than full load to ride through normal inrushes and (2) they are unable to distinguish between normal currents and low-magnitude fault currents which may be less than full-load currents.

Dangerous temperatures and magnetic forces are proportional to current for overloads and short circuits; therefore, overcurrent protective devices usually are adequate to protect against such faults. However, the temperatures of arcing faults are generally independent of current magnitude; and arcs of great and extensive destructive capability can be sustained by currents not exceeding the overcurrent device settings. Other means of protection are therefore necessary. A ground detection device which "sees" only ground-fault current can be coupled to an automatic switching device to open all three phases when a line-to-ground fault exists on the circuit.

Section 230-95 requires ground-fault protection equipment be provided for each service *disconnecting means* rated 1000 A or more in a solidly grounded-wye electrical service that operates with its ungrounded legs at more than 150 V to ground. Note that this applies to the rating of the disconnect, not to the rating of the overcurrent devices or to the capacity of the service-entrance conductors.

The wording of the first sentence of this section makes clear that service GFP (ground-fault protection) is required under specific conditions: only for grounded-wye systems that have voltage over 150 V to ground and less than 600 V phase to phase. In effect, that means the rule applies only to 480/277-V grounded-wye and *not* to 120/208-V systems or any other commonly used systems (Fig. 5-18). And GFP is *not* required on any systems operating over 600 V phase to phase.

In a typical GFP hookup as shown in Fig. 5-19, part (a) of the section specifies that a ground-fault current of 1200 A or more must cause the disconnect to open all ungrounded conductors. Thus the maximum ground-fault pickup setting permitted is 1200 A, although it may be set lower.

With a GFP system, at the service entrance a ground fault anywhere in the system is immediately sensed in the ground-relay system, but its action to open the circuit usually is delayed to allow some normal overcurrent device near the point of fault to open if it can. As a practical procedure, such time delay is designed to be only a few cycles or seconds, depending on the voltage of the circuit, the time-current characteristics of the overcurrent devices in the system, and the location of the ground-fault relay in the distribution system. Should any of the conventional short-circuit overcurrent protective devices fail to operate in the time predetermined to clear the circuit, and if the fault continues, the ground-fault protective relays will open the circuit. This provides added overcurrent protection not available by any other means.

The rule requiring GFP for any service disconnect rated 1000 A or more (on 480/277-V services) specifies a maximum *time delay of 1 s for ground-fault currents of 3000 A or more* (Fig. 5-20).

The maximum permitted setting of a service GFP hookup is 1200 A, but the time-current trip characteristic of the relay must ensure opening of the disconnect in not more than 1 s for any ground-fault current of 3000 A or more. This change in the Code was made to establish a specific level of protection in GFP equipment by setting a maximum limit on i^2t of fault energy.

The reasoning behind this change was explained as follows:

The amount of damage done by an arcing fault is directly proportional to the time it is allowed to burn. Commercially available GFP systems can easily meet the 1-s limit. Some users are requesting time delays up to 60 s so all downstream overcurrent devices can have plenty of time to trip thermally before the GFP on the main disconnect trips. However, an arcing fault lasting 60 s can virtually destroy a service equipment installation. Coordination with downstream overcurrent devices can and should be achieved by adding GFP on feeder circuits where needed. The Code should require a reasonable time limit for GFP. Now, 3000 A is 250 percent of 1200 A, and 250 percent of setting is a calibrating point specified in ANSI 37.17. Specifying a maximum time delay starting at this current value will allow either flat or inverse time-delay characteristics for ground-fault relays with approximately the same level of protection.

Selective coordination between GFP and conventional protective devices (fuses and CBs) on service and feeder circuits is now a very clear and specific task as a result of rewording of Sec. 230-95(a) that calls for a maximum time delay of 1 s at any ground-fault current value of 3000 A or more.

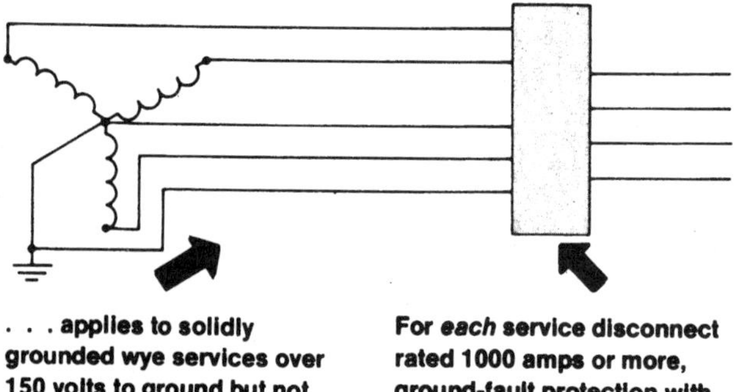

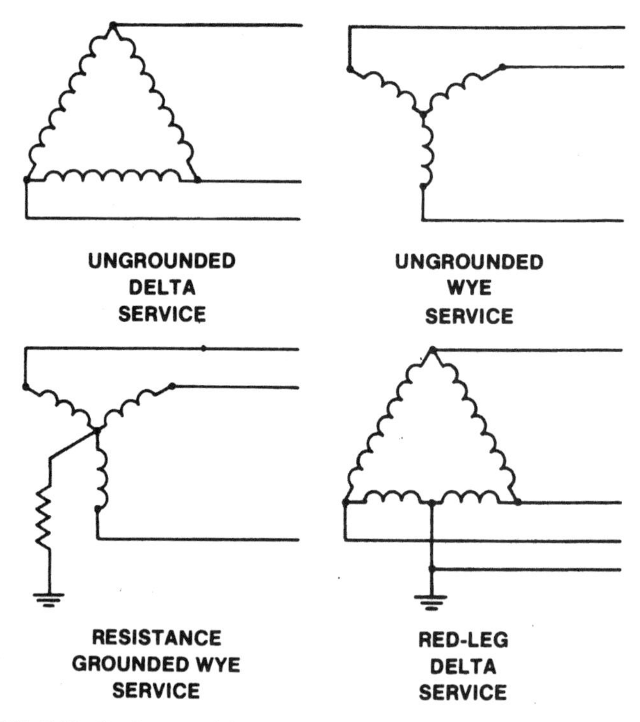

FIG. 5-18 Service ground-fault protection is mandatory.

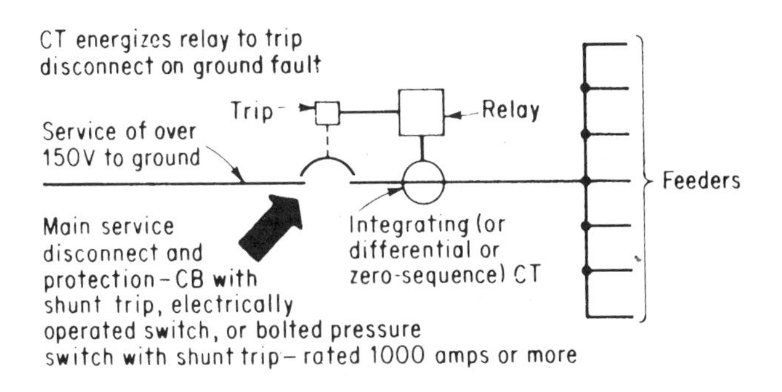

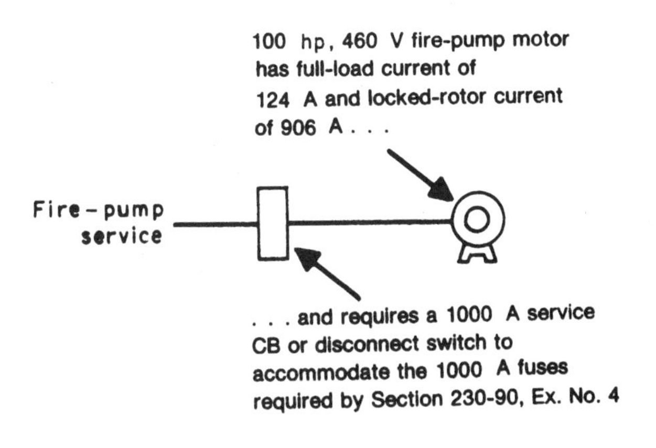

FIG. 5-19 GFP is required for each disconnect rated 1000 A or more, but not for a fire-pump disconnect.

For applying the rule of Sec. 230-95, the rating of any service disconnect means shall be determined as shown in Fig. 5-21.

Because the rule on required service GFP applies to the rating of each service disconnect, there are many instances where GFP would be required if a single service main disconnect is used but *not* if the service subdivision option of Sec. 230-71(a) were taken, as shown in Fig. 5-22.

By Exception 1 to part (a), continuous industrial process operations are exempted from the GFP rules of Sec. 230-95(a) where the electrical system is under the supervision of qualified persons who will effect orderly shutdown of the system and thereby avoid hazards, greater than ground fault itself, that would result from the nonorderly, automatic interruption that GFP would produce in the supply to such critical continuous operations. Exception 1 excludes GFP requirements where a nonorderly shutdown will introduce additional or increased hazards. The idea behind that is to provide maximum protection against service outage for such industrial processes. With highly trained personnel at such locations, design and maintenance of the electrical system can often

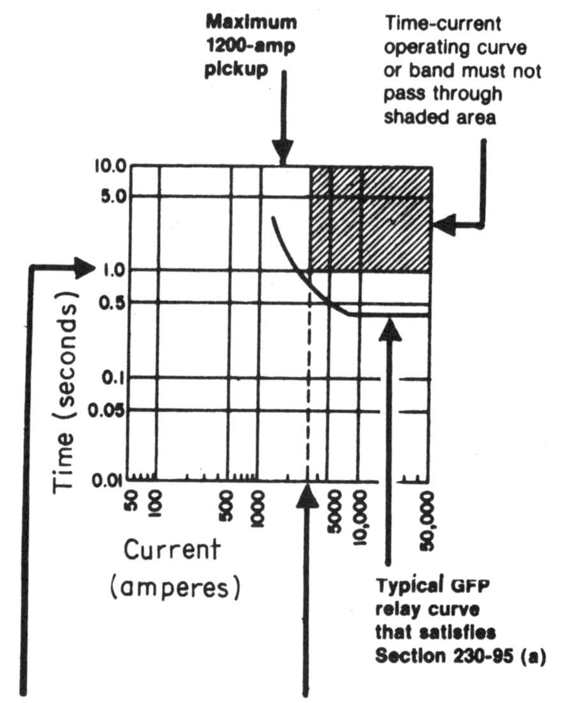

FIG. 5-20 The rule specifies maximum energy let-through for GFP operation.

accomplish safety objectives more readily without GFP on the service. Electrical design can account for any danger to personnel resulting from loss of process power versus damage to electrical equipment.

Exception 2 excludes fire-pump service disconnects from the basic rule that requires ground-fault protection on any service disconnect rated 1000 A or more on a grounded-wye 480/277-V system.

Because fire pumps are required by Sec. 230-90, Exception 4, to have overcurrent protection devices large enough to permit locked-rotor current of the pump motor to flow without interruption, larger fire pumps (100 hp and more) would have disconnects rated 1000 A or more. Without the exception, those fire-pump disconnects would be subject to the basic rule and would have to be equipped with ground-fault protection. But GFP on any fire pump is objectionable on the same basis that Sec. 230-90, Exception 4, wants nothing less than protection rated for locked rotor. The intent is to give the pump motor every chance to operate when it functions during a fire, to prevent opening of the motor circuit or any overload up to and including stalling or seizing of the shaft or bearings. For the same reason, Sec. 430-31 exempts fire pumps from the need for overload protection, and Sec. 430-72, Exception 4, requires overcurrent protection to be omitted from the control circuit of a starter for a fire pump.

FUSED SWITCH (bolted pressure switch, service protector, etc.)

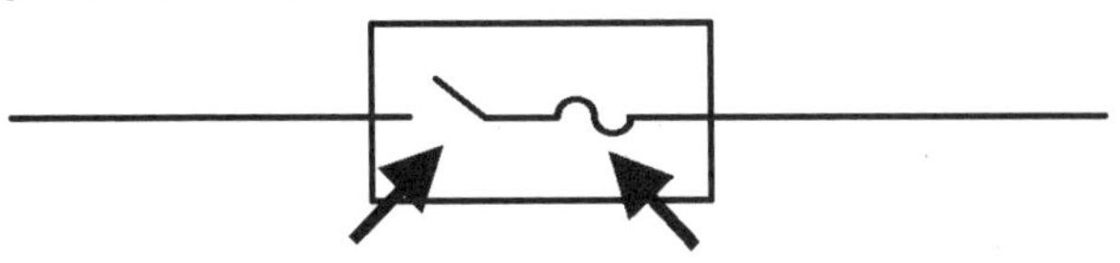

Rating of switch is taken as ... ... the amp rating of the largest fuse that can be installed in the switch fuseholders.

EXAMPLE

If 900-amp fuses are used in this service switch, ground-fault protection would be required because the switch can take fuses rated 1200 amps–which is above the 1000-amp level at which GFP becomes mandatory.

CIRCUIT BREAKER

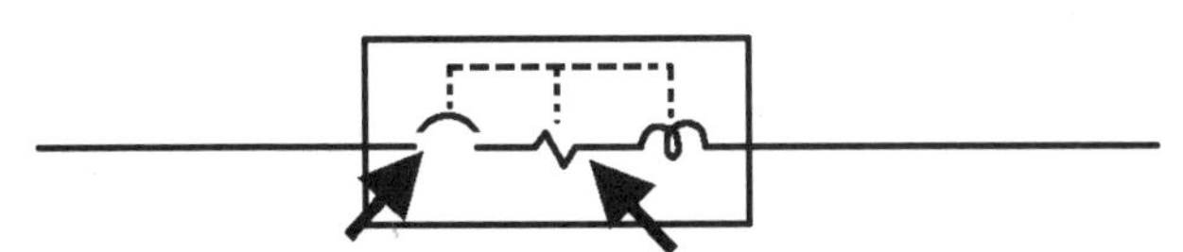

Rating of breaker is taken as ... ... the maximum continuous current rating (pickup of long time-delay) for which the trip device in the breaker is set or can be adjusted.

Example: GFP would be required for a service CB with, say, an 800-amp trip setting if the CB had a trip device that can be adjusted to 1000 amps or more.

FIG. 5-21 Determining the rating of service disconnect for GFP rule.

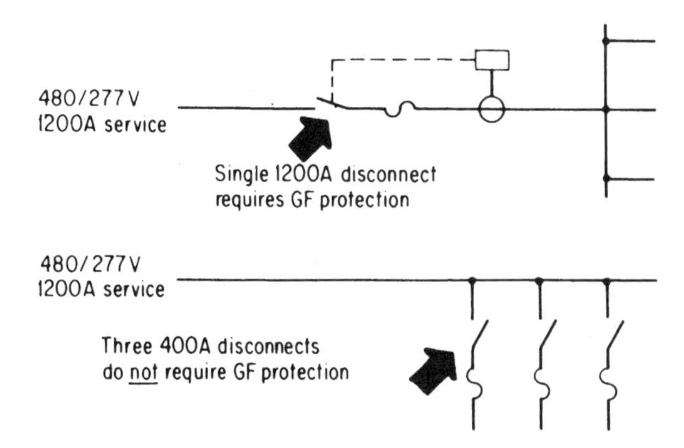

FIG. 5-22 Subdivision option on disconnects affects GFP rule.

And note that Exception 2 in Sec. 230-95 says that ground-fault protection *"shall not apply"* to fire-pump motors, making omission of GFP mandatory.

Important considerations are given in fine-print notes in this section. Obviously, the selection of ground-fault equipment for a given installation merits a detailed study. The option of subdividing services discussed under *six service entrances from one drop* (Sec. 230-2, Exception 7) should be evaluated. A 4000-A service, for example, could be divided using five 800-A disconnection means, and in such cases GFP would not be required.

One very important note (FPN 3) in Sec. 230-95(b) warns about potential desensitizing of ground-fault sensing hookups when an emergency generator and transfer switch are provided in conjunction with the normal service to a building. The note applies to those cases where a solid neutral connection from the normal service is made to the neutral of the generator through a 3-pole transfer switch. With the neutral grounded at the normal service and the neutral bonded to the generator frame, ground-fault current on the load side of the transfer switch can return over two paths, one of which will escape detection by the GFP sensor, as shown in Fig. 5-23. Such a hookup can also cause nuisance tripping of the GFP due to normal neutral current. Under normal (nonfaulted) conditions, neutral current due to normal load unbalance on the phase legs can divide at the common neutral connection in the transfer switch, with some current flowing toward the genera-

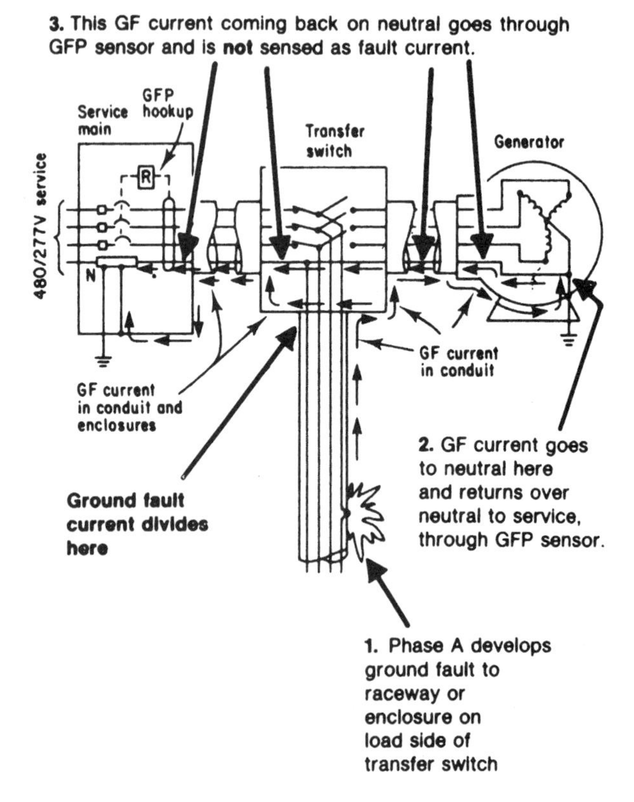

FIG. 5-23 Improper operation of GFP equipment can result from emergency system transfer switch.

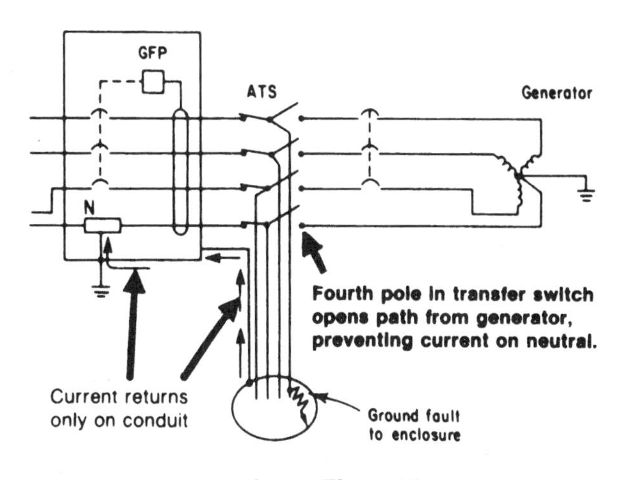

FIG. 5-24 Four-pole transfer switch is one way to avoid desensitizing GFP.

tor and returning to the service main on the conduit—indicating falsely that a ground fault exists and causing nuisance tripping of GFP. The note points out that "means or devices" (such as a 4-pole, neutral-switched transfer switch) "may be needed" to ensure proper, effective operation of the GFP hookup (Fig. 5-24).

Very important! Because of so many reports of improper and/or unsafe operation (or failure to operate) of ground-fault protective hookups, part (c) of Sec. 230-95 *requires* (a mandatory rule) that *every* GFP hookup be "performance tested when first installed." And the testing *must be done on the job site!* Factory testing of a GFP system does not satisfy this Code rule. This rule requires that such testing be done according to "approved instructions. . .provided with the equipment." A written record must be made of the test and must be available to the inspection authority.

Figure 5-25 shows two basic types of GFP hookup used at service entrances.

MAIN AND EQUIPMENT BONDING JUMPERS

Part (a) of Sec. 250-79 calls for use of copper or other "corrosion-resistant" conductor material, which does include aluminum and copper-clad aluminum. Part (c) demands use of connectors, lugs, and other fittings that have been designed, tested, and listed for the particular application.

Part (d) covers sizing of any bonding jumper within the service equipment enclosure or on the line or supply side of that enclosure. Refer to the definition of "bonding jumper, main" in Art. 100.

Figure 5-26 shows examples of sizing bonding jumpers in accordance with the first sentence of part (d) of this section.

At A, the bonding bushing and jumper are used to comply with part (d) of Sec. 250-72. Referring to Table 250-94, with 500-kcmil copper as the "largest service-entrance conductor," the minimum permitted size of grounding electrode conductor is No. 1/0

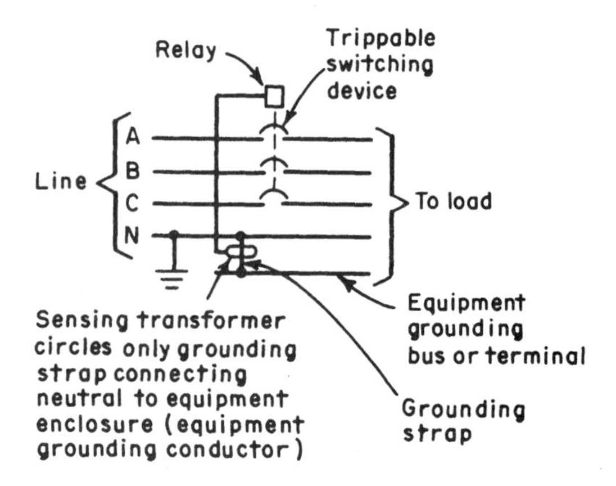

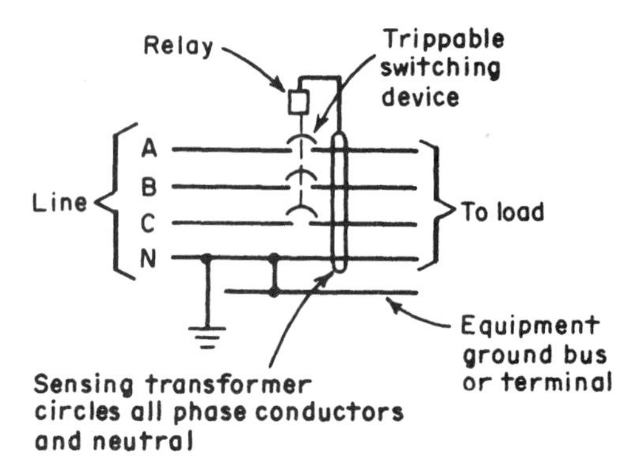

FIG. 5-25 Types of ground-fault detection that may be selected for use at services.

copper (or No. 3/0 aluminum). That therefore is the minimum permitted size of the required bonding jumper.

At B, with each service phase leg made up of two 500-kcmil copper conductors in parallel, the left-hand column heading in Table 250-94 refers to the "equivalent for parallel conductors." As a result, the phase leg is taken at $2 \times 500 = 1000$ kcmil, which is the physical equivalent of the makeup. Then Table 250-94 requires a minimum bonding jumper of No. 2/0 copper (or No. 4/0 aluminum).

Figure 5-27 shows an example of sizing a service bonding jumper in accordance with the second sentence of part (d) of this section. In this sketch, the jumper between the neutral bus and the equipment ground bus is defined by the NEC as a "main bonding jumper," and the minimum required size of this jumper for this installation is determined by calculating the size of one service phase leg. With three 500-kcmil per phase, that works out to 1500-kcmil copper per phase. Because that value is in excess of 1100-

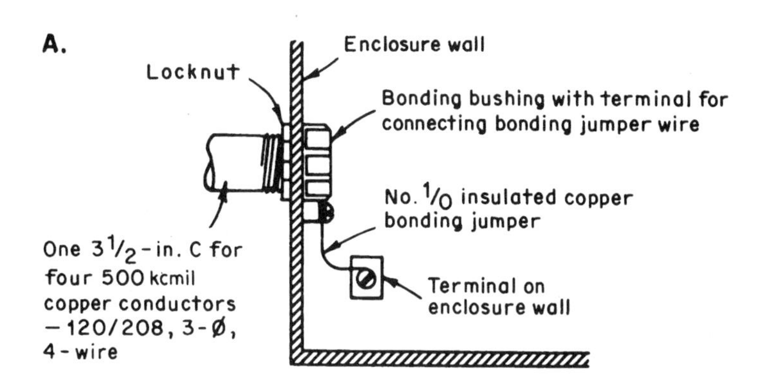

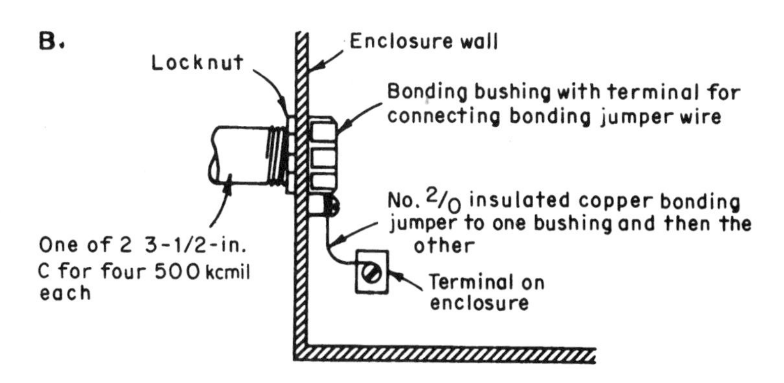

NOTE: Bushing with jumper is acceptable bonding for a clean KO or one with punched rings still in place.

FIG. 5-26 Examples of the basic sizing of service bonding jumpers.

kcmil copper, as noted in the Code rule, the minimum size of the main bonding jumper must equal at least $12\frac{1}{2}$ percent of the phase leg cross-sectional area. Then

$$12\tfrac{1}{2}\% \times 1500 \text{ kcmil} = 0.125 \times 1500 = 187.5 \text{ kcmil}$$

Referring to Table 8 in Chap. 9 in the back of the Code book, the smallest conductor with at least that cross-sectional area (csa) is No. 4/0 with a csa of 211,600 CM or 211.6 kcmil. Note that a No. 3/0 has a csa of only 167.8 kcmil. Thus No. 4/0 copper with any type of insulation would satisfy the Code.

The jumper shown in Fig. 5-27 running from one conduit bushing to the other and then to the equipment ground bus is defined by the NEC as an "equipment bonding jumper." It is sized the same as a main bonding jumper (above). In this case, therefore, the equipment bonding jumper would have to be not smaller than No. 4/0 copper. And

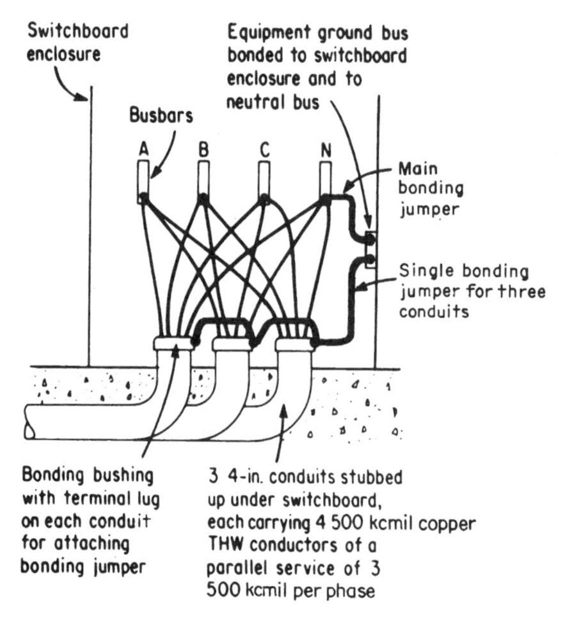

FIG. 5-27 Sizing of main bonding jumper and other jumpers at service equipment.

with the calculation that uses the 12½ percent value, if the jumper conductor is to be aluminum instead of copper, a calculation must be made as described below.

In the sketch of Fig. 5-27, if each of the three 4-in conduits has a separate bonding jumper connecting each one individually to the equipment ground bus, the next-to-last sentence of part (d) may be applied to an individual bonding jumper for each separate conduit (Fig. 5-28). The size of a separate bonding jumper for each conduit in a parallel service must be not less than the size of the grounding electrode conductor for a service of the size of the phase conductor used in each conduit. Referring to Table 250-94, a 500-kcmil copper service calls for at least a No. 1/0 grounding electrode conductor. Therefore, the bonding jumper run from the bushing lug on each conduit to the ground bus must be at least a No. 1/0 copper (or 3/0 aluminum).

The third sentence of part (d) *requires* separate bonding jumpers when the service is made up of multiple conduits and the equipment bonding jumper is run within each raceway (such as plastic pipe) for grounding service enclosures. According to the third sentence of part (d), when service-entrance conductors are paralleled in two or more raceways, an equipment bonding jumper that is routed within the raceways must also be run in parallel, one in each raceway, as at the right in Fig. 5-28. This clarifies application of nonmetallic service raceway where parallel conduits are used for parallel service-entrance conductors. As worded, the rule applies to both nonmetallic and metallic conduits where the bonding jumper is run within the raceways rather than from lugs on bonding bushings on the conduit ends. But for metallic conduits stubbed up under service equipment, if the conduit ends are to be bonded to the service equipment enclosure by jumpers from lugs on the conduit bushings, either a single large common bonding jumper may be used—from one lug, to another lug, to another, etc., and then to the ground bus—

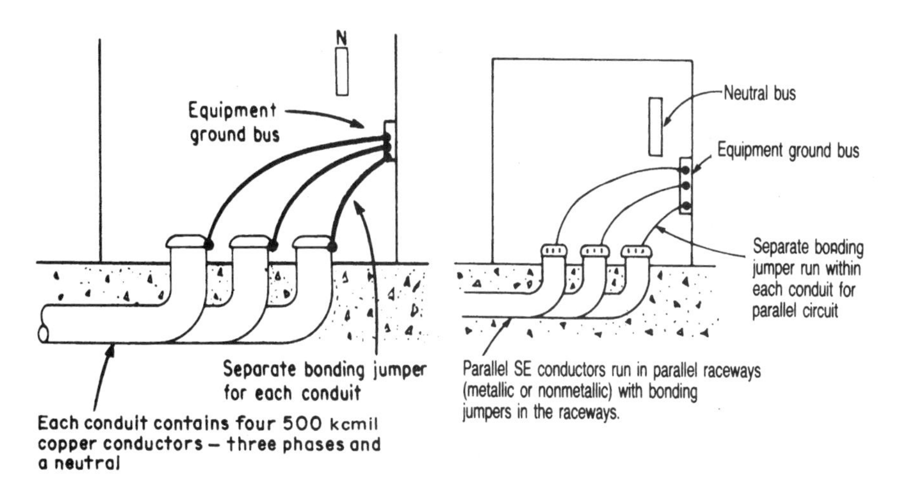

FIG. 5-28 An individual bonding jumper may be used for each conduit (left) and *must* be used as shown at right.

or an individual bonding jumper (of smaller size from Table 250-94, based on the size of conductors in each conduit) may be run from each bushing lug to the ground bus.

The second sentence of part (d) sets minimum sizes of copper *and* aluminum service-entrance conductors above which a service bonding jumper must have a cross-sectional area "not less than 12½ percent of the area of the largest phase conductor." And the rule states that if the service conductors and the bonding jumper are of different materials (that is, service conductors are copper, say, and the jumper is aluminum), the minimum size of the jumper shall be based on the assumed use of phase conductors of the same material as the jumper and with an ampacity equivalent to that of the installed phase conductors (Fig. 5-29).

The last sentence in part (d) covers the sizing of a bonding jumper used to bond raceway that contains a grounding electrode conductor. Such a raceway is required to provide mechanical protection for a grounding electrode conductor smaller than No. 6, as noted in Sec. 250-92(a). And protection is frequently provided for larger grounding electrode conductors.

At service equipment, a bonding jumper for a raceway containing a grounding electrode conductor only has to be at least the same size as the required grounding electrode conductor, as shown in Fig. 5-30. The last sentence makes clear that the bonding jumper for a grounding electrode conductor conduit does *not* have to be sized at 12½ percent of the cross-sectional area of the largest phase conductor of the service—as required by the foregoing text of Sec. 250-79(d), when the largest service phase conductor is larger than 1100-kcmil copper or 1750-kcmil aluminum. That requirement for 12½ percent of the service phase size applies only to bonding jumpers used with conduits containing ungrounded service phase conductors, but not with raceway sleeves for grounding electrode conductors.

Part (b) of Sec. 250-92 covers details on the use of "metal enclosures" (such as conduit or EMT sleeves) for grounding electrode conductors. The grounding conductor must be connected to its protective conduit at both ends so that any current that might flow over the conductor will also have the conduit as a parallel path. The regulation pre-

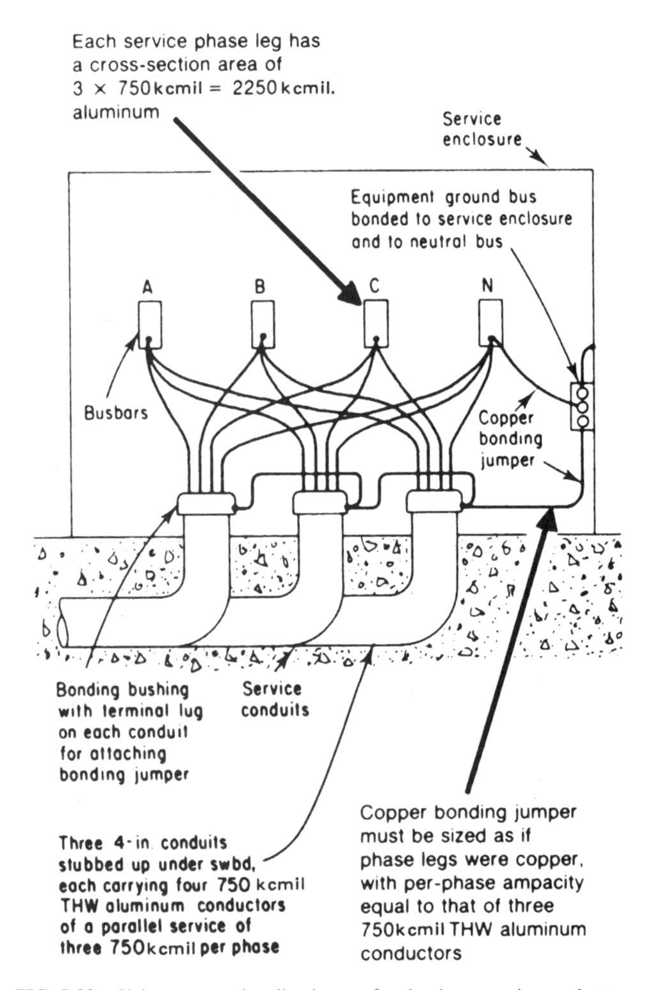

FIG. 5-29 Sizing a copper bonding jumper for aluminum service conductors.

sented is actually a performance description of the rule of Sec. 250-71(a)(3), which specifically and simply requires that any conduit or armor enclosing a grounding electrode conductor be electrically parallel with the conductor. Bonded connections at both ends of an enclosing raceway must be used for any grounding electrode arrangement at a service and for grounding of a separately derived system, such as a generator or transformer secondary. Any bonding jumper for that application simply has to be at least the same size as required for the grounding electrode conductor run inside the conduit.

Part (e) requires a bonding jumper on the load side of the service to be sized as if it were an equipment grounding conductor for the largest circuit with which it is used. And sizing would have to be done from Table 250-95, as follows.

Figure 5-31 shows a floor trench in the switchboard room of a large hotel. The conductors are feeder conductors carried from circuit breakers in the main switchboard (just

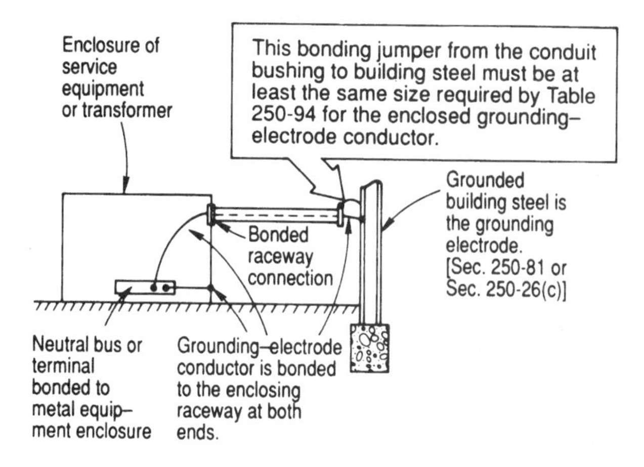

FIG. 5-30 A typical example of the rule on bonding jumpers for raceways enclosing grounding electrode conductors.

FIG. 5-31 Conduits in trench carry feeder conductors from switchboard at right (arrow) out to various panels and control centers. A single, common bonding jumper—run continuously from bushing to bushing—may be used to bond all conduits to the switchboard ground bus.

visible in upper right corner of photograph) to feeder conduits going out at left, through the concrete wall of the trench, and under the slab floor to the various distribution panels throughout the building. Because the conduits themselves are not metallically connected to the metal switchboard enclosure, bonding must be provided from the conduits to the switchboard ground bus to ensure electrical continuity and conductivity as required by NEC Secs. 250-33, 250-42(e), 250-51, and 250-57.

1. The single, common, continuous bonding conductor that bonds all the conduits to the switchboard must be sized in accordance with NEC Table 250-95, based on the highest rating of CB or fuses protecting any one of the total number of circuits run in all the conduits.

2. Sizing of the single, common bonding jumper would be based on the highest rating of overcurrent protection for any one of the circuits run in the group of conduits. For instance, some of the circuits could be 400-A circuits made up of 500 kcmils in individual 3-in conduits, and others could be parallel-circuit makeups in multiple conduits, such as 800-A circuits, with two conduits per circuit, and 1200-A circuits, with three conduits. If, for instance, the highest rated feeder in the group were protected by a 2000-A circuit breaker, then the single, common bonding jumper for all the conduits would have to be 250-kcmil copper or 400-kcmil aluminum—determined readily from Table 250-95, by simply going down the left column to the value 2000 and then reading across. The single conductor is run through a lug on each of the conduit bushings and then to the switchboard ground bus.

In the case shown in Fig. 5-31, however, because the bonding jumper from the conduit ends to the switchboard is much longer than a jumper would be if the conduits stubbed up under the switchboard, better engineering design might dictate that a separate equipment grounding *conductor* (rather than a "jumper") be used for each individual circuit in the group. If one of the conduits is a 3-in conduit carrying three 500-kcmil conductors from a 400-A CB in the switchboard, the minimum acceptable size of bonding jumper (or equipment grounding conductor) from a grounding bushing on the conduit end to the switchboard ground bus will be No. 3 copper or No. 1 aluminum or copper-clad aluminum, as shown opposite the value of 400 A in the left column of Table 250-95. If another two of the 3-in conduits are used for a feeder consisting of two parallel sets of three 500-kcmil conductors (each set of three 500-kcmils in a separate conduit) for a circuit protected at 800 A, a single bonding jumper could be used, run from one grounding bushing to the other grounding bushing and then to the switchboard ground bus. This single bonding jumper will have to be a minimum No. 1/0 copper, from NEC Table 250-95 on the basis of the 800-A rating of the feeder overcurrent protective device.

With such a long run for a jumper, as shown in Fig. 5-31, Code rules could be interpreted to require that the bonding jumper be subject to the rules of Sec. 250-95; that is, use of a bonding jumper must conform to the requirements for equipment grounding conductors. As a result, bonding of conduits for a parallel-circuit makeup would have to comply with the second sentence in Sec. 250-95, which requires equipment grounding conductors to be run in parallel "where conductors are run in parallel in multiple raceways. . . ." That would then be taken to require that bonding jumpers *also* be run in parallel for multiple-conduit circuits. And that concept is supported by the next-to-last sentence of part (d) of Sec. 250-79. In the case of the 800-A circuit above, instead of a single No. 1/0 copper jumper from one bushing lug to the other bushing lug and then to the ground bus, it would be necessary to use a separate No. 1/0 copper from each bushing to the ground bus, so the jumpers are run in parallel, as required for equipment grounding conductors. Figure 5-32 shows the two possible arrangements. The wording

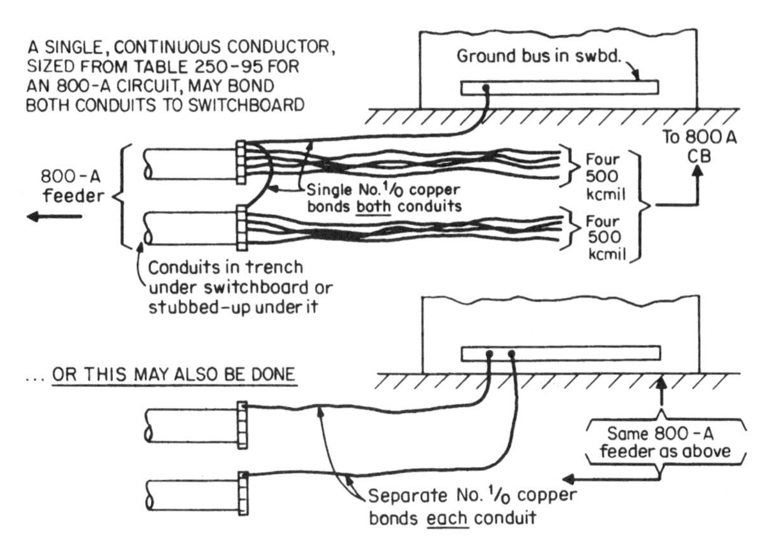

FIG. 5-32 One jumper may be used to bond two or more conduits on the load side of the service.

of Secs. 250-79(e) and 250-95 can be used to support either method. Bonding jumpers on the load side of service equipment are sized and routed the same as equipment grounding conductors because such bonding jumpers and equipment grounding conductors serve identical functions. And note that Sec. 250-95 requires the equipment grounding conductor for each of the conduits for a parallel circuit to be the full size determined from the circuit rating. In the case here, a No. 1/0 copper for each conduit is required, based on the 800-A rating of the feeder protective device.

An exception notes that an equipment bonding jumper never has to be larger than the circuit conductors within a conduit being bonded.

Part (f) of Sec. 250-79 follows the thinking that was described in Sec. 250-58(a) for external grounding of equipment attached to a properly grounded metal rack or structure. A short length of flexible metal conduit, liquidtight flex, or any other raceway may, if the raceway itself is not acceptable as a grounding conductor, be provided with grounding by a "bonding jumper" (note: *not* an "equipment grounding conductor") run *either* inside or *outside* the raceway or enclosure *provided that* the *length* of the *equipment bonding jumper* is *not more* than 6 ft and the jumper is routed with the raceway or enclosure.

Where an equipment bonding conductor is installed within a raceway, it must comply with all the Code rules on identification of equipment grounding conductors. A bonding jumper installed in flexible metal conduit or liquidtight flex serves essentially the same function as an equipment grounding conductor. For that reason, a bonding jumper should comply with the identification rules of Sec. 310-12(b)—on the use of bare, green-insulated or green-taped conductors for equipment grounding.

Note that this application has limited use for the conditions specified and is a special variation of the concept of Sec. 250-57(b), which requires grounding conductors run inside raceways. Its big application is for external bonding of short lengths of liquidtight or standard flex, under those conditions where the particular type of flex itself is not

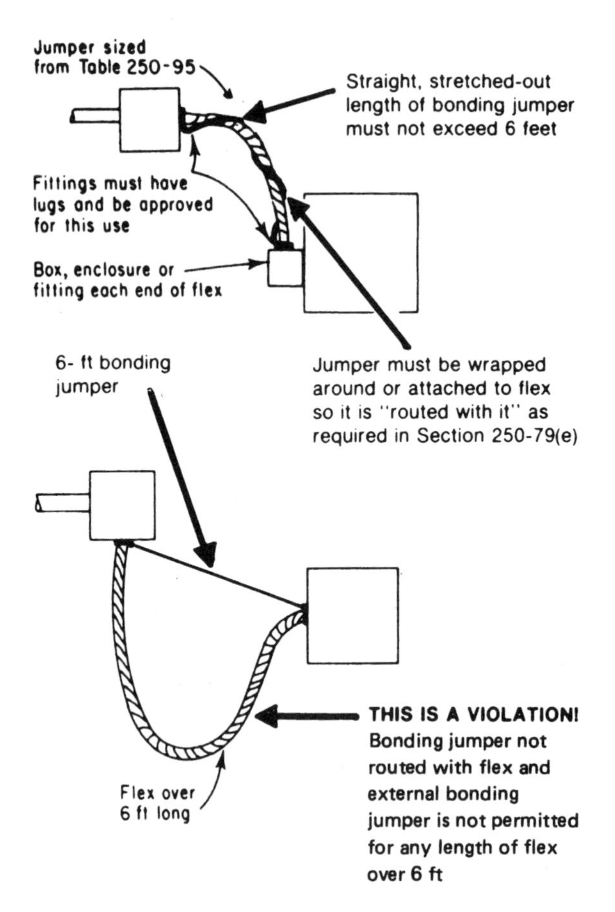

Bonding jumper required for flex may be outside the flex.

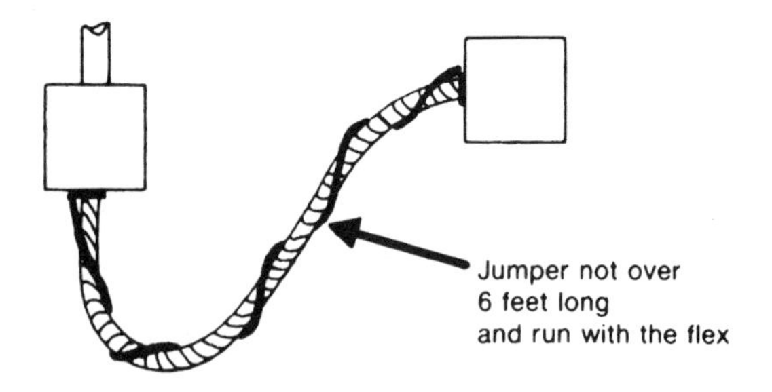

Liquidtight flex larger than 1¼ in. size must have bonding jumper — inside or outside — in any length up to 6 ft.

FIG. 5-33 Bonding jumper rules for standard flex and liquidtight flex.

suitable for providing the grounding continuity required by Secs. 350-5 and 351-9. Refer also to Sec. 250-91(b), Exceptions 1 and 2.

The top of Fig. 5-33 shows how an external bonding jumper may be used with standard flexible metallic conduit (so-called Greenfield). If the length of the flex is not over 6 ft but the conductors run within the flex are protected at more than 20 A, a bonding jumper *must* be used either inside or outside the flex. An outside jumper must comply as shown. For a length of flex not over 6 ft, containing conductors that are protected at not more than 20 A and used with conduit termination fittings that are approved for grounding, a bonding jumper is *not* required, as covered in Sec. 250-91(b), Exception 1.

The bottom of Fig. 5-33 shows use of an external bonding jumper with liquidtight flexible metallic conduit. If liquidtight flex is not over 6 ft long *but* is larger than 1¼-in trade size, a bonding jumper must be used, installed *either* inside or outside the liquidtight. An outside jumper must comply as shown. If a length of liquidtight flex larger than 1¼ in is short enough to permit an external bonding jumper that is not more than 6 ft long between external grounding-type connectors at the ends of the flex, an external bonding jumper may be used. *But watch out!* The rule says the *jumper,* not the flex must not exceed 6 ft in length *and* the jumper "shall be routed with the raceway," that is, run along the flex surface and not separated from the flex.

Figure 5-34 shows the typical use of copper, aluminum, or copper-clad aluminum conductor to connect the bonded neutral and equipment ground terminal of service equipment to each of the one or more grounding electrodes used at a service (Sec. 250-91). Controversy has been common on the permitted color of an insulated (or covered) grounding electrode conductor. Section 200-7 limits use of white or natural gray color for any conductor other than a "grounded conductor," such as the grounded neutral or phase leg, as described in the definition of "grounded conductor." Green color is reserved for equipment grounding conductors, although there is no Code rule clearly prohibiting a green grounding electrode conductor. Refer to Sec. 310-12(b).

Exception 2 of Sec. 250-91(a) covers the installation of the grounding electrode conductor for a service layout consisting of two to six service disconnects in separate enclosures with a separate set of service-entrance conductors run down to each disconnect. Such an arrangement is covered in Sec. 230-40, Exception 2. Previous Code editions required (and it still satisfies the Code to use) either a separate grounding electrode conductor run from each enclosure to the grounding electrode or a single, unspliced conductor looped from enclosure to enclosure. And a single grounding electrode conductor used to ground all the service disconnects has to be without splice [the last line in Sec.

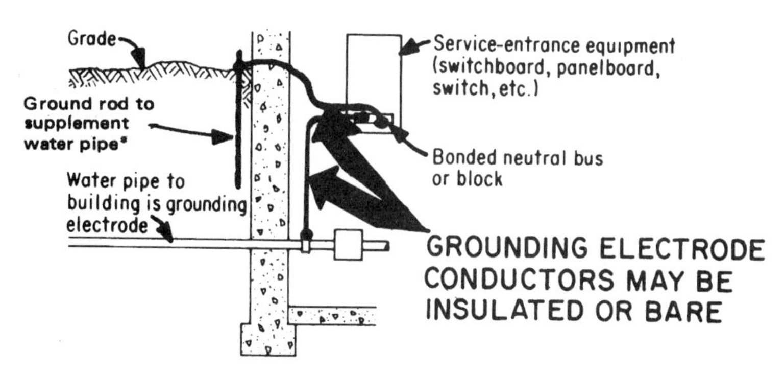

FIG. 5-34 An insulated grounding electrode conductor may be any color other than white, natural gray, or green.

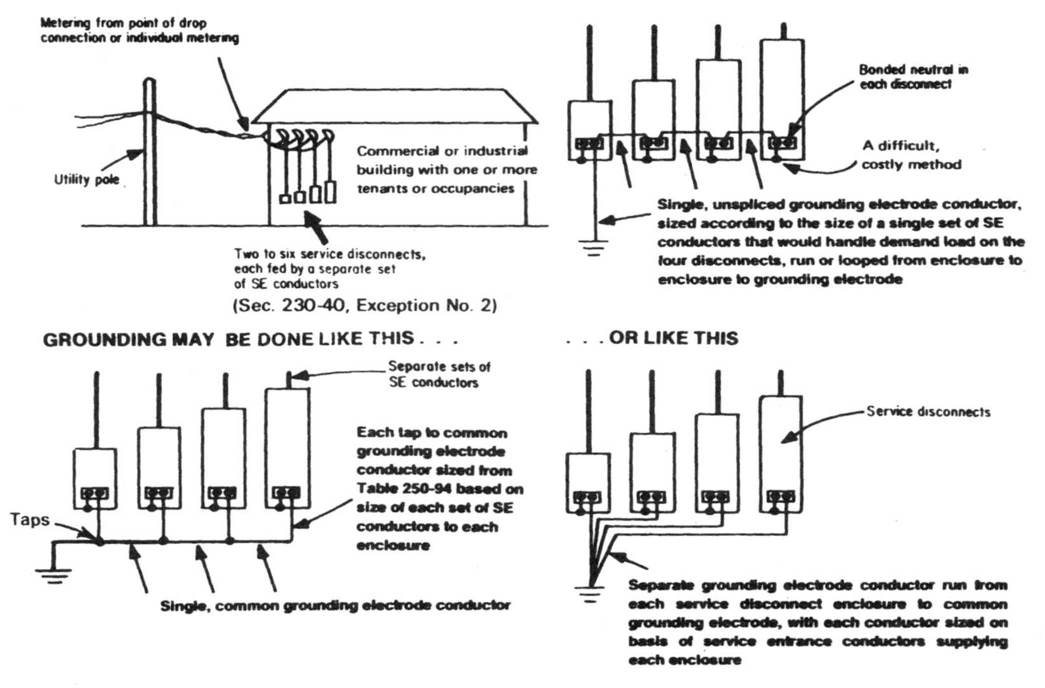

FIG. 5-35 Grounding electrode conductor may be tapped for multiple service disconnects.

250-91(a)]—run from one enclosure to the other and then to the water pipe or other grounding electrode. The specific rule of Exception 2 says that when a service disconnect consists of two to six switches or CBs in separate enclosures, a tap from each enclosure may be made to a single grounding electrode conductor that *must* be without splice or joint. For such an arrangement (shown at the bottom left in Fig. 5-35), the last sentence of Exception 2 clarifies that there must be no break in the single, common grounding electrode conductor used to connect the taps to ground two or more service disconnects in individual enclosures.

The wording of Exception 2 is clear on the sizing of the main, common grounding electrode conductor. Wording of the exception requires the main grounding electrode conductor to be sized from Table 250-94, which has a footnote that covers multiple sets of SE conductors. The main grounding electrode conductor must be sized for the sum of the cross-sectional areas of the total number of conductors connected to one hot leg of the service drop. All considerations of function, adequacy, and safety would be satisfied by running the single, common, unspliced grounding electrode conductor from the disconnect for the largest of the several sets of service-entrance conductors directly to the grounding electrode. That single, common conductor could be sized directly from Table 250-94, based on the size of SE conductors as determined from the note under Table 250-94. Each other service-entrance disconnect would then have a tap conductor that taps to the unspliced common grounding electrode conductor.

The wording of the rule makes clear that the size of the grounding electrode tap to each separate enclosure may be determined from Table 250-94 on the basis of the largest service hot leg serving each enclosure, as shown in Fig. 5-36. Although that illustration shows an overhead service to the layout, the two to six service disconnects

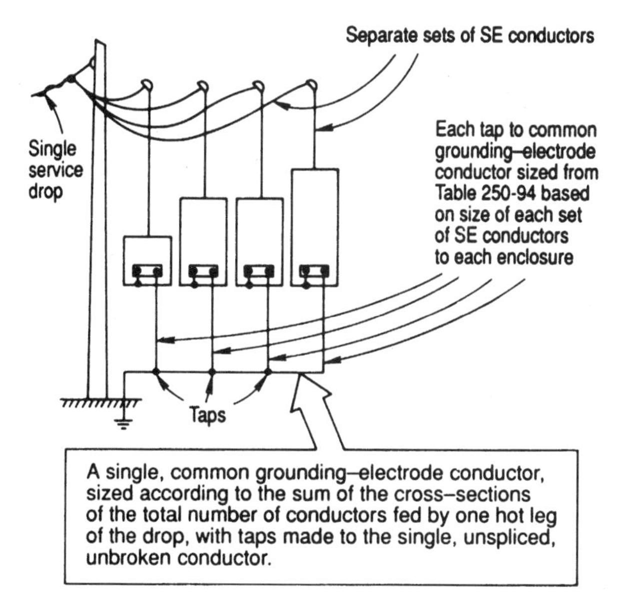

FIG. 5-36 Rule covers sizing main and taps of grounding electrode conductor at multiple-disconnect services. A single common grounding electrode conductor must be "without splice or joint," with taps made to the grounding electrode conductor.

could be fed by individual sets of underground conductors, making up a "single" service lateral as permitted by Exception 7 to Sec. 230-2.

Part (b) describes the various types of conductors and metallic cables or raceways that are considered suitable for use as equipment grounding conductors. And the Code recognizes cable tray as an equipment grounding conductor as permitted by Art. 318.

Exception 1 recognizes flexible metal conduit or flexible metallic tubing (see Art. 349) or liquidtight flexible metal conduit with termination fittings UL-listed for use as a grounding means (without a separate equipment grounding wire) if the length of the flex is not over 6 ft and the contained circuit conductors are protected by overcurrent devices rated at 20 A or less.

Standard flexible metal conduit (also known as Greenfield) is not listed by UL as suitable for grounding in itself. However, Sec. 350-5 of the NEC as well as Exception 1 in part (b) permit flex to be used without any supplemental grounding conductor when any length of flex in a ground return path is not over 6 ft and the conductors contained in the flex are protected by overcurrent devices rated not over 20 A (Fig. 5-37). Use of standard flex with the permission given in Sec. 250-79(e) for either internal or external bonding must be as follows:

1. When conductors within a total length of flex up to 6 ft are protected at more than 20 A, equipment grounding may not be provided by the flex, but a separate conductor must be used for grounding. If a length of flex is short enough to permit a bonding jumper not over 6 ft long to be run between external grounding-type connectors at the flex ends, while keeping the jumper *along* the flex, such an external jumper may be used where equipment grounding is required—as for a short length of flex with circuit conductors in it protected at more than 20 A. Of course, such short lengths of flex may also be "bonded" by a bonding jumper inside the flex, instead of external. Refer to Sec. 250-79(e).

2. Any length of standard flex that would require a bonding jumper longer than 6 ft may not use an external jumper. In the Code sense, when the length of such a grounding conductor exceeds 6 ft, it is *not* a bonding jumper but *is* an equipment grounding conductor and must be run only *inside* the flex, as required by Sec. 250-57(b). Combining UL data with the rule of Sec. 250-79(e) and Exception 1 to Sec. 350-5, *every* length of flex that is over 6 ft must contain an equipment grounding conductor run *only* inside the flex (Fig. 5-38).

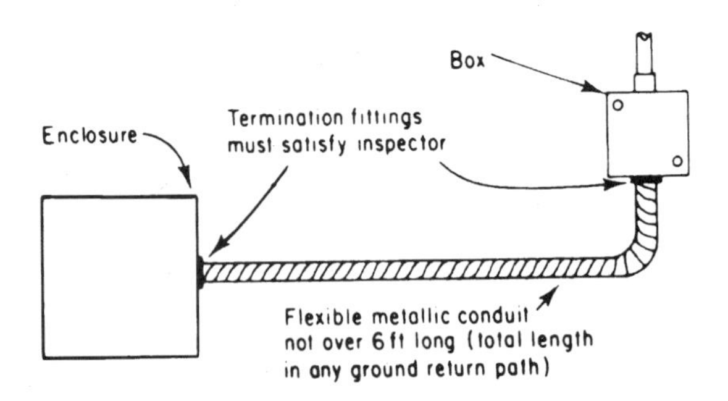

Flex not over 6 ft long is suitable as a grounding means (without a separate ground wire) if the conductors in it are protected by OC devices rated not more than 20 amps.

FIG. 5-37 Standard flex is limited in use without an equipment ground wire.

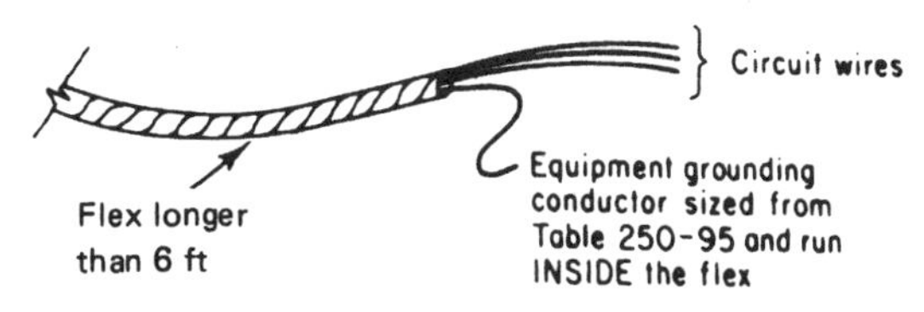

FIG. 5-38 Internal equipment grounding is required for any flex over 6 ft long.

In part (a) of Exception 1, note that exemption from the need for an equipment grounding conductor applies only to flex where there is not over 6 ft of "total length in any ground return path." That means that from any branch-circuit load device—lighting fixture, motor, etc.—all the way back to the service ground, the total permitted length of flex without a ground wire is 6 ft. In the total circuit run from the service to any outlet, there could be one 6-ft length of flex, or two 3-ft lengths, or three 2-ft lengths, or a 4-ft and 2-ft length, where the flex lengths are in series as equipment ground return paths. In any circuit run—feeder to subfeeder to branch circuit—any length of flex that would make the total series length over 6 ft would have to use an internal or external bonding jumper, regardless of any other factors.

In all cases, sizing of bonding jumpers for all flex applications is made according to Sec. 250-79(d), which requires the same minimum size for bonding jumpers as is required for equipment grounding conductors. In either case, the size of the conductor is selected from Table 250-95, based on the maximum rating of the overcurrent devices protecting the circuit conductors that are within the flex.

In part (c) of Exception 1, the term *listed for grounding* as applied to termination fittings will require the authority having jurisdiction to verify the grounding capabilities of fittings by requiring only "listed" fittings to be used with these short conduit lengths. See also Secs. 350-5 and 351-9.

Exception 2 presents conditions under which *liquidtight* flexible metal conduit may be used without need for a separate equipment grounding conductor:

1. Both Exception 2 and the UL's *Electrical Construction Materials Directory* (the Green Book) note that any listed liquidtight flex in 1¼-in and smaller trade size, in a length not over 6 ft, may be satisfactory as a grounding means through the metal core of the flex, without need of a bonding jumper (or equipment grounding conductor) either internal or external (Fig. 5-39). Liquidtight flex in 1¼-in and smaller trade size may be used without a bonding jumper inside *or* outside provided that the "total length" of that flex "in any ground return path" is not over 6 ft. Thus, two or more separate 6-ft lengths installed in a raceway run would not be acceptable with the bonding jumper omitted from all of them. In such cases, one 6-ft length or more than one length that does not total over 6 ft may be used with a bonding jumper, but any additional lengths 6 ft or less in the same raceway run must have an external bonding jumper sized from Table 250-95.

The required conditions for use of liquidtight flex without need of a separate equipment bonding jumper (or equipment grounding conductor) are as follows: Where terminated in fittings investigated for grounding and where installed with not more than 6 ft (total length) in any ground return path, liquidtight flexible metal conduit in the ⅜- and ½-in trade sizes is suitable for grounding where used on circuits rated 20 A or less, and the ¾-, 1-, and 1¼-in trade sizes are suitable for grounding where used on circuits rated 60 A or less. See the category "conduit fittings" (DWTT) with respect to fittings suitable as a grounding means.

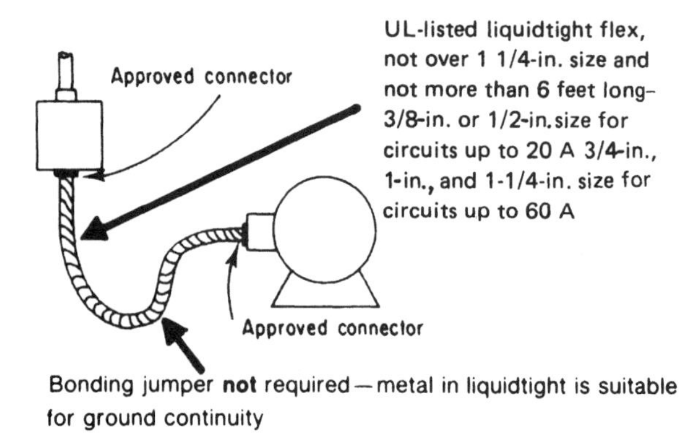

FIG. 5-39 Liquidtight flex may be used with a separate ground wire.

The following are not considered to be suitable as a grounding means:

- The 1½-in and larger trade sizes.
- The ⅜- and ½-in trade sizes where used on circuits rated higher than 20 A, or where the total length in the ground return path is greater than 6 ft.
- The ¾-, 1-, and 1¼-in trade sizes where used on circuits rated higher than 60 A, or where the total length in the ground return path is greater than 6 ft.

Although UL gives the same grounding recognition to its "listed" liquidtight flex, this Code rule covers liquidtight that the UL does not list, such as high-temperature type.

2. For liquidtight flex over 1¼ in, UL does not list any as suitable for equipment grounding, thereby requiring use of a separate equipment grounding conductor installed in *any* length of the flex, as required by the Code. If a length of liquidtight flex larger than 1¼ in is short enough to permit an external bonding jumper not more than 6 ft long between external grounding-type connectors at the ends of the flex, an external bonding jumper may be used. *But watch out!* The rule says the *jumper,* not the flex, must not exceed 6 ft in length *and* the jumper "shall be routed with the raceway," that is, run along the flex surface and not separated from the flex.

3. If any length of flex is *over 6 ft,* then the flex is not a suitable grounding conductor, regardless of the trade size of the flex, whether it is larger or smaller than 1¼ in. In such cases, an *equipment grounding conductor* (not a *bonding jumper*—the phrase reserved for short lengths) must be used to provide grounding continuity and it must be run *inside* the flex, not external to it, in accordance with Sec. 250-57(b).

Part (c) of Sec. 250-91 is an extremely important rule that has particular impact on the use of electrical equipment outdoors. The first part of the rule accepts the use of "supplementary grounding electrodes," such as a ground rod, to "augment" the equipment grounding conductor; *but* an equipment grounding conductor must always be used where needed, and the connection of outdoor metal electrical enclosures to a ground rod is never a satisfactory alternative to the use of an equipment grounding conductor because use of just ground-rod grounding would have the earth as "the sole equipment grounding conductor" and that is expressly prohibited by the last clause of part (c).

This whole matter of earth ground usually comes up as follows:

Question: When direct-burial or nonmetallic conduit circuits are run underground to supply lighting fixtures or other equipment mounted on metal standards or poles or fed by metal conduit run up a pole or building wall, is it necessary to run an equipment grounding conductor to ground the metal standard or pole or conduit if a ground rod has been driven for the same purpose?

Answer: Yes. An equipment grounding conductor is necessary to provide low impedance for ground-fault current return to ensure fast, effective operation of the circuit protective device when the circuit is derived from a grounded electrical system (such as 240/120 V, single-phase or 208Y/120 or 480Y/277 V, 3-phase). Low impedance of a grounding path "to facilitate the operation of the circuit protective devices in the circuit" is clearly and specifically required by NEC Sec. 250-51. When a ground rod is used to ground an outdoor metal standard or pole or outdoor metal conduit and no other grounding connection is used, ground-fault current must attempt to return to the grounded system neutral by flowing through the earth. Such an earth return path has impedance that is too high, limiting the current to such a low value that the circuit protective device does not operate. In that case, a conductor that has faulted (made conductive contact) to a metal standard, pole, or conduit will put a dangerous voltage on the metal, exposing persons to shock or electrocution hazard as long as the fault exists. The basic concept of this problem—and Code violation—is revealed in Fig. 5-40.

The same undesirable condition would exist where a direct-burial or nonmetallic conduit circuit fed up through a metal conduit outdoors. As shown in Fig. 5-41, fault current would have to return through the high impedance of the earth.

The hazard of arrangements using only a ground rod as shown arises from the chance that a person might make contact with the energized metal standard or conduit and have a high enough voltage across the person to produce a dangerous current flow through the person's body. The actual current flow through the body will depend upon the contact resistances and the body resistance in conjunction with the voltage gradient (potential difference) imposed across the body. As shown in Fig. 5-42, a person contacting an energized metal pole can complete a circuit to earth or to some other pipe or metal that is grounded back to the system neutral.

It is important to note that part (c) of Sec. 250-91 applies to these situations where outside metal standards or conduits are grounded by means of a ground rod at the standard or conduit. The installations shown in Figs. 5-40, 5-41, and 5-42 are in violation of the last clause of Sec. 250-91(c) as well as Sec. 250-51(c).

Compliance with the letter and spirit of those Code sections—and other sections on equipment grounding—will result from use of an equipment grounding conductor run with the circuit conductors—either closely placed in the same trench with direct-burial conductors (type UF or type USE) or pulled into nonmetallic conduit with the circuit conductors. Such arrangement is also dictated by Sec. 250-57(b), which requires that an equipment grounding conductor be within the same raceway, cable, or cord or otherwise run with the circuit conductors.

[Although NEC Art. 338 on "Service-Entrance Cable" does not say that type USE cable may be used as a feeder or branch circuit (on the load side of service equipment), the UL listing on type USE cable says it "is suitable for all of the underground uses for which Type UF cable is permitted by the National Electrical Code."]

If an underground circuit to a metal standard or pole is run in metal conduit, a separate equipment grounding conductor is not needed in the conduit if the conduit end within the standard is bonded to the standard by a bonding jumper. Section 250-91(b) recognizes metal conduits as suitable equipment grounding conductors in themselves.

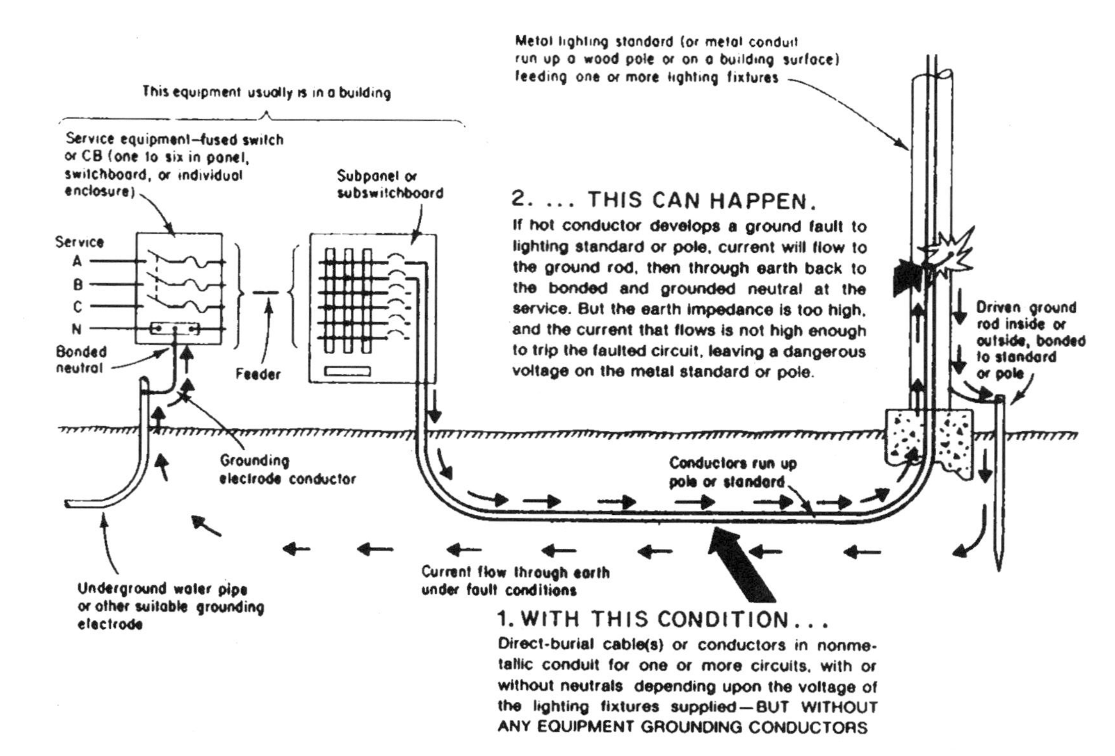

FIG. 5-40 This Code violation produces a dangerous condition at a metal standard.

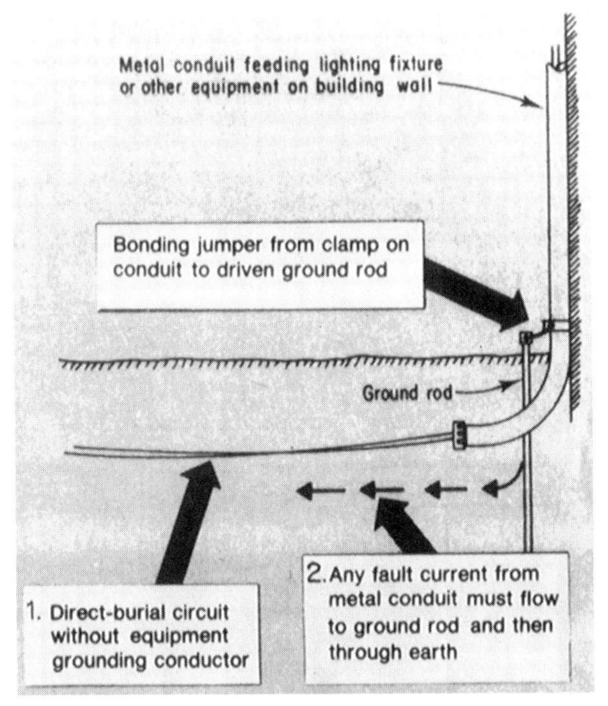

FIG. 5-41 This is also a dangerous condition for conduit up a pole or building.

SIZING GROUNDING ELECTRODE CONDUCTOR

For copper wire, a minimum size of No. 8 is specified by Sec. 250-94 in order to provide sufficient carrying capacity to ensure an effective ground and sufficient mechanical strength to be permanent. Where one of the service conductors is a grounded conductor, the same grounding electrode conductor is used for grounding both the system and the equipment. Where the service is from an ungrounded 3-phase power system, a grounding electrode conductor of the size given in Table 250-94 is required at the service.

If the sizes of service-entrance conductors for an ac system are known, the minimum acceptable size of grounding electrode conductor can be determined from NEC Table 250-94. Where the service consists of only one conductor for each hot leg or phase, selection of the minimum permitted size of electrode conductor is a relatively simple, straightforward task. If the largest phase leg is, say, a 500-kcmil copper THW, Table 250-94 shows No. 1/0 copper or No. 3/0 aluminum (reading across from "Over 350 kcmil through 600 kcmil") as the minimum size of a grounding electrode conductor.

But use of the table for services with multiple conductors per phase leg (e.g., four 500 kcmil for each of three phase legs of a service) is more involved. The heading over the left-hand columns of this table is "Size of Largest Service-Entrance Conductor or Equivalent for Parallel Conductors." To make proper use of this table, the meaning of the word *equivalent* must be clearly understood. *Equivalent* means that parallel conductors per phase are to be converted to a single conductor per phase that has a cross-

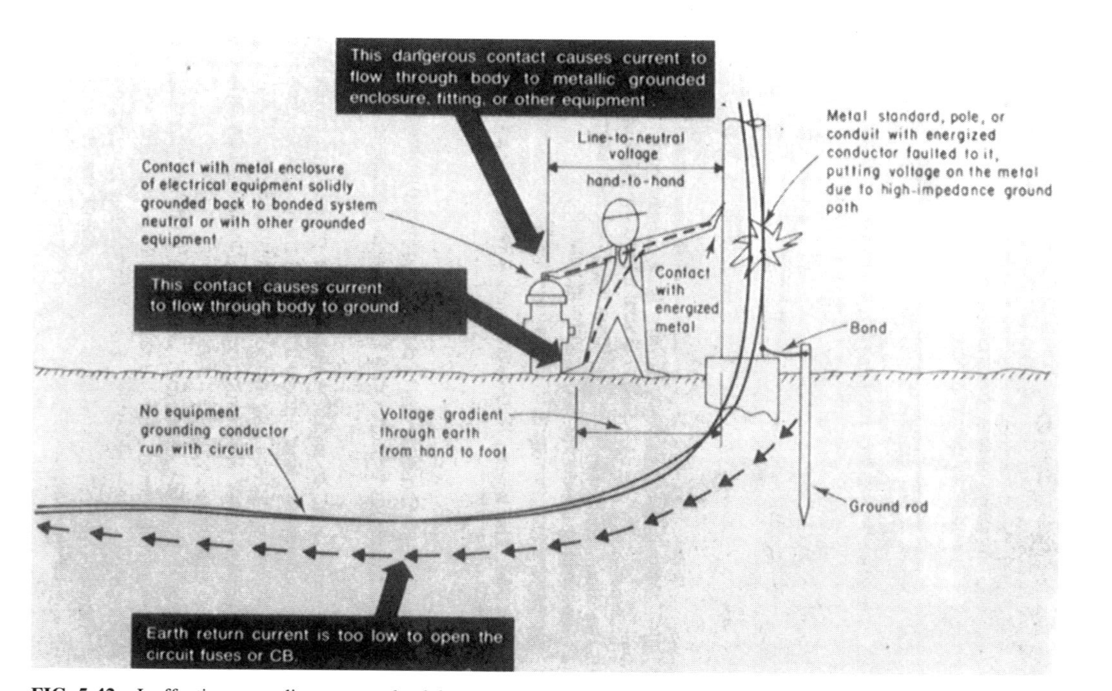

FIG. 5-42 Ineffective grounding creates shock hazards.

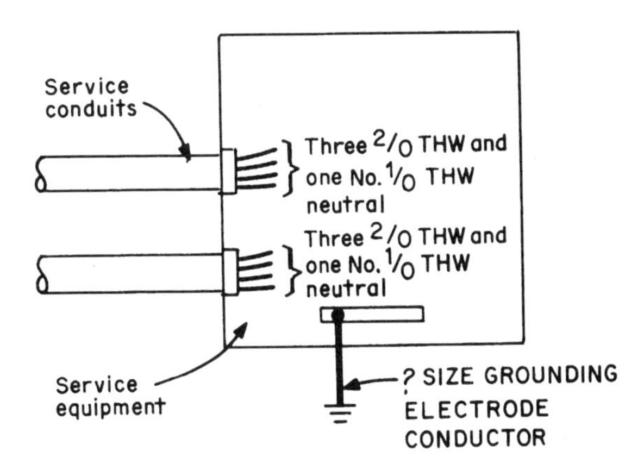

FIG. 5-43 Typical task of sizing the conductor to the grounding electrode.

sectional area of its conductor material at least equal to the sum of the cross-sectional areas of the conductor materials of the two or more parallel conductors per phase. (The cross-sectional area of the insulation must be excluded.)

For instance, two parallel 500-kcmil copper RHH conductors in separate conduits would be equivalent to a single conductor with a cross-sectional area of 500 + 500 = 1000 kcmil. From Table 250-94, the minimum size of grounding electrode conductor required is shown to be No. 2/0 copper or No. 4/0 aluminum—opposite the left column entry "Over 600 kcmil through 1100 kcmil." Note that use of this table is based solely on the size of the conductor material itself, regardless of the type of insulation. No reference is made at all to the kind of insulation.

Figure 5-43 shows a typical case where a grounding electrode conductor must be sized for a multiple-conductor service. A 208/120-V, 3-phase, 4-wire service is made up of two sets of parallel copper conductors of the sizes shown in the sketch. The minimum size of grounding electrode conductor which may be used with these service-entrance conductors is determined by first adding the physical sizes of the two No. 2/0 conductors which make up each phase leg of the service:

1. From NEC Table 8 in Chap. 9 in the back of the Code book, which gives physical dimensions of the conductor material itself (excluding insulation cross-sectional area), each of the phase conductors has a cross-sectional area (csa) of 133,100 kcmil. Two such conductors per phase have a total csa of 266,200 kcmil.

2. The same table shows that the single conductor which has a csa at least equal to the total csa of the two conductors per phase is a 300-kcmil conductor. That conductor size is then located in the left-hand column of Table 250-94 to determine the minimum size of grounding electrode conductor, which turns out to be No. 2 copper or No. 1/0 aluminum or copper-clad aluminum.

Figure 5-44 shows another example of conductor sizing, as follows:

1. The grounding electrode conductor A connects to the street side of the water meter of a metallic water supply to a building. The metallic pipe extends 30 ft underground outside the building.

2. Because the underground metallic water piping is at least 10 ft long, the underground piping system is a grounding electrode and must be used as such.
3. Based on the size of the service-entrance conductors (5 × 500 kcmil = 2500 kcmil per phase leg), the minimum size of grounding electrode conductor to the water pipe is No. 3/0 copper or 250-kcmil aluminum or copper-clad aluminum.
4. The connection to the ground rod at B satisfies the rule of Sec. 250-81(a) requiring a water-pipe electrode to be supplemented by another electrode.
5. But the minimum size of grounding electrode conductor B required between the neutral bus and the made electrode is No. 6 copper or No. 4 aluminum, as covered by part (a) of Exception 1 and part (a) of Exception 2, in this section. Although the Code does not require the conductor to a made electrode to be larger than No. 6, regardless of the ampacity of the service phases, a larger size of conductor is commonly used for mechanical strength, to protect it against breaking or damage. As discussed under Sec. 250-81(a), the conductor at B in Fig. 5-44 can be considered to be a bonding jumper, as covered by the last paragraph of Sec. 250-81(a), which also says that the conductor to the ground rod need not be larger than No. 6 copper or No. 4 aluminum.

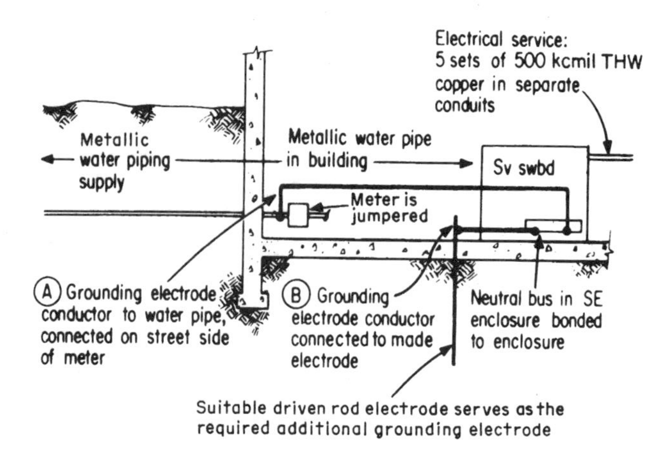

FIG. 5-44 Two different sizes of grounding electrode conductors are required for installations like this one.

Parts (b) and (c) of Exceptions 1 and 2 make clear that a grounding electrode conductor does not have to be larger than a conductor-type electrode to which it connects. Section 250-81(c) recognizes a "concrete-encased" electrode—which must be at least 20 ft of one or more ½-in-diameter steel reinforcing bars or rods in the concrete or at least 20 ft of bare No. 4 copper conductor (or a larger conductor), concrete-encased in the footing or foundation of a building or structure. Section 250-81(d) recognizes a "ground-ring" electrode made up of at least 20 ft of No. 2 bare copper conductor (or larger), buried directly in the earth at a depth of at least 2½ ft. Because each of those electrodes is described under Sec. 250-81, they are *not* "made electrodes," which are described under Sec. 250-83. As electrodes from Sec. 250-81, such electrodes would normally be subject to the basic rule of Sec. 250-94, which calls for connection to any such electrode by a grounding electrode conductor sized from Table 250-94—requiring up to No. 3/0 copper for use on high-capacity services. *But* that is not required, as explained in these paragraphs of the two exceptions.

Parts (b) and (c) recognize that there is no reason to use a grounding electrode conductor that is larger than a conductor electrode to which it connects. The grounding electrode conductor need not be larger than No. 4 copper for a No. 4 concrete-encased electrode and need not be larger than No. 2 copper if it connects to a ground-ring electrode, as in part (c) of the exceptions. Where Table 250-94 would permit a grounding/electrode conductor smaller than No. 4 or No. 2 (based on size of service conductors), the smaller conductor may be used—but the electrode itself must not be smaller than No. 4 or No. 2.

The first note under Table 250-94 correlates to Sec. 230-40, Exception 2, and Sec. 250-91(a), Exception 2, as follows:

When two to six service disconnects in separate enclosures are used at a service, with a separate set of SE conductors run to each disconnect, the size of a single common grounding electrode conductor must be based on the largest sum of the cross sections of the same phase leg of each of the several sets of SE conductors. When using multiple service disconnects in separate enclosures, with a set of SE conductors run to each from the drop or lateral (Sec. 230-40, Exception 2) and using a single, common grounding electrode conductor, either run continuous and unspliced from one disconnect to another and then to the grounding electrode, or with taps from each disconnect to a common grounding electrode conductor run to the electrode—as in Sec. 250-91(a), Exception 2, this note is used to determine the size of the grounding electrode conductor from Table 250-94. The "equivalent area" of the size of SE conductors is the largest sum of the cross-sectional areas of one ungrounded leg of each of the several sets of SE conductors.

CHAPTER 6
MOTOR CIRCUITS

MOTOR BRANCH CIRCUITS

Motor loads vary widely in size and electrical characteristics, but all motor circuits require careful wiring and the protection of conductors and equipment to ensure safe and reliable operation. In any plant or building, of course, the problem of providing maximum safety and reliability must be solved along with other problems—minimizing voltage drop, avoiding excessive copper loss, providing sufficient flexibility for changes in the locations of equipment, designing for ease and economy of maintenance of the motors and equipment, and providing spare capacity for future load increases.

Although the effective application of motor controllers is based primarily on thorough engineering analysis, careful consideration should also be given to the National Electrical Code (NEC), which sets forth minimum safety provisions for the control of motors.

In no way is the NEC a substitute for the intelligent design of motor control circuits suited to the particular characteristics of each individual application. However, because it does represent the accumulation of years of experience with motor circuits, it presents an excellent general outline of motor circuit design. Within this basic framework, designers can add specific equipment features and circuit techniques to meet their own needs.

The minimum requirements that must be observed in designing branch circuits for motor loads are set out in two NEC articles which are directed specifically to motor applications.

1. Article 430 of the NEC covers the application and installation of motor circuits and motor control hookups, including conductors, short-circuit and ground-fault protection, starters, disconnects, and running overload protection.
2. Article 440, "Air-Conditioning and Refrigerating Equipment," contains provisions for such motor-driven equipment and for branch circuits and controllers for this equipment, taking into account the special considerations involved with the use of sealed (hermetic-type) motor-compressors, in which the motor operates under the cooling effect of the refrigeration.

It must be noted that the rules of Art. 440 apply in addition to or are amendments to the rules given in Art. 430 for motors in general. The basic rules of Art. 430 apply to air-conditioning and refrigerating equipment unless exceptions are indicated in Art. 440. Article 440 further clarifies the application of NEC rules to air-conditioning and refrigeration equipment as follows:

1. Air-conditioning and refrigerating equipment which does not incorporate a sealed (hermetic-type) motor-compressor must satisfy the rules of Art. 422 (appliances), Art.

424 (space-heating equipment), or Art. 430 (conventional motors), as they apply. For instance, where refrigeration compressors are driven by conventional motors, the motors and controls are subject to Art. 430, not Art. 440. Furnaces with air-conditioning evaporator coils installed must satisfy Art. 424. Other equipment in which the motor is not a sealed compressor and which must be covered by Art. 422, 424, or 430 includes fan coil units, remote forced-air-cooled condensers, remote commercial refrigerators, and similar equipment.

2. Room air conditioners are covered in Part G of Art. 440 (Secs. 440-60 through 440-64) but must also comply with the rules of Art. 422.

3. Household refrigerators and freezers, drinking water coolers, and beverage dispensers are considered by the NEC to be appliances; their application must comply with Art. 422 and must satisfy the rules of Art. 440, because such devices contain sealed motor-compressors.

Figure 6-1 shows the six basic elements which the NEC requires the designer to account for in any motor circuit. Although these elements are shown separately here, in certain cases the NEC permits a single device to serve more than one function. For instance, in some cases, one switch can serve as both disconnecting means and controller. In other cases, short-circuit protection and overload protection can be combined in a single circuit breaker (CB) or set of fuses.

Throughout this discussion of motor and control circuits, it should be clearly understood that the NEC rules provide only the foundation for modern design. Many equipment applications, circuit layouts, and features must be incorporated in the design, over and above NEC rules, to provide convenience, flexibility, and fulfillment of the objectives of a specific installation.

Motor Circuit Conductors

The basic NEC rule says that the conductors carrying load current to a single-speed motor used for continuous duty must have a current-carrying capacity of not less than 125 percent of the motor full-load current rating. In the case of a multispeed motor, the selection of branch-circuit conductors on the line side of the controller must be based on the highest of the full-load current ratings shown on the motor nameplate. The selection of branch-circuit conductors between the controller and the motor, which are energized for a particular speed, must be based on the current rating for that speed.

Figure 6-2 shows the sizing of branch-circuit conductors to four different motors fed from a panel. The full-load current for each motor is taken from NEC Table 430-150.

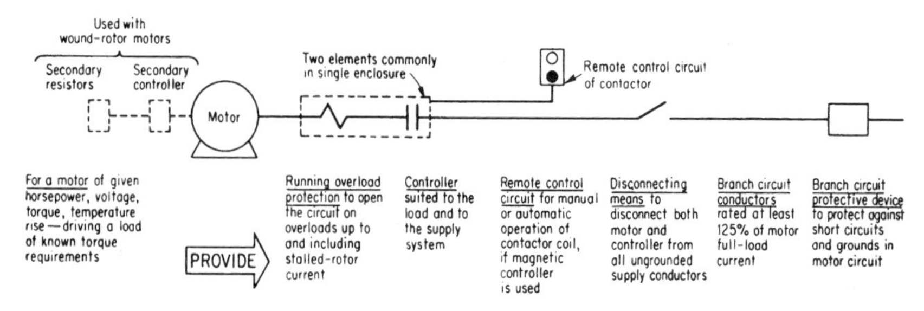

FIG. 6-1 Design of any motor branch circuit must account for each of these elements.

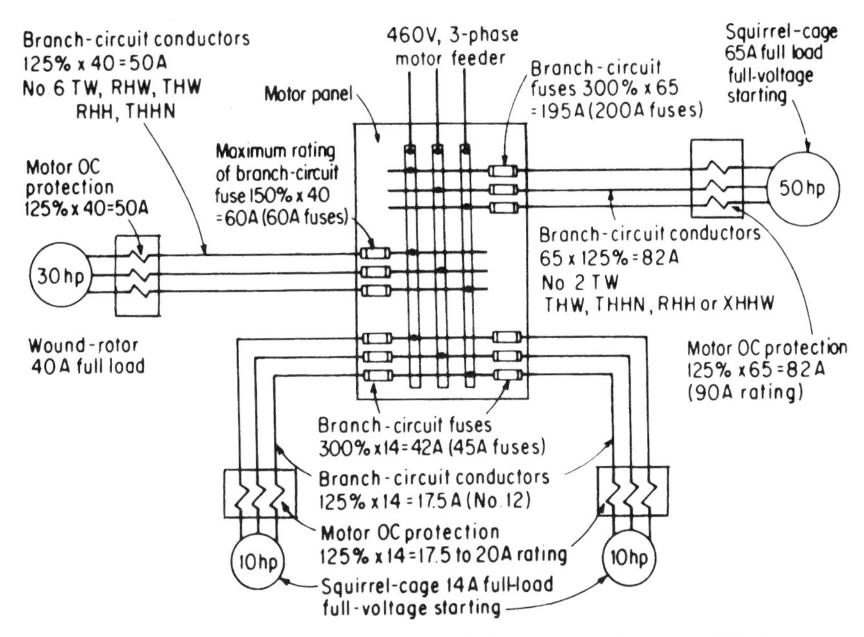

FIG. 6-2 Basic design steps in sizing motor branch-circuit conductors and overcurrent devices.

Running overload protection is sized on the basis that nameplate values of motor full-load currents are the same as the values in Table 430-150. If the nameplate and table values are not the same, overload protection is sized according to the nameplate value. The conductor sizes shown are for copper. The given current values and NEC Table 310-16 may be used to select the correct sizes of aluminum conductors. (Figure 6-2 also shows the sizing of branch-circuit protection and running overload protection, as discussed below. Refer to NEC Table 430-150 for motor full-load currents and Table 430-152 for maximum ratings of fuses.)

Although it is not shown in Fig. 6-2, the NEC addresses the concern related to application of the so-called design E motors, which are high-efficiency motors. Prior to the 1996 NEC standards, application of such motors was nearly impossible with certain short-circuit and ground-fault protective devices because of the "maximum" ratings permitted for those devices by the Code. The significantly higher inrush current that is inherent in design E motors—due to lower internal impedance—caused instantaneous-trip CBs in listed starters to "open" on start-up. To solve this problem, the "maximum" ratings or settings for an instantaneous-trip CB may be higher when used with design E motors.

Conductors supplying two or more motors must have a current rating of not less than 125 percent of the full-load current rating of the largest motor supplied plus the sum of the full-load current ratings of all other motors supplied.

It is important to note that these are minimum conductor ratings based on temperature rise only; they do not take into account voltage drop or power loss in the conductors. Such considerations frequently require an increase in the size of the branch-circuit conductors.

Section 430-22(a) includes requirements for sizing individual branch-circuit wires serving motors used for short-time, intermittent, periodic, or other varying duty. For such types of duty, the frequency of starting and duration of operating cycles impose varying heat loads on conductors. Conductor sizing, therefore, varies with the application. But note that any motor is considered to be wired for continuous duty unless the nature of the apparatus that it drives is such that the motor cannot operate continuously with load under any condition of use.

Conductors connecting the secondary of a wound-rotor induction motor to the controller must have a carrying capacity at least equal to 125 percent of the motor's full-load secondary current if the motor is used for continuous duty. If the motor is used for less than continuous duty, the conductors must have a capacity not lower than the percentage of full-load secondary nameplate current given in NEC Table 430-22(a). Conductors from the controller of a wound-rotor induction motor to its starting resistors must have an ampacity in accordance with Table 430-23(c). (See Fig. 6-3.)

What Current to Use? The NEC contains definite provisions for determining current-carrying capacities:

Section 430-6. For general motor applications (excluding applications of torque motors and sealed hermetic-type refrigeration motor-compressors), whenever the current rating of a motor is used to determine the current-carrying capacities of conductors, switches, fuses, or circuit breakers, the values given in Tables 430-147 through 430-150 must be used instead of the current rating marked on the motor nameplate. However, the selection of separate motor running overcurrent protection must be based on the actual motor nameplate current rating.

For torque motors, shaded-pole motors, permanent split-capacitor motors, and ac adjustable-voltage motors, refer to Sec. 430-6.

Section 440-5. For sealed (hermetic-type) refrigeration motor-compressors, the actual nameplate full-load running current of the motor must be used in determining the current rating of the disconnecting means, controller, branch-circuit conductors, short-

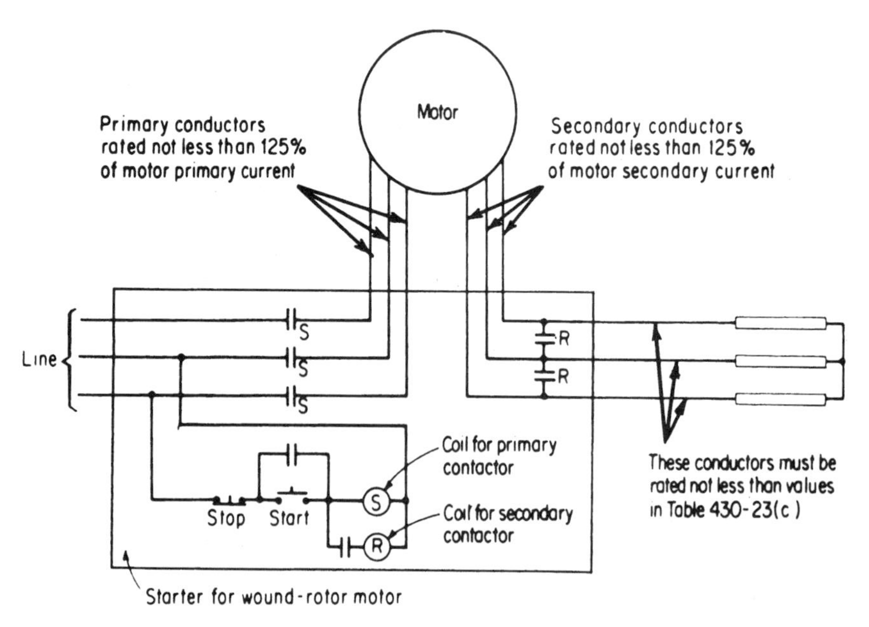

FIG. 6-3 Circuit design for wound-rotor motors must include the sizing of both primary and secondary conductors.

circuit and ground-fault protective devices, and motor running overload protection. When such equipment is marked with a branch-circuit selection current, it must be used instead of the rated load current to determine the rating or ampacity of the disconnecting means, branch-circuit conductors, controller, and branch-circuit, short-circuit, and ground-fault protection. But the nameplate rated load current must still be used for sizing separate running overload protection.

Example. The circuit components for a sealed hermetic motor-compressor with a nameplate current rating of 10 A are sized as follows:

1. The full-load current used for all calculations for a sealed (hermetic) motor-compressor is the current marked on the unit's nameplate [NEC Sec. 440-5(a)].
2. For continuous operation, the branch-circuit conductors must have a rating at least equal to 125 percent of 10 A, or 13 A (NEC Sec. 440-32).
3. Conductors for the branch circuit must, therefore, be not smaller than No. 14 copper or No. 12 aluminum (type TW, THW, RHW, RHH, THHN, or XHHW).
4. Running overload protection from overload relays in the starter must have a rating or setting not over 140 percent of the full-load (nameplate) current rating. This calls for an overload relay set to trip at 1.4 × 10 A, or 14 A [NEC Sec. 440-52(a)].
5. The maximum fuse rating for short-circuit protection for a hermetic motor-compressor is 175 percent of the motor nameplate full-load current rating, which is 1.75 × 10 A, or 17.5 A. Section 440-22 does not indicate that the next-higher standard rating of fuse may be used, but it does permit up to 225 percent of the motor current where needed for starting of the motor. That would indicate the use of 20-A fuses.

Short-Circuit Protection

The NEC requires that the branch-circuit protection for motor circuits protect the circuit conductors, the control apparatus, and the motor itself against overcurrent due to short circuits or grounds (Secs. 430-51 through 430-58).

The first, and obviously necessary, rule is that the branch-circuit protective device for an individual branch circuit to a motor must be capable of carrying the motor starting current without opening the circuit. Given this condition, the NEC places maximum values on the ratings or settings of such overcurrent devices. It says that such devices must not be rated in excess of values determined from Table 430-152. If such values do not correspond to standard sizes or ratings of fuses, nonadjustable CBs, or thermal devices or to possible settings of adjustable CBs adequate to carry the load, the next-lower size, rating, or setting may be used. If this is not adequate to carry the load, the next-higher standard size, rating, or setting is permitted. Where these values do not permit the motor to start, the device may be rated as follows:

1. The rating of a non-time-delay fuse not exceeding 600 A may be increased but may in no case exceed 400 percent of the full-load current.
2. The rating of a time-delay (dual-element) fuse may be increased but must never exceed 225 percent of the full-load current.
3. The setting of an instantaneous-trip circuit breaker (without time delay) may be increased, but never over 1300 percent of the motor full-load current.
4. The rating of an inverse-time (time-delay) circuit breaker may be increased but must not exceed 400 percent of a full-load current of 100 A or less and must not exceed 300 percent of a full-load current over 100 A.

5. The rating of a fuse rated 601 to 6000 A may be increased but must not exceed 300 percent of the full-load current.
6. Torque motors must be protected at the motor nameplate rating; if a standard overcurrent device is not made in that rating, the next-higher standard rating of protective device may be used.

For a multispeed motor, a single short-circuit and ground-fault protective device may be used for one or more windings of the motor, provided the rating of the protective device does not exceed the above applicable percentage of the nameplate rating of the smallest winding protected.

The NEC establishes maximum values for branch-circuit protection, setting the limit of safe application. However, the use of branch-circuit protective devices of smaller sizes is obviously permitted, and this does offer the opportunity for substantial economy in the selection of circuit breakers, fuses and the switches used with them, panelboards, etc. However, a branch-circuit device with a lower rating than the maximum permitted rating must have sufficient time delay in operation to permit the motor starting current to flow without opening the circuit. But a circuit breaker for branch-circuit protection must have a continuous-current rating of not less than 115 percent of the motor full-load current.

Where maximum protective-device ratings are shown in the manufacturer's heater table for use with a marked controller or are otherwise marked with the equipment, they must not be exceeded, even if higher values are allowed by the rules given above. (See Fig. 6-2 for the sizing of overcurrent protection in accordance with Table 430-152.)

Instantaneous-Trip CBs. The NEC recognizes the use of an instantaneous-trip CB (without time delay) for short-circuit protection of motor circuits. Such breakers—also called *magnetic-only breakers*—may be used only if they are adjustable and if combined with motor starters in combination assemblies. An instantaneous-trip CB or a motor short-circuit protector (MSCP) may be used *only* as part of a "listed" (such as by UL) combination motor controller. A combination motor starter using an instantaneous-trip breaker must have running overload protection in each conductor (Fig. 6-4). Such a

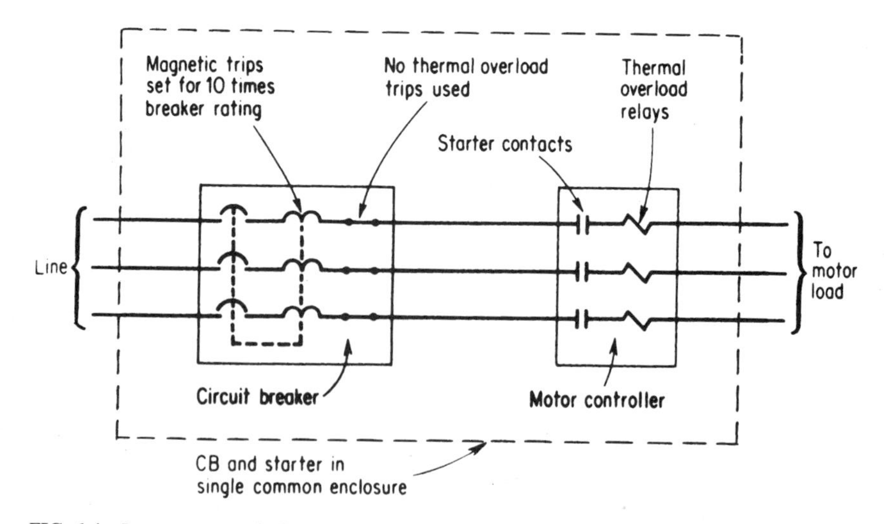

FIG. 6-4 Instantaneous-only CBs are permitted only in combination starters.

combination starter offers use of a smaller CB than would be possible if a standard thermomagnetic CB were used. And the smaller CB offers faster operation for greater protection against grounds and short circuits, in addition to offering greater economy.

A combination motor starter, as shown in Fig. 6-4, is based on the characteristics of the instantaneous-trip CB, which is covered by the third percent column from the left in Code Table 430-152. Molded-case CBs with only magnetic instantaneous-trip elements in them are available in almost all sizes. Use of such a device requires careful accounting for the absence of overload protection in the CB, up to the short-circuit trip setting. Such a CB is designed for use as shown in Fig. 6-4. The circuit conductors are sized for at least 125 percent of motor current. The thermal overload relays in the starter protect the entire circuit and all equipment against operating overloads up to and including stalled rotor current. They are commonly set at 125 percent of motor current. In such a circuit, a CB with an adjustable magnetic trip element can be set to take over the interrupting task at currents above stalled rotor and up to the short-circuit duty of the supply system at that point of installation. The magnetic trip in a typical unit might be adjustable from 3 to 17 times the breaker current rating; that is, a 100-A trip can be adjusted to trip anywhere between 300 and 1700 A. Thus the CB serves as motor circuit disconnect and short-circuit protection.

Selection of such a listed assembly with an instantaneous-only CB is based on choosing a nominal CB size with a current rating at least equal to 115 percent of the motor full-load current to carry the motor current and to qualify under Secs. 430-58 and 430-110(a) as a disconnected means. Then the adjustable magnetic trip is set to provide the short-circuit protection—the value of current at which instantaneous circuit opening takes place, which should be just above the starting current of the motor involved—using a multiplier of something like 1.5 on locked-rotor current to account for asymmetry in starting current. Asymmetry can occur when the circuit to the motor is closed at that point on the alternating voltage wave where the inrush starting current is going through the negative maximum value of its alternating wave. That is the same concept as asymmetry in the initiation of a short-circuit current. Where supplying design E motors, a greater inrush can be anticipated on start-up. As a result, higher initial and maximum settings are recognized.

Listed equipment using an instantaneous CB type is available with very simple instructions by the manufacturer to make proper selection and adjustment of the instantaneous-trip CB combination a quick, easy matter. The following describes the concept behind the application of listed combination starters with instantaneous-only CBs.

Given: A 30-hp, 230-V, 3-phase, squirrel-cage motor marked with the code letter M, indicating that the motor has a locked-rotor current of 10 to 11.19 kVA/hp, from Code Table 430-7(b). A full-voltage controller is combined with the CB, with running overload protection in the controller to protect the motor within its heating damage curve on overload in a listed unit.

Required: Select the maximum setting and minimum rating for the CB which will provide short-circuit protection and will qualify as the motor circuit disconnect means.

Solution: The motor has a full-load current of 80 A (Code Table 430-150). A CB suitable for use as disconnect must have a current rating at least 115 percent of 80 A. As covered in Sec. 430-52(c)(3), for instantaneous-trip CBs, the initial setting from Table 430-152 would be limited to 800 percent of the 80-A full-load current. The maximum setting—for other than the high-efficiency design E motors—is 1300 percent. For design E motors, the initial setting may be 1100 percent of motor full-load current with a maximum setting of 1700 percent of the motor full-load current.

Note that settings above 800 or 1100 percent of the motor's full-load current are only permitted if nuisance tripping occurs on starting *or* if evaluation of the motor's starting characteristics and the time-current trip curve of the breaker indicates that a greater setting is needed. Although not completely clear, the trip value established through the "engineering evaluation" should be considered as the maximum setting.

Because the use of a magnetic-only CB does not protect against low-level grounds and shorts in the circuit conductors on the line side of the starter running overload relays, the NEC rule permits such application only where the CB and starter are part of a *listed* combination starter in a single enclosure.

MSCPs. A motor short-circuit protector, as referred to in the second paragraph of Sec. 430-52, is a fuselike device designed for use only in its own type of fusible-switch combination motor starter. The combination offers short-circuit protection, running overload protection, disconnect means, and motor control—all with assured coordination between the short-circuit interrupter (the motor short-circuit protector) and the running overload devices. It involves the simplest method of selection of the correct MSCP for a given motor circuit. This packaged assembly is a third type of combination motor starter—added to the conventional fusible-switch and CB types.

The NEC recognizes motor short-circuit protectors in Secs. 430-40 and 430-52 provided the combination is a "listed" assembly. This means a combination starter equipped with motor short-circuit protectors and listed by Underwriters Laboratories Inc., or another nationally recognized testing laboratory, as a package called an MSCP starter.

One Circuit for Two or More Motors. A single branch circuit may be used to supply two or more motors as follows.

Small Motors. Two or more motors, each rated not more than 1 hp and each drawing not over 6-A full-load current, may be used on a branch circuit protected at not more than 20 A at 125 V or less, or 15 A at 600 V or less. The rating of the branch-circuit protective device marked on any of the controllers must not be exceeded. [See Sec. 430-53(a).] Individual running overload protection is necessary in such circuits, unless the motor (1) is not permanently installed, or (2) is manually started and is within sight of the controller location, or (3) has sufficient winding impedance to prevent overheating due to stalled-rotor current, or (4) is part of an approved assembly which does not subject the motor to overloads and which incorporates protection for the motor against stalled rotor, or (5) cannot operate continuously under load.

Motors of Any Rating. Two or more motors of any rating, each having individual running overload protection, may be connected to a branch circuit which is protected by a short-circuit protective device selected in accordance with the maximum rating or setting of a device which could protect an individual circuit to the motor of the smallest rating. This may be done only where it can be determined that the branch-circuit device so selected will not open under the most severe normal conditions of service which might be encountered [Sec. 430-53(b)].

This permission of Sec. 430-53(b) offers wide application of more than one motor on a single circuit, particularly small, integral-horsepower motors installed on 440-V, 3-phase systems. The application primarily concerns the use of small, integral-horsepower 3-phase motors in 208-, 220-, and 440-V industrial and commercial systems. Only such 3-phase motors have full-load operating currents low enough to permit more than one motor on circuits fed from 15-A protective devices.

Figure 6-5 covers the use of more than one motor on a branch circuit, where small integral-horsepower and/or fractional-horsepower motors are used in accordance with Sec. 430-53(b).

In case I, with a 3-pole CB used as the branch-circuit protective device, application is made in accordance with Sec. 430-53(b) as follows:

CASE I—USING A CIRCUIT BREAKER FOR PROTECTION

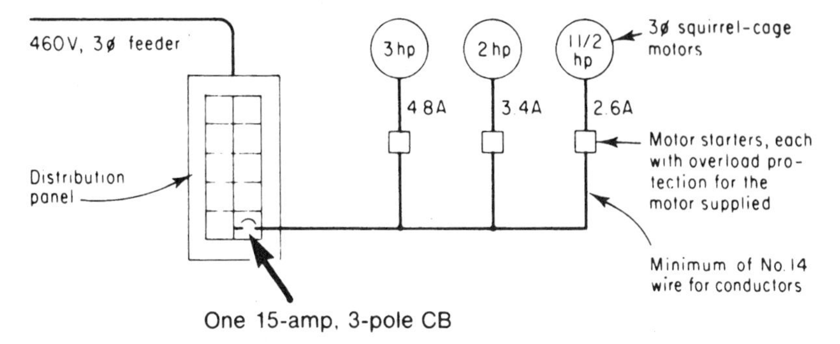

HERE IS THE KEY: A 15-amp, 3-pole CB is used, based on Section 430-52 and Table 430-152. This is the "next higher size" of standard protective device above 250% × 2.6 amps (the required rating for the smallest motor of the group). The 15-amp CB makes this application possible, because the 15-amp CB is the smallest standard rating of CB and is suitable as the branch-circuit protective device for the 1½-hp motor.

CASE II—USING A CIRCUIT BREAKER FOR PROTECTION

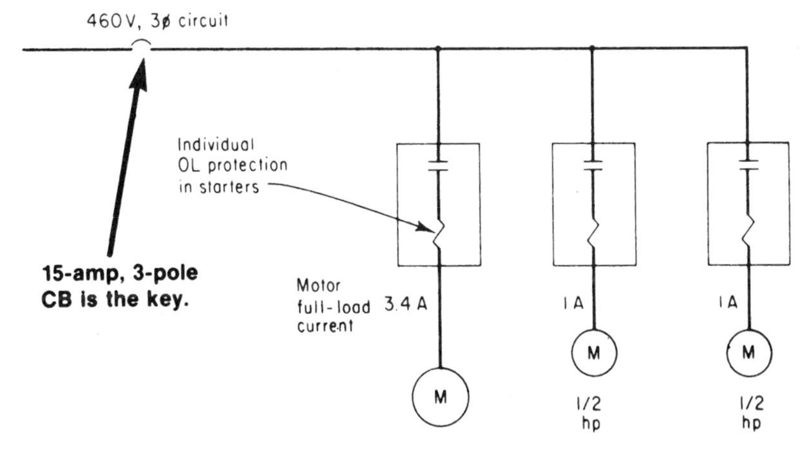

FIG. 6-5 Branch circuit supplying two or more motors must be carefully designed.

1. The full-load current for each motor is taken from NEC Table 430-150 [as required by Sec. 430-6(a)].
2. A circuit breaker is used instead of fuses for branch-circuit protection. The rating of the branch-circuit protective device, 15 A, does not exceed the maximum short-circuit protection required by Sec. 430-52 and Table 430-152 for the smallest motor of the group, which is the $1\frac{1}{2}$-hp motor. Although 15 A is greater than the maximum value of 250 percent of the motor full-load current (2.5×2.6 A = 6.5 A) set by Table 430-152, the 15-A breaker is the "next higher size, rating or setting" for a standard circuit breaker as permitted by Sec. 430-52. A 15-A circuit breaker is the smallest standard rating recognized by the NEC in Sec. 240-6.
3. The total load of motor currents is

$$4.8 \text{ A} + 3.4 \text{ A} + 2.6 \text{ A} = 10.8 \text{ A}$$

This is well within the rating of the 15-A CB, which has sufficient time delay in its operation to permit starting of any one of these motors with the other two already operating. The torque characteristics of the loads on starting are not high, and it was determined that the CB will not open under the most severe normal service.
4. Each motor is provided with individual running overload protection in its starter.
5. The branch-circuit conductors are sized in accordance with Sec. 430-24:

$$4.8 \text{ A} + 3.4 \text{ A} + 2.6 \text{ A} + 0.25 \times 4.8 \text{ A} = 12 \text{ A}$$

The conductors must have an ampacity at least equal to 12 A, and No. 14 THW, TW, RHW, RHH, THHN, or XHHW conductors will fully satisfy this requirement.

In case II, a similar hookup is used to supply three motors—also with a CB for branch-circuit protection:

1. Section 430-53(b) requires the branch-circuit protection rating be not higher than the maximum current set by Sec. 430-52 for the lowest-rated motor of the group.
2. From Sec. 430-52 and Table 430-152, that maximum protection rating for a circuit breaker is 250 percent of 1 A (the rating of the lowest-rated motor), or 2.5 A. But 2.5 A is not a "standard rating" of circuit breaker from Sec. 240-6; the third paragraph of Sec. 430-52 permits use of the "next higher size, rating or setting" of standard protective device.
3. Because 15 A is the lowest standard rating of CB, it is the next-higher device rating above 2.5 A and satisfies NEC rules on the rating of the branch-circuit protection.

The applications shown in cases I and II permit the use of several motors up to circuit capacity, based on Secs. 430-24 and 430-53(b) and on the starting-torque characteristics, the operating duty cycles of the motors and their loads, and the time delay of the CB. Such applications greatly reduce the number of CB poles, number of panels, and amount of wire used in the system. However, one limitation is placed on this practice in the next-to-last sentence of Sec. 430-52(c)(2) as follows:

> Where maximum branch-circuit, short-circuit and ground-fault protective device ratings are shown in the manufacturer's overload relay table for use with a motor controller or are otherwise marked on the equipment, they shall not be exceeded even if higher values are allowed as shown above.

BUT, WATCH OUT!!!

CASE III—USING FUSES FOR CIRCUIT PROTECTION

Interpretation of NE Code rules of Section 430-53(b) in conjunction with the "standard" ratings of fuses in Section 240-6 may require different circuit makeup when fuses are used to protect the branch circuit to several motors.

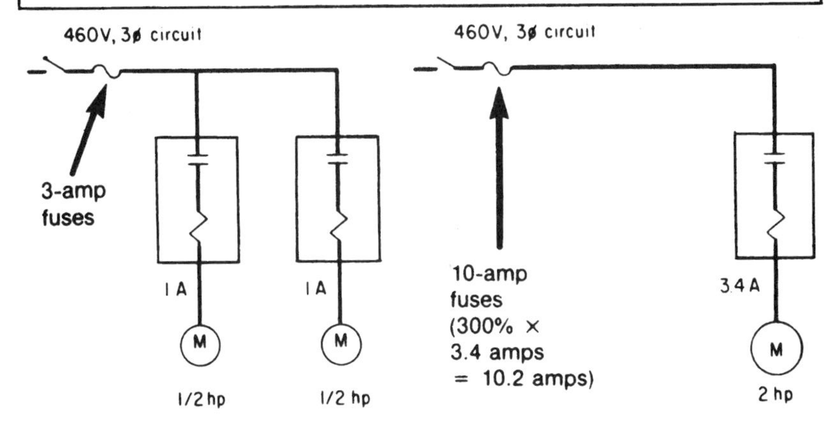

FIG. 6-6 Fuses, instead of CBs, alter the design characteristics of a multimotor branch circuit.

In case III, shown in Fig. 6-6, the three motors of case II would be hooked up differently to comply with the rules of Sec. 430-53(B) if fuses, instead of a circuit breaker, were used for branch-circuit protection:

1. To comply with Sec. 430-53(b), fuses used as branch-circuit protection must have a rating not in excess of the value permitted by Sec. 430-52 and Table 430-152 for the smallest motor of the group—here, one of the ½-hp motors.
2. Table 430-152 shows that the maximum permitted rating for non-time-delay fuses is 300 percent of the full-load current for 3-phase squirrel-cage motors. Applying that rule to one of the ½-hp motors gives a maximum fuse rating of 3.0 × 1 A = 3 A.
3. But there is no permission for the fuses to be rated higher than 3 A, because 3 A is a "standard" rating of fuse (but not a standard rating of circuit breaker). Section 240-6 considers fuses rated at 1, 3, 6, and 10 A to be "standard."
4. The maximum branch-circuit protection permitted by Sec. 430-53(b) for a ½-hp motor is therefore 3 A.
5. The two ½-hp motors may be fed from a single branch circuit with three 3-A fuses in a 3-pole switch.
6. Following the same NEC rules, the 2-hp motor would require fuse protection rated not over 10 A (3.0 × 3.4 A = 10.2 A).

Note: Because there are standard fuse ratings below 15 A, fuses have a different relationship to the applicable NEC rules than CBs; interpretation of the rules will be required to resolve the question of acceptable application in case II versus case III. Such interpretation will be necessary to determine if circuit breakers are excluded as circuit protection in those cases where the use of fuses, in accordance with the precise wording of the NEC, provides lower rated protection than circuit breakers when the rule of the third paragraph of Sec. 430-52 is applied. And if the motors of case I are fed from a circuit protected by fuses, the literal effect of the rules is to require different circuiting for those motors.

Figure 6-7 shows one way of combining cases II and III to satisfy Secs. 430-53(b), 430-52, and 240-6; however, the 15-A CB would then be feeder protection, because the fuses would be serving as the "branch-circuit protective devices" required by Sec. 430-53(b). Those fuses might be acceptable in each starter, without a disconnect switch, in accordance with Sec. 240-40, which allows the use of cartridge fuses at any voltage without an individual disconnect for each set of fuses, provided only qualified persons have access to the fuses. But Sec. 430-112 would have to be satisfied if the single CB were to be used as a disconnect for the group of motors. And part (b) of the exception to that section recognizes one common disconnect in accordance with Sec. 430-53(a) but not with Sec. 430-53(b). Certainly, the use of a fusible-switch-type combination starter for each motor would satisfy all rules.

Figure 6-8 shows another hookup of several motors on one branch circuit—an actual job installation which was based on application of Sec. 430-53(b). The installation was studied as follows.

Example. A factory has one hundred 1½-hp, 3-phase motors with individual motor starters incorporating overcurrent protection, rated for 460 V. Provide circuits.

Solution. Prior to 1965, the NEC would not permit several motors on one branch circuit fed from a 3-pole CB. Each of the 100 motors would have had to have its own individual 3-phase circuit fed from a 15-A, 3-pole CB in a panel. As a result, a total of 300 CB poles would have been required, calling for seven panels of 42 circuits each plus a smaller panel (or special panels of more than 42 poles per panel).

Under present rules, depending upon the starting-torque characteristics and operating duty of the motors and their loads, with each motor rated for 2.6 A, three or four motors could be connected on each 3-phase, 15-A circuit; this greatly reduces the number of panel boards and overcurrent devices and the amount of wire involved in the system. The time delay of the CB influences the number of motors on each circuit. However, an extremely important requirement is given in the next-to-last sentence of Sec. 430-52(c)(2), which says:

> Where maximum branch-circuit short-circuit and ground-fault protective device ratings are shown in the manufacturer's overload relay table for use with a motor controller or are otherwise marked on the equipment, they shall not be exceeded even if higher values are allowed as shown above.

Because Sec. 430-53(b) includes the condition that "the branch-circuit short-circuit and ground-fault protective device is selected not to exceed that allowed by Sec. 430-52," it is clear that all of Sec. 430-52 (including the sentence quoted above) must be fully satisfied.

Motors for "Group Installation." Two or more motors of any rating may be connected to one branch circuit if each motor has running overcurrent protection, if the overload devices and controllers are listed for group installation, and if the branch-circuit fuse or inverse-time CB rating is in accordance with Sec. 430-52 for the largest motor plus the sum of the full-load current ratings of the other motors. The branch-circuit fuses or circuit breaker must not be larger than the rating or setting of short-circuit protection permitted by Sec. 430-52 for the smallest motor of the group, unless this value is less than the ampacity of the supply conductors, in which case the fuse or CB

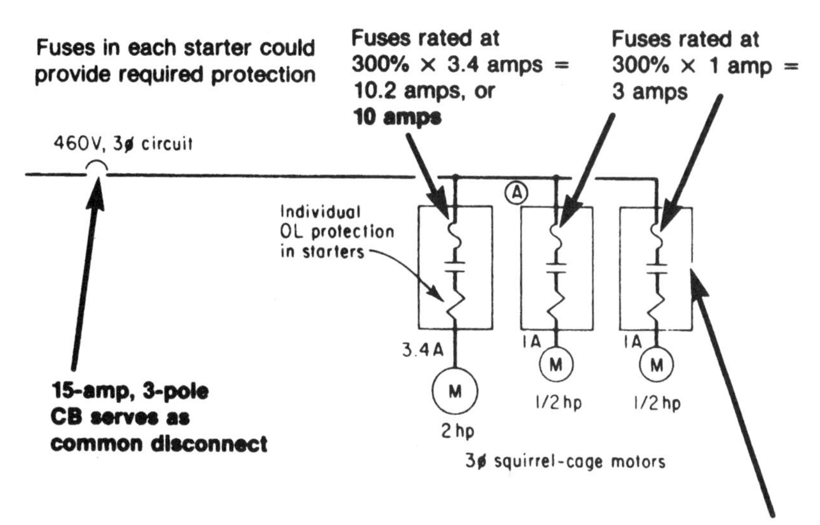

. . . AND A HOOKUP LIKE THIS MIGHT BE REQUIRED TO MAKE CASE 1 OF FIG. 6.5 COMPLY WITH NE CODE RULES

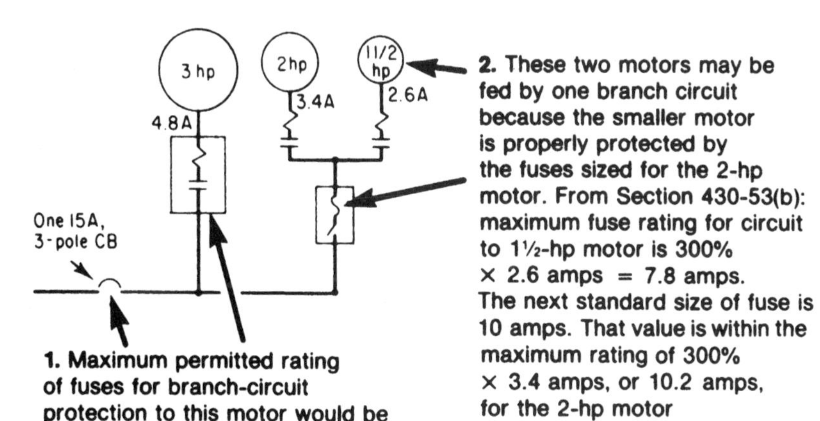

FIG. 6-7 Other designs for multimotor circuits.

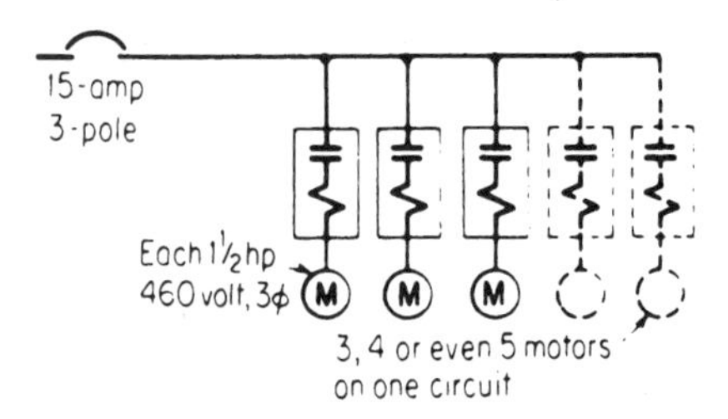

FIG. 6-8 Economy often dictates the use of multiple fractional-horsepower or small integral-horsepower motors on a single circuit.

may be increased to the same ampere value as the supply conductors' ampacity. But in such cases, the fuses or CBs may not be rated more than the overload device indicated by Sec. 430-40. [See Secs. 430-53(c) and 430-40.]

Air Conditioning and Refrigeration

NEC Sec. 440-22(a) covers the rating or setting of the branch-circuit short-circuit and ground-fault protective device for a circuit to an individual sealed hermetic motor-compressor. The rule says that the device "shall be capable of carrying the starting current of the motor." The required protection is considered to be obtained when the device has a rating or setting not exceeding 175 percent of the compressor rated load current or branch-circuit selection current, whichever is greater (with a 15-A size minimum); however, where the specified protection is not sufficient for the starting current of the motor, it may be increased but may not exceed 225 percent of the motor rated load current or branch-circuit selection current, whichever is greater.

Section 440-22(b) covers sizing of the short-circuit and ground-fault protective device for a branch circuit to equipment which incorporates more than one sealed hermetic motor-compressor or one sealed motor-compressor and other motors or other loads. This extensive coverage of branch-circuit protection for such complicated motor loads should be carefully studied to ensure effective compliance with the NEC on such work. In all such cases, where more than one motor is supplied by a single branch circuit, the rules of Sec. 450-53 ("Several Motors or Loads on One Branch Circuit") must be applied.

Section 440-62(a) points out that a room air conditioner must be treated as a single motor unit in determining its branch-circuit requirements when all the following conditions are met:

1. The unit is cord-and-plug connected.
2. Its total rating is not more than 40 A and 250 V, single-phase.
3. The total rated load current is shown on the unit nameplate, rather than individual motor currents.
4. The rating of the branch-circuit, short-circuit, and ground-fault protective device does not exceed the ampacity of the branch-circuit conductors or the rating of the receptacle, whichever is less.

Section 440-60 describes a room air conditioner as an ac appliance of the air-cooled window, console, or in-wall type, with or without provisions for heating,

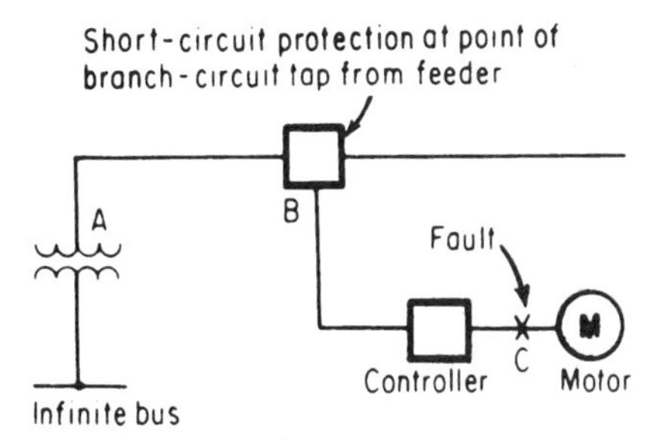

FIG. 6-9 Branch-circuit protection must be properly rated for the maximum current drawn by a fault anywhere on its load side.

installed in the conditioned room, and incorporating one or more hermetic refrigerant motor-compressors.

Figure 6-9 calls attention to the fact that branch-circuit protection must always be capable of interrupting the amount of short-circuit current which might flow through it. In the diagram, a short-circuit fault at *C* will draw current until the circuit is opened by the protection at *B*. The value of the short-circuit current available at *C* depends upon the apparent-power (kVA) rating of the supply transformer *A*, percentage reactance of the transformer, secondary voltage, and effective impedance of the current path from the transformer to the point of fault. In application, therefore, motor controllers must be carefully coordinated with the characteristics of the branch-circuit protective device, which must be able to safely interrupt the short-circuit current. And not only must the device be rated to interrupt the fault current, but also it must act quickly enough to open the circuit before let-through current can damage the controller. The speed in clearing the circuit must be compared to the abilities of the various circuit elements to withstand the damaging effects of short-circuit current flow during the time it takes the protective device to operate. Figure 6-10 shows a typical example of the problem that can arise when the operating speed of the branch-circuit overcurrent device does not protect other circuit elements from the effects of let-through current. The damage to the motor controller could have been avoided by use of a faster-opening CB or fast-acting fuses which would limit the fault current to a value that the controller could withstand.

Motor Controllers

As used in the NEC, the term *controller* includes any switch or device normally used to start and stop a motor, in addition to motor controllers as such. The basic requirements of Secs. 430-81 and 430-83 regarding sizes and types of motor controllers are as follows (refer to Fig. 6-11, where controller types are numbered to correspond to the following list):

1. The *branch-circuit protective device* may serve as the controller for a stationary motor rated ⅛ hp or less which is normally left running and is not subject to damage from overload or failure to start. (Clock motors are typical of this application.)
2. A *plug and receptacle connection* may serve as the controller for portable motors up to ⅓ hp.

3. A *controller* must be capable of starting and stopping the motor which it controls, must be able to interrupt the stalled-rotor current of the motor, and must have a horsepower rating not lower than the rating of the motor, with the following exceptions
4. For the high-efficiency design E motors the controller must be listed for use with design E motors, in which case it must have at least the same horsepower rating as the motor. Nondesign E listed controllers must have a horsepower rating of 1.4 or 1.3 times the motor's rating, for motors rated 3 to 100 hp or more than 100 hp, respectively.
5. A *general-use switch* rated at not less than twice the full-load motor current may be used as the controller for stationary motors up to 2 hp, rated 300 V or less. On ac circuits, a general-use snap switch suitable only for use on alternating current may be used to control a motor having a full-load current rating not over 80 percent of the ampere rating of the switch.
6. A *branch-circuit circuit breaker,* rated in amperes only, may be used as a controller. If the same circuit breaker is used both as a controller and to provide overload protection for the motor circuit, it must be rated accordingly.
7. Exception 3 of NEC Sec. 430-83 covers controllers for torque motors. The controller must have a continuous-duty current rating that is at least equal to the nameplate motor current. Or the "equivalent current rating" can be calculated from the data given in Tables 430-147 to 430-150.

Although Sec. 430-83 permits the use of horsepower-rated switches as controllers and the UL lists horsepower-rated switches up to 500 hp, the UL states in its Green Book that "enclosed switches rated higher than 100 hp are restricted to use as motor disconnect means and are not for use as motor controllers." But a horsepower-rated switch up to 100 hp may be used as both a controller and a disconnect if it breaks all ungrounded legs to the motor.

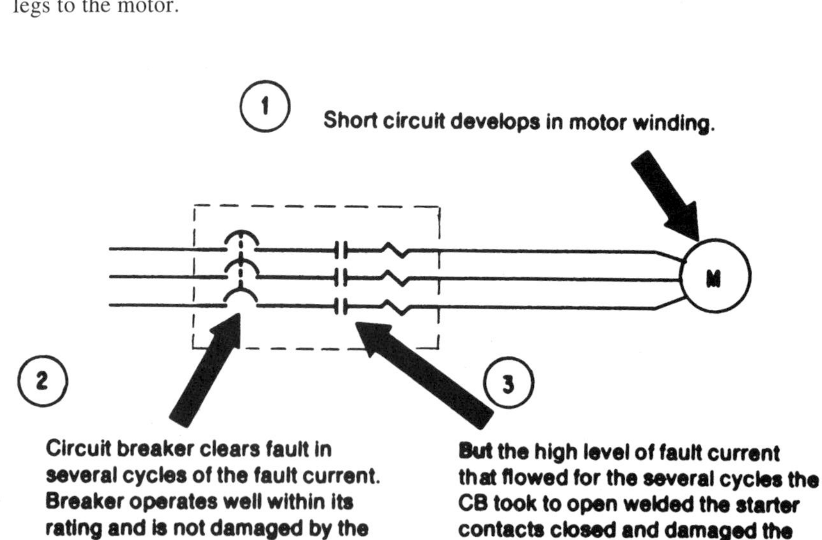

FIG. 6-10 Effective design can prevent this kind of common fault.

CONTROLLERS FOR FRACTIONAL HP MOTORS*

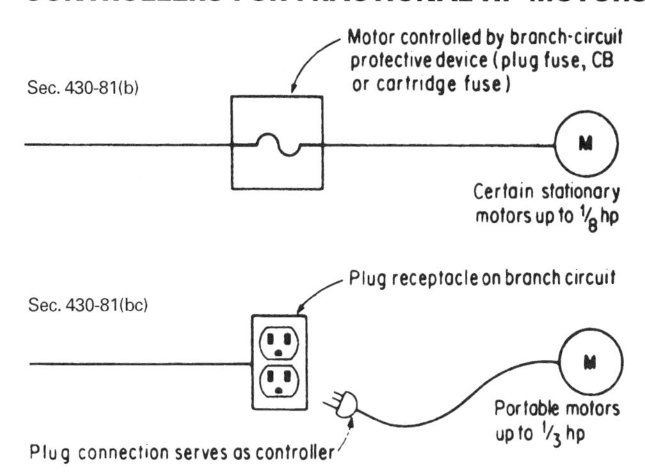

FOR OTHER MOTORS, THIS IS THE BASIC RULE

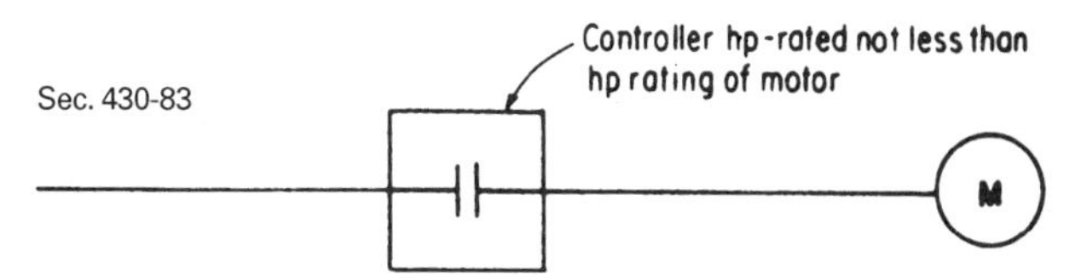

THESE ARE EXCEPTIONS TO THE BASIC RULES

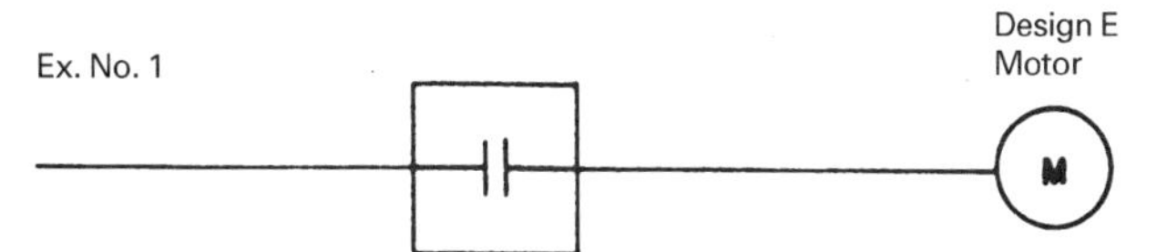

Same hp if listed for use with design E motor, nondesign E listing requires at least 1.4 times hp of design E motor (3 hp – 100), or at least 1.3 times motor rating if over 100 hp

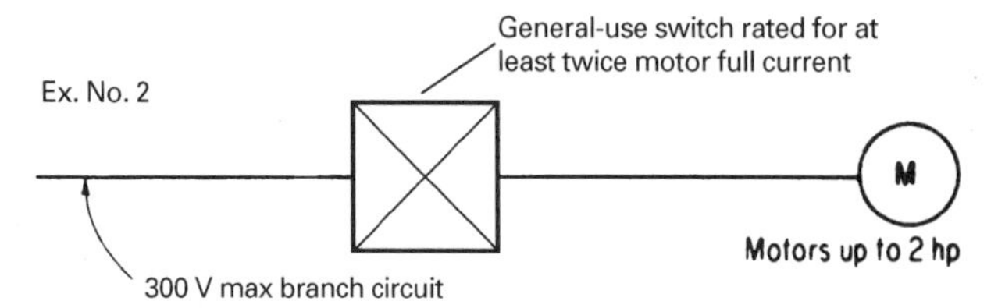

*** Not required to be horsepower-rated**

FIG. 6-11 Some type of motor controller must be provided.

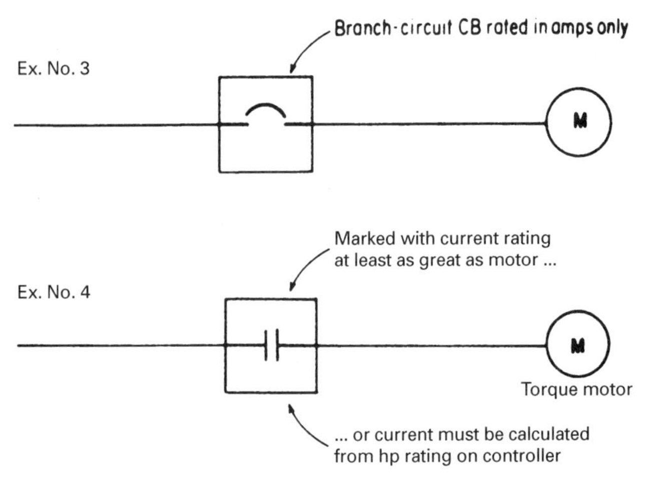

FIG. 6-11 (*Continued*) Some type of motor controller must be provided.

For sealed (hermetic-type) refrigeration motor-compressors, selection of the size of controller is slightly more involved than for standard applications. Because of their low-temperature operating conditions, hermetic motors can handle heavier loads than general-purpose motors of equivalent size and rotor-stator construction. And because the capabilities of such motors cannot be accurately defined in terms of horsepower, they are rated in terms of full-load current and locked-rotor current for polyphase motors and larger single-phase motors. Accordingly, the selection of controller size for a hermetic motor is different from that for a general-purpose motor, for which horsepower ratings must be matched.

NEC rules on controllers for motor-compressors are covered in Sec. 440-41. The controller must have a continuous-duty full-load current rating and a locked-rotor current rating that are not less than the full-load and locked-rotor currents of the motor.

For controllers rated in horsepower, the size required for a particular hermetic motor can be selected after the nameplate rated load current or branch-circuit selection current (whichever is greater) and locked-rotor current of the motor have been converted to an equivalent horsepower rating. To get this equivalent horsepower rating, which is the required size of controller, the tables in Art. 430 must be used. First, the nameplate full-load current of the motor is located in one of Tables 430-148 to 430-150, and the horsepower rating which corresponds to it is noted. Then the nameplate locked-rotor current of the motor is found in Table 430-151, and again the corresponding horsepower is noted. In all tables, if the exact value of current is not listed, the next-higher value should be used. If the two horsepower ratings obtained in this way are not the same, the larger value is taken as the required size of controller.

Example. A 230-V, 3-phase, squirrel-cage induction motor in a compressor has a nameplate full-load current of 25.8 A and a nameplate locked-rotor current of 90 A. What size of horsepower-rated controller should be used?

Solution. From Table 430-150, 28 A is the next-higher current than the nameplate current of 25.8 A, and the corresponding horsepower rating for a 230-V, 3-phase motor is 10 hp.

From Table 430-151, a locked-rotor current rating of 90 A for a 230-V, 3-phase motor requires a controller rated at 5 hp. The two values of horsepower obtained are not the same, so the higher rating is selected as acceptable for the given conditions. A 10-hp motor controller must be used.

Some controllers may be rated in full-load current and locked-rotor current rather than in horsepower. For use with a hermetic motor, such a controller must have current ratings equal to or greater than the nameplate full-load current and locked-rotor current of the motor.

Starter Poles. It is interesting to note that the NEC says that a controller need not open all conductors to a motor, except when the controller serves also as the required disconnecting means. For instance, a 2-pole starter could be used for a 3-phase motor if running overload protection is provided in all three circuit legs by devices separate from the starter. The controller must interrupt only enough conductors to start and stop the motor.

However, when the controller is a manual (nonmagnetic) starter or is a manually operated switch or CB (as permitted by the NEC), the controller itself may also serve as the disconnect means if it opens all ungrounded conductors to the motor. This eliminates the need for another switch or CB to serve as the disconnecting means. But note that only a manually operated switch or circuit breaker may serve such a dual function. A magnetic starter cannot also serve as the disconnecting means, even if it does open all ungrounded conductors to the motor. These conditions are shown in Figs. 6-12 and 6-13.

The word *ungrounded* above refers to the condition that none of the circuit conductors is grounded. They may be the ungrounded conductors of grounded systems.

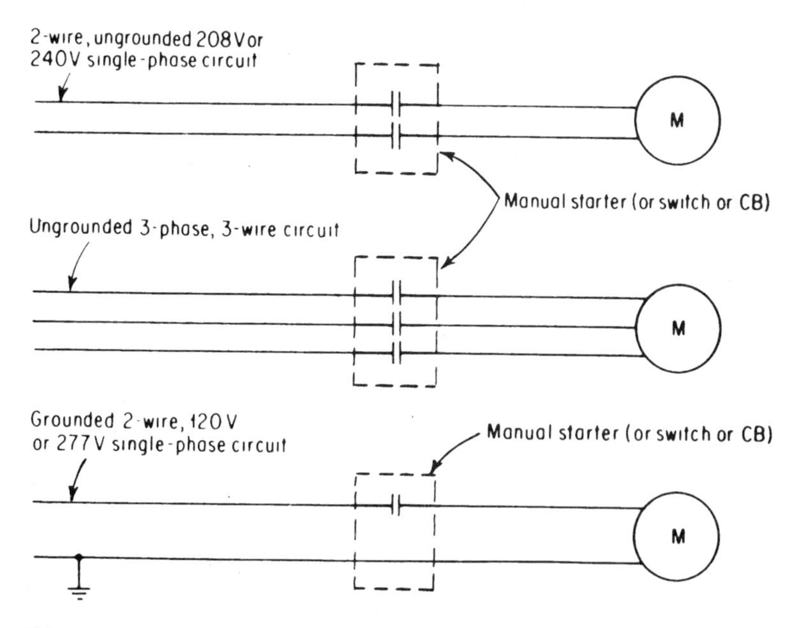

FIG. 6-12 Manual starter that opens all ungrounded conductors to a motor may serve as *both* controller and disconnect.

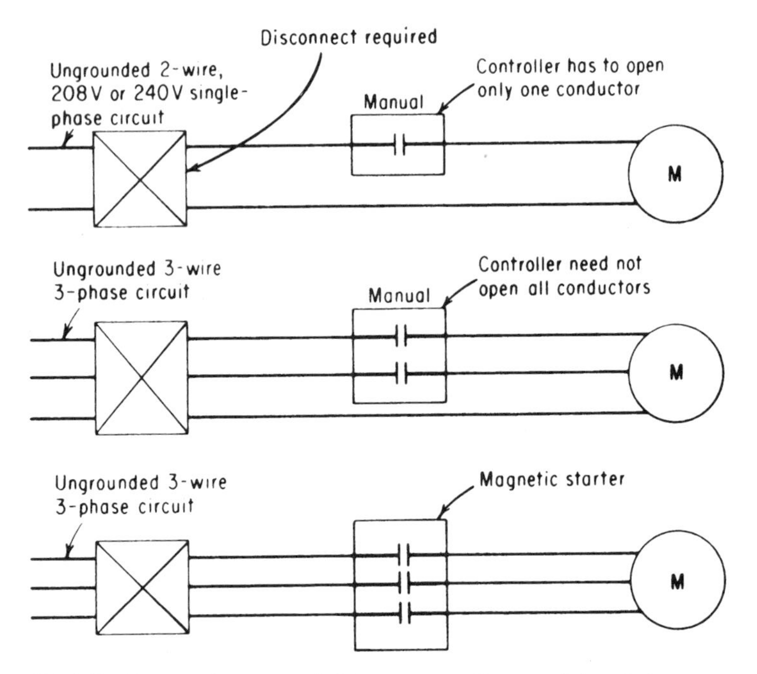

FIG. 6-13 If a manual starter or switch does not open all ungrounded conductors, a separate disconnect is required.

Overload Protection

The NEC contains specific requirements regarding motor running overcurrent (overload protection); these are intended to protect the elements of the branch circuit—the motor itself, the motor control apparatus, and the branch-circuit conductors—against excessive heating due to motor overloads. Such an overload is considered to be an operating overload up to and including stalled-rotor current. When the overload persists for a sufficient time, it will cause damage or dangerous overheating of the apparatus. This overload type does not include fault current due to shorts or grounds.

Typical NEC requirements for running overload protection, shown in Fig. 6-14, are as follows:

1. Running overload protection must be avoided for motors of more than 1 hp, if used for continuous duty. This protection may be an external overcurrent device actuated by the motor running current and set to open at not more than 125 percent of the motor full-load current for motors marked with a service factor of not less than 1.15 and for motors with a temperature rise not over 40°C. Sealed (hermetic-type) refrigeration motor-compressors must be protected against overload and failures to start, as specified in Sec. 440-52, by one of the following:
   - **a.** An overload relay set at up to 140 percent of the motor full-load current
   - **b.** An approved, integral thermal protector
   - **c.** A branch-circuit fuse or CB rated at not over 125 percent of the rated load current

1. MOTORS RATED MORE THAN 1 HP

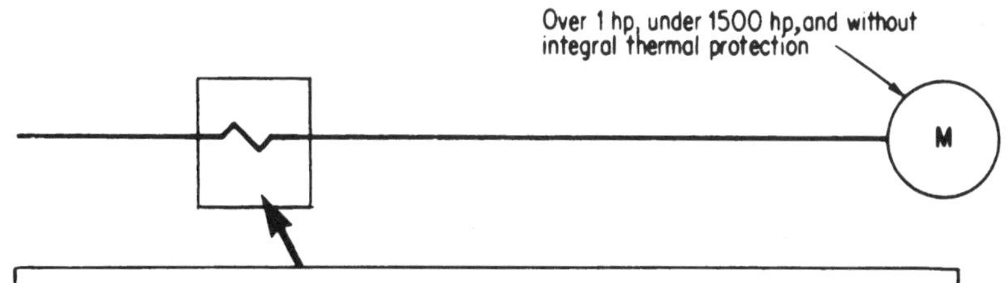

A separate overload device that is responsive to motor current. This device shall be selected to trip or shall be rated at no more than the following percent of the motor nameplate full-load current rating.

Motors with a marked service factor not less than 1.15	125%
Motors with a marked temperature rise not over 40°C....................	125%
All other motors..	115%

Modification of this value shall be permitted as provided in Section 430-34. For a multispeed motor, each winding connection shall be considered separately.

Where a separate motor overload device is so connected that it does not carry the total current designated on the motor nameplate, such as for wye-delta starting, the proper percentage of nameplate current applying to the selection or setting of the overload device shall be clearly designated on the equipment, or the manufacturer's selection table shall take this into account.

(FPN): Where power factor correction capacitors are installed on the load side of the motor overload device, see Section 460-9.

2. MOTORS RATED 1 HP OR LESS

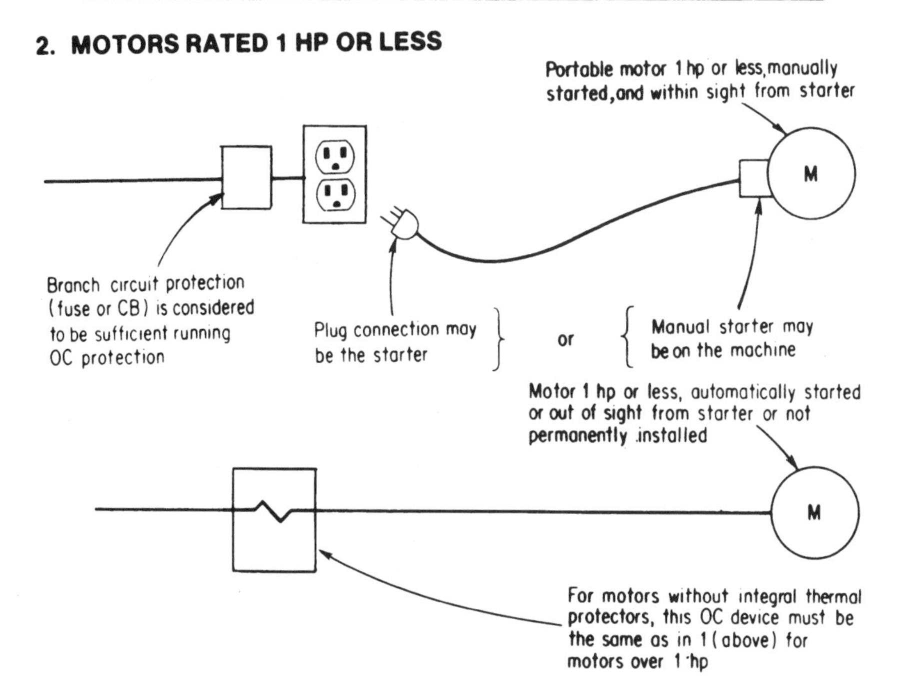

FIG. 6-14 Circuit design must include some form of protection against overloads, up to locked-rotor current.

d. A special protective system

The overload device must be rated or set to trip at not more than 115 percent of the motor full-load current for all other motors, such as motors with a 1.0 service factor or a 55°C rise.

Section 430-32 presents a thorough guide to the application of motor and branch-circuit overcurrent (overload) protection for conventional motors of the nonhermetic type. Section 440-52 gives detailed regulations on running overload protection for motor-compressors.

2. Motors of 1 hp or less which are not permanently installed and are manually started are considered protected against overload by the branch-circuit protection if the motors are within sight of the starter. Running overload devices are not required in such cases. A distance over 50 ft is considered out of sight.

Any motor of 1 hp or less which is not portable, is not manually started, and/or is not within sight of its starter location must have specific running overload protection. Automatically started motors of 1 hp or less must be protected against running overload in the same way as motors rated over 1 hp. That is, a separate or integral overload device must be used.

Basic NEC requirements concern the rating or setting of overcurrent devices separate from motors. However, the rules permit the use of thermal protectors integral with motors, provided that such devices are approved for their particular applications and that they prevent dangerous overheating of the motors. Exceptions to the basic rules on providing specific running overload protective devices are as follows:

1. Where the values specified for motor running overload protection do not permit the motor to start or to carry the load, the next-higher size of overload relay may be used, but not higher than the following percentages of motor full-load current rating:

- Motors with a marked service factor not less than 1.5: 140 percent
- Motors with a marked temperature rise not over 40°C: 140 percent
- Sealed (hermetic-type) motor-compressors: 140 percent
- All other motors: 130 percent

Fuses or circuit breakers may be used for running overload protection but may not be rated or set up to those values. Fuses and CBs must have a maximum rating as listed above for motors of more than 1 hp. If the value determined as indicated there does not correspond to a standard rating of fuse or CB, the next-smaller size must be used. A rating of 125 percent of the full-load current is the absolute maximum for fuses and circuit breakers.

2. Under certain conditions, no specific running overload protection need be used: The motor is considered to be properly protected if it is part of an approved assembly which does not normally subject the motor to overloads and which has controls to protect against stalled rotor. Or if the impedance of the motor windings is sufficient to prevent overheating due to failure to start, the branch-circuit protection is considered adequate.

3. A motor used for service which is inherently short-time, intermittent, periodic, or varying duty (see NEC Table 430-22) is considered as protected against overcurrent by the branch-circuit overcurrent device. A motor is considered to be wired for continuous duty unless the motor cannot operate continuously with load under any condition of use.

Complete data on the required number and location of overcurrent devices are given in Table 430-37. Table 430-37 requires three running overload devices (trip coils, relays, thermal cutouts, etc.) for all 3-phase motors unless protected by other approved

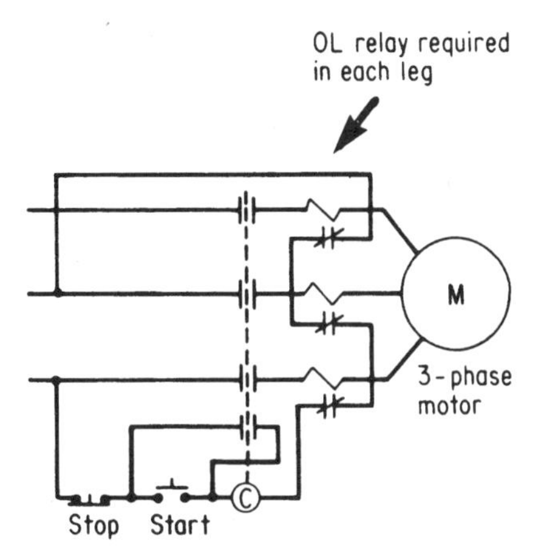

FIG. 6-15 Three-phase starter must have three overload devices.

means, such as specially designed embedded detectors with or without supplementary external protective devices (Fig. 6-15).

The usual acceptable rating of the overload protection in a motor starter is 125 percent of the motor full-load current (or as specifically required in NEC Secs. 430-32 and 430-34). But when a power-factor capacitor is installed on the load side of the motor starter or at the motor itself, a correction must be made in the rating or setting of the overload device. The device must be rated to take into account the fact that the magnetizing current for the motor is being supplied by the capacitor, and the total current flowing from the supply circuit through the starter is lower than it would be without the power-factor capacitor.

Figure 6-16 explains the need for correcting the size of the running overload protection in motor controllers when power-factor capacitors are used on the load side of the controller, as required by Sec. 460-9 on the use of capacitors to correct power factor. And, as required by the NEC, the rating of the capacitors should not exceed the value required to raise the no-load power factor of the motor to unity. Capacitors of these maximum ratings usually result in a full-load power factor of 95 to 98 percent.

Although the NEC contains all those requirements on the running overload protection of motors, Sec. 430-44 does recognize that there are cases when automatic opening of a motor circuit due to overload may be objectionable from a safety standpoint. In recognition of the circumstances of many industrial applications, Sec. 430-44 permits alternatives to automatic opening of a circuit in the event of overload. This permission to eliminate overload protection is similar to the permission given in Sec. 240-12 to eliminate overload protection when automatic opening of the circuit on an overload would constitute a more serious hazard than the overload itself. As the rule notes, "if immediate automatic shutdown of a motor by a motor overload protective device(s) would introduce additional or increased hazard(s) to a person(s) and continued motor operation is necessary for safe shutdown of equipment or process," then automatic overload opening is not required. However, as shown in Fig. 6-17, the circuit must be provided with a motor overload-sensing device conforming to the NEC requirement on overload protection, to indicate the presence of the overload by means of a supervised alarm. Such overload indication (instead of automatic opening) will alert personnel to the objectionable condition and will permit corrective action for an orderly shutdown,

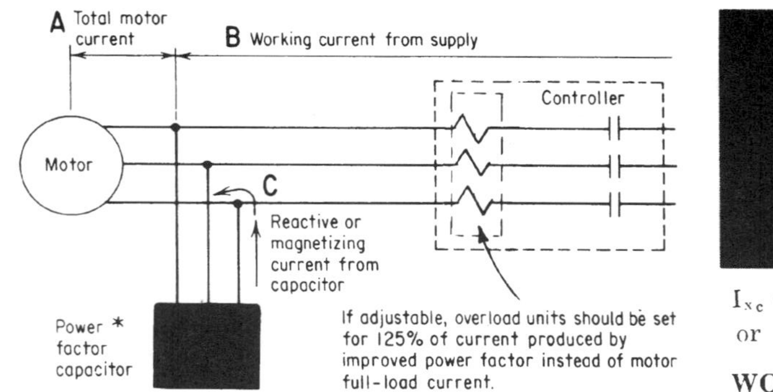

A (total motor current) = vector sum of B and C
where B = in-phase, working current
C = reactive, magnetizing current

EXAMPLE:

A motor with 70% power factor has a full-load current rating of 143 amps. Normally, the OL relay would be set for, say, 125% of 143 or 179 amps. *BUT*, because a PF capacitor is installed at the motor, the OL relay no longer will have 143 amps flowing through it at full load. If the capacitor corrects to 100% PF, the effect will be as shown at right.

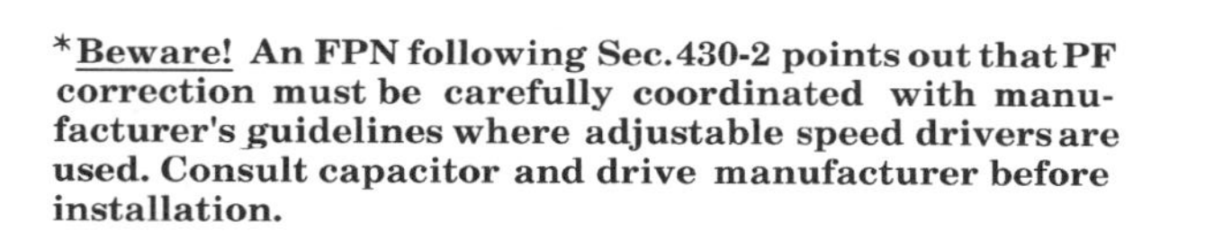

***Beware!* An FPN following Sec. 430-2 points out that PF correction must be carefully coordinated with manufacturer's guidelines where adjustable speed drivers are used. Consult capacitor and drive manufacturer before installation.**

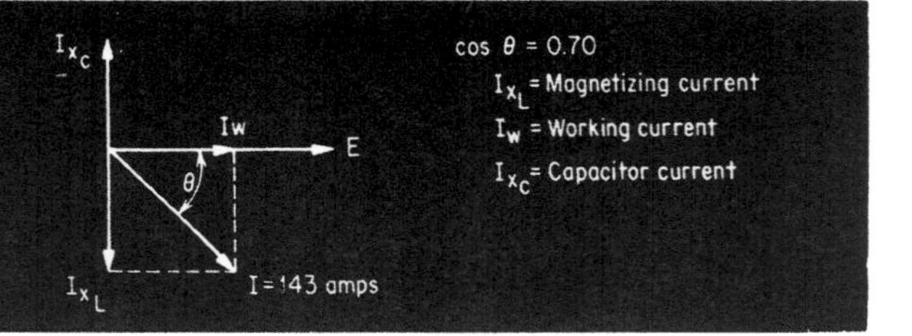

I_{XC} cancels I_{XL} at 100% PF, leaving only the working or in-phase current to be supplied from the circuit.

WORKING CURRENT =
TOTAL MOTOR CURRENT × POWER FACTOR
= 143 × 0.70 = 100 AMPS
THAT IS THE CURRENT THAT WILL BE FLOWING THROUGH THE OL RELAY AT FULL LOAD. THE OL RELAY, THEREFORE, MUST BE SET AT 125% × 100 AMPS = 125 AMPS

FIG. 6-16 Setting of overload protection must be corrected for reduced line current due to capacitors.

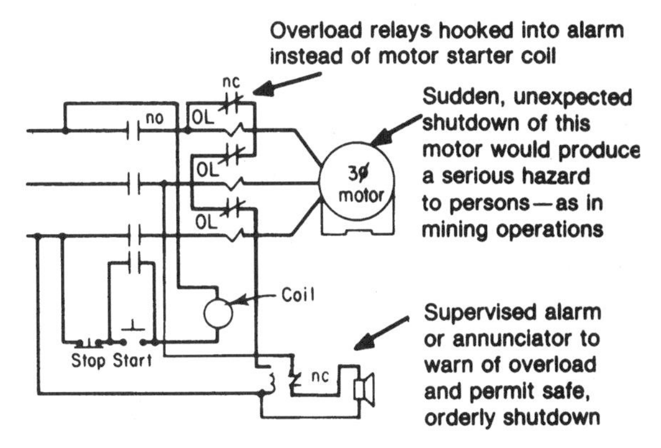

FIG. 6-17 Overload protection may be eliminated under these conditions.

either immediately or at some more convenient time, to resolve the difficulty. But, as is required in Sec. 240-12, short-circuit protection on the motor branch circuit must be provided to protect against those high-level ground faults and short circuits that would be more hazardous than a simple overload.

Motor Disconnect

The NEC specifically requires that a "listed" disconnecting means—basically, a motor circuit switch rated in horsepower or a CB or a molded-case switch—be provided in each motor circuit to disconnect the motor and its controller from all ungrounded supply conductors. In a motor branch circuit, every switch in the circuit in sight of the controller must satisfy the requirements regarding the type and rating of the disconnect means. And the disconnect switch or CB generally must be rated to carry at least 115 percent of the nameplate current rating of the motor for circuits up to 600 V. Figure 6-18 sets forth the basic requirements on types of disconnects.

The NEC includes a basic requirement that the disconnecting means for a motor and its controller be a motor circuit switch rated in horsepower. For motors rated up to 500 hp, this rule is readily complied with, inasmuch as the UL lists motor circuit switches up to 500 hp and manufacturers mark switches to conform. But for motors rated over 100 hp, the NEC does not require that the disconnect have a horsepower rating. An exception to the basic rule permits the use of ampere rated switches or isolation switches, provided the switches have a carrying capacity of at least 115 percent of the nameplate current rating of the motor. The UL notes that a horsepower rated switch that is rated over 100 hp must not be used as both a disconnect *and* a controller. But switches rated up to 100 hp may serve the dual function of disconnect and controller.

For design E motors rated from 3 to 100 hp, the disconnect must be "listed" for use with design E motors or have a rating that is 1.4 times the motor rating. Design E motors rated over 100 hp may be used with nondesign E listed disconnects that are rated at least 1.3 times the motor's horsepower rating.

As noted above, although Exception 4 to Sec. 430-109 sets the maximum horsepower rating required for motor circuit switches at 100 hp, higher-rated switches are now available, will provide additional safety, and should be used to ensure adequate interrupting ability.

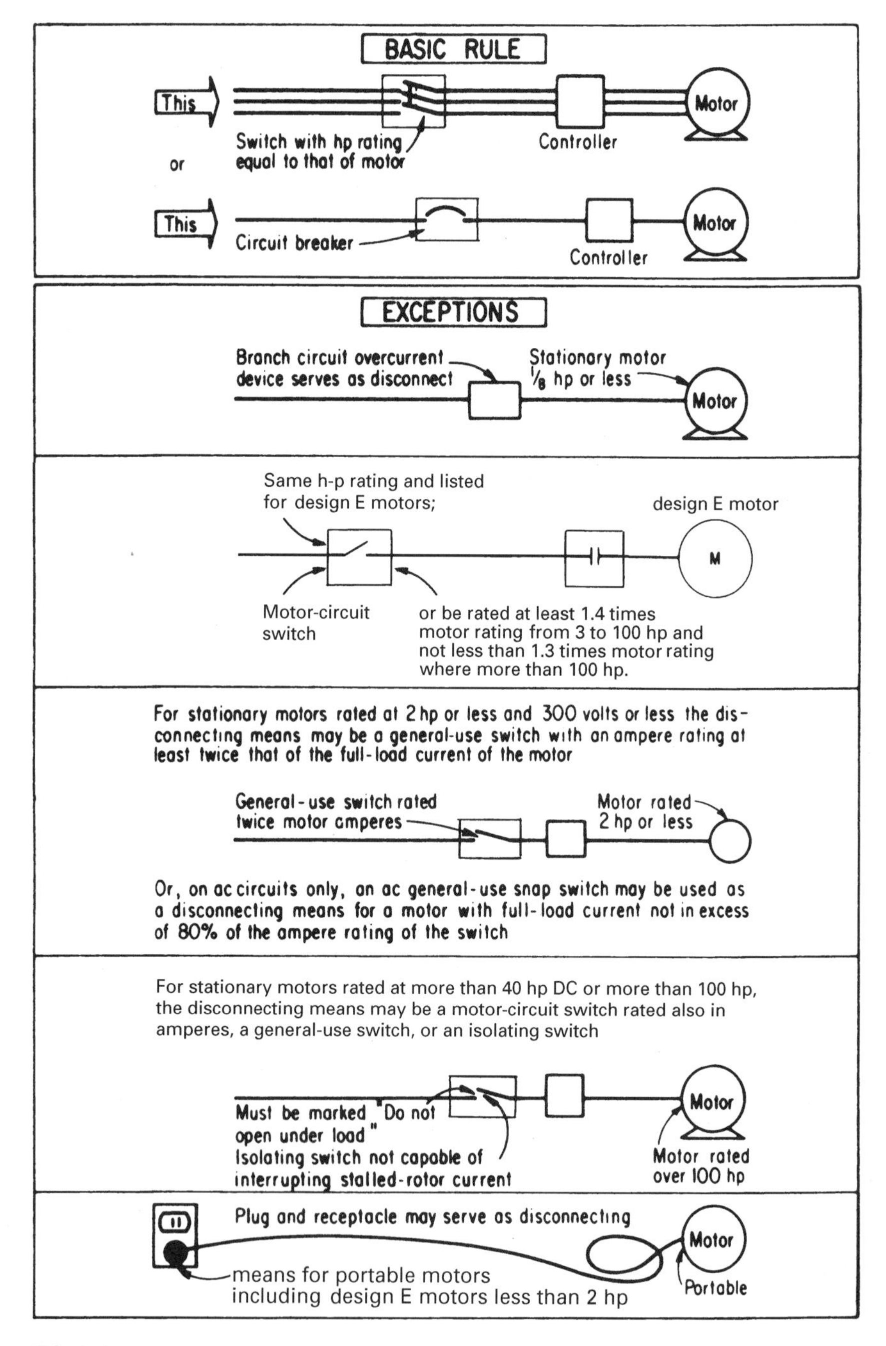

FIG. 6-18 Design of motor branch circuit must include a suitable disconnect means.

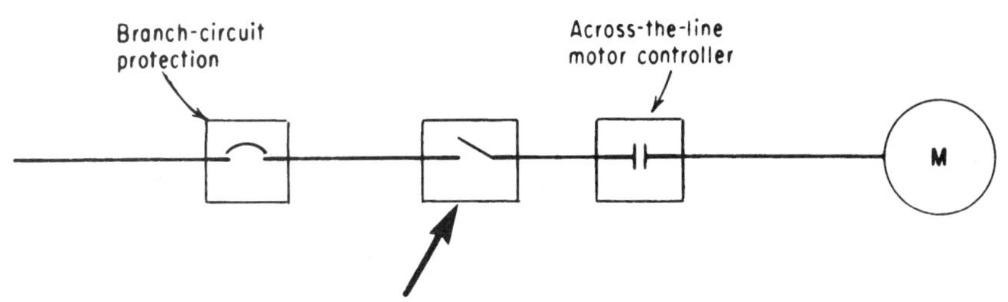

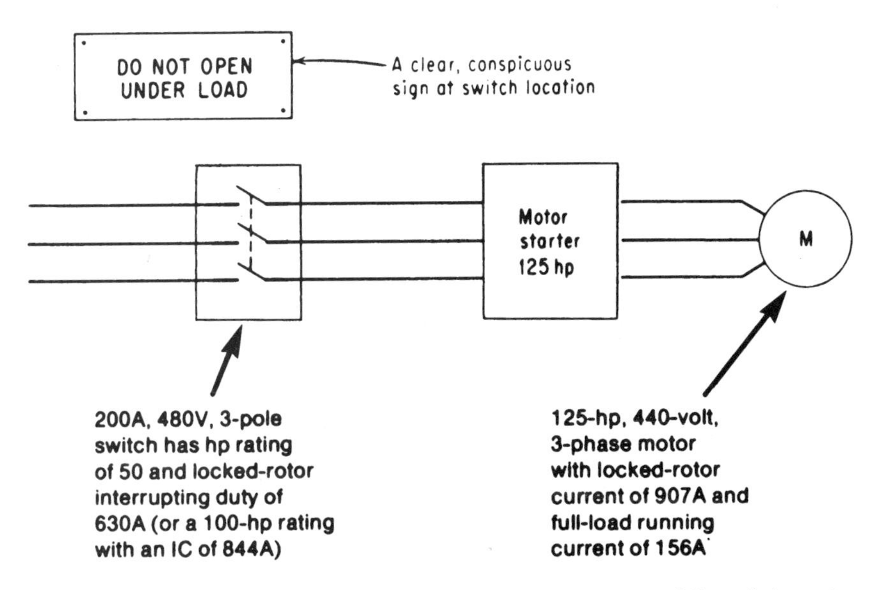

FIG. 6-19 Motor disconnect selected on the basis of ampere rating must be carefully applied.

Isolation switches for motors over 100 hp, if they are not capable of interrupting stalled-rotor currents, must be plainly marked "Do not open under load" (Fig. 6-19).

Example. Provide a disconnect for a 125-hp, 3-phase, 460-V motor. Use a nonfusible switch, inasmuch as short-circuit protection is provided at the supply end of the branch circuit.

Solution. The full-load running current of the motor is 156 A. A suitable disconnect must have a continuous carrying capacity of 156 A × 1.15 = 179 A. This calls for a 200-A, 3-pole switch rated for 480 V. The switch may be a general-use switch, a current-and-horsepower marked motor circuit switch, or an isolation switch. A motor circuit switch with the required current and voltage rating for this case would be marked for 50 hp, but the horsepower rating is of no concern because the switch does not have to be horsepower-rated for motors larger than 100 hp.

If the 50-hp switch were of the heavy-duty type, it would have an interrupting rating of 10 × 65 A (the full-load current of a 440-V, 50-hp motor), or 650 A. But the locked-rotor current of the 125-hp motor might run as high as 900 A. In such a case, the switch should be marked "Do not open under load."

If a fusible switch had to be provided for this motor for disconnect and short-circuit protection, the size of the switch would be determined by the size and type of fuses used. For a fuse rating of 250 percent of the motor current (which does not exceed the 300 percent maximum in Table 430-152) for standard fuses, the application would call for 400-A fuses in a 400-A switch. This switch would certainly qualify as the motor disconnect. However, if time-delay fuses were used, a 200-A switch would be large enough to take the time-delay fuses and could be used as the disconnect (because it is rated at 115 percent of motor current).

In the foregoing, the 400-A switch might have an interrupting rating high enough to handle the locked-rotor current of the motor. Or the 200-A switch might be of the type that has an interrupting rating up to 12 times the rated load current of the switch itself. In either case, there would probably be no need to mark the switch "Do not open under load."

Up to 100 hp, a switch which satisfies the NEC rule on rating for use as a motor controller may also provide the required disconnect means—the two functions being performed by the one switch—provided it opens all ungrounded conductors to the motor, is protected by an overcurrent device (which may be the branch-circuit protection or fuses in the switch itself), and is a manually operated air-break switch or an oil switch not rated over 600 V or 100 A.

As described under "Controllers," a manual starting switch capable of starting and stopping a given motor, capable of interrupting the stalled-rotor current of the motor, and having the same horsepower rating as the motor may serve the functions of the controller and disconnecting means in many motor circuits, if the switch opens all ungrounded conductors to the motor. A single circuit breaker may also serve as controller and disconnect; this is permitted by Sec. 420-111. However, an autotransformer-type controller, even if manual, may not also serve as the disconnecting means. Such controllers must be provided with a separate means for disconnecting the controller and motor.

The acceptability of a single switch used as both the controller and disconnecting means is based on the single switch satisfying NEC requirements for a controller and for a disconnect. This "combination" finds application where general-use switches or horsepower-rated switches are used, as permitted by the NEC in conjunction with time-delay fuses which are rated low enough to provide both running overload protection and branch-circuit (short-circuit) protection. In such cases, a single fused switch may serve a total of four functions: (1) controller, (2) disconnect, (3) branch-circuit protection, and (4) running overload protection. And it is also possible for a single circuit breaker to serve these four functions.

For sealed refrigeration compressors, Sec. 440-12 gives the procedure for determining the disconnect rating, based on the nameplate rated load current or branch-circuit selection current (whichever is greater) and the locked-rotor current of the motor-compressor. As an example, suppose a 3-phase, 460-V hermetic motor rated at 11-A branch-circuit selection current and 60-A locked-rotor current is to be supplied with a disconnect switch rated in horsepower. The first step is to determine the equivalent horsepower rating of the motor by referring to NEC Table 430-150. This table lists $7\frac{1}{2}$ hp as the horsepower size of a 460-V, 11-A motor. To ensure adequate interrupting capacity, Table

430-151 is now used. This table shows $7\frac{1}{2}$ hp as the equivalent horsepower rating for any locked-rotor current over 45 A and up to 66 A for a 460-V motor. Both tables thus establish a $7\frac{1}{2}$-hp disconnect as adequate for the given motor. If the ratings obtained from the two tables had been different, the higher rating would have been chosen.

In general, each motor is provided with a separate disconnecting means. However, a single disconnect sometimes may serve a group of motors. Such a disconnect must have a rating sufficient to handle a single load equal to the sum of the horsepower ratings or current ratings of all the motors it serves. The single disconnect may be used for a group of motors driving different parts of a single piece of apparatus, for several motors on one branch circuit, or for a group of motors in a single room within sight of the disconnect location (Sec. 430-112).

An important rule in NEC Sec. 430-102 states simply and clearly that "a disconnecting means shall be located in sight from the controller location." This applies always, for all motor loads rated up to 600 V—even if an "out-of-sight" disconnect can be locked in the open position (Fig. 6-20). There are, however, two exceptions to this important rule:

Exception 1 permits the disconnect for a high-voltage (over 600 V) motor to be out of sight of the controller location if the controller is marked with a warning label giving the

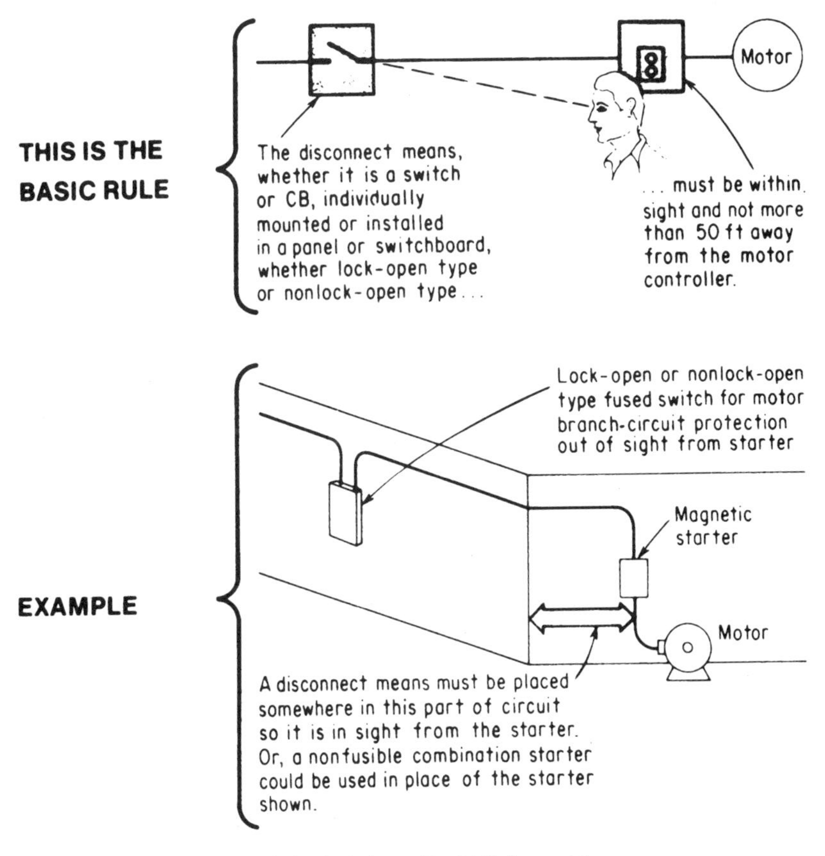

FIG. 6-20 Basic motor-circuit design includes an "in-sight" disconnect.

location and the identification of the disconnecting means to be locked in the open position.

Exception 2 is aimed at permitting practical, realistic disconnect means for large and complex industrial machinery utilizing a number of motors to power various interrelated machine parts. This exception recognizes that a single common disconnect for a number of controllers, as permitted by part (a) of the exception to Sec. 430-112, often cannot be installed "within sight" of all the controllers, even though the controllers are "adjacent one to the other." On much industrial process equipment, the components of the overall structure obstruct the view of many controllers. Exception 2 permits the single disconnect to be technically out of sight of some of or even all the controllers if the disconnect is simply "adjacent" to them, that is, nearby on the equipment structure.

Voltage Regulation

To ensure the proper and efficient operation of motors, the matter of voltage regulation must be considered carefully. Many factors (for example, number of motors, sizes and types of motors, duty cycles, load densities, type of distribution system, loading of various feeders, and power factor) are related to the design problem of ensuring the necessary level and stability of motor voltages. Voltage regulation follows through every step in design and must be accounted for in sizing conductors.

The voltage drop from the sources of voltage to any motor in the system must not exceed 5 percent. Normally, the proper power feeder design will limit the voltage drop to 3 percent, leaving a maximum permissible voltage drop of 2 percent in any motor branch circuit under full-load conditions. However, a 1 percent maximum circuit drop is recommended.

The power factor should be taken into consideration in all calculations; the known power-factor value or an 80 percent assumed value should be used. By raising the power factor, for the same actual power delivered, the current is decreased in the generator, transformers, and lines, up to the point where the capacitor is connected. Power-factor capacitors can be connected into motor circuits to neutralize the effect of lagging-power-factor loads, thereby reducing the current drawn for a given kilowatt load. In a distribution system, small capacitor units may be connected at the individual loads, or the total required capacitance (in kilovars) may be grouped at one point and connected to the main. Although the kilovar total is the same in both cases, the use of small capacitors at the individual loads reduces the current all the way from the loads back to the source; this method therefore has a greater power-factor corrective effect than the use of one big unit on the main, which reduces the current only from the point of installation back to the source.

Figure 6-21 shows the basic relationship involved in power-factor correction. A typical calculation is as follows:

Example. A 3-phase, 460-V, 50-hp motor has a power factor of 70 percent. What rating of capacitor is needed to improve the power factor to 90 percent?

Solution. From NEC Table 430-150, the full-load current of the motor is 65 A. That gives a kilowatt load of

$$L = \frac{460 \times 65 \times 1.73}{1000} \mathrm{PF}_1$$

$$= \frac{51{,}740}{1000} \times 0.7 = 36.2 \text{ kW}$$

If $\cos \theta_1 = 0.70$, then $\theta_1 = 45.58°$ and $\tan \theta_1 = 1.0203$; and if $\cos \theta_2 = 0.90$, then $\theta_2 = 25.84°$ and $\tan \theta_2 = 0.4844$. Then

TO IMPROVE THE POWER FACTOR OF A CIRCUIT FROM PF_1 TO PF_2 THE REQUIRED RATING OF A CAPACITOR MUST BE

$$KVAR_R = KW(\tan\theta_1 - \tan\theta_2)$$

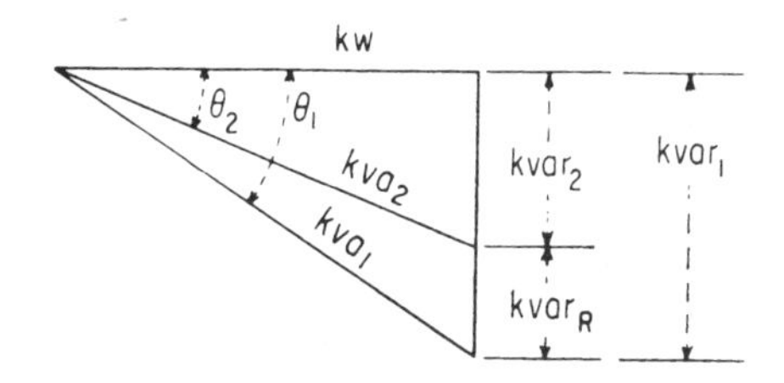

where, in the sketch

$KVAR_R$ = required kilovars rating of capacitor to change from PF_1 to PF_2

KW = the kilowatt value of the circuit load

θ_1 = original phase angle

θ_2 = improved phase angle

PF_1 = power factor before correction

PF_2 = power factor after correction

$KVAR_1$ = reactive kilovolt-amperes at PF_1

$KVAR_2$ = reactive kilovolt-amperes at PF_2

NOTE: The phase angles θ_1 and θ_2 can be determined from a table of trigonometric functions using the following relationships:

θ_1 = The angle which has its cosine equal to the decimal value of the original power factor (e.g., 0.70 for 70% PF; 0.65 for 65%; etc.)

θ_2 = The angle which has its cosine equal to the decimal value of the improved power factor

FIG. 6-21 Motor circuit power factor is raised with this procedure.

$$\begin{aligned}KVAR_R &= L(\tan\theta_1 - \tan\theta_2)\\ &= 36.2(1.0203 - 0.4844)\\ &= 36.2 \times 0.5359 = 19.4 \text{ kvar}\end{aligned}$$

Depending upon the relationship between the voltage drop and the power factor, measures for improving the power factor may be designed into motor branch circuits. At individual motor locations, power-factor-correcting capacitors offer improved voltage regulation. As shown in Fig. 6-22, power-factor capacitors installed at terminals of motors provide maximum relief from reactive currents, reducing the required current-carrying capacities of conductors from their point of application all the way back to the

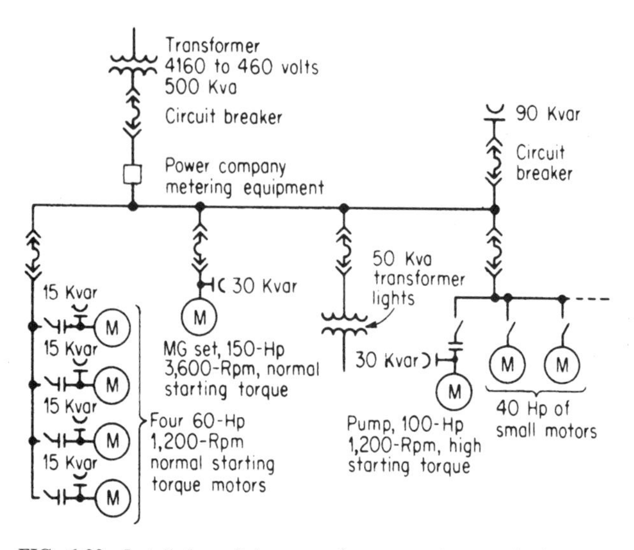

FIG. 6-22 Installation of the power-factor capacitors at the individual motors offers maximum benefit from the correction.

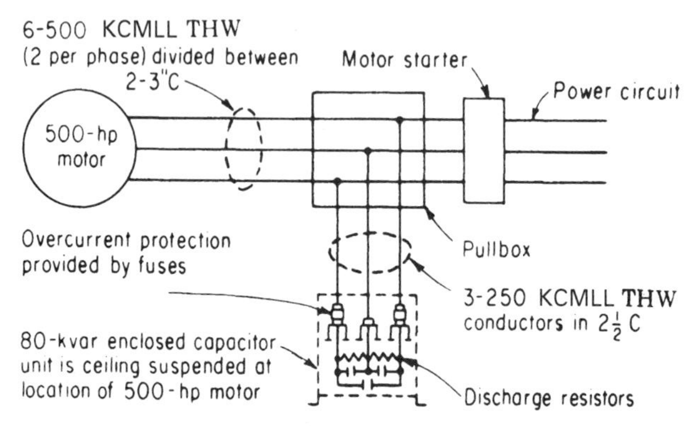

FIG. 6-23 Design must provide a place for connecting a capacitor unit between motor and starter.

supply system. Such application also eliminates extra switching devices, since each capacitor can be switched with the motor it serves. Figure 6-23 shows details included on a typical set of electrical plans to convey the designer's concept for power-factor correction of a large motor. For small, numerous motors that are operated intermittently, however, it is often economically more desirable to install the required capacitor kilovars at the motor load center, as at the right in Fig. 6-22.

In most applications, power-factor capacitors are installed to raise the system power factor for increased circuit or system current-carrying capacity, reduce power losses, and lower reactive-power charges (most utility companies include a power factor penalty clause in their industrial billing). Additional benefits derived from power-factor capacitor installation are reduced voltage drop and increased voltage stability. Capacitor manufacturers provide various tables and graphs to facilitate the selection of the proper capacitor for a given motor load.

The NEC limits power-factor correction to unity (100 percent, or 1.0) when there is no load on the motor. This results in a power factor of 95 percent or better when the motor is fully loaded. Section 460-9 recognizes the use of capacitors either sized at the value that will produce 100 percent power factor when the motor is running at no load or sized at a value equal to 50 percent of the kilovoltampere rating of the motor input (as noted in the exception), for motors up to 50 hp and 600 V, as shown in Fig. 6-24. Most motor-associated capacitors are used with low-voltage, 5- to 50-hp, 1800- and 1200-rpm, across-the-line-start motors. In this range, the no-load rule for determining the maximum capacitor kilovars restricts capacitors to the equivalent of less than 50 percent of the horsepower. Noticeable economy can be effected by applying larger capacitors, up to 50 percent of the motor input power, to such motors. This has been done for years, with excellent results and no field trouble.

The no-load power factor of a motor is a design constant for the motor; it may be obtained from the manufacturer of the motor, or it may be measured or calculated. With reference to Fig. 6-21, the known no-load power factor PF of the motor may be used to calculate the required power in kilowatts as $\text{kW} = \text{PF} \times \text{kVA}_1$, where kVA_1 is calculated from the circuit voltage and current, as measured with a clamp-on ammeter. Then the kilovar rating of the capacitor required to raise the no-load power factor to 100 percent equals the square root of $(\text{kVA}_1)^2 - (\text{kW})^2$.

Capacitors installed as a group or bank at some central point, such as a switchboard, load center, busway, or outdoor substation, usually serve only to reduce the utility company penalty charges. However, in many instances, installation costs also will be lowered.

Capacitor installations may consist of an individual unit connected as close as possible to the inductive load (for example, at the terminals of a motor) or of a bank of many units connected in multiple across a main feeder. Units are available in specific kilovar and voltage ratings. Standard low-voltage capacitor units are rated from about 0.5 to 25

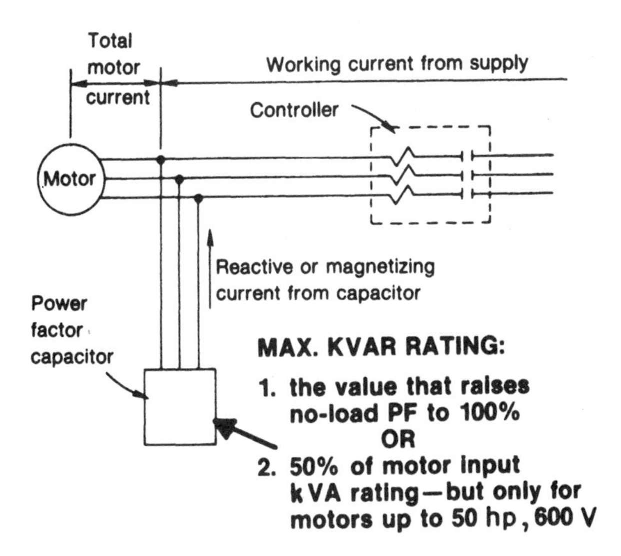

FIG. 6-24 The NEC sets a maximum value on capacitor rating for safety purposes.

Desired power factor in percentage

Original power factor in percentage	80%	81	82	83	84	85	86	87	88	89	90	91	92	93	94	95	96	97	98	99	100
50%	.982	1.008	1.034	1.060	1.086	1.112	1.139	1.165	1.192	1.220	1.248	1.276	1.303	1.337	1.369	1.402	1.441	1.481	1.529	1.590	1.732
51	.936	.962	.988	1.014	1.040	1.066	1.093	1.119	1.146	1.174	1.202	1.230	1.257	1.291	1.320	1.357	1.395	1.435	1.483	1.544	1.686
52	.894	.920	.946	.972	.998	1.024	1.051	1.077	1.104	1.132	1.160	1.188	1.215	1.249	1.281	1.315	1.353	1.393	1.441	1.502	1.644
53	.850	.876	.902	.928	.954	.980	1.007	1.033	1.060	1.088	1.116	1.144	1.171	1.205	1.237	1.271	1.309	1.349	1.397	1.458	1.600
54	.809	.835	.861	.887	.913	.939	.966	.992	1.019	1.047	1.075	1.103	1.130	1.164	1.196	1.230	1.268	1.308	1.356	1.417	1.559
55	.769	.795	.821	.847	.873	.899	.926	.952	.979	1.007	1.035	1.063	1.090	1.124	1.156	1.190	1.228	1.268	1.316	1.377	1.519
56	.730	.756	.782	.808	.834	.860	.887	.913	.940	.968	.996	1.024	1.051	1.085	1.117	1.151	1.189	1.229	1.277	1.338	1.480
57	.692	.718	.744	.770	.796	.822	.849	.875	.902	.930	.958	.986	1.013	1.047	1.079	1.113	1.151	1.191	1.239	1.300	1.442
58	.655	.681	.707	.733	.759	.785	.812	.838	.865	.893	.921	.949	.976	1.010	1.042	1.076	1.114	1.154	1.202	1.263	1.405
59	.618	.644	.670	.696	.722	.748	.775	.801	.828	.856	.884	.912	.939	.973	1.005	1.039	1.077	1.117	1.165	1.226	1.368
60	.584	.610	.636	.662	.688	.714	.741	.767	.794	.822	.849	.878	.905	.939	.971	1.005	1.043	1.083	1.131	1.192	1.334
61	.549	.575	.601	.627	.653	.679	.706	.732	.759	.787	.815	.843	.870	.904	.936	.970	1.008	1.048	1.096	1.157	1.299
62	.515	.541	.567	.593	.619	.645	.672	.698	.725	.753	.781	.809	.836	.870	.902	.936	.974	1.014	1.062	1.123	1.265
63	.483	.509	.535	.561	.587	.613	.640	.666	.693	.721	.749	.777	.804	.838	.870	.904	.942	.982	1.030	1.091	1.233
64	.450	.476	.502	.528	.554	.580	.607	.633	.660	.688	.716	.744	.771	.805	.837	.871	.909	.949	.997	1.058	1.200
65	.419	.445	.471	.497	.523	.549	.576	.602	.629	.657	.685	.713	.740	.774	.806	.840	.878	.918	.966	1.027	1.169
66	.388	.414	.440	.466	.492	.518	.545	.571	.598	.626	.654	.682	.709	.743	.775	.809	.847	.887	.935	.996	1.138
67	.358	.384	.410	.436	.462	.488	.515	.541	.568	.596	.624	.652	.679	.713	.745	.779	.817	.857	.905	.966	1.108
68	.329	.355	.381	.407	.433	.459	.486	.512	.539	.567	.595	.623	.650	.684	.716	.750	.788	.828	.876	.937	1.079
69	.299	.325	.351	.377	.403	.429	.456	.482	.509	.537	.565	.593	.620	.654	.686	.720	.758	.798	.840	.907	1.049
70	.270	.296	.322	.348	.374	.400	.427	.453	.480	.508	.536	.564	.591	.625	.657	.691	.729	.769	.811	.878	1.020
71	.242	.268	.294	.320	.346	.372	.399	.425	.452	.480	.508	.536	.563	.597	.629	.663	.701	.741	.783	.850	.992
72	.213	.239	.265	.291	.317	.343	.370	.396	.423	.451	.479	.507	.534	.568	.600	.634	.672	.712	.754	.821	.963
73	.186	.212	.238	.264	.290	.316	.343	.369	.396	.424	.452	.480	.507	.541	.573	.607	.645	.685	.727	.794	.936
74	.159	.185	.211	.237	.263	.289	.316	.342	.369	.397	.425	.453	.480	.514	.546	.580	.618	.658	.700	.767	.909
75	.132	.158	.184	.210	.236	.262	.289	.315	.342	.370	.398	.426	.453	.487	.519	.553	.591	.631	.673	.740	.882
76	.105	.131	.157	.183	.209	.235	.262	.288	.315	.343	.371	.399	.426	.460	.492	.526	.564	.604	.652	.713	.855
77	.079	.105	.131	.157	.183	.209	.236	.262	.289	.317	.345	.373	.400	.434	.466	.500	.538	.578	.620	.687	.829
78	.053	.079	.105	.131	.157	.183	.210	.236	.263	.291	.319	.347	.374	.408	.440	.474	.512	.552	.594	.661	.803
79	.026	.052	.078	.104	.130	.156	.183	.209	.236	.264	.292	.320	.347	.381	.413	.447	.485	.525	.567	.634	.776
80	.000	.026	.052	.078	.104	.130	.157	.183	.210	.238	.266	.294	.321	.355	.387	.421	.459	.499	.541	.608	.750
81	—	.000	.026	.052	.078	.104	.131	.157	.184	.212	.240	.268	.295	.329	.361	.395	.433	.473	.515	.582	.724
82	—	—	.000	.026	.052	.078	.105	.131	.158	.186	.214	.242	.269	.303	.335	.369	.407	.447	.489	.556	.698
83	—	—	—	.000	.026	.052	.079	.105	.132	.160	.188	.216	.243	.277	.309	.343	.381	.421	.463	.530	.672
84	—	—	—	—	.000	.026	.053	.079	.106	.134	.162	.190	.217	.251	.283	.317	.355	.395	.437	.504	.645
85	—	—	—	—	—	.000	.027	.053	.080	.108	.136	.164	.191	.225	.257	.291	.329	.369	.417	.478	.620

FIG. 6-25 Multipliers for calculating capacitor kilovars required to raise the power factors of motor circuits.

kvar at voltages from 216 to 600 V. For high-voltage applications, standard ratings are 15, 25, 50, and 100 kvar. Available in single-, 2-, or 3-phase configurations, power capacitors may be supplied either unfused or equipped with current-limiting or high-capacity fuses (single-phase units are furnished with one fuse; 3-phase capacitors usually have two fuses). On low-voltage units, fuses may be mounted on the capacitor bushings inside the terminal compartment.

Figure 6-25 is a table for calculating the capacitor kilovar rating required to raise the power factor from an original value to a desired higher value. Its use is illustrated in the following example.

Example. A circuit supplies a load that has a 100-kW value (obtained either by calculation or from a wattmeter) and operates at a power factor of 68 percent. What capacitor kilovar rating is needed to raise the power factor to 96 percent?

Solution. Find the required multiplier in the table of Fig. 6-25 by locating 68 percent in the vertical column at the left and then moving across the column of multipliers headed 96 percent. The multiplier is 0.788. Then the required rating is

$$100 \text{ kW} \times 0.788 = 78.8 \text{ kvar}$$

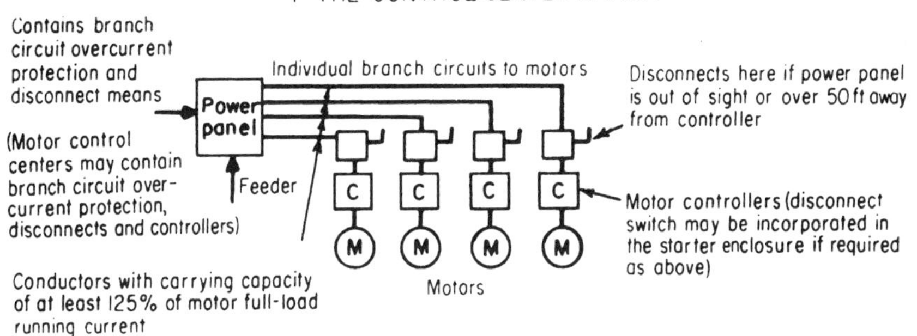

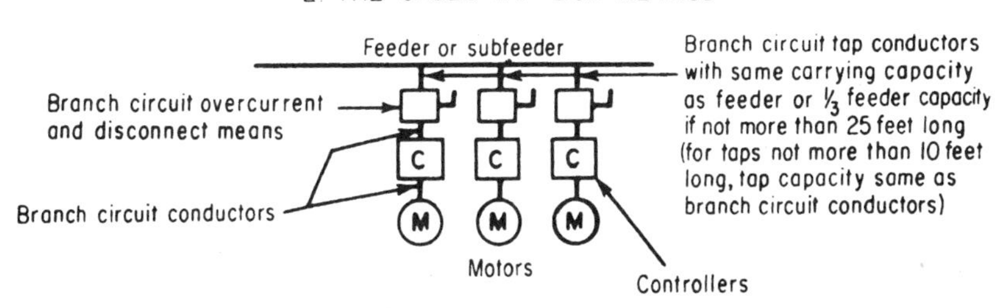

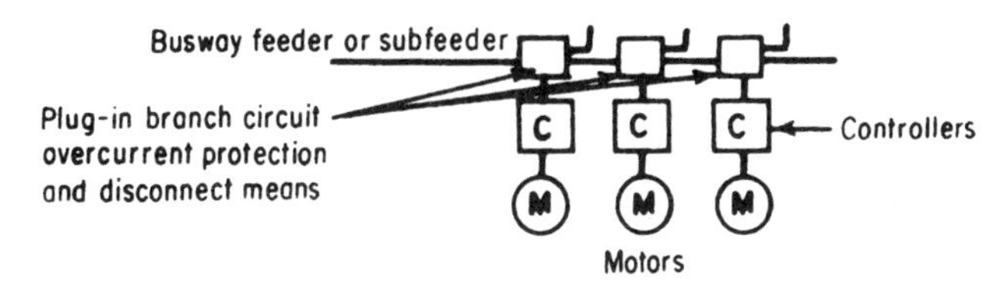

FIG. 6-26 Motor circuit layout must suit the particular needs of a given installation.

MOTOR FEEDERS

When a load consists only of motors or of motors plus lighting, heating, and/or other types of loads, the feeder conductors and overcurrent protection must be carefully selected on the basis of all applicable design considerations (see Fig. 6-26).

The sizing of feeder conductors must satisfy applicable NEC rules and, among other considerations, ensure that voltage drop and copper loss are kept to reasonable values. The initial design procedure is based on the following requirements:

1. The current-carrying capacity of conductors supplying several motors must be at least equal to 125 percent of the full-load current of the highest-rated motor plus the sum of the full-load current of the other motors supplied by the feeder (NEC Sec. 430-24).

2. The current-carrying capacity of feeder conductors supplying a single motor plus other loads must be at least equal to 125 percent of the full-load current of the motor plus other load current.
3. The current-carrying capacity of feeder conductors supplying a motor load and a lighting and/or appliance load must be sufficient to handle the lighting and/or appliance load as determined from the procedure for sizing lighting feeders, plus the motor load as determined from items 1 and 2 above.

The NEC permits inspectors to authorize the use of demand factors for motor feeders—based on reduced heating of conductors that supply motors operating intermittently, motors on duty cycles, or motors not operating together. Where necessary, the authority enforcing the NEC should be consulted to ensure that the conditions and operating characteristics are deemed suitable for reduced-capacity feeders.

Sizing Conductors

The NEC allows the sizing of motor feeders (and feeders supplying combination power and lighting loads) on the basis of maximum demand running current, calculated as follows:

$$\text{Running current} = (1.25 \times I_f) + (\text{DF} \times I_t)$$

where I_f = full-load current of largest motor, A
DF = demand factor
I_t = sum of full-load currents of all motors except largest, A

But modern design must include provisions for limiting the voltage drop under conditions of high motor current inrush and therefore dictates the use of the maximum-demand starting current in sizing conductors for improved voltage stability on the feeder. Thus current is calculated as follows:

$$\text{Starting current} = I_s + \text{DF} \times I_t$$

where I_s is the average starting current of the largest motor. (Use the percentage of motor full-load current given for fuses in Table 430-152.)

Voltage Drop and Power Factor

Voltage drop and I^2R loss must be carefully considered in sizing motor feeders. The design voltage drop percentage may vary with the particular operating conditions and layout of motor loads; however, it should never exceed a 3 percent drop from the service entrance to the point of origin of motor branch circuits. And a maximum voltage drop of only 2 percent from the service to the motor branch circuit is widely used for motor feeder design.

Both the reactance and the resistance of the feeder conductors must be included in voltage drop calculations, as both contribute to the drop. Power factor must also be accounted for in these calculations. And voltage drop in a feeder must be analyzed in terms of the number of motors supplied and the size and operating duty of each motor.

When a number of motors may be starting simultaneously, or when several motors driving sluggish loads may be started at or near the same time, the voltage drop in the feeder could be large, unless it is sized to account for high load currents. Of course, such conditions, when analyzed, will often clearly indicate the need for further subdivision or adjustment of feeder loads, as well as the best types of motors and/or controllers to select for use. The initial value of the starting current—the locked-rotor current—must be used in studying the effect of motor loads on voltage drop.

The I^2R loss in a motor feeder—the power (watts) lost in the conductors due to heat developed by current flow through the conductors—is equal to the square of the total current drawn through the conductors times the total resistance of the conductors. This loss may frequently be substantial, even when the voltage drop in the feeder is within recommended limits. All voltage-drop studies should include the consideration of an increase in conductor size beyond that necessary to limit the voltage drop, for the purpose of limiting I^2R losses and the cost of the power they waste.

The power factor of a motor feeder is improved by placing a capacitor of the proper kilovar rating at the supply end of the feeder; this reduces both the required current flow to a given load and the voltage drop, by eliminating the reactive current from the feeder conductors.

The power factor of a circuit supplying several motors is determined as follows:

1. For each motor, multiply its horsepower by its power factor at 75 percent of rated load.
2. Add these products for all the motors.
3. Divide the sum obtained in step 2 by the total horsepower connected to the circuit, to obtain the approximate (but accurate enough for most calculations) power factor of the circuit.

The circuit can then be corrected to raise the power factor to a desired level by selecting a capacitor of the proper kilovar rating for the given load, as directed by the literature of capacitor manufacturers.

Figure 6-27 is a handy table commonly used for computing the capacitor kilovars (reactive kilovoltamperes) required to raise the power factor from one value to another for a given kilowatt load.

Figure 6-28 presents data on the application of power-factor capacitors on plug-in busway supplying power loads. The busways were used in a modern industrial plant to supply plug-in circuits to motor loads, and the data were used as follows: The total horsepower of the connected motor load was determined for each busway run. A demand factor was applied, based on past experience and metering, reducing the demand on the busway run. The horsepower ratings were then corrected for the efficiency of the motors, which was assumed to be 95 percent. This corrected horsepower was converted to kilowatts, and a corresponding kilovoltampere value was established for an 80 percent power factor, for each busway. Reactive kilovoltampere (kilovar) values were also determined. Then kilovoltampere and kilovar values were calculated for the given kilowatt values, based on an assumed power factor of 95 percent. The difference in kilovars was noted for each busway, to establish the kilovars required to bring the power factor of each busway up to about 95 percent. The last column in Fig. 6-28 shows the amounts of capacitor equipment that were required, based on the use of 15-kvar unit capacitors rated at 575 V for the 12 sections of 600-A, 480-V, 3-phase, 4-wire busway. The capacitor equipment was mounted on the busway sections.

Protecting Motor Feeders

Overcurrent protection for a feeder to several motors must have a rating or setting not greater than the largest rating or setting of the branch-circuit protective device for any motor of the group, based on the maximum permitted value from Sec. 430-52,

Desired Power Factor in Percentage

Original Power Factor in Percentage	80%	81	82	83	84	85	86	87	88	89	90	91	92	93	94	95	96	97	98	99	100
50%	.982	1.008	1.034	1.060	1.086	1.112	1.139	1.165	1.192	1.220	1.248	1.276	1.303	1.337	1.369	1.40[illegible]	1.441	1.481	1.529	1.590	1.732
51	.936	.962	.988	1.014	1.040	1.066	1.093	1.119	1.146	1.174	1.202	1.230	1.257	1.291	1.32[illegible]	1.357	1.395	1.435	1.483	1.544	1.686
52	.894	.920	.946	.972	.998	1.024	1.051	1.077	1.104	1.132	1.160	1.188	1.215	1.249	1.281	1.315	1.353	1.393	1.441	1.502	1.644
53	.850	.876	.902	.928	.954	.980	1.007	1.033	1.060	1.088	1.116	1.144	1.171	1.205	1.237	1.271	1.309	1.349	1.397	1.458	1.600
54	.809	.835	.861	.887	.913	.939	.966	.992	1.019	1.047	1.075	1.103	1.130	1.164	1.196	1.230	1.268	1.308	1.356	1.417	1.559
55	.769	.795	.821	.847	.873	.899	.926	.952	.979	1.007	1.035	1.063	1.090	1.124	1.156	1.190	1.228	1.268	1.316	1.377	1.519
56	.730	.756	.782	.808	.834	.860	.887	.913	.940	.968	.996	1.024	1.051	1.085	1.117	1.151	1.189	1.229	1.277	1.338	1.480
57	.692	.718	.744	.770	.796	.822	.849	.875	.902	.930	.958	.986	1.013	1.047	1.079	1.113	1.151	1.191	1.239	1.300	1.442
58	.655	.681	.707	.733	.759	.785	.812	.838	.865	.893	.921	.949	.976	1.010	1.042	1.076	1.114	1.154	1.202	1.263	1.405
59	.618	.644	.670	.696	.722	.748	.775	.801	.828	.856	.884	.912	.93[illegible]	.973	1.005	1.039	1.077	1.117	1.165	1.226	1.368
60	.584	.610	.636	.662	.688	.714	.741	.767	.794	.822	.849	.878	[illegible]	.939	.971	1.005	1.043	1.083	1.131	1.192	1.334
61	.549	.575	.601	.627	.653	.679	.706	.732	.759	.787	.815	.843	.870	.904	.936	.970	1.008	1.048	1.096	1.157	1.299
61	.515	.541	.567	.593	.619	.645	.672	.698	.725	.753	.781	.809	.836	.870	.902	.936	.974	1.014	1.062	1.123	1.265
63	.483	.509	.535	.561	.587	.613	.640	.666	.693	.721	.749	.777	.804	.838	.870	.904	.942	.982	1.030	1.091	1.233
64	.450	.476	.502	.528	.554	.580	.607	.633	.660	.688	.716	.744	.771	.805	.837	.871	.909	.949	.997	1.058	1.200
65	.419	.445	.471	.497	.523	.549	.576	.602	.629	.657	.685	.713	.740	.774	.806	.840	.078	.918	.966	1.027	1.169
66	.388	.414	.440	.466	.492	.518	.545	.571	.598	.626	.654	.682	.709	.743	.775	.809	.847	.887	.935	.996	1.138
67	.358	.384	.410	.436	.462	488	.515	.541	.568	.596	.624	.652	.679	.713	.745	.779	.817	.857	.905	.966	1.108
68	.329	.355	.381	.407	.433	.459	.486	.512	.539	.567	.595	.623	.650	.684	.716	.750	.788	.828	.876	.937	1.079
69	.209	.325	.351	.377	.403	.429	.456	.482	.509	.537	.565	.593	.620	.654	.686	.720	.758	.798	.840	.907	1.049
70	.270	.296	.322	.348	.374	.400	.427	.453	.480	.508	.536	.564	.591	.625	.657	.691	.729	.769	.811	.878	1.020
71	.242	.268	.294	.320	.346	.372	.399	.425	.452	.480	.508	.536	.563	.597	.629	.663	.701	.741	.783	.850	.992
72	.213	.239	.265	.291	.317	.343	.370	.396	.423	.451	.479	.507	.534	.568	.600	.634	.672	.712	.754	.821	.963
73	.186	.212	.238	.264	.290	.316	.343	.369	.396	.424	.452	.480	.507	.541	.573	.607	.645	.685	.727	.794	.936
74	.159	.185	.211	.237	.263	.289	.316	.342	.369	.397	.425	.453	.480	.514	.546	.580	.618	.658	.700	.767	.909
75	.132	.158	.184	.210	.236	.262	.289	.315	.342	.370	.398	.426	.453	.487	.519	.553	.591	.631	.673	.740	.882
76	.105	.131	.157	.183	.209	.235	.262	.288	.315	.343	.371	.399	.426	.460	.492	.526	.564	.604	.652	.713	.855
77	.079	.105	.131	.157	.183	.209	.236	.262	.289	.317	.345	.373	.400	.434	.466	.500	.538	.578	.620	.687	.829
78	.053	.079	.105	.131	.157	.183	.210	.236	.263	.291	.319	.347	.374	.408	.440	.474	.512	.552	.594	.661	.803
79	.026	.052	.078	.104	.130	.156	.183	.209	.236	.264	.292	.320	.347	.381	.413	.447	.485	.525	.567	.634	.776
80	.000	.026	.052	.078	.104	.130	.157	.183	.210	.238	.266	.294	.321	.355	.387	.421	.459	.499	.541	.608	.750
81	—	.000	.026	.052	.078	.104	.131	.157	.184	.212	.240	.268	.295	.329	.361	.395	.433	.473	.515	.582	.724
82	—	—	.000	.026	.052	.078	.105	.131	.158	.186	.214	.242	.269	.303	.335	.369	.407	.447	.489	.556	.698
83	—	—	—	.000	.026	.052	.079	.105	.132	.160	.188	.216	.243	.277	.309	.343	.381	.421	.463	.530	.672
84	—	—	—	—	.000	.026	.053	.079	.106	.134	.162	.190	.217	.251	.283	.317	.355	.395	.437	.504	.645
85	—	—	—	—	—	.000	.027	.053	.080	.108	.136	.164	.191	.225	.257	.291	.329	.369	.417	.478	.620

Example: Total kw input of load from wattmeter reading 100 kw at a power factor of 60%. The leading reactive kva necessary to raise the power factor to 90% is found by multiplying the 100 kw by the factor found in the table, which is .849. Then 100 kw × 0.849 = 84.9 kva. Use 85 kva.

FIG. 6-27 Capacitor kilovars required to improve a load to a desired power factor.

Exception 2, plus the sum of the full-load currents of the other motors supplied by the feeder. It is possible for motors of different horsepower ratings to have the same rating of branch-circuit protective device, depending upon the type of motor and the type of protective device. If two or more motors in the group are of different horsepower ratings but the rating or setting of the branch-circuit protective device is the same for both motors, then one of the protective devices should be considered as the largest for the calculation of feeder overcurrent protection.

Because NEC Table 430-152 recognizes many different ratings of branch-circuit protective devices (based on the use of fuses or circuit breakers and depending upon the particular type of motor), it is possible for two motors of equal horsepower rating to have vastly different ratings of branch-circuit protection. For instance, for a 25-hp motor protected by non-time-delay fuses, Table 430-152 gives 300 percent of the full-load motor current as the maximum rating or setting of the branch-circuit device. Thus the nearest standard fuse that does not exceed this value is one of 225 A, which would be used for a motor with a 78-A full-load rating for branch-circuit protection. But for feeder protection the exception to Sec. 430-62 allows the maximum rating from Sec. 430-52, Exception 2, which permits 400 percent of full-load current when a non-time-delay fuse

Bus duct	Total hp	40 % Demand	95 % Eff.	Kw	Kva 80%	Kvar 80%	Kva 95%	Kvar 95%	Kvar diff.	Kvar use
1 A	364	146	153	109	136	82	115	35	47	45
1 B	374	150	157	117	142	85	124	38.4	46.6	45
1 C	461	185	194	145	181	109	153	47	62	60
2 A	735	(30%) 220	230	172	215	130	181	56	74	75
2 B	1300	(30%) 390	410	306	382	230	322	100	130	135
3 A	729	291	307	229	286	172	241	75	97	105
3 B	275	110	116	87	109	65	92	29	36	45

FORMULAS: Power Factor = $\frac{KW}{KVA}$ KW output of motor = HP x .746 $\cos \emptyset_1 = .8$ $\cos \emptyset_2 = .95$

$KVA = \frac{1.73 \times E \times I}{1000}$ KW input to motor = $\frac{HP \times .746}{Efficiency}$ $\sin \emptyset_1 = .6$ $\sin \emptyset_2 = .312$

KVA input to induction motor = $\frac{HP \times .746}{P.F. \times Eff.}$ $KVA = \frac{KW}{\cos \emptyset}$ KVAR = KVA x sin Ø

Power Factor = cos Ø

Power factor can be increased with capacitors or with synchronous motors. The following shows how a 75 kVA synchronous motor load (which draws current that "leads" the voltage instead of current that "lags" the voltage, as with an induction motor) was used to provide 45 kvar (reactive kilovolt amperes) to offset part of the 135 kvar of the induction motor load. The kW components add up to 290 kW, and the lagging kvar is reduced to 90 kvar.

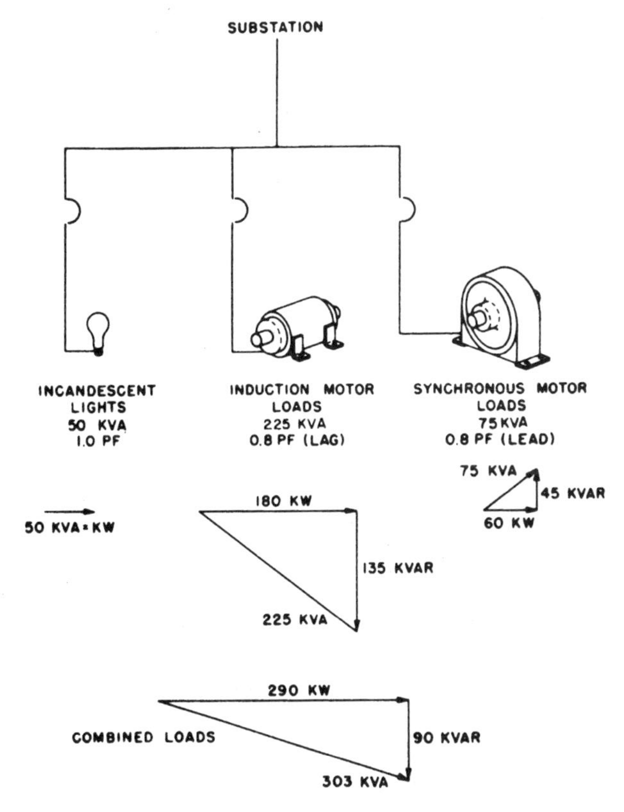

FIG. 6-28 In an industrial system, these design data were used to determine capacitor power-factor correction for plug-in busway to motor loads.

does not exceed 600 A. Thus, 400 percent × 78 A = 312 A; the nearest standard fuse that does not exceed this value is a 300-A short-circuit ground-fault protective device. Another motor of the same horsepower (and even of the same type) using time-delay fuses must be rated at not more than 225 percent of 78 A, which would mean using a 175-A fuse for calculating feeder short-circuit ground-fault protection. If the two 25-hp motors were of different types—one being a wound-rotor motor and the other a squirrel-cage induction motor—it would still be necessary to base the selection of the feeder protection on the maximum permitted value from Sec. 430-52, Exception 2, for the specific type being used, regardless of the horsepower rating of the motor (Fig. 6-29).

In large-capacity installations where extra feeder capacity is provided for load growth or future changes, the feeder overcurrent protection may be calculated on the basis of the rated current-carrying capacity of the feeder conductors. In some cases, such as where two or more motors on a feeder may be started simultaneously, feeder conductors may have to be larger than usually required for feeders to several motors.

The NEC calculation for selecting the size of a feeder overcurrent protective device is concerned with establishing the maximum setting or rating of the CB or fuse. If a lower value of protection is suitable, it may be used.

Example. In Fig. 6-30, 300-A fuses are used to protect the feeder conductors to a group of four motors; the fuses were selected as follows: The four motors supplied by

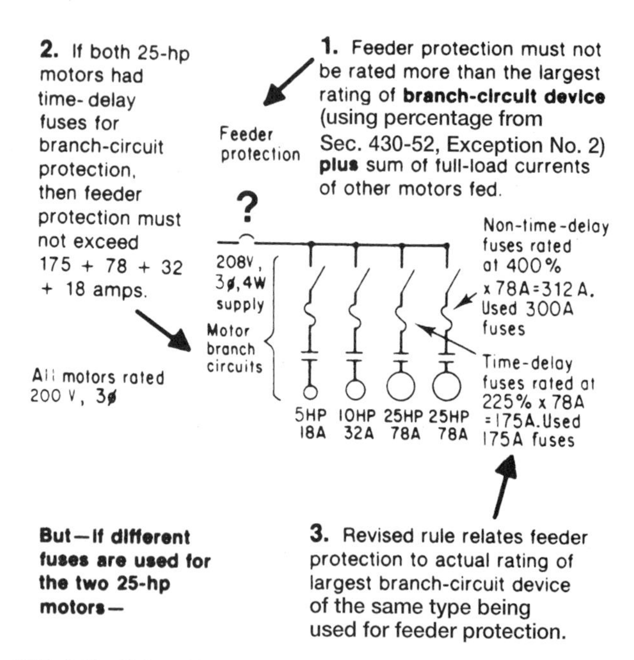

FIG. 6-29 Sizing of feeder protective device is based on the largest type of branch-circuit protective device, regardless of the actual motor horsepower ratings. *Example:* When time-delay fuse is used for feeder, 175 A + 78 A + 32 A + 18 A = 303 A; the nearest standard fuse that does not exceed this value is one of *300 A.* When non-time-delay fuse is used for feeder, 300 A + 78 A + 32 A + 18 A = 428 A; the nearest standard fuse that does not exceed this value is one of *400 A.*

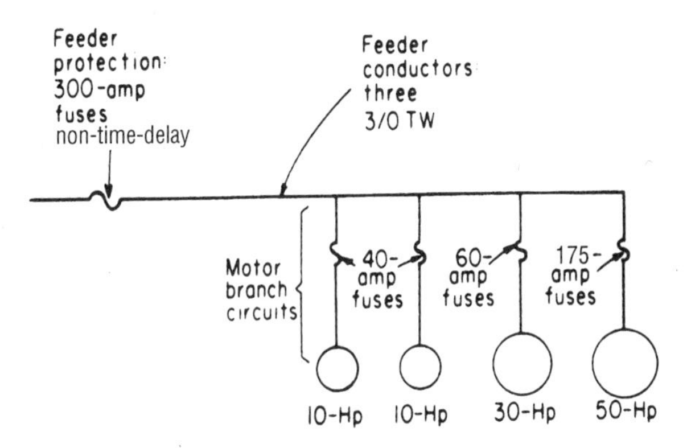

FIG. 6-30 Rating of feeder protection and size of feeder conductors are based on motor full-load currents.

the 3-phase, 440-V, 60-Hz feeder are not marked with a code letter (NEC Table 430-152) and are

- One 50-hp squirrel-cage induction motor (full-voltage starting)
- One 30-hp wound-rotor induction motor
- Two 10-hp squirrel-cage induction motors (full-voltage starting)

Step 1: Branch-Circuit Loads. From NEC Table 430-150, the motors have full-load current ratings as follows:

50-hp motor: 65 A
30-hp motor: 40 A
10-hp motors: 14 A

Step 2: Conductors. The feeder conductors must have a carrying capacity that is computed as follows:

$$1.25 \times 65 \text{ A} = 81 \text{ A}$$

$$81 \text{ A} + 40 \text{ A} + (2 \times 14 \text{ A}) = 149 \text{ A}$$

The feeder conductors must be at least No. 3/0 TW, No. 1/0 THW, or No. 1/0 RHH, XHHW, or THHN (copper) wires.

Step 3: Branch-Circuit Protection. The required overcurrent (branch-circuit) protection (from NEC Table 430-152 and Sec. 430-52) using non-time-delay fuses is found as follows:

1. The 50-hp motor must be protected at not more than 175 A (since 300 percent of 65 A is 3 × 65 A = 195 A and the next-lower standard size of fuse is 175 A).
2. The 30-hp motor must be protected at not more than 60 A (150 percent of 40 A).
3. Each 10-hp motor must be protected at not more than 40 A (300 percent of 14 A = 42 A).

Step 4: Feeder Protection. Using non-time-delay fuses, the maximum rating or setting for the overcurrent device protecting such a feeder must not be greater than the largest rating or setting of branch-circuit protective device from Sec. 430-52, Exception 2, for one of the motors of the group plus the sum of the full-load currents of the other motors. From the above, then, the maximum allowable size of feeder fuses is 250 + 40 + 14 + 14 = 318 A.

This calls for a maximum rating of 300 A for the motor feeder fuses, since 300 A is the nearest standard fuse rating that does not exceed the maximum allowable value of 318 A.

Note: There is no provision in NEC Sec. 430-62 for the use of the next-higher size, rating, or setting of protective device for a motor feeder when the calculated maximum rating does not correspond to a standard size of device.

Example. Figure 6-31 shows a feeder that supplies five motors as follows:

- One 40-hp wound-rotor induction motor
- Two 25-hp squirrel-cage induction motors (full-voltage start)
- One 10-hp squirrel-cage induction motor (full-voltage start)
- One 5-hp squirrel-cage induction motor (full-voltage start)

The calculation procedure involves these steps:

Step 1. The NEC full-load running currents of these motors are as follows (motors rated 208 V, 3-phase):

40-hp motor:	114.4 A
25-hp motors:	74.8 A
10-hp motor:	30.8 A
5-hp motor:	16.7 A

These current values are derived from NEC Table 430-150.

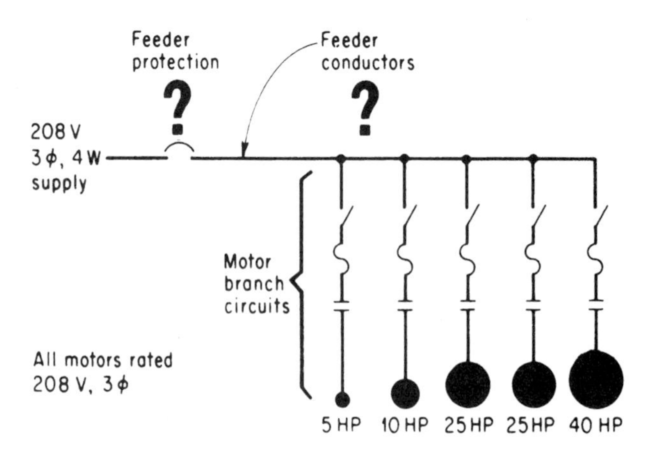

FIG. 6-31 Rating of a circuit breaker for motor feeder protection is determined by the largest rating of a branch-circuit circuit breaker in a case like this plus the sum of the full-load currents of the other motors.

Step 2. The minimum permissible ampacity of the feeder conductors supplying this group of continuously operating motors is the sum of the full-load running currents of all the motors plus 25 percent of the full-load current of the motor with the highest rated running current. (However, if some of the motors were operating on short-time, intermittent, periodic, or varying duty instead of all operating continuously, the sizing of the feeders would be modified as explained in NEC Sec. 430-24.)

Step 3. Based on the full-load running currents of these continuous-duty motors, the feeder conductors here must have an ampacity of not less than 340.1 A, that is, 114.4 A + 74.8 A + 74.8 A + 30.8 A + 16.7 A + (0.25 × 114.4 A).

Step 4. If THW aluminum conductors are used in conduit for the feeder, the smallest conductor size that will properly satisfy the need here is 700 kcmil (375 A).

Step 5. The maximum permissible rating of a standard inverse-time CB used for the overcurrent protection of this feeder is 300 percent of the full-load current rating of the 40-hp motor (which is 300 percent × 114.4 A = 343.2 A; the nearest standard rating that does not exceed 343.2 A is 300 A) plus the sum of the full-load currents of the other four motors of the group.

Step 6. Adding the largest overcurrent protective device of 300 A to the full-load currents of the other motors (74.8, 74.8, 30.8, and 16.7 A) gives 497.1 A.

Step 7. The maximum setting or rating of a standard, time-delay-type, molded-case CB that may be used for the protection of this feeder is 450 A, since Sec. 430-62 does not recognize the use of the next-larger size of CB.

Power and Lighting Feeders

The conductors and protection for a feeder to both a motor load and a lighting and/or appliance load must be sized on the basis of both loads. The rating or setting of the overcurrent device must be sufficient to carry the lighting and/or appliance load plus the rating or setting of the motor branch-circuit protective device if only one motor is supplied, *or* plus the highest rating or setting of branch-circuit protective device for any one motor plus the sum of the full-load currents of the other motors, if more than one motor is supplied.

Example. Figure 6-32 shows a feeder arrangement for a combination power and lighting load. As shown, the feeder supplies a 277-V lighting panel and four 3-phase motors. Such considerations as voltage drop, I^2R loss, spare capacity, and lamp dimming on motor starting would have to be accounted for in determining the actual wire size and overcurrent protection to use for the job. The circuiting shown would be safe, but perhaps not as efficient or effective as it could be.

The feeder conductors that will carry both the lighting-load current and the motor-load current must have sufficient capacity for the total load. The basic design steps are as follows:

Step 1. The total load is the sum of the motor load and the lighting load:

$$\text{Motor load} = 65\text{ A} + 40\text{ A} + 14\text{ A} + 14\text{ A} + (0.25 \times 65\text{ A}) = 149\text{ A per phase}$$

$$\text{Lighting load} = 120\text{ A} \times 1.25 = 150\text{ A per phase}$$

$$\text{Total load} = 149\text{ A} + 150\text{ A} = 299\text{ A per phase}$$

Note that, because the lighting load of 120 A per phase is a "continuous" load, NEC Sec. 220-10(b) requires that feeder conductors carrying that load have ampacity at least equal to 125 percent of the load current. The multiplier 1.25 used above satisfies the requirement. But only the phase conductors of the feeder are subject to that 125 percent factor;

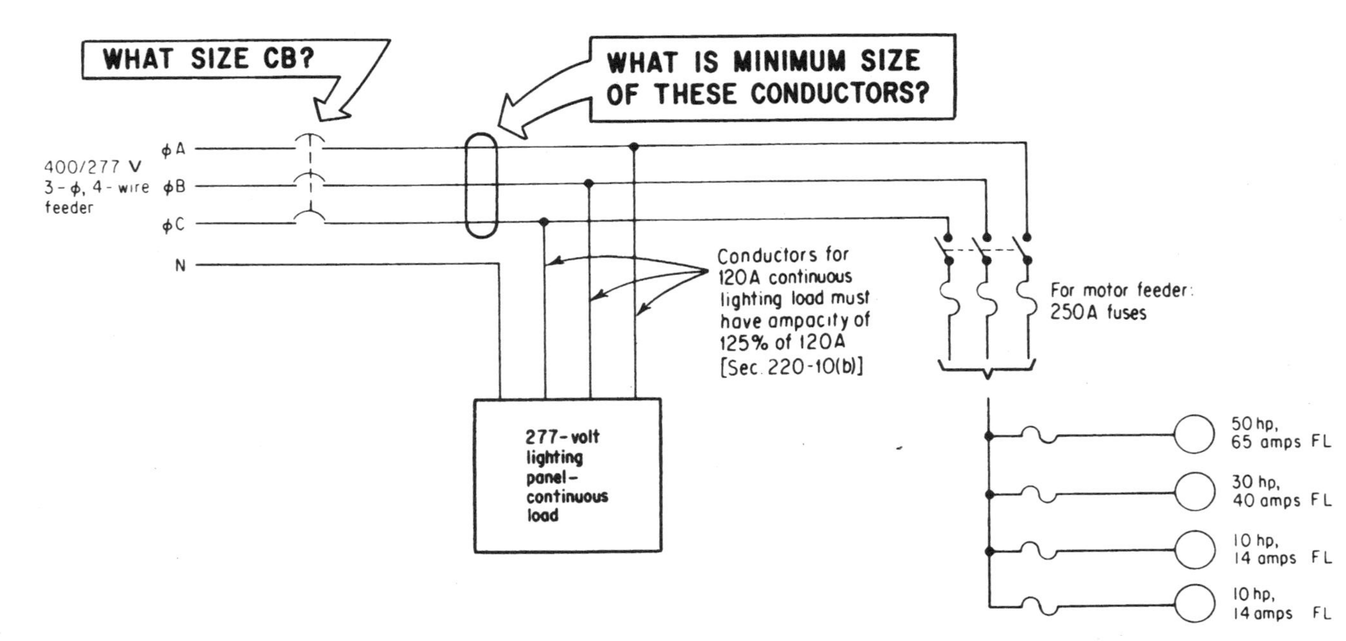

FIG. 6-32 Conductors and protection for a power and lighting feeder must be sized to satisfy both loads.

the neutral conductor of the feeder requires only an ampacity of 120 A. The 125 percent continuous-load factor is not applied to the feeder neutral because the neutral does not connect to the terminal of a switch, CB, or other device for which heating would be a problem under continuous load.

Step 2. NEC Table 310-16 shows that a combined power and lighting load of 299 A can be served by the following copper conductors:

500-kcmil TW (320 A)

350-kcmil THW (310 A)

Table 310-16 shows that this load can be served by the following aluminum or copper-clad aluminum conductors:

700-kcmil TW (310 A)

500-kcmil THW, RHH, or THHN (310 A)

Step 3. The protective device for a feeder supplying a combined motor and lighting load may not have a rating greater than the sum of the maximum rating of the motor feeder protective device and the lighting load; these are obtained as follows:

1. The maximum rating of the motor feeder protective device is the rating or setting of the largest branch-circuit device of any motor of the group being served plus the sum of the full-load currents of the other motors; this is

$$200 \text{ A (50-hp motor)} + 40 \text{ A} + 14 \text{ A} + 14 \text{ A} = 268 \text{ A}$$

This calls for a maximum standard rating of 250 A for the motor feeder fuses (the nearest standard fuse rating that does not exceed the maximum permitted value of 268 A).

2. For the lighting load, $120 \text{ A} \times 1.25 = 150 \text{ A}$.

A CB for the combined load must be rated at $268 \text{ A} + 150 \text{ A} = 418 \text{ A}$ maximum. This calls for a 400-A CB (the nearest standard rating that does not exceed the 418-A maximum).

Again, the NEC *does not* have a provision which permits the use of the next-higher size, rating, or setting of protective device for a motor feeder when the calculated maximum rating does not correspond to a standard size of device.

Example. Figure 6-33 shows a layout in which a main feeder is run to a point where it is tapped (such as at an auxiliary gutter) by two subfeeders—one for a lighting load and one for a motor load. The main switch for the combined feeder might be in a switchboard or panelboard, or it could be a main service switch (disconnect and protection at a service entrance). The conductors for the combined feeder and the fuses in the 200-A combined feeder switch are sized on the basis of the following loads: The motor load is four 3-phase, 230-V, squirrel-cage induction motors, designed for 40°C temperature rise, marked with code letter H, started across the line, and rated as follows:

- One 10-hp motor
- One $7\frac{1}{2}$-hp motor
- Two $1\frac{1}{2}$-hp motors

The lighting load is a 20-kW, single-phase, 115-V load.

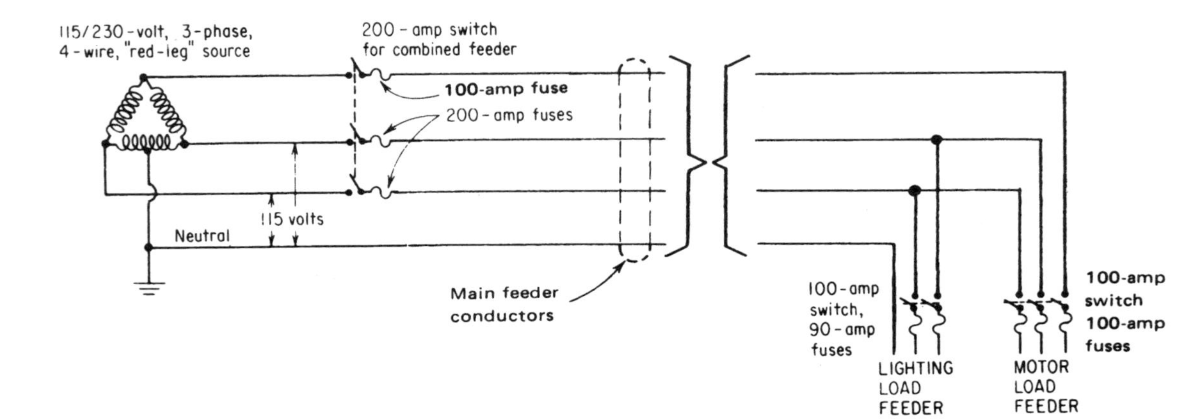

FIG. 6-33 Sizing of conductors and protection must be based on actual phase loadings.

Step 1. The average full-load motor currents (from NEC Table 430-150) are

10-hp motor: 28 A
$7\frac{1}{2}$-hp motor: 22 A
$1\frac{1}{2}$-hp motors: 5.2 A each

Step 2. The conductors for the motor load feeder must be selected to have an ampacity based on the following:

10 hp: 1.25 × 28 A	35 A
$7\frac{1}{2}$ hp	22 A
$1\frac{1}{2}$ hp: 2 × 5.2 A	10.4 A
Total	67.4 A

The required ampacity of 67.4 could be provided by

Three No. 4 TW in 1-in conduit, or
Three No. 4 THW in 1-in conduit, or
Three No. 4 RHH (without covering) in 1-in conduit, or
Three No. 4 THHN in 1-in conduit

Step 3. The motor feeder protection (Sec. 430-62), using circuit breakers for the branch circuits but fuses in the feeder, is sized as follows:

10 hp: 28 A × 2.5	70 A (rating of branch CB)
$7\frac{1}{2}$ hp	22 A
$1\frac{1}{2}$ hp: 2 × 5.2 A	10.4 A
Total	102.4 A

The maximum size of the feeder fuses is 102.4 A. Therefore, a 100-A switch with 100-A non-time-delay fuses or smaller time-delay fuses would be used.

Note: NEC Sec. 430-62 does not permit the motor feeder fuses to be upsized to the next standard rating above 102.4 A, or 110 A. The fuses must be selected at 100 A, the next-*lower* rating.

Step 4. For the lighting load, the full-load current would be

$$\frac{20 \times 1000\ \text{W}}{230\ \text{V}} = 87\ \text{A}$$

A 100-A switch with two 90-A fuses would be used for a lighting load that is not continuous, that is, one where the full load is not on for 3 h or more. [If the 87-A load were continuous, a fused switch would be selected so that the load was not over 80 percent of the fuse rating. This means that the minimum fuse size would be 87 A × 1.25 = 108.75 A, so a 110-A fuse would be used for each phase in a 200-A switch (per NEMA standards).]

Thus, three No. 2 TW, THW, RHH, or THHN in $1\frac{1}{4}$-in conduit or in 1-in conduit (for THHN), would be used. (For a continuous 87-A load, where the fuse size would be increased to 110 A, the feeder conductors would have to be rated for at least 110 A, as are No. 1 TW and No. 2 THW, RHH, or THHN.)

Step 5. Steps 1 to 4 cover the sizing of the individual feeders—one for the motor load and one for the lighting load. A single main feeder which supplies the combined motor and lighting load will have the 3-phase motor load on each of its three phase conductors and the lighting load on only two of its phase conductors and a neutral. A 3-phase, 4-wire, "red-leg" circuit would be designed as follows:

Step 6. The two phase conductors which supply both the motor load and a noncontinuous lighting load must have a minimum capacity of

Motor load	67.4 A
Lighting load	87 A
Total	154.4 A

Each of these two conductors must be at least No. 3/0 TW (165 A) or No. 2/0 THW (175 A) or RHH or THHN (185 A). [If the lighting load were continuous, a value of 108.75 A (125 percent of 87 A) would be used instead of 87 A in the previous calculation.]

Step 7. The third phase conductor of the main combined feeder serves only the motor and must have a rating of at least 67.4 A, which calls for No. 4 TW, THW, RHH, or THHN as noted in step 2.

Step 8. Under the assumption that the 20-kW single-phase lighting load is all 115-V loads, equally balanced from hot legs to neutral, the neutral must be rated for the maximum unbalance; this is 87 A, the same as the load on each hot leg. This neutral conductor would then be a No. 2, TW, THW, RHH, or THHN, as shown in step 4. (If the lighting load were continuous, the neutral could still be sized at 87 A, although the phase legs would require an ampacity at least equal to 12.5 $\times$ 87 A in such a case.)

Step 9. Based on the loads on the three hot legs—154, 154, and 67.4 A—a 3-pole, 200-A switch is needed for this combined feeder.

Step 10. Overcurrent protection for the two hot legs that carry the motor load and the lighting load is sized according to NEC Sec. 430-63:

Motor load	102.4 A setting
Lighting load	87 A
Combined load	189.4 A

Thus, the two fuses in the 200-A switch protecting the combined load must have a rating not greater than 200 A. (NEC Sec. 430-63 may be interpreted as setting a maximum fuse rating of 190 A—that is, 100 A for the rating of the motor feeder fuses plus 90 A for the lighting feeder fuses—thereby requiring the use of the next-smaller size fuse, or 175 A.)

Step 11. The fuse which protects the third hot leg—the one serving only the motor load—must be rated not over 100 A.

Note: This sizing calculation would not change at all if the feeder constituted the service-entrance conductors to a building with the given load, rather than the feeder to the combined load within the building.

CHAPTER 7
TRANSFORMERS (TO 600 VOLTS)

The wide availability of economical, efficient, and highly effective dry-type transformers has strongly promoted the use of such transformers to carry electric power, at higher-than-utilization voltages, as close as possible to the center of a layout of utilization equipment—lighting, appliances, and receptacle loads.

Widespread acceptance and application of industrial medium-voltage systems (above 600 V and usually up to 35 kV) has stimulated the use of higher voltages in electrical systems for commercial and industrial buildings. Not too many years ago, 3-wire, single-phase, 120/240-V distribution systems and 3-phase, 4-wire, 120/208-V distribution systems were standards for nonindustrial buildings. Today, although these lower-voltage systems still find wide and efficient application, the 480Y/277-V, 3-phase, 4-wire system is established as the preferred secondary voltage system for use where substantial motor loads and general-area fluorescent or other electric-discharge lighting must be supplied.

Tremendous growth in the size of electrical loads in commercial buildings was responsible for the development of the 480Y system, which has many of the characteristics of industrial-type systems. Air-conditioning loads and business and other machine loads increased the ratio of power loads to lighting loads, calling for the type of load center circuit treatments and layouts which are used in industrial plants to economically and efficiently serve heavy motor loads.

Higher-voltage feeders to motor loads and to step-down transformers for lighting and receptacle circuits proved to be the ideal solution. Less copper is needed to distribute the heavy power requirements, and voltage drop and other losses are effectively minimized. The wide availability of extensive lines of lighting equipment which operates on 277-V circuits has also contributed greatly to the success of this system.

Studies have shown that a typical 480Y system can provide savings of more than $40 per kilovoltampere demand compared to a 120/208-V system handing the same load. Many installations over the past years—in schools, office buildings, industrial plants, shopping centers, and sports arenas—have convincingly proved the economic and operating advantages of the 480/277-V system.

The 480Y system was developed to meet the requirements of commercial building load conditions. Usually, the general lighting is fluorescent and can be served by 277-V circuits. Motors for air-conditioning compressors, circulating fans, elevators, and pumps make up an average load of about 4 VA/ft^2. These motors can be more efficiently and economically supplied at 480 V than at lower voltages, and 480-V motors are less expensive than lower-voltage motors of the same horsepower rating. The combined power and general lighting loads average between 5 and 15 W/ft^2. Receptacle and miscellaneous loads—desk lamps, local lights, business machines, appliances, water coolers, etc.—average only about 0.5 to 2 W/ft^2.

As can be seen from the above, about 80 percent of a building load may be served directly by 480/277-V feeders. The 120-V circuits may be provided either by using separate 120/208-V substations (fed by medium-voltage feeders of 4160 V, 13.2 kV, etc.) or by using dry-type transformers (480 to 120/240-V or 480-V, 3-phase to 120/208-V) installed locally at the center of each concentration of 120-V loads. The latter method offers greater economy, even when the 120-V loading is as high as 5W/ft^2.

The design of electrical systems utilizing dry-type transformer applications must include careful consideration of a number of transformer characteristics. These are discussed in the next several paragraphs.

VOLTAGE RATING

The primary and secondary voltage ratings of a transformer are determined very simply from the voltage rating of the supply to the primary winding and the required voltage level on the secondary. Both single- and 3-phase transformers are made with a wide variety of voltage ratings to afford ready application in standard system configurations and even in nonstandard arrangements.

The two basic selection factors are secondary voltage and secondary current. Together they determine the kilovoltampere capacity of the transformer, which is its load-handling ability. The two terms are defined as follows:

- The *rated secondary voltage* of a constant-potential transformer is the voltage at which the transformer secondary is designed to deliver the rated kilovoltampere capacity.
- The *rated secondary current* of a constant-potential transformer is the secondary current obtained by dividing the rated kilovoltampere capacity by the rated secondary voltage.

Transformer selection must first focus on those two basic ratings, depending on the load to be supplied. Then the primary voltage rating of the transformer must be matched to the voltage of the primary feeder circuit to the transformer.

To match actual system conditions, transformers are available that allow some variation from the nominal primary voltage but still produce the required secondary voltages. Figure 7-1 shows some typical transformer primary tap connections for varying input voltages. On general-purpose transformers, the primary is commonly provided with a number of taps for input voltages that vary over a range of 15 percent; for instance, two 2½ percent taps above normal voltage and four 2½ percent taps below normal voltage may be provided. On a transformer with a 480-V primary, then, the lowest of these taps would be connected to a primary feeder which operated at 432 V to obtain the rated secondary output of 240/120 V (since 480 V $-$ 4 $\times$ 2.5 percent of 480 V $=$ 432 V. If the 432-V primary supply were connected to the 480-V transformer terminals or to any of the other taps, the secondary voltage would be lower than 240/120 V. The taps thus compensate for system voltage drops.

General-purpose transformers have no-load tap changers, requiring that the transformer be deenergized when tap changes are made. Invariably, suitable taps for the given primary voltage are set once and left alone, unless some change in load conditions alters the voltage level at the transformer primary. On large transformers serving fluctuating demand loads, automatic tap changers which operate under load can be used on the transformer to maintain a constant voltage level at the transformer secondary terminals or at a remote point on the secondary output feeder.

The selection of transformer taps should be based on the no-load or maximum-voltage conditions at the transformer. Taps above or below the nominal rated voltage should

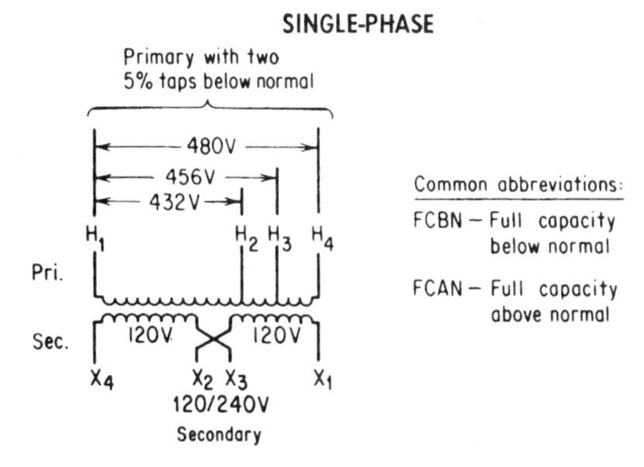

FIG. 7-1 Transformer primary taps provide for matching the transformer input voltage rating to the voltage of the supply circuit.

be selected to accommodate anticipated variations in supply voltage and, thereby, provide the required no-load secondary voltage. This requires that voltage drops in secondary circuits be taken into account to arrive at proper voltage levels at utilization devices.

Remember that transformer taps change the voltage transformation by changing the turns ratio of the transformer. Because of this, the kilovoltampere rating of a transformer is not changed when taps are changed.

For a given voltage rating, a transformer winding has a certain current rating, which determines the kilovoltampere rating of the transformer. The current (ampere) rating is a maximum value determined by the impedance of the winding and the capability of the transformer to dissipate heat. The conditions of use must not place a higher current on either the primary or the secondary winding. If the voltage input to a set of primary transformer terminals of given voltage rating is reduced below that voltage rating, the maximum kilovoltampere rating is also reduced, because the current cannot be increased to maintain constant kilovoltamperes. And if an overvoltage is applied, overheating will result and will shorten the life of the transformer.

Single-phase transformer fundamentals

CIRCUIT CONDITIONS

E_p = Primary voltage
I_p = Primary current
E_s = Secondary voltage
I_s = Secondary current
E_{vd} = Voltage drop in circuit conductors
E_L = Voltage across load
T_p = Number of turns on primary
T_s = Number of turns on secondary

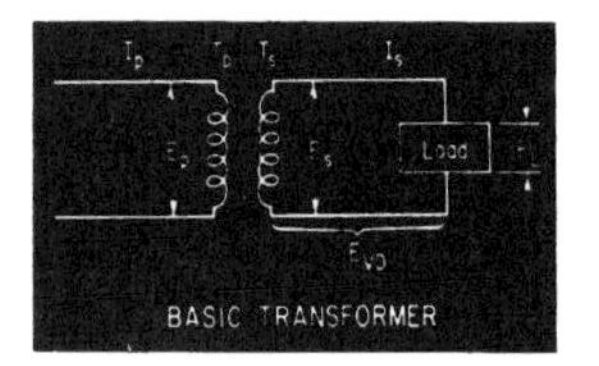

PRACTICAL RELATIONSHIPS

$\frac{E_p}{E_s} = \frac{T_p}{T_s}$ $T_p \times I_p = T_s \times I_s$ $\frac{I_p}{I_s} = \frac{T_s}{T_p}$ $\frac{I_p}{I_s} = \frac{E_s}{E_p}$ $I_p \times E_p = I_s \times E_s$ Transformer kva rating $= \frac{I_p \times E_p}{1000} = \frac{I_s \times E_s}{1000}$

Three-phase transformers currents and voltages

For any delta connection (pri. or sec.):

$$I_L = \frac{\text{Rated transformer kva} \times 1000}{1.73 \times E_L}$$

For any wye connection (pri. or sec.):

$$I_L = \frac{\text{Rated transformer kva} \times 1000}{1.73 \times E_L}$$

Where, I_L = Line current (pri. or sec.) of transformer
I_w = Rated current of each transformer winding
E_L = Rated voltage (phase-to-phase) of transformer
E_w = Rated voltage of each transformer winding

NOTE: This data applied to 3-phase transformers and to single-phase transformers connected for 3-phase use.

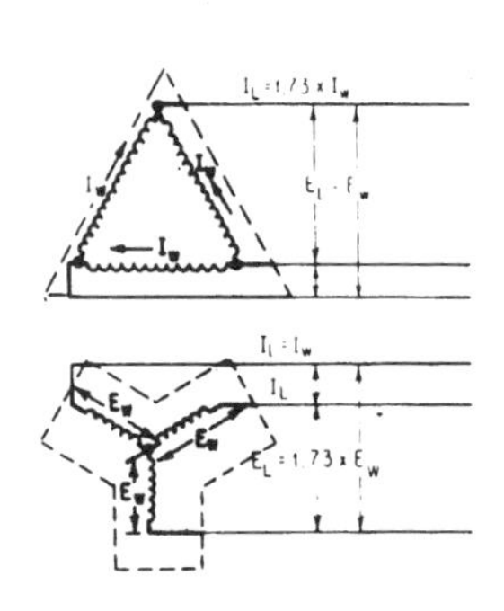

FIG. 7-2 Transformer circuit design is based on these relationships.

Figure 7-2 shows the basic current, voltage, and kilovoltampere relationships for single-phase and 3-phase transformers. These relationships are used in design calculations as shown in the following examples.

Example. What are the primary and secondary current ratings of a 50-kVA transformer for stepping 480 V down to 240 V?

First use

$$I_P = \frac{\text{transformer kVA rating} \times 1000}{E_P}$$

$$= \frac{50 \times 1000}{480} = 104 \text{ A}$$

Then

$$I_s = I_p \frac{E_p}{E_s} = 104 \left(\frac{480}{240}\right) = 208 \text{ A}$$

Example. What are the primary and secondary current ratings of a 500-kVA, 3-phase transformer stepping 480-V delta down to 120/208-V wye?

On the delta primary, with E_L = 480 V,

$$I_L = \frac{500 \times 1000}{1.73 \times 480} = \frac{500{,}000}{830} = 603 \text{ A}$$

On the wye secondary, with $E = 208$ V,

$$I_L = \frac{500 \times 1000}{1.73} \times 208 = \frac{500{,}000}{360} = 1388 \text{ A}$$

Note: The above formulas do not take transformer losses into consideration and, therefore, provide only approximate determinations. However, because transformers have extremely high efficiencies (for example, 98 percent), the calculations shown are generally acceptable for practical applications of transformers.

One consideration in selecting a transformer is the *impedance* of the unit, which is indicative of its ability to supply short-circuit currents into faults on the load side of the transformer. The impedance of a transformer is the opposition which the transformer presents to the flow of short-circuit current through it.

Every transformer has an impedance, which is generally expressed as a percentage—the percentage of the normal rated primary voltage which must be applied to the transformer to cause rated full-load current to flow in the short-circuited secondary. For instance, if a 480- to 120-V transformer has an impedance of 5 percent, then 5 percent of 480 V, or 24 V, applied to the primary will cause rated load current to flow in the short-circuited secondary. If 5 percent of the primary voltage will cause such a current, then 100 percent of the primary voltage will cause 100/5 (or 20) times the rated full-load current to flow through a solid short circuit on the secondary terminals.

From the foregoing, it can be seen that the lower the impedance of a transformer of given kilovoltampere rating, the higher the short-circuit current which it can deliver. Consider two transformers, both rated at 500 kVA. Assume the rated secondary load current is the same for both transformers. If one transformer is rated at 10 percent impedance, it can supply 100/10 (or 10) times the rated secondary current into a short circuit on its secondary terminals. If the other is rated at 2 percent impedance, it can supply 100/2 (or 50) times the rated secondary current into a short circuit on its secondary terminals. Thus, the second transformer can supply 5 times as much short-circuit current as the first, even though both have the same kilovoltampere load-handling ability. Common impedance values for general-purpose transformers are between 3 and 6 percent.

Voltage regulation is another transformer consideration, and impedance is related to it. Voltage regulation is a measure of how the secondary voltage of a transformer varies as the load on the transformer varies from full load to zero, with the primary voltage held constant.

Voltage regulation is expressed as a percentage that is calculated as the no-load voltage minus the full-load voltage divided by the full-load voltage times 100 percent. If a transformer has a no-load secondary voltage of 240 V, and the voltage drops to 220 V when the transformer is fully loaded (supplying the rated current output), then the regulation percentage is

$$(240 - 220) \div 220 \times 100 \text{ percent} = 9 \text{ percent}$$

This indicates that 9 percent of the secondary voltage is being dropped across the internal impedance of the transformer. It is obvious, then, that the higher the impedance of a transformer, the greater the drop from no-load to full-load voltage and the higher the percentage regulation. It is generally desirable to keep regulation as low as possible, to minimize variations in voltage as load-current demand varies. Typical regulation values are between 2 and 4 percent.

Low regulation becomes very important when the transformer supplies varying load demands for utilization equipment which is sensitive to voltage changes. In cases where the load current on the transformer is relatively constant, however, the transformer taps or the primary supply conditions can be adjusted to provide fixed compensation for the voltage drop in the transformer. The voltage level can be set as required at the utilization devices, and it will not change because the load current is not changing. In such cases, acceptable voltage conditions at the utilization devices can be achieved with transformers of either low or high regulation.

But it should be pointed out that the high-regulation transformer provides a high internal impedance, which reduces the level of possible short circuits on the secondary. As a result, fuses, circuit breakers, and other protective devices need not be rated for as high interrupting capacities as they would need if a low-regulation (and low-impedance) transformer were used.

KILOVOLTAMPERE RATING

The rated kilovoltampere output of a transformer is that output which it can deliver for a specified time, at the rated secondary voltage and rated frequency, without exceeding a specified temperature rise based on insulation life and ambient temperature. The output that a transformer can deliver without objectionable deterioration of the insulation may be more or less than the rated output, depending upon ambient temperature and load cycles.

Transformers may be loaded above their kilovoltampere ratings with no sacrifice of life expectancy only in accordance with prescribed temperature testing or short-time loading data from the manufacturer. In addition, the load that can be carried by self-cooled transformers may be increased considerably by the use of fans or other auxiliary cooling equipment. Such an application may be advantageous for short-time peak loading, although the design characteristics of the transformer, including increased voltage regulation, must be considered.

The basic loading conditions, for which the normal life expectancy of a transformer is determined, are as follows:

1. The transformer is continuously loaded at its rated kilovoltamperes and rated voltage.
2. The average temperature of the cooling air during any 24-h period is 30°C.
3. The temperature of the cooling air at no time exceeds 40°C.

The rated kilovoltamperes, rated voltage, and rated primary and secondary currents are considered to be maximum values. The kilovoltampere rating is based on drawing the rated current at the rated voltage; but, as noted previously, the kilovoltampere handling capacity cannot be kept constant by, say, operating at a lower voltage and a higher current. This capacity is determined by the maximum current the transformer can carry without reaching a dangerous temperature, and that current cannot be exceeded. Moreover, note that transformers are rated in *kilovoltamperes,* so that their load-handling ability is dependent only upon current and voltage, without regard for the effect of circuit power factor. The power-handling capacity (rated kilowatt load) of a transformer, therefore, is greater for high-power-factor loads than for low-power-factor loads.

A transformer selected for a particular application must have a kilovoltampere rating at least equal to the kilovoltampere rating of load to be supplied. But modern engineering practice, based on experience with load growth in electrical systems, demands the inclusion of spare capacity in every transformer installation. Of course, this spare capacity will be provided automatically if the feeders and branch circuits supplied by the transformer have spare capacity and the transformer is sized to accommodate that extra capacity.

In selecting transformers to supply motor loads, several general rules are used to determine the required capacity. One such rule says that a transformer should have at least 1 kVA of capacity for each horsepower of motor load. Another guide suggests at least $1\frac{1}{4}$ kVA per horsepower of motor load for motors rated 5 hp and above. Motors over 50 hp may be satisfactorily served by more or less than $1\frac{1}{4}$ kVA/hp, depending upon the torque characteristics (NEMA design) of the particular motor, the horsepower rating of the motor, the load characteristics, whether or not reduced-voltage starting is used, and what type of reduced-voltage starter is used. Of course, greater accuracy can be achieved in matching transformer capacity to a number of motors if the nameplate current ratings of the motors are added and then multiplied by the rated voltage to obtain the required kilovoltamperes. Effective, economical, and reliable operation of large motors may often depend upon selection of a transformer with the correct kilovoltampere rating and internal impedance for the particular conditions of motor starting, duty cycle, and loading. Early engineering assistance from a transformer manufacturer can prevent many problems in motor applications.

Understanding the *K Factor*

The *K* factor already plays a significant role in the work of the professional engineer, electrical contractor, building management, and maintenance personnel. However, it is not, at present, clearly understood by very many people. The application of the *K* factor to electrical distribution systems must be understood first as it relates to the rated device (the transformer in this case) and the load connected to that transformer.

The following is one effort to explain the *K* factor, what it is and how it is used. We all know that nonlinear loads cause harmonic currents to flow in the circuit conductors. These currents can constitute a significant amount of the total load current and can adversely affect other components of the electrical network.

The *K* factor is computed by multiplying the square of the individual harmonic current by the square of that current's harmonic number. These products are then summed to produce the *K* factor. In this case, because we are using percentage values rather than actual current values, the sum of the products of harmonic and current number is divided by the sum of the percentage units squared (Fig. 7-3).

Using the values shown in Fig. 7-3, the *K* factor is computed as follows: The harmonic number (order) is shown in column (a), and the harmonic current, as a percentage of the fundamental for each harmonic order is expressed as a decimal value in column (c). Both values—those from columns (a) and (c)—are squared, and the results are shown in columns (b) and (d), respectively. The values shown in columns (b) and (d) are then multiplied, and the product is given in column (e). Next, the individual values in column (e) are added, which in this case equals 8.47. But, as indicated, because percentage values for the harmonic currents are used instead of the actual current values of a specific harmonic order, the total of 8.47 must be divided by the sum of values shown in column (d) to give the *K* factor for that particular loading. Therefore, in this case the *K* factor equals 8.47/1.61, or 5.25

The chart in Fig. 7-4 illustrates the computed *K* factor for varying individual percentages of harmonic currents at the odd harmonic orders, which are more commonly observed in the real world of electrical distribution. Single-phase nonlinear loads produce large amounts of third harmonic current and moderate amounts of fifth (and sometimes seventh) harmonic currents.

Harmonic	Harmonic Squared	Current % Fund.	Current % Fund. Squared	Product (b)*(d)
(a)	(b)	(c)	(d)	(e)
1	1	1.00	1.00	1.00
3	9	0.70	0.49	4.41
5	25	0.35	0.12	3.06
7	49	0.00	0.00	0.00
9	81	0.00	0.00	0.00
11	121	0.00	0.00	0.00
13	169	0.00	0.00	0.00
			1.61	8.47
		Total Harmonic Distortion		78.3%
		"K" Factor		5.25

FIG. 7-3 Typical *K* factor computation.

Percent of Fundamental	Harmonic					
	3	5	7	9	11	13
10	1.08	1.24	1.48	1.79	2.19	2.66
20	1.31	1.92	2.85	4.08	5.62	7.46
30	1.66	2.98	4.96	7.61	10.9	14.9
40	2.10	4.31	7.62	12.0	17.6	24.2
50	2.60	5.80	10.6	17.0	25.0	34.6
60	3.12	7.35	13.7	22.2	32.8	45.5
70	3.63	8.89	16.8	27.3	40.5	56.2
80	4.12	10.4	19.7	32.2	47.8	66.6
90	4.58	11.7	22.5	36.8	54.7	76.2
100	5.00	13.0	25.0	41.0	61.0	85.0

FIG. 7-4 The *K* factor computation for various harmonics.

The most interesting aspect shown in the chart (Fig. 7-4), is that the same percentage of fundamental current at higher harmonic orders will produce a higher *K* factor. For example, 50 percent of the fundamental as third harmonic produces a *K* factor of approximately 2.6, while 50 percent of the fundamental as seventh harmonic produces a *K* factor of 10.6. Careful examination of the chart and the accompanying graph (Fig. 7-5) will provide insight into the effect that different values of harmonic current at specific harmonic orders have on the *K* factor. It becomes immediately clear that higher-order harmonics can produce very high *K* factors. For example, 3-phase nonlinear loads, such as six-pulse variable-frequency drives, can produce *K* factors of 16 and higher.

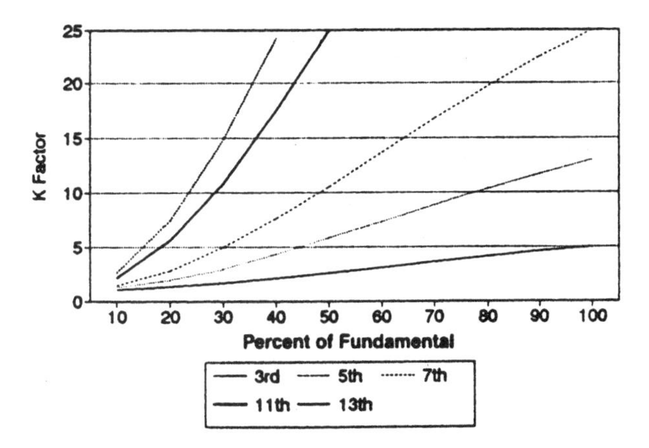

FIG. 7-5 Effects of different harmonic orders on the *K* factor.

TRANSFORMER PROTECTION

An important element in the design of transformer layouts for electrical distribution systems is the provision of effective overcurrent protection for each transformer. With regard to transformer protection, careful observance of the rules of NEC Sec. 430-3(b) will ensure both compliance with the Code and complete design adequacy for transformers rated up to 600 V on their primaries.

Figure 7-6 shows the *basic* rule as applied to protection of the dry-type transformers used for power step-down to 120/240 V, single-phase or to 120/208 V, 3-phase, 4-wire: Overcurrent protection for a dry-type transformer is provided by fuses or CBs rated at not more than 125 percent of the transformer primary full-load current (TPFLC) to protect the circuit conductors that supply the transformer primary. These circuit conductors must have an ampacity that is properly protected by the rating of the overcurrent protection.

In the layout in Fig. 7-6, a CB or set of fuses rated not over 125 percent of the transformer rated primary full-load current provides all the overcurrent protection required by the NEC for the transformer. This overcurrent protection is in the feeder circuit to the transformer, and it is logically placed at the supply end of the feeder so it may also provide the overcurrent protection required for the primary feeder conductors. There is no limit on the distance between the primary protection and the transformer. When the correct maximum rating of transformer protection is selected and installed at any point on the supply side of the transformer (either near to or far from the transformer), the feeder circuit conductors must be sized so that the CB or fuses selected will provide the proper protection as required for the conductors. The ampacity of the feeder conductors must be at least equal to the ampere rating of the CB or fuses unless Exception 1 of Sec. 240-3 is satisfied. That is, when the rating of the overcurrent protection selected is not more than 125 percent of the rated primary current, the primary feeder conductor may have an ampacity such that the overcurrent device is the next-higher standard rating.

The rules for protecting a 600-V transformer with a CB or set of fuses in its primary circuit are given in Fig. 7-7. Note that, for transformers with rated primary current of 9 A or more, "the next higher standard" rating of protection may be used, if needed. Figure 7-7 shows the absolute maximum values of protection for smaller transformers.

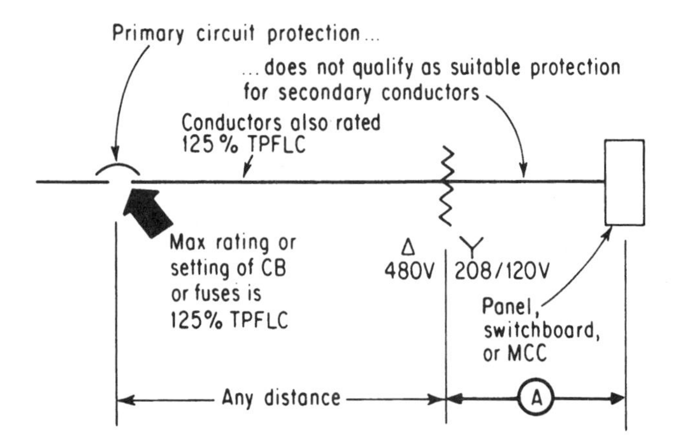

Distance "A" from transformer to panel is limited to 10 or 25 ft, subject to the requirements of Section 240-21 (b) and (d). If overcurrent protection is placed at the transformer secondary connection to protect secondary conductors, the circuit can run any distance to the panel. If the panel fed by the transformer secondary is a lighting panel and requires main protection, the protection must be on the secondary [Section 384-16(d)].

FIG. 7-6 Most common layout for protection of dry-type 600-V transformers.

When the factor 1.67 (167 percent) or 3 (300 percent) is used, if the resultant current value is not exactly equal to a standard rating of fuse or CB, then the next-*lower* standard rated fuse or CB must be selected.

It is important to note that transformer primary circuit protection is not acceptable as suitable protection for the secondary circuit conductors—even if the secondary conductors have an ampacity equal to the ampacity of the primary conductors times the primary-to-secondary voltage ratio. On 3- and 4-wire transformer secondaries, it is possible that an unbalanced load may greatly exceed the secondary conductor ampacity that is selected assuming balanced conditions. Because of this, the NEC does not permit the protection of secondary conductors with overcurrent devices operating from the primary through a transformer having a 3-wire or 4-wire secondary. For other than 2-wire–to–2-wire transformers, protection for secondary conductors has to be provided completely separately from any primary-side protection. Section 384-16(d) states that required main protection for a lighting panel on the secondary side of a transformer must be located on the secondary side. However, Exception 5 of Sec. 240-3 permits the secondary circuit from a transformer to be protected by means of fuses or a CB in the primary circuit to the transformer—if the transformer has no more than a 2-wire primary circuit and a 2-wire secondary. As shown in Fig. 7-8, the 2-to-1 primary-to-secondary turns ratio of the transformer allows 20-A primary protection to protect against any secondary current in excess of 40 A, thereby protecting, say, secondary No. 8 TW wires rated at 40 A. The protection on the primary (here, 20 A) must not exceed the product of the secondary conductor ampacity (40 A) and the secondary-to-primary transformer voltage ratio ($120 \div 240 = 0.5$). Here, $40\text{ A} \times 0.5 = 20\text{ A}$, and the rule is satisfied.

A transformer with rated primary current of *9 amps or more* . . .

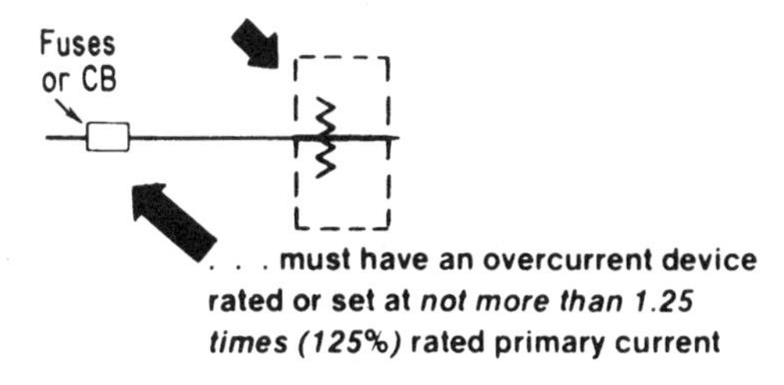

NOTE: Where 1.25 times primary current does not correspond to a standard rating of protective device, the next higher standard rating from Section 240-6 is permitted.

A transformer with rated primary current of *less than 9 amps* . . .

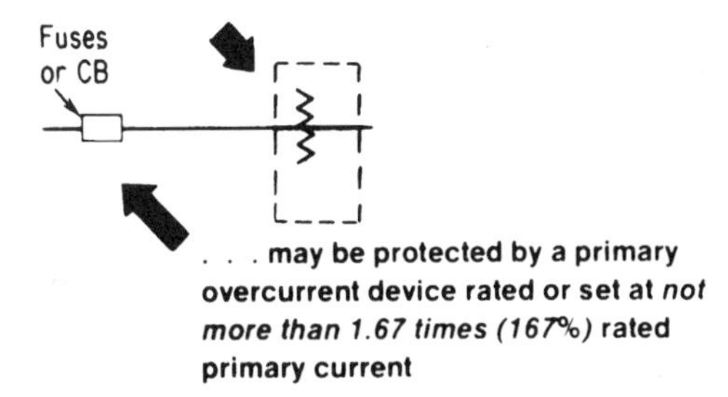

A transformer with rated primary current of *less than 2 amps* . . .

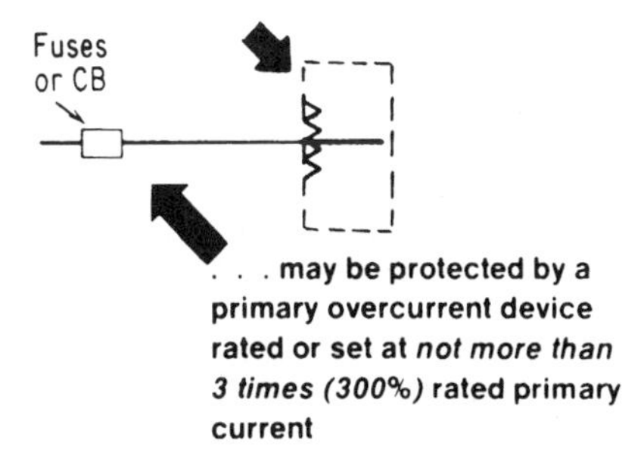

FIG. 7-7 Basic rules on protecting a transformer with an overcurrent device on the primary.

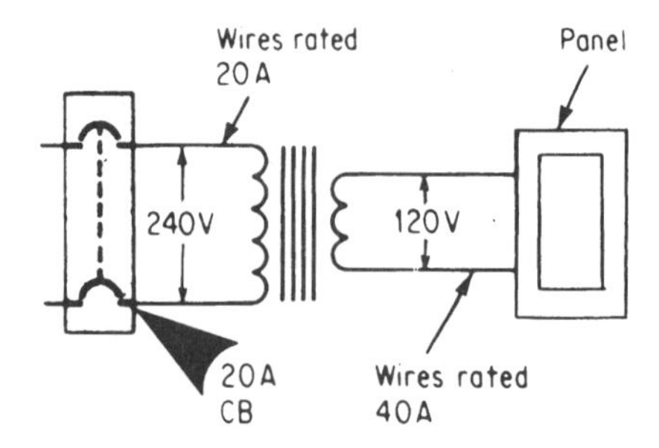

FIG. 7-8 This kind of protection of secondary conductors is permitted.

When the foregoing overcurrent protection rules are observed, the transformer itself is properly protected and the primary feeder conductors, sized to correspond, are provided with the required protection. But the secondary side must be evaluated separately and independently for any transformer with a 3-wire or 4-wire secondary. When a transformer is provided with primary-side overcurrent protection, a whole range of design and installation possibilities are available for a secondary arrangement that satisfies the NEC. The basic approach is to provide the required overcurrent protection for the secondary conductors right at the transformer—as with a fused switch or CB attached to the transformer enclosure, as shown in Fig. 7-9. Or, 10- to 25-ft taps may be made, as covered under "Feeder Taps" in Chapter 4 of this handbook.

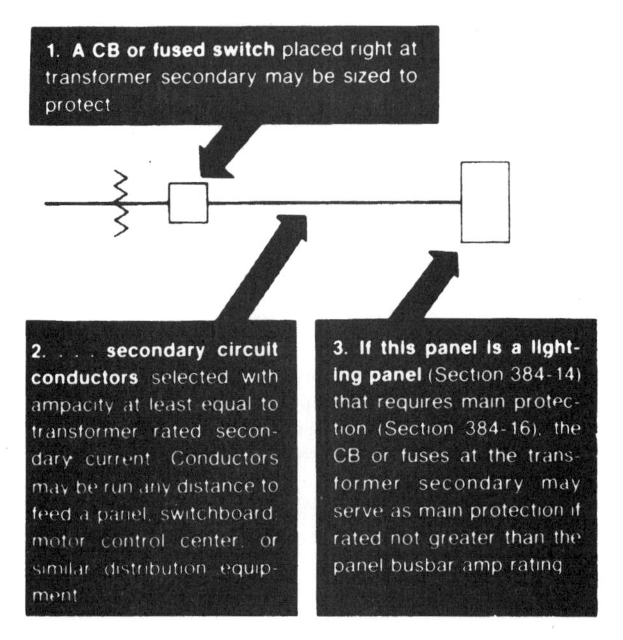

FIG. 7-9 Protection of secondary circuit conductors must be independent of primary protection.

In designing transformer circuits, the rules of NEC Sec. 450-3 on overcurrent protection for transformers can be coordinated with Sec. 240-21(d), which provides special rules for tap conductors used with transformers. This exception would be used mainly where the primary overcurrent devices were rated at 250 percent of rated transformer primary current, according to Sec. 450-3(b)(2), or where a secondary tap would exceed 10 ft. (Refer to the discussion of "Feeder Taps.")

Where secondary feeder taps do not exceed 10 ft in length, the requirements of Sec. 240-21(b) could apply, as in the case of any other feeder tap, with no restriction on the size of the feeder overcurrent device ahead of the tap. In applying the tap rules in Exceptions 2 and 8, the requirements of Sec. 450-3 on transformer overcurrent protection must always be satisfied.

Another acceptable way to protect a 600-V transformer is described in NEC Sec. 450-3(b)(2). In this method, the transformer primary may be fed from a circuit which has overcurrent protection (and circuit conductors) rated up to 250 percent (instead of 125 percent, as above) of the rated primary current; but in such cases there must be a protective device on the secondary side of the transformer, and that device must be rated or set at not more than 125 percent of the transformer's rated secondary current as shown in Fig. 7-10. This secondary protective device must be located right at the transformer secondary terminals or not more than the length of a 10- or 25-ft tap away from the transformer, and the rules on tap conductors must be fully satisfied.

The secondary protective device rated at not over 125 percent of the rated secondary current may readily be incorporated as part of other required provisions on the secondary side of the transformer, such as protection for a secondary feeder from the transformer to a panel or switchboard or a motor control center fed from the switchboard. And a single secondary protective device rated not over 125 percent of the rated secondary current may serve as required panelboard main protection as well as the required transformer secondary protection, as shown at the bottom of Fig. 7-11.

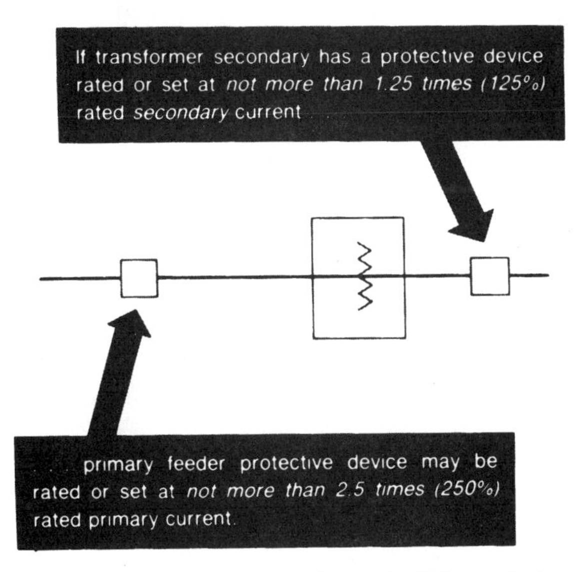

FIG. 7-10 Use of secondary protection permits higher-rated primary protection and circuit.

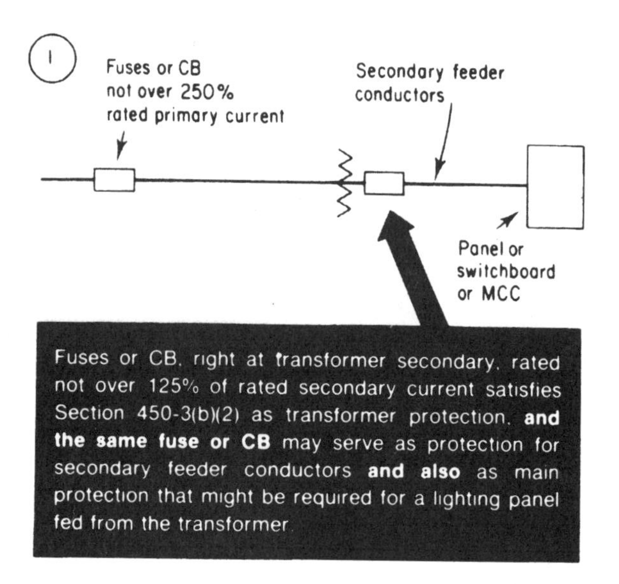

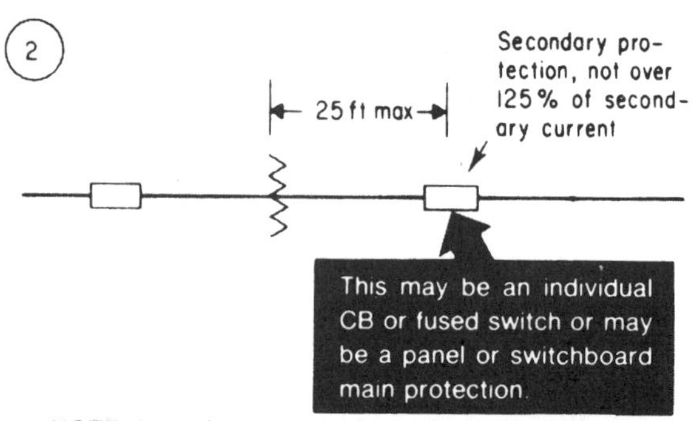

NOTE: In both cases, primary and secondary feeder conductors must be sized to be properly protected by the fuses or CB in both the primary and secondary circuit, which will give them more than adequate ampacity for the transformer full-load current.

FIG. 7-11 Secondary protection rated at 125 percent of the secondary current may be placed at one of several points.

The use of a transformer circuit with primary protection rated up to 250 percent of the rated primary current offers the opportunity to avoid situations in which a set of primary fuses or CB rated at only 125 percent would cause nuisance tripping or opening of the circuit on transformer inrush current. But the use of 250 percent rated primary protection has a more widely applicable advantage: It makes possible the feeding of two or more transformers from the same primary feeder. The number of transformers that might be used in any case would depend on the amount of continuous load on all the transformers. But in all such cases, the primary protection must be rated not more than 250 percent of any one transformer, if they are all the same size, or 250 percent of the

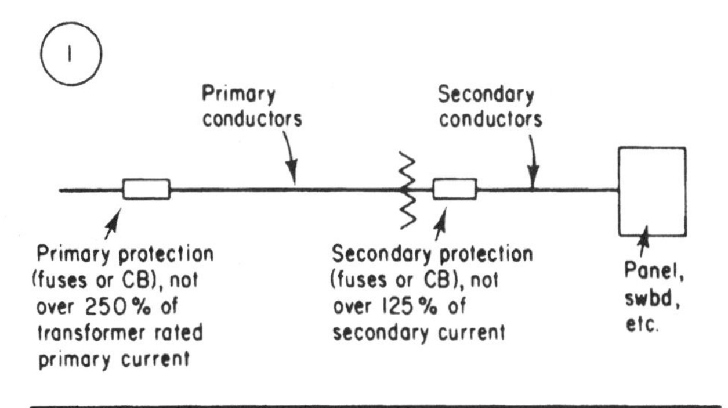

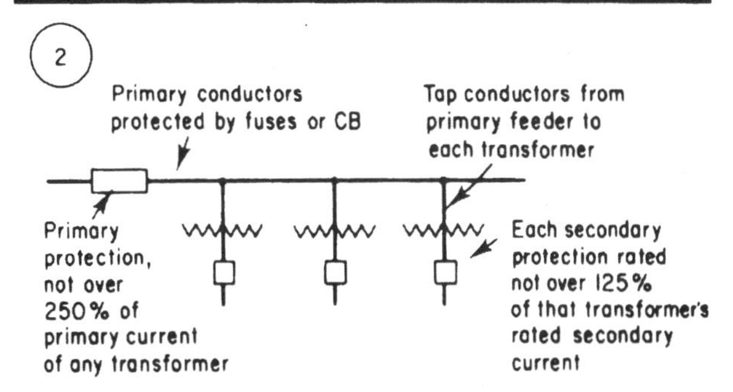

NOTE: Each set of tap conductors from primary feeder to each transformer may be same size as primary feeder conductors **OR** may be smaller than primary conductors if sized in accordance with Section 240-21(d), which permits a 25-ft tap from a primary feeder to be made up of both primary and secondary tap conductors. The 25-ft tap may have any part of its length on the primary or secondary but must not be longer than 25 ft and must terminate in a single CB or set of fuses.

FIG. 7-12 Transformer primary protection rated at 250 percent protects one or more transformers.

smallest transformer, if they are of different sizes. And for each transformer fed, there must be a set of fuses or CB on the secondary side rated at not more than 125 percent of the rated secondary current, as shown in Fig. 7-12.

Figure 7-13 shows an application of 250 percent primary protection to a feeder supplying three transformers (such as at the bottom of Fig. 7-12). This example indicates how the rules of NEC Sec. 450-3(b) must be carefully related to Sec. 240-21 and to other NEC rules.

Section 240-21(d) covers the use of a 25-ft unprotected tap from feeder conductors, with a transformer inserted in the 25-ft tap. This rule does not eliminate the need for secondary protection; it sets a special condition for placement of the secondary protective device. It is a restatement of Sec. 240-21(c), as applied to a tap containing a transformer, and applies to both single-phase and 3-phase transformer feeder taps.

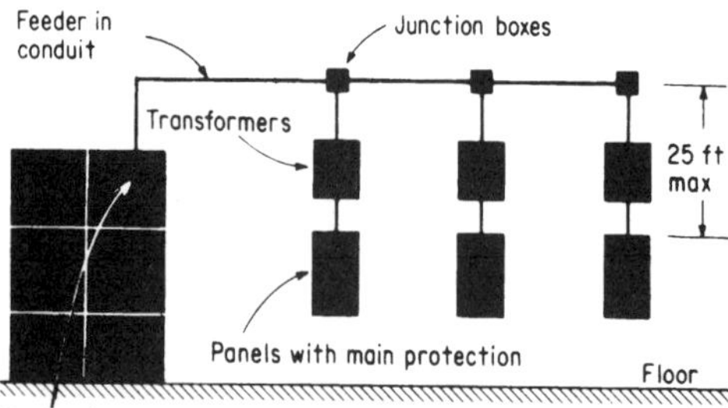

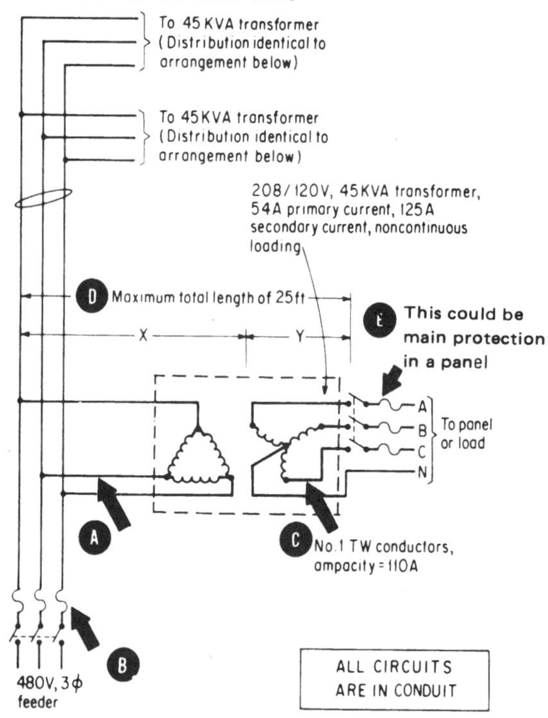

FIG. 7-13 This layout of feeder and transformer taps provides economical application with NEC compliance.

Figure 7-13 shows a feeder supplying three 45-kVA transformers; each transformer is fed through a 25-ft feeder tap that conforms to Exception 8 of Sec. 240-21. Although each transformer has a rated primary current of 54 A at full load, the demand load on each transformer primary was calculated to be 41 A, based on secondary loading. Thus, No. 1 THW copper feeder conductors were considered adequate for the total noncontinuous demand load of the three transformer primaries, which is equal to 3 × 41 A, or 123 A. In the following step-by-step analysis of this system, the letters refer to the circled letters on the sketch:

A. The primary circuit tap conductors are No. 6 TW, rated at 55 A, which gives them "an ampacity at least ⅓ that of the conductors or overcurrent protection from which they are tapped," because these conductors are tapped from the feeder conductors protected at 125 A. The use of No. 6 TW conductors is permissible for the 41-A primary current.

B. The 125-A fuses in the feeder switch properly protect the No. 1 THW feeder conductors, which are rated at 130 A and feed the taps for the three transformers.

C. According to Sec. 240-21(d), the conductors supplied by the transformer secondary must have "an ampacity that, when multiplied by the ratio of the secondary-to-primary voltage, is at least ⅓ the ampacity of the conductors or overcurrent protection from which the primary conductors are tapped." The secondary-to-primary voltage ratio of the transformer is

$$\frac{208 \text{ V}}{480 \text{ V}} = 0.433$$

(Note that phase-to-phase voltages must be used to determine this ratio.) Then, for the secondary conductors, according to Exception 8b,

$$\text{Minimum ampacity } 0.433 = \tfrac{1}{3} \times 5 \text{ A} = 41.67 \text{ A}$$

and

$$\text{Minimum ampacity} = \frac{41.67 \text{ A}}{0.433} = 96 \text{ A}$$

The No. 1 TW secondary conductors, rated at 110 A, are above the 96-A minimum and are, therefore, satisfactory.

D. The total length of the unprotected tap—the primary conductor length plus the secondary length ($x + y$) for any circuit leg—must not be greater than 25 ft.

E. The secondary tap conductors from the transformer must terminate in a single CB or set of fuses that will limit the load on those conductors to their rated ampacity from Table 310-16. Note that there is no exception to that requirement, and the next-higher standard device rating *may not be used* if the conductor ampacity does not correspond to the rating of a standard device.

The overcurrent protection required at E, the load end of the 25-ft tap conductors, must not be rated higher than the ampacity of the No. 1 TW conductors:

$$\text{Maximum rating of fuses or CB at E} = 110 \text{ A}$$

But a 100-A main would satisfactorily protect the 96-A secondary load.

Note: The overcurrent protective device required at E could be the main protective device required for a lighting and appliance panel fed from the transformer.

Watch out for this trap!

The foregoing calculation shows how unprotected taps may be made from feeder conductors by satisfying Exception 8 of Sec. 240-21. However, the rules of Sec. 240-21 are concerned with the *protection of conductors only.* Consideration must now be given to transformer protection, as follows:

1. Note that Sec. 240-21 makes no reference to transformer protection. But Sec. 450-3 calls for the protection of transformers, and no exception is made for the conditions of Exception 8 to Sec. 240-21.
2. It is clear that each transformer shown in Fig. 7-13 is *not* protected by a primary-side overcurrent device rated not more than 125 percent of the rated primary current (54 A), as required by Sec. 450-3(b)(1), because 1.25×54 A = 68 A.
3. But Sec. 450-3(b)(2) does offer a way to provide the required protection. The 110-A protection at E is secondary protection rated not over 125 percent of the rated secondary current (1.25×125 A secondary current = 156 A). With that secondary protection, a primary feeder overcurrent device rated not more than 250 percent of the rated primary current would satisfy Sec. 450-3(b)(2). That would call for fuses in the feeder switch (or a CB), at B in the diagram, rated at not over

$$2.5 \times 54 \text{ A (primary current)} = 135 \text{ A}$$

Because the fuses in the feeder switch are rated at 125 A (which is not in excess of 250 percent of the transformer primary rated current), those fuses satisfy Sec. 450-3(b)(2).

In addition to the two basic methods described above for protecting transformers, Sec. 450-3(b)(2) provides for protection with built-in thermal overload protection, as shown in Fig. 7-14.

Typical Applications

The following examples are based on the foregoing design considerations and procedures.

Example. In Fig. 7-15, the feeder to the 3-phase dry-type transformer supplies 480-V power for step-down to 120/208 V for a lighting panel with a continuous load of 30 A on each phase. What are the required minimum sizes of conductors and protective devices?

Note: Knowing these minimums under conditions of continuous use, the designer can effectively provide spare capacity without depriving the conductors of the transformer of the NEC-required levels of protection.

Step 1. The kilovoltampere rating of the 30-A, 120/208-V load on the panel is

$$30 \times 208 \times 1.73 \div 1000 = 10.18 \text{ kVA}$$

That load will require at least a 15-kVA transformer, assuming a larger one is not needed for even greater possible load growth.

Step 2. Assuming eventual use of the entire transformer capacity, the rated primary current is computed as

$$\frac{15{,}000}{408 \times 0.732} = \frac{15{,}000}{831} = 18 \text{ A}$$

That value of primary current would require conductors with a minimum current-carrying capacity of 20 A: No. 12 copper or No. 10 aluminum. Because the footnote to NEC

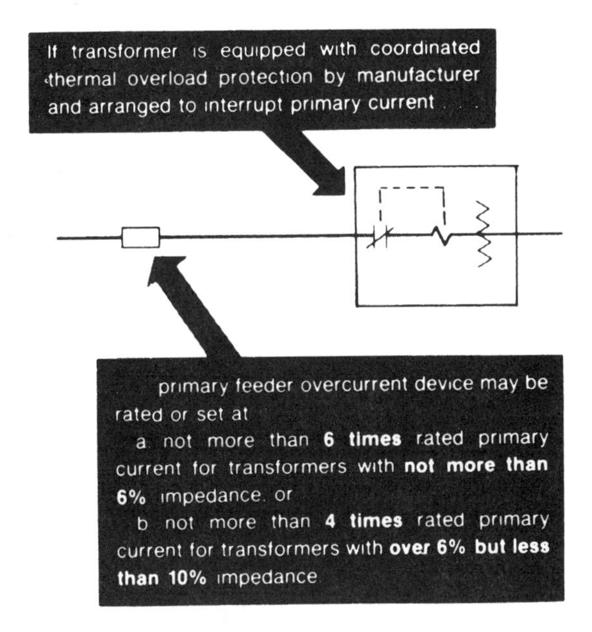

FIG. 7-14 "Built-in" overload protection is an alternative to the other methods shown for transformer protection.

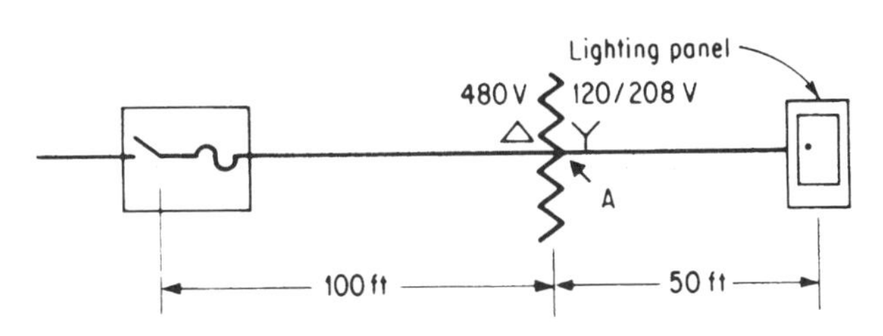

FIG. 7-15 Sizing of conductors and overcurrent protection must satisfy all transformer circuit conditions.

Table 310-16 says that No. 14 copper wire must not have a load-current rating over 15 A, the 18-A current prohibits the use of No. 14 conductors.

Step 3. NEC Sec. 450-3(b)(1) says that primary overcurrent protection for the transformer must be rated or set at not more than 125 percent of the rated primary current. In this example,

$$1.25 \times 18 \text{ A} = 22.5 \text{ A}$$

Because Exception 1 of Sec. 450-3(b)(1) says that the next-higher rated device may be used here, this would call for a 25-A, 3-pole CB, or 25-A fuses. This transformer overcurrent protection may be installed at the supply end of the primary circuit feeder.

Step 4. Section 220-10(b) says that the feeder overcurrent device must be rated at least 125 percent of the continuous load. Since

$$1.25 \times 18 \text{ A} = 22.5 \text{ A}$$

the 25-A CB protection computed in step 3 above for an 18-A full-load transformer current would not violate the limitations set by Sec. 220-10(b).

Step 5. The primary circuit conductors to the transformer are also required by Sec. 220-10(b) to have an ampacity of not less than 125 percent of 18 A, or 22.5 A. This would call for No. 12 copper or No. 10 aluminum circuit conductors. From NEC Table 310-16, No. 12 TW copper wire has an ampacity of 25 A, and No. 10 aluminum has that same ampacity. And No. 12 copper has a load rating of 20 A (from the footnote to Table 310-16), which exceeds the 18-A load current.

Step 6. The rated secondary current of a 15-kVA transformer would be

$$\frac{15{,}000}{208 \times 1.732} = 41.7 \text{ A}$$

Section 220-10(b) requires that the secondary circuit conductors from the transformer to the panel have an ampacity of at least 125 percent of that value, or 52 A. To carry this load, the following secondary conductors could be used in conduit:

Copper: No. 6 TW (55 A), THW (65 A), or THHN (75 A)

Aluminum: No. 4 TW (55 A), THW (65 A), or THHN (75 A)

Step 7. But the secondary conductors cannot be protected "through" the transformer. Exception 5 of Sec. 240-3 permits the protection of transformer secondary conductors by overcurrent devices placed on the primary side of the transformer only for single-phase transformers with 2-wire primary and 2-wire secondary circuits. That is not the case here.

Therefore, protection must be provided for the secondary feeder conductors which run 50 ft from the transformer to the panel. Such protection must be located right at the transformer, that is, "at the point where the conductor to be protected receives its supply," as covered in Sec. 240-21. Of course, secondary conductors from the transformer can be used as 10- or 25-ft tap conductors provided they fully comply with Exception 2, 3, or 8 of Sec. 240-21. But where the length from the transformer to a panel is over 25 ft, as in this case, separate overcurrent protection must be inserted to protect the secondary conductors at their prescribed capacity.

A 60-A CB or 60-A fuses would protect any of the secondary conductors listed in step 6 above in accordance with its ampacity. An enclosure containing this 60-A protection could be placed right on the transformer case. Or, it could be placed so that 10-ft tap conductors or 25-ft tap conductors could reach it.

Note: The protection for the secondary conductors must never be placed farther from the transformer than the distance at which 25-ft tap conductors could be used. Overcurrent protection must be placed at A in Fig. 7-15, and the protection for the secondary conductors may also provide required protection for the lighting panel supplied—if the rating of the protection is not higher than the main rating of the panel.

Thus, in this example, a 60-A device, for instance, could protect, say both No. 6 THW secondary conductors and a 60-A lighting panel fed from the transformer.

Example. The circuit components in Fig. 7-16 are sized as follows:

Step 1. The lighting load will be served by 15-A, 2-wire circuits. Because the lighting load will be operating continuously to supply fixtures in a store, each circuit must

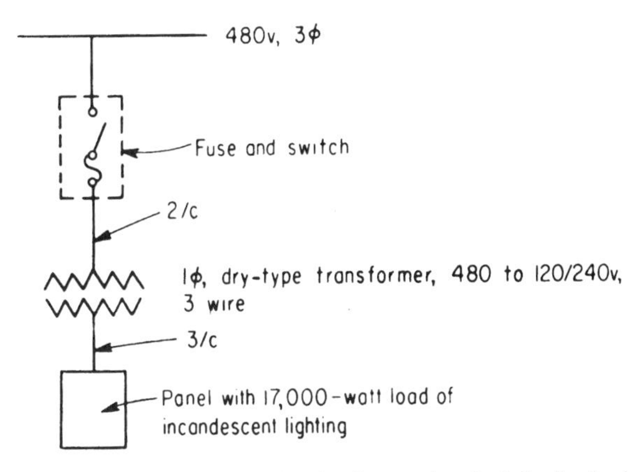

FIG. 7-16 Transformer feeder circuits may be sized for the load capacity or for the full capacity of the transformer selected.

not be loaded to more than 80 percent of its rating. At 15 A and 120 V, each circuit has a maximum capacity of 15 × 120 = 1800 W. The 80 percent limit gives 1800 W × 0.8 = 1440 W. The required number of circuits is determined as follows:

$$\text{No. of circuits} = \frac{\text{total load in watts}}{\text{watts per circuit}} = \frac{17{,}000}{1440} = 12 \text{ circuits}$$

Step 2. The feeder to the 120/240-V panel must have sufficient capacity for the 17-kW load, which operates continuously. The required feeder conductor ampere loading is found as follows:

$$I = \frac{\text{load in watts}}{2 \times \text{phase-to-neutral voltage}} = \frac{17{,}000}{240} = 71 \text{ A}$$

Although the feeder circuit is described as being rated at 120/240 V, these are high values of voltage that usually will not occur during normal operation. It is better to use 115/230 V as more likely and typical of the voltage conditions that will exist. And since the current is determined by dividing wattage by voltage, it is safer to use the lower voltage value, which yields a higher current value to be used in sizing the feeder conductors. This will provide a higher-rated conductor and reduce the chance of the circuit's being overloaded. Thus,

$$I = \frac{17{,}000}{230} = 74 \text{ A}$$

The demand factors which may be applied to an overall feeder load, as covered in NEC Sec. 220-11, are not applicable because this is a store building and all the lights will be operating simultaneously and constantly.

Step 3. But because the load on the feeder is continuous (operates for longer than 3 h), the value of 74 A must be multiplied by 125 percent to satisfy NEC Sec. 220-10(b) for continuous loading. The feeder must be sized for

$$1.25 \times 74 \text{ A} = 93 \text{ A}$$

To meet that ampacity requirement, the secondary feeder could be made up of three No. 2 TW copper conductors (rated at 95 A); or, No. 2 THW, XHHW, or THHN copper conductors could be used.

Step 4. According to NEC Sec. 384-13, the bus bars of a panel must have a current rating not lower than the minimum feeder capacity required for the load. This means that, in this case, the 120/240-V panel must have a rating of at least 93 A—the load on the feeder. A 100-A panel would satisfy this requirement.

Step 5. Based on a total load of 17 kW, the smallest standard size of transformer which could be used to continuously supply the load would be rated at 25 kVA. If the load were not continuously at the maximum value, it might be possible to use a 15-kVA transformer and allow it to operate overloaded during peak-load times. But a 25-kVA transformer is properly required for a continuous 17-kW load.

Step 6. To obtain data for sizing the transformer's overcurrent protection, its primary and secondary full-load currents should be calculated using the formula for a single-phase transformer:

$$\text{Primary current} = \frac{\text{transformer kVA rating} \times 1000}{\text{primary voltage}} = \frac{25 \times 1000}{480} = 52 \text{ A}$$

Its full-load secondary current is found with the same formula, applied to the secondary:

$$\text{Secondary current} = \frac{\text{transformer kVA rating} \times 1000}{\text{secondary voltage}} = \frac{25{,}000}{240} = 104 \text{ A}$$

Step 7. Under the load conditions given, the load current on the transformer secondary will be the 74-A load current drawn by the lighting panel. The load current drawn by the transformer primary is found from the formula which relates the currents and voltages on the primary and secondary sides of the transformer.

$$\frac{I_p}{I_s} = \frac{E_s}{E_p}$$

where I_p = primary current, A
I_s = secondary current, A
E_s = secondary voltage, V
E_p = primary voltage, V

For a secondary current of 74 A, the primary current is calculated as

$$I_p = I_s \frac{E_s}{E_p} = 74 \left(\frac{240}{480}\right) = 37 \text{ A}$$

To handle a continuous load of 37 A, the primary circuit conductors must have an ampacity of

$$1.25 \times 37 \text{ A} = 46 \text{ A}$$

Two No. 6 TW copper conductors (rated at 55 A) could be used, as could No. 6 THW, THHN, XHHW, or RHH conductors.

But because a 25-kVA transformer has been selected, good design practice calls for sizing the entire feeder for the capacity of the transformer—even though the present loading is lower than that capacity.

Step 8. In an application of this sort, the transformer must be protected by a primary CB or fuses in a switch, rated or set at not more than 125 percent of the transformer rated primary current. The rated primary current is computed as 25,000 VA ÷ 480 V = 52 A, and 1.25 × 52 A = 65 A. That rating of protection would not properly protect No. 6 TW conductors (55 A) but would protect No. 6 THW (65-A) or No. 6 THHN or XHHW (75-A) conductors. However, to obtain the full output of the transformer for future use, primary conductors rated at 1.25 × 52 A = 65 A would have to be used. And because equipment rated up to 100 A is required by UL to be wired with 60°C conductors (TW)—or with higher-temperature conductors that are sized as if they were TW conductors—the primary circuit would require at least No. 4 TW (70 A) or THW, THHN, or XHHW conductors.

Step 9. If the primary feeder is sized to allow the full 52 A of rated primary current for a 25-kVA transformer, the secondary conductors must be resized at 25,000 VA ÷ 240 V × 1.25 = 130 A to allow for future load increase up to the transformer's rated 25-kVA capacity. That calls for No. 1 THW copper conductors (130 A) or No. 2/0 TW copper conductors (145 A). If THHN or XHHW conductors are used, they must be sized as if they were 75°C conductors (THW), which means No. 1 copper must be used.

Step 10. The 100-A lighting panel must then be equipped with 100-A main protection. That protection could be placed right at the transformer secondary, or it could be placed at the load end of unprotected tap conductors that run no more than 25 ft from the transformer secondary to the main 100-A CB or fused switch that serves as main protection in the panel.

Example. Figure 7-17 shows a single-phase transformer with a 2-wire primary circuit and a 2-wire secondary, which is used to feed ungrounded circuits for a hospital operating room.

Step 1. In such a circuit, the secondary conductors may be protected by the overcurrent device protecting the primary circuit to the transformer, if the size of the conductors and the rating of the protective device are calculated on the basis of the transformer turns ratio.

Step 2. In the hookup shown, the 30-A, 2-pole CB protecting the primary circuit to the transformer is acceptable as protection for the transformer if the transformer rated primary current is more than 20 A [since 1.25 × 21 A = 26 A, and a 30-A CB is the

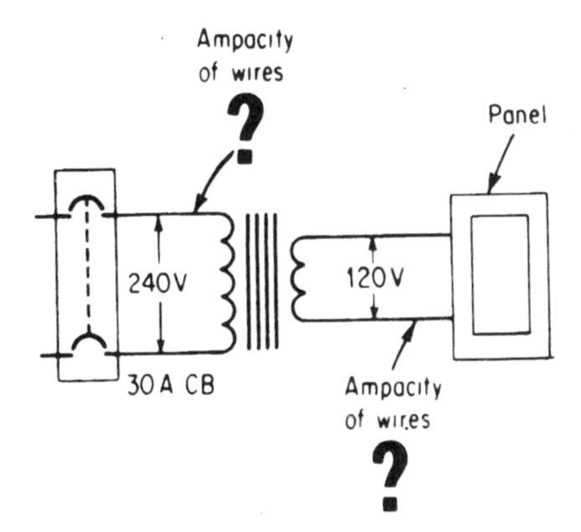

FIG. 7-17 Primary protection will protect feeder circuit conductors.

next-higher standard rating of protective device, as permitted by NEC Sec. 450-3(b)(1)]. The minimum size of copper conductors acceptable for use as the primary circuit to the transformer is No. 10 TW. The minimum acceptable size of secondary circuit conductor is No. 6 copper (55 A), for which the 30-A primary CB acts as a 60-A protective device through the 2-to-1 current step up from primary to secondary. And such 60-A protection is the "next higher standard rating" of CB above 55 A, as permitted by Exception 1 of Sec. 240-3.

Step 3. If the transformer shown in the sketch were rated at 5 kVA, it would have a rated primary current (at 240 V) of 5000 VA ÷ 240 V, or 20.8 A. The maximum rating of overcurrent device for such a transformer would be 125 percent of 20.8 A, or 26 A. Because Exception 1 of Sec. 480-3(b)(1) permits the next-higher standard rating of protective device from Sec. 240-6 when 1.25 times the rated primary current does not correspond to a standard rating, the maximum standard rating of CB acceptable as protection for this transformer is 30 A, which is the next-higher standard size of CB above 26 A.

Example. Calculations of the various loads supplied by the 208/120-V switchboard shown in Fig. 7-18 reveal that the switchboard would have a continuous demand load of 762 A per phase.

Step 1. The minimum transformer kilovoltampere rating that can handle that load is determined with the formula

$$\text{kVA rating} = \text{line-to-line voltage} \times \text{current per phase} \times 1.732 \div 1000$$

$$= 208\text{ V} \times 762\text{ A} \times 1.732 \div 1000 = 274\text{ kVA}$$

Step 2. The nearest standard rating of transformer, which must be used to handle the load, is 300 kVA.

Step 3. A transformer with a 300-kVA rating has a rated primary current of 300 × 1000 ÷ (480 × 1.732), or 361 A.

Step 4. The CB or fuses used to protect the primary feeder circuit to the transformer must have a rating or setting of not more than 125 percent of the transformer rated pri-

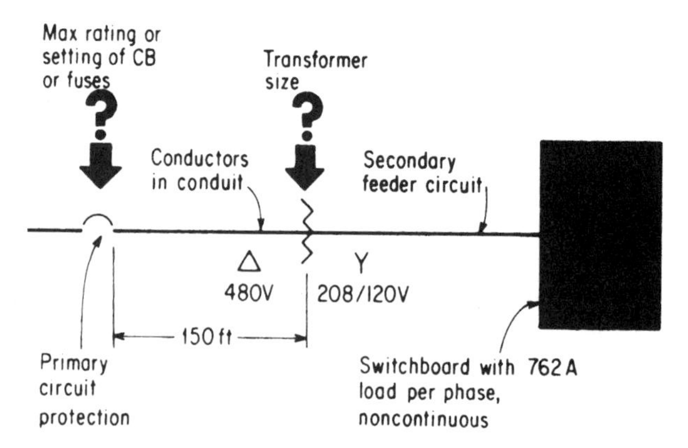

FIG. 7-18 Switchboard loading dictates transformer size, which then dictates conductor and protection ratings.

mary current if the primary circuit protection is to qualify by itself as the required overcurrent protection for the transformer.

Step 5. Therefore, the primary overcurrent device must have a rating of 1.25 × 361 A = 451 A.

Step 6. The maximum standard rating of protective device permitted by Exception 1 to NEC Sec. 450-3(b) is 450 A. (Actually, Exception 1 would permit the use of 500-A protection as the next-higher rating above 451 A, as calculated in step 5 above.)

Step 7. Based on the use of primary overcurrent protection of the rating established in step 5 above, the primary circuit conductors may be of the same ampacity or of such ampacity that the primary overcurrent device in step 5 is of the "next higher standard rating or setting" above the ampacity rating of the conductors. For this circuit, that condition is achieved if each phase leg of the primary circuit has an ampacity of over 400 A; either a single conductor per phase leg of that ampacity may be used, or two conductors per leg may be used, with each conductor rated over 200 A.

Step 8. If the primary circuit to the transformer is made up of two parallel sets of three conductors, with all six conductors in the same conduit, each conductor could be a 300-kcmil THHN or XHHW aluminum conductor (in a dry location), which has a normal rating of 255 A (from NEC Table 310-16); but these conductors must be derated to 80 percent of their ampacity (0.8 × 255 A, or 204 A), as required by Note 8 to Table 316-16, because there are more than three current-carrying conductors in the conduit.

Step 9. With two parallel sets of conductors as described in step 8 above, the ampacity of each phase leg of the primary circuit is 2 × 204 A, or 408 A, and the overcurrent protection sized in step 6 is the next-higher rating above the ampacity of the circuit phase legs.

Step 10. The minimum size of conduit required for the six 300-kcmil THHN or XHHW circuit conductors of step 8 above is 3 in.

Note: Although 90°C conductors are being used (THHN or XHHW), they will be operating at only 361 A when the transformer is loaded to its capacity. Because 361 A is lower than the 368-A rating of two 300-kcmil THWs in parallel (2 × 230 A × 0.8 = 68 A), the equipment terminal lugs will be within the 75°C operating limit that UL places on conductor terminal operations.

Step 11. If the load on the switchboard fed from the transformer secondary were continuous instead of noncontinuous, the calculations given above for sizing the primary circuit overcurrent protection and the primary conductors would have to be evaluated in light of NEC Sec. 220-10(b), which requires feeder conductors and overcurrent protection to have an ampacity at least equal to 125 percent of the load current. The 274-kVA load on the transformer draws a primary current equal to 274 × 1000 ÷ (480 × 1.732), or 330 A of primary current. Then 1.25 × 330 A = 412 A. The primary circuit conductors selected in step 9 above have a derated capacity of 408 A per phase which is not quite up to the required 412-A level. The 300-kcmil conductors are, therefore, not fully adequate for the continuous load. If 350-kcmil THHN aluminum conductors (280 A) are used instead of 300-kcmil conductors, their derated ampacity of 2 × 280 A × 0.8 = 448 A will satisfy the need for an ampacity of 412 A. And the 450-A rating of the primary protection is acceptable because it is greater than 125 percent of the continuous load of 330 A.

Step 12. The above calculations cover the primary protection, primary circuit makeup, and transformer protection. The conductors and other components on the secondary side of the transformer must be sized separately, on the basis of other, independent calculations; the primary protection does not affect the required secondary protection or the sizing of the secondary feeder conductors.

Step 13. The secondary conductors from the transformer to the switchboard must be sized on the basis of the demand load on the panel. For the noncontinuous load indicated, the secondary conductors must have an ampacity of not less than 762 A.

Step 14. The secondary circuit conductors are selected on the basis of the load on the conductors, the distance of the switchboard from the transformer, and consideration of the need for overcurrent protection for the conductors.

Step 15. If the transformer is located adjacent to the switchboard, as shown at the top of Fig. 7-19, so that secondary conductors only 9 ft long can be used in conduit to connect the transformer secondary terminals to the main lugs in the switchboard, those conductors need only have an ampacity that is not less than the main current rating of the switchboard; in addition, they qualify as 10-ft tap conductors and may be used without overcurrent protection.

Step 16. If the secondary conductors are run in conduit from the transformer to the switchboard located 13 ft away, as at the bottom of Fig. 7-19, they may be run without overcurrent protection at the transformer secondary to a single CB or set of fuses in the switchboard, provided that the conductors have an ampacity of at least 762 A (for a non-continuous load), they are not more than 25 ft long, and the protective device in which they terminate protects them at their rated ampacity.

Step 17. If the secondary conductors are run in conduit to the switchboard located 50 ft from the transformer, they must be supplied with overcurrent protection at the transformer (their supply end); this overcurrent protection must protect the secondary conductors at their rated ampacity, which must be not less than 762 A.

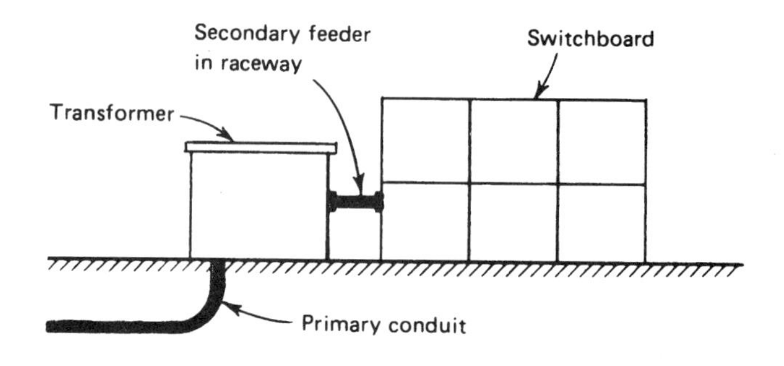

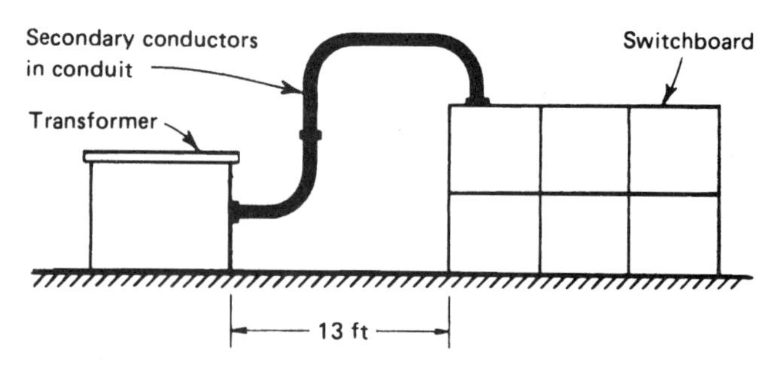

FIG. 7-19 Distance between a transformer and the equipment it feeds can alter calculations.

Step 18. If the load on the switchboard consisted of 500 A of continuous load and 262 A of noncontinuous load, the rating of the overcurrent protection at the transformer in step 17 above would have to be not less than 125 percent of 500 A plus 262 A, or 887 A. And the current rating of the switchboard would then have to be at least 1000 A.

Step 19. The condition described in step 18 would call for a standard secondary overcurrent device (CB or set of fuses) rated not below 1000 A. Then the secondary circuit conductors would have to have at least that ampacity and not a lower ampacity, because the overcurrent device would be rated over 800 A.

TRANSFORMER GROUNDING

With the great popularity of distribution systems using transformer load centers, particularly 480/277-V systems, effective grounding of transformer secondaries has become an important consideration in modern electrical system design. In a 480/277-V system, 120-V circuits for incandescent lighting and receptacle outlets may be provided either by using single-phase transformers to step 480 V down to 240/120-V, single-phase secondaries or by using 3-phase transformers to step the 480-V, 3-phase primary supply down to 208/120-V, 3-phase, 4-wire secondaries. The ac systems supplied from the secondaries of such transformers are considered to be "separately derived ac wiring systems" and are subject to the grounding requirements of NEC Sec. 250-26. The same rules on grounding and bonding apply to the output circuit of an engine generator used for emergency or standby power supply and to that of a generator used for on-site supply of the normal power to a building or facility.

Details on the grounding of transformer secondaries are as follows:

1. Any system that operates at over 50 V but not more than 150 V to ground must be grounded; that is, it must be operated with one of its circuit conductors connected to a grounding electrode.
2. This requires the grounding of secondaries of dry-type transformers serving 208/120-V, 3-phase or 240/120 V, single-phase circuits for lighting and appliance outlets and receptacles at load centers throughout a building, as shown in Fig. 7-20. And 480//277-V, wye-connected secondaries must also be grounded if indoor lighting is to be supplied—with loads connected from line to neutral and/or from line to

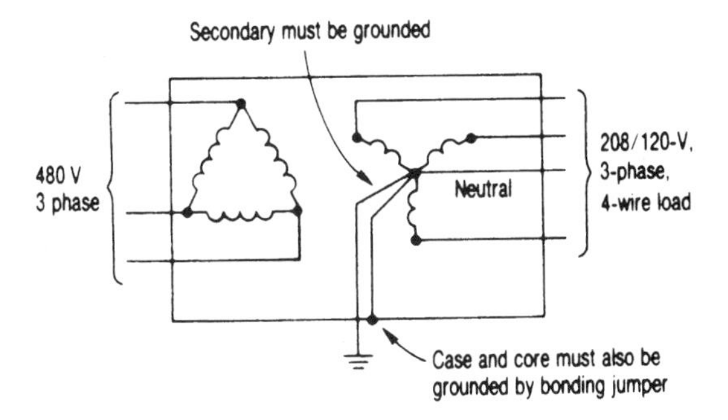

FIG. 7-20 Transformer secondary neutral must be both grounded and bonded to the case.

line. Electric-discharge lighting that is supplied by 480-V ballasts indoors *must* be fed from a 480/277-V transformer secondary with a grounding connection to the neutral point to provide a voltage that is not over 300 V to ground, as required by Exceptions 1 and 2 of NEC Sec. 210-6(a).

3. All rules applying to both system and equipment grounding must be satisfied in such installations.

The wiring system conductor to be grounded in compliance with items 1 and 2 above depends upon the type of system. In 3-wire, single-phase systems, the midpoint of the transformer windings—the point from which the system neutral is derived—is grounded. In grounded 3-phase wiring systems (either 3-wire or 4-wire systems), the neutral point of the wye-connected transformer or generator is the point connected to ground. In delta-connected transformer hookups, grounding can be effected by grounding one of the three phase legs, by grounding a center-tap point on one of the transformer windings (as in the 3-phase, 4-wire "red-leg" delta system), or by using a special grounding transformer which establishes a neutral point of a wye connection which is grounded.

The steps involved in designing transformer secondary connections are as follows (see Fig. 7-21):

Step 1

A bonding jumper must be installed between the transformer secondary neutral terminal and the metal case of the transformer. The size of this bonding conductor is obtained from NEC Table 250-94, based on the size of the transformer secondary phase conductors (as if they were service-entrance conductors), and is the same as the size of the required grounding-electrode conductor. For cases where the transformer secondary circuit conductors are larger than 1100-kcmil copper or 1750-kcmil aluminum per phase leg, the size of the bonding jumper must not be less than 12½ percent of the cross-sectional area of the secondary phase leg.

Example. A 75-kVA transformer has a 208/120-V, 3-phase, 4-wire secondary, with current of

$$75{,}000 \div (208 \times 1.732) = 209 \text{ A}$$

If No. 4/0 THW copper conductors (with a 230-A rating) were used for the secondary phase legs, then the size of the required bonding jumper would be obtained from Table 250-94 as if No. 4/0 service conductors were being used. The table shows that No. 4/0 copper service conductors require a minimum of No. 2 copper or No. 1/0 aluminum for a grounding-electrode conductor. The bonding jumper would have to be of either of those two sizes.

If the transformer were a 500-kVA unit with a 120/208-V secondary, its rated secondary current would be

$$\frac{500 \times 1000}{1.732 \times 208} = 1388 \text{ A}$$

If, say, THW aluminum conductors were used, then each secondary phase leg would be made up of four 700-kcmil aluminum conductors in parallel (each 700-kcmil THW aluminum is rated at 375 A, and four are rated at 4 × 375 A, or 1500 A, which suits the 1388-A load). Then, because 4 × 700 kcmil equals 2800 kcmil per phase leg, which is

STEP 1 – BONDING JUMPER

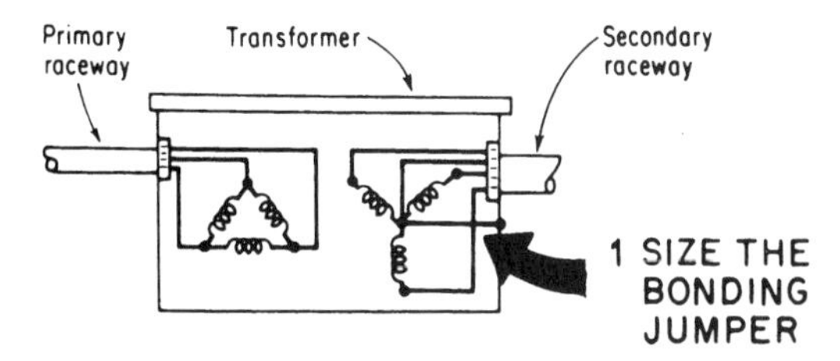

STEP 2 – GROUNDING ELECTRODE CONDUCTOR

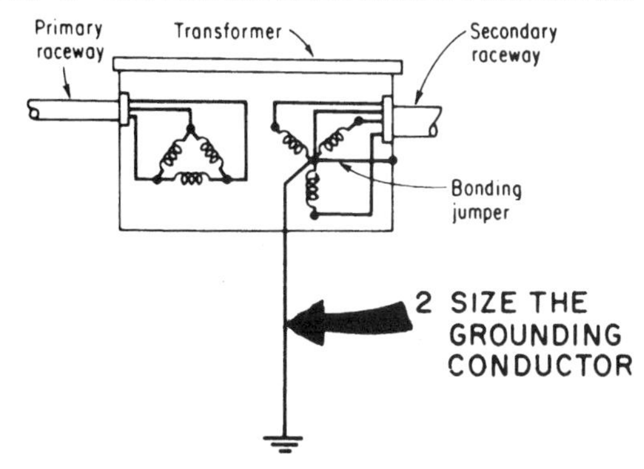

STEP 3 – GROUNDING ELECTRODE

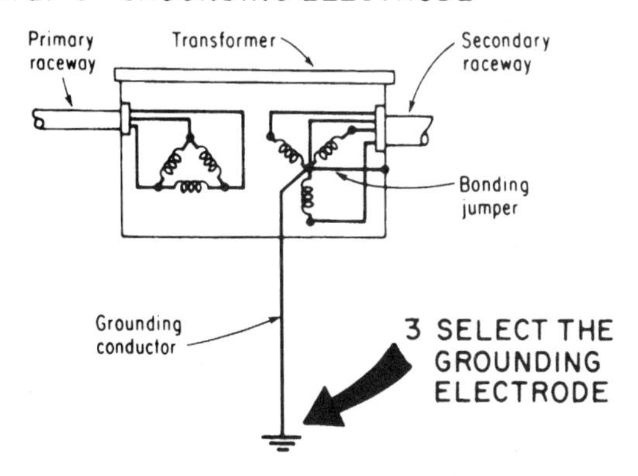

FIG. 7-21 These design steps provide for transformer secondary connections.

in excess of 1750 kcmil, Sec. 250-79(c) would require the bonding jumper from the case to the neutral terminal to be at least equal to 12½ percent of 2800 kcmil (0.125 × 2800 kcmil), or 350-kcmil aluminum.

Step 2

A grounding-electrode conductor must be installed from the transformer secondary neutral terminal to a suitable grounding electrode. This grounding conductor is sized the same as the required bonding jumper in the above example. That is, the grounding-electrode conductor is sized according to Table 250-94 as if it were a grounding-electrode conductor for a service with service-entrance conductors equal in size to the phase conductors used on the transformer secondary side. But this grounding-electrode conductor does not have to be larger than 3/0 copper or 250-kcmil aluminum when the transformer secondary phase legs are over 1100-kcmil copper or 1750-kcmil aluminum.

Example. For the 75-kVA transformer in step 1, the grounding-electrode conductor must be not smaller than the required minimum size shown in Table 250-94 for 4/0 phase legs, which makes it the same size as the bonding jumper, that is, No. 2 copper or No. 1/0 aluminum. For the 500-kVA transformer, the grounding-electrode conductor is sized directly from Table 250-94, which requires 3/0 copper or 250-kcmil aluminum where the phase legs are over 1100-kcmil copper or 1750-kcmil aluminum. Note that the grounding-electrode conductor for this 500-kVA transformer is smaller than the bonding jumper sized in step 1.

In the 1978 NEC, this rule called for the bonding and grounding of a grounded transformer secondary or generator output (for example, 208/120-V wye) "at the source" of the separately derived system. That phrase was frequently interpreted very rigidly to mean "only at the transformer itself." In the present NEC, the last sentence of Sec. 250-26(a) and (b) permits the bonding and grounding connections to be made either right at the transformer or generator or at the first disconnect or overcurrent device fed from the transformer or generator, as in Fig. 7-22.

The last sentence of Sec. 250-26(a) and (b) does, however, say that the bonding and grounding "shall be made at the source"—which appears to mean right at the transformer or generator, and not any other point—where the transformer supplies a system that "has no disconnecting means or overcurrent device." A local transformer that has no disconnect or overcurrent devices on its secondary is one that supplies only a single circuit and has overcurrent protection on its primary, such as a control transformer to supply motor starter coils. But in such applications, the transformer does not supply a separately derived "system" in the usual sense that a system consists of more than one circuit. However, for a transformer supplying only one circuit, any required bonding and grounding of a secondary grounded conductor—such as the grounded leg of a 2-wire, 120-V control circuit—would normally have to be done right at the transformer.

Another interpretation that might be put on that last phrase is that it refers to a hookup in which the transformer secondary feeds main lugs only in a panelboard, switchboard, or motor control center. In such cases, the absence of a main CB or fused switch would mean that there is no disconnect means or overcurrent device for the overall "system" fed by the transformer, even though there are disconnects and overcurrent protection for the individual "circuits" that make up the "system." That interpretation would require bonding and grounding of the secondary right at the source (the transformer itself).

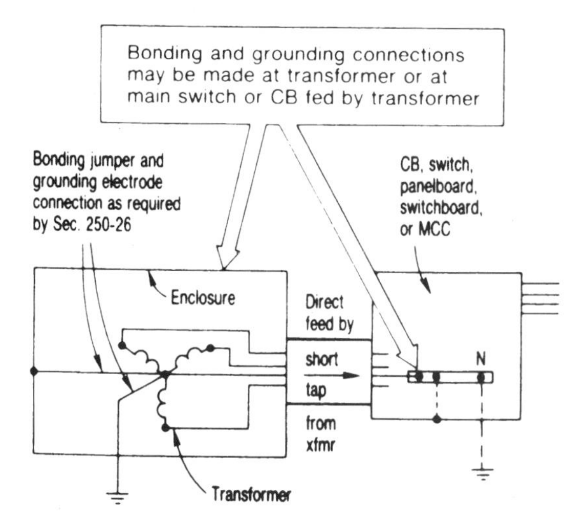

FIG. 7-22 Secondary neutral grounding and bonding may be accomplished at either of two places.

Step 3

The grounding-electrode conductor, installed and sized as in step 2 above, must be properly connected to a grounding electrode that must be "as near as practicable to and preferably in the same area as the grounding conductor connection to the system." That is, the grounding electrode must be as near as possible to the transformer itself. In order of preference, the grounding electrode must be one of the following:

1. The nearest available structural steel of the building, provided it is established that such building steel is effectively grounded
2. The nearest available water pipe, provided it is effectively grounded

Section 250-112 clarifies the term *effectively grounded* by noting that the grounding connection to a grounding electrode must "assure a permanent and effective ground." Thus, when a nearby cold-water pipe is used as a grounding electrode, it would appear to be necessary to bond around any unions or valves that might be opened and thereby might break the piping connection to the earth (Fig. 7-23). And Sec. 250-18(a) requires a bonding jumper to be used around all indoor water meters to ensure continuity to earth or through the interior water pipe system. Such bonding jumpers must be at least the same size as the grounding conductor from the transformer to the water pipe and other electrodes.

At least 10 ft of the metal water pipe must be buried in earth outside the building for the water pipe system to qualify as a grounding electrode for the transformer secondary neutral. However, even when there is not 10 ft of metal pipe in the earth, there must always be a connection between an interior metal water pipe system and the service-entrance grounded conductor (the neutral of the system that feeds the primary of transformers in the building). The grounding connection for the service neutral or other system grounded conductor must be made at the service. Where a metallic water pipe system in a building is fed from a nonmetallic underground piping system or has less

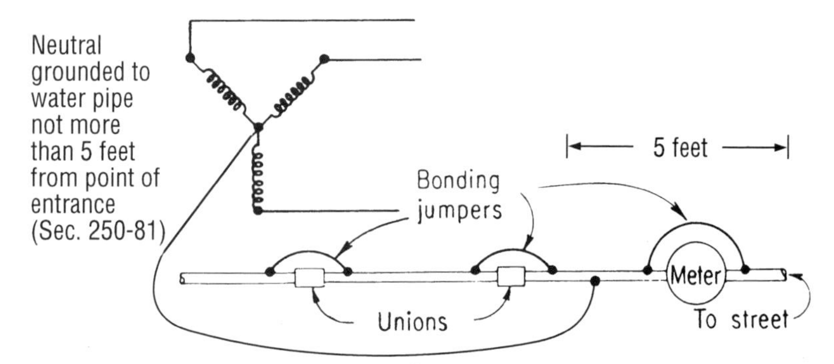

FIG. 7-23 Reliability and conductivity of grounding-electrode connections must be assured.

than 10 ft of metal pipe underground, the water pipe is not a grounding electrode, and the service or other service grounded conductor must have a connection to a ground rod or other electrode in addition to the connection to the interior metal water pipe system.

Note that Sec. 259-26 accepts an "effectively grounded" metal water pipe as a suitable grounding electrode for a "separately derived system" such as a transformer secondary. But Sec. 250-81 *does not* permit the use of a water pipe electrode (with at least 10 ft in the ground) as the *sole* grounding electrode for the service to a building. Any such water pipe electrode must be supplemented by bonding to at least one other grounding electrode, such as grounded building steel, reinforcing bars in the footing or foundation, or a driven ground rod.

Figure 7-24 shows the various elements involved in grounding and bonding at a transformer. Transformer housings must be grounded by connection to grounded cable

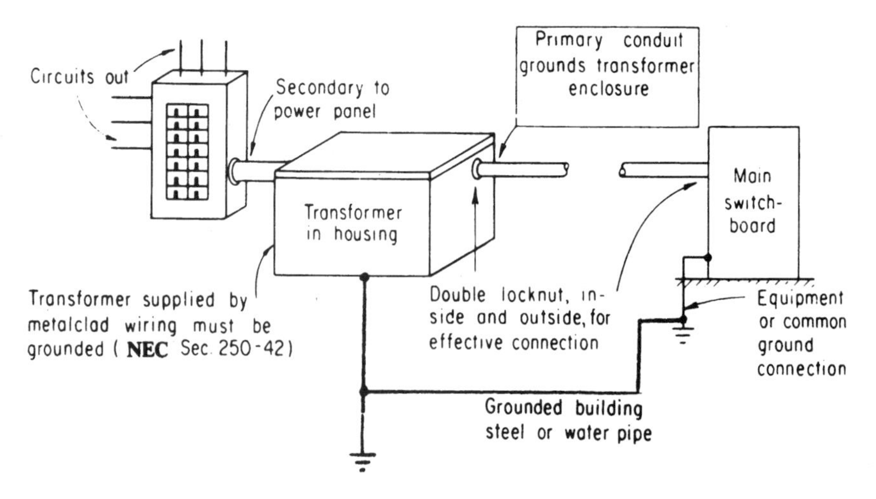

FIG. 7-24 All the elements of transformer grounding and bonding.

armor or metal raceway, by use of a grounding conductor run with the primary circuit conductors (either a bare conductor or a conductor with green covering), or by a separate equipment grounding conductor. Such an effective grounding conductor (metal raceway or a separate conductor) is necessary to provide for fault-current return in case an energized conductor of a grounded primary supply circuit (such as a 3-phase, 480-V primary circuit derived from a grounded wye, 480/277-V system) faults to the metal transformer housing. Even though the secondary neutral is bonded to the case and grounded to building steel, a low-impedance ground-return path must be provided for primary circuit faults in the metal housing to clear a primary protective device, as required by NEC Sec. 250-57(b). But the clearing of ground faults on the *secondary* size of the transformer is facilitated by the bonding that Sec. 250-26 requires between the secondary neutral and the metal case of the transformer, as shown in Fig. 7-25.

Figure 7-26 shows an important detail for effective grounding of a transformer case and secondary neutral. A common technique for protecting a bare or insulated system

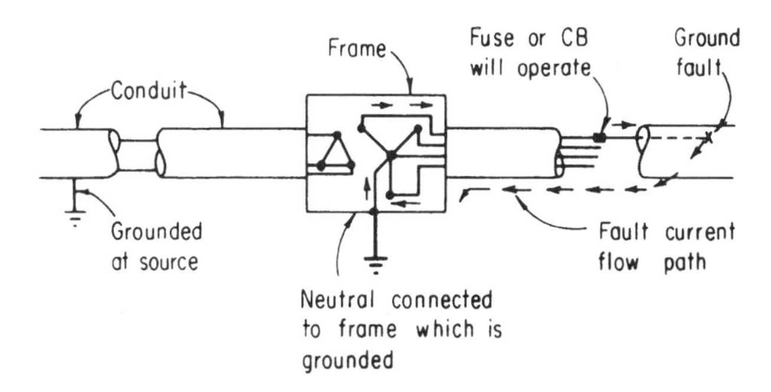

FIG. 7-25 Secondary neutral bonding provides for fault clearing on secondary faults, but not on primary faults.

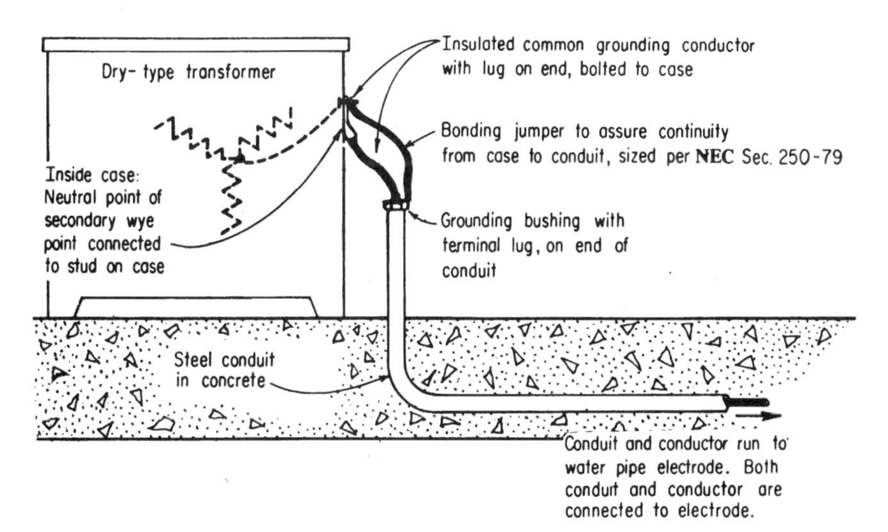

FIG. 7-26 Protective metallic conduit sleeving for a grounding-electrode conductor *must* be bonded to the conductor at both ends.

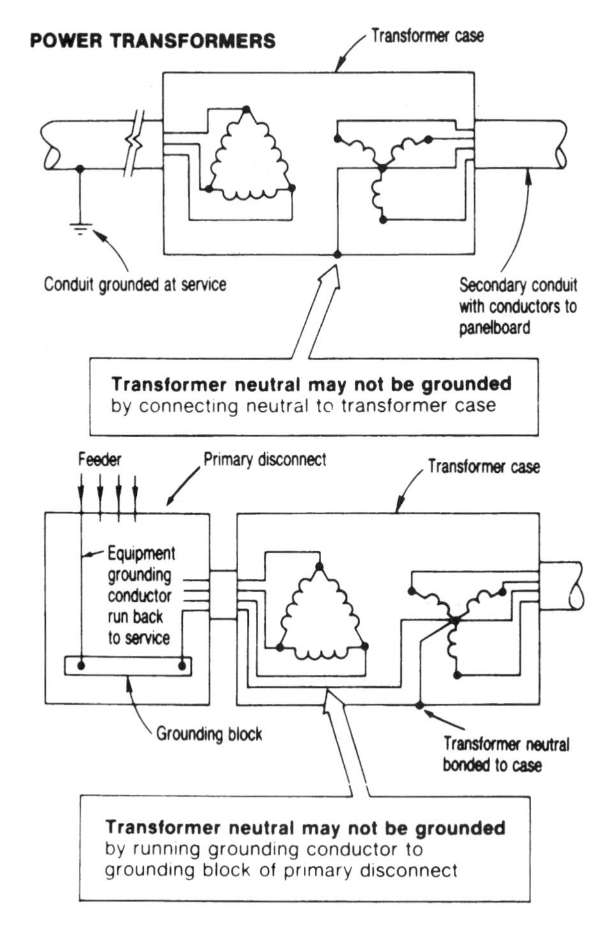

FIG. 7-27 These grounding methods are *not* acceptable.

grounding conductor (one which grounds the wiring system and equipment cases) and for protecting equipment-only grounding conductors makes use of a metal conduit sleeve, run open or installed in concrete. In all such cases, the NEC says that "metallic enclosures for grounding conductors shall be electrically continuous from the point of attachment to cabinets or equipment to the grounding electrode, and shall be securely fastened to the ground clamp or fitting." This means that the grounding connector must be connected to its protective conduit at both ends so that any current which might flow over the conductor will also have the conduit as a parallel path.

The necessity for making a grounding conductor electrically parallel with its protective conduit applies to any grounding conductor. If the protective conduit were arranged so that the conductor and conduit were not acting as parallel conductors—such as would be the case in Fig. 7-26 if there were no bonding jumper from the conduit bushing to the conductor lug—the presence of magnetic metal conduit (steel) would serve to greatly increase the inductive reactance of the grounding conductor in limiting the flow of current to ground. The steel conduit would act as the core of a "choke" to restrict current flow.

Where building steel or a metal water pipe is not available for the grounding of in-building transformers (or generators), other electrodes may be used, based on NEC Sec. 250-81 or 250-83.

Figure 7-27 shows techniques of transformer grounding that have been used in the past but are no longer acceptable.

As shown in Fig. 7-28, the exception to Sec. 250-26(b) exempts small control and signaling transformers from the basic requirement for a grounding-electrode conductor run from the bonded secondary grounded conductor (such as a neutral) to a grounding electrode (nearby building steel or a water pipe). A class 1 remote-control or signaling transformer that is rated not over 1000 VA simply has to have a grounded secondary conductor bonded to the metal case of the transformer; no grounding-electrode conductor is needed—provided that the metal transformer case itself is properly grounded by grounded metal raceway supplying its primary or by means of a suitable equipment-grounding conductor that ties the case back to the grounding electrode for the primary system. The exception to Sec. 250-26(a) permits the use of a No. 14 copper conductor to bond the grounded leg of the transformer secondary to the transformer frame, leaving the supply conduit to the transformer to provide the path to ground back to the main service ground, but depending on the connection between neutral and frame to provide an effective return for clearing faults, as shown. Transformer housings must be grounded by connection to grounded cable armor or metal raceway or by use of a grounding conductor run with circuit conductors (either a bare conductor or a conductor with green covering).

Because the rule on bonding jumpers for the secondary neutral point of a transformer refers to Sec. 250-79(c) and therefore ties into Table 250-94, the smallest size that may be used is No. 8 copper, as shown in that table. But for small transformers (such as those used for class 1 remote-control or signaling circuits) a bonding jumper that large is not necessary and is not suited to termination provisions. For that reason, the exception to

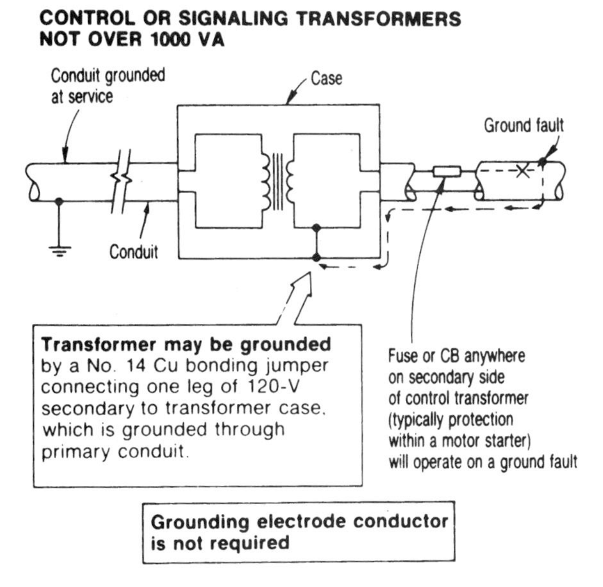

FIG. 7-28 Grounding-electrode connection is *not* required here.

Sec. 250-26(a) permits the bonding jumper for such transformers rated not over 1000 VA to be smaller than No. 8 wire. The jumper simply has to be at least the same size as the secondary phase legs, but in no case smaller than No. 14 copper or No. 12 aluminum.

OTHER DESIGN EXAMPLES

In Fig. 7-29, for a feeder of a 480/277-V, 3-phase, 4-wire system, a 140-kVA demand load of incandescent lighting and receptacle circuits must be fed from a transformer which will feed two 120/208-V, 3-phase, 4-wire panelboards for the required branch circuits to the loads:

1. What size of 3-phase, dry-type transformer must be used for the total 120/208-V load?
2. What size of primary and secondary conductors must be used to carry the transformer current?
3. What size of bonding jumper and grounding-electrode conductors must be used at the transformer?
4. What ratings and arrangements of primary and secondary overcurrent protection are required?

These questions are answered in the next several paragraphs, for both noncontinuous and continuous loading.

Transformer

The total demand load to be supplied from the 120/208-V, 3-phase, 4-wire secondary is 140 kVA, and the next-highest standard size of dry-type transformer is 150 kVA. If the extra 10 kVA of capacity is sufficient to accommodate anticipated load growth, a 150-kVA transformer will do the job.

Conductors

The full-load primary current for a 150-kVA transformer with a 480-V, 3-phase, 4-wire delta primary hookup is determined using the standard formula:

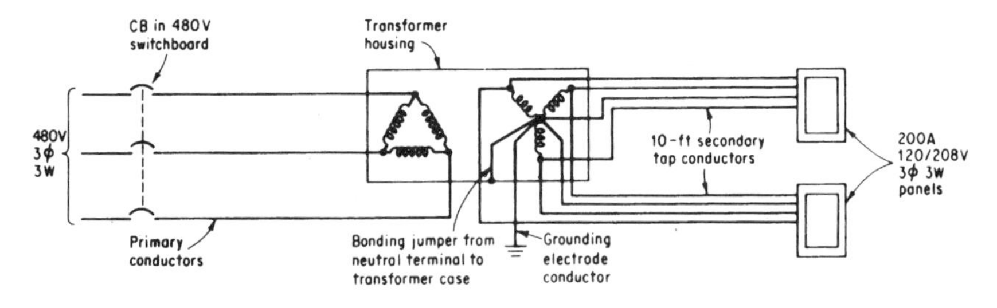

FIG. 7-29 Sizing of all circuit components must be carefully coordinated for all conditions.

$$\text{Rated primary current} = \frac{\text{rated transformer kVA} \times 1000}{1.73 \times \text{rated primary phase-to-phase voltage}}$$

$$= \frac{150 \times 1000}{1.73 \times 480} = \frac{150{,}000}{831} = 180 \text{ A}$$

This means that the ampacity of the primary circuit conductors from the CB in the 480-V switchboard to the transformer must be at least 180 A to provide full use of the 150-kVA rating of the transformer.

Note: If the load fed by the circuit is continuous, a 125 percent multiplier must be used in sizing all circuit conductors, as explained later.

Conductors are selected with NEC Table 310-16. If the primary conductors are to be copper and in raceway, then No. 4/0 type TW conductors (rated 195 A) or No. 3/0 type THW conductors (rated 200 A) or No. 3/0 type THHN or XHHW conductors (rated 225 A) appear to be acceptable. All these conductors would be operating within the ampacity and heating limitation for 75°C (THW) conductors. If aluminum conductors are to be used, Table 310-16 shows that three No. 4/0 TW in 2-in conduit, or three No. 3/0 THW in 2-in conduit, or three No. 3/0 THHN or XHHW in $1\frac{1}{2}$-in conduit would be acceptable.

As shown in Fig. 7-29, the primary circuit run to the transformer from the 480-V switchboard is long, but the two panels fed by the transformer are located immediately adjacent to the transformer, and each is fed by tap conductors not over 10 ft long from the transformer to each panel.

We know that the rated secondary current of the transformer is

$$\text{Rated secondary current} = \frac{\text{rated transformer kVA} \times 1000}{1.73 \times \text{rated secondary phase-to-phase voltage}}$$

$$= \frac{150 \times 1000}{1.73 \times 208} = \frac{150{,}000}{360} = 417 \text{ A}$$

But in this case, the transformer capacity will be divided equally between the two 200-A panels; that is, a set of secondary conductors rated at 200 A will be used for the 10-ft tap to each panel (assuming a separate raceway to each panel).

From NEC Table 310-16, each 10-ft feeder tap could be made up of four of any of the following copper conductors:

250-kcmil type TW (rated 215 A)
No. 3/0 type THW (rated 200 A)
No. 3/0 type THHN or XHHW (rated 225 A)

Note: The 225-A rating of the THHN or XHHW and the 215-A rating of the 250-kcmil TW make the full secondary capacity of the transformer available, so that two conductors of any of these types, when paralleled, have a current-carrying capacity greater than 417 A, the full secondary current rating. If a single set of secondary conductors had been taken from the transformer to supply, say, a single panel, the conductors would have had to be rated at least 417 A (as are 600-kcmil type THW, rated 420 A). However, in this example the secondary capacity is divided between two panels.

Taps from the transformer, if not more than 10 ft long, are acceptable, just as taps from a feeder with no change in voltage level would be. It is only necessary that all design factors be accounted for:

- The primary circuit conductors to the transformer must be sized and protected properly.
- The transformer must be protected.
- The tap conductors must be sized and installed in accordance with Exception 2 of Sec. 240-21:
  1. The tap conductors must not be over 10 ft long.
  2. The ampacity of the tap conductors must not be less than the combined computed loads on the panel fed, and not less than the rating of the panel mains or not less than the rating of fuses or a CB in which they terminate.
  3. The tap conductors must be enclosed in raceway from the transformer case to each panel.

Protection of the panel mains must also be considered. If either of these panels is a "lighting and appliance" panel (more than 10 percent of its CB poles or fuses rated 30 A or less, with a neutral terminal block provided in the panel), each such panel must have main protection as described in Sec. 384-16. Basically, this means that a lighting and appliance panel must be protected by a main CB or fused switch with a rating not greater than the rating of the panel mains (that is, not over 200 A).

Grounding

The bonding jumper and grounding-electrode conductor required by NEC Sec. 250-26 for the transformer would be sized as follows:

The bonding jumper used to connect the system neutral of the transformer to the metal housing of the transformer must be sized in accordance with Sec. 250-79(c), which says that Table 250-94 must be used just as in sizing the grounding-electrode conductor that runs to a water pipe or building steel.

The required size of the bonding jumper, either copper or aluminum, is selected by referring to Table 250-94 and using the size of the transformer secondary conductors as if they were service-entrance conductors. In this example, with the secondary circuit made up of two sets of, say, No. 3/0 copper THW, it is necessary to convert the two No. 3/0 conductors to an equivalent single copper conductor with a cross section equal to that of the two No. 3/0 THWs.

Note: In NEC Table 250-94, the phrase "or Equivalent for Parallel Conductors" in the heading of the left-hand column must be taken as meaning equivalent in cross-sectional area, not in ampacity. And the type of insulation on the service-entrance conductors is of no concern in this calculation. Physical size of conductor material—effective cross section per phase—is the determinant of the required size of grounding-electrode conductor (and bonding jumper).

The cross-sectional area of a single No. 3/0 stranded conductor is 167,800 cmil (from Table 9 of NEC Chap. 9). The single conductor "equivalent" to two No. 3/0 conductors would have a cross-sectional area of 2 × 167,800, or 335,600 cmil. From Table 8, the single conductor having a cross-sectional area closest to this figure is the 350-kcmil, which has a cross-sectional area of 350,000 cmil.

Table 250-94 shows that for 350-kcmil service conductors, a minimum of No. 2 copper or No. 1/0 aluminum must be used for a grounding-electrode conductor. For the transformer in this example, then, the bonding jumper must not be smaller than No. 2 copper or No. 1/0 aluminum.

The grounding-electrode conductor used to connect the transformer neutral (and the bonded equipment enclosure) to the grounding electrode must be sized according to Sec. 250-26, which says in effect that the grounding-electrode conductor must be sized the same as the bonding jumper, using Table 250-94 as described above. Thus, for this transformer application, the grounding-electrode conductor must not be smaller than No. 2 copper or No. 1/0 aluminum (again, from Table 250-94).

Important: The size of a grounding-electrode conductor and bonding jumper for a dry-type transformer application is based on the conductor material and cross-sectional area of the secondary phase conductors. Thus, the use of higher-temperature-rated secondary phase conductors to obtain the required capacity in a smaller conductor (say, THW instead of TW) will result in a smaller required grounding-electrode conductor and bonding jumper.

In this example, if TW conductors had been used for the secondary phase conductors rather than THW, they would have had to be at least 250-kcmil conductors. Two per phase would have been equivalent in size to one 500-kcmil, which would require a minimum of No. 1/0 copper or No. 3/0 aluminum for the grounding-electrode conductor and transformer bonding jumper (Table 250-94).

When transformer secondary conductors are selected at 125 percent of the "continuous" load current, they are still larger than those required for noncontinuous operation—and thereby often require larger grounding-electrode conductors and bonding jumpers.

If the 150-kVA transformer of this example had been used to supply say, one 400-A panel with a single set of conductors, each secondary phase conductor might have been a single 600-kcmil THW, rated at 420 A. The minimum required size of grounding-electrode conductor and bonding jumper would then have been No. 1/0 copper or No. 3/0 aluminum.

Overcurrent Protection

Overcurrent protection for this 150-kVA dry-type transformer must conform to NEC Sec. 450-3(b). According to subparagraph (1) of that section, a dry-type transformer must be protected by an overcurrent device rated or set at not more than 125 percent of the rated primary current. This device may be right at the transformer primary or at the supply end of the circuit that feeds the transformer primary.

In this example, say, three No. 3/0 THW conductors are used for the primary supply circuit to carry the 180 A of rated primary current (assuming future full use of the transformer). These conductors must then be protected at not more than their 200-A ampacity. A 200-A CB or 200-A fuses would provide such protection.

Because Sec. 450-3(b)(1) sets a basic limit of 125 percent of the primary current on the rating of the transformer protective device, a CB or fuses used to protect this transformer must be rated at not more than 1.25 × 180 A, or 225 A. Any rating or setting below this value would be acceptable according to Sec. 450-3(b)(1), so that a 200-A CB or 200-A fuse could be used.

Could these No. 3/0 THW primary conductors be used, then, for the circuit to the transformer, with a 200-A CB or 200-A fuses in a 200-A switch for both transformer and circuit protection? Before this question can be answered, other rules must be accounted for. If the 120/208-V load fed by the transformer is a 180-A "continuous" load (the full load operates continuously for 3 h or more), the following must be considered.

1. When a feeder supplies a continuous load, the rating of the overcurrent device protecting the feeder and the ampacity of the primary feeder conductors must be not less than 125 percent of the continuous load [NEC Sec. 220-10(b)].
2. If the CB or fused switch feeding the transformer is located in a panelboard, the load on either one of them (180 A, under full-load conditions) must not exceed 80 percent of the rating of the CB or fuses [NEC Sec. 384-16(c)].

3. A fused switch used for other than a motor circuit must not be loaded in excess of 80 percent of the rating of the fuses in the switch (UL requirements). And a general UL rule limits the continuous loading on any molded-case circuit to not over 80 percent of the CB rating.

If the full load on the transformer is, say, lighting that will operate for 3 h periods or longer, or if it is assumed that the load will be continuous, the foregoing rules would prohibit the use of a 200-A CB or fuses for this transformer circuit. Certainly the circuit to the transformer is a feeder which must have protection complying with item 1 above. Then the use of a 200-A CB or 200-A fuses is not acceptable.

If the transformer's rated primary current of 180 A must not exceed 80 percent of the rating of the overcurrent protection (that is, the rating of the device is not less than 125 percent of 180 A), then the minimum acceptable rating of overcurrent protection is 180 A × 1.25 = 225 A. Thus, two conditions must be met:

- The primary circuit protection device (CB or fuses) must not be rated below 225 A [Sec. 220-10(b)].
- The same protective device, if it is also to provide protection for the transformer, must not be rated above 225 A.

But when 225-A overcurrent protection is substituted for the initial selection of a 200-A CB or 200-A fuses, the size of the primary circuit conductors must be increased from No. 3/0 THW, because the 225-A protection does not protect these 200-A conductors within their allowable ampacity. It is important to note that Sec. 220-10(b) also requires the circuit conductors to be rated not less than 180 A × 1.25, or 225 A, to correspond with the standard fuse or CB ratings.

Primary circuit conductors which are acceptable for use with 225-A protection must be selected from Table 310-16. The conductors could be as follows:

Copper: 300-kcmil TW (rated 240 A), No. 4/0 THW (rated 230 A), or No. 4/0 THHN or XHHW (rated 260 A)

Aluminum: 400-kcmil TW (rated 225 A), 300-kcmil THW (rated 230 A), or 300-kcmil THHN or XHHW (rated 255 A)

Note: For other than continuous loads, where conductors have a current rating that does not correspond to a standard fuse or CB rating, the next-larger size of fuse or CB may be used. However, for continuous loading, Sec. 220-10(b) does not provide that type of exception to the rule that the conductors be rated at 125 percent of the load.

But for transformer protection, it must be understood that exceptions to Sec. 450-3(b) do give permission to use the next-larger standard overcurrent device rating when the transformer full-load current does not correspond to a standard rating of CB or fuse. However, in some cases, the next-smaller size must be used to avoid exceeding the maximum value. (Permission for use of the "next higher size" is also given in Sec. 430-52, covering motor branch-circuit protection.)

Example. In Fig. 7-30, a 500-kVA transformer is used to step 480-V, 3-phase power down to 208/120 V for incandescent lighting, electric heating, and receptacle outlets. No motor loads are fed by the secondary. The total load on the transformer is 422 kVA, of which 70 percent is continuous.

Step 1. The secondary current rating of the transformer is determined from the formula

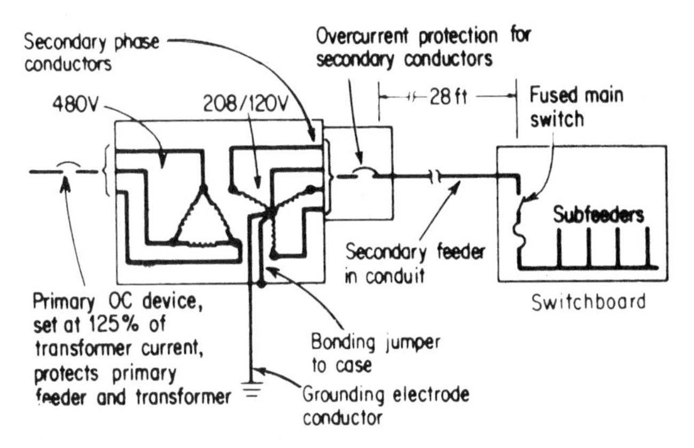

FIG. 7-30 Loading, conductor sizing, and the relative locations of equipment influence design calculations.

$$I = \text{kVA rating} \times 1000 \div (1.732 \times 208)$$

Step 2. The rated secondary current is

$$\frac{500 \times 1000}{1.732 \times 208} = 1388 \text{ A}$$

Step 3. To use the full secondary capacity of the transformer, a circuit makeup of four 700-kcmil THW aluminum conductors per phase leg (each rated at 375 A) would be acceptable for the load (4 × 375 A = 1500 A). Of course, to maintain that load ampacity, each set of three phases and a neutral must be installed in a separate conduit.

Step 4. Based on the use of four 700-kcmil aluminum conductors per phase, the bonding jumper must be sized for a phase leg size of 4 × 700 kcmil, or 2800 kcmil. The minimum acceptable size of aluminum bonding jumper would be $12\frac{1}{2}$ percent of 2800 kcmil, or 350 kcmil, which would require the use of a standard 350-kcmil jumper. If a copper bonding jumper were used, it would have to be sized at $12\frac{1}{2}$ percent of 2000 kcmil [four 500-kcmil THW copper conductors, which have a total ampacity of 4 × 380 A, or 1520 A, and comprise the copper equivalent of four 700-kcmil THW aluminum conductors per phase, as required by NEC Sec. 250-79(c)]. That would call for a 250-kcmil copper jumper.

Step 5. The minimum acceptable size of grounding-electrode conductor to connect this transformer neutral to a nearby grounded structural-steel column is 3/0 copper or 250-kcmil aluminum.

Step 6. The secondary conductors from the transformer to the switchboard are provided with overcurrent protection right at the transformer, and the protective device must have a rating or setting that is at least equal to 125 percent of the continuous load plus the noncontinuous load. The total load on the secondary is equal to 422 kVA ÷ (208 V × 1.732), or 1171 A. The continuous part of that load is 70 percent of 1171 A, or 820 A. The secondary conductors and their protection must be rated for at least 1.25 × 820 A + 351 A, or 1376 A.

Step 7. With such loading, the circuit breaker protecting the secondary would have to be at least a 1400-A standard size (the next rating above 1376 A). The 1500-A rating of the secondary circuit of four 700-kcmil THW aluminum conductors fully satisfies the need for an ampacity of at least 1376 A. And the 1400-A CB is acceptable protection for conductors that have an ampacity of 1500 A.

Note: Under those conditions, the entire circuit and transformer are loaded to within 24 A of the maximum permissible load on a 1400-A circuit breaker that is *not* UL-listed for continuous operation at full-load rating (1400 A − 1376 A). Up to 24 A of noncontinuous load may be added. If it is added, the load on the transformer will then be 1171 A + 24 A, or 1195 A. But the additional transformer capacity of 193 A (1388 A rated less 1195-A load) could not be utilized. This example makes a strong case for use of 100 percent continuous rated circuit breakers or 100 percent fusible switching equipment, in which case a CB or fused switch rated at 1200 A could be used for the 1171-A secondary load, with conductors having an ampacity of at least 1200 A.

Step 8. If the switchboard were located 5 ft closer to the transformer, there would be no need for overcurrent protection at the transformer, provided the secondary conductors were tap conductors not over 25 ft long, terminating in a single main CB or fused switch in the switchboard with protection rated at not more than the ampacity of the conductors.

AUTOTRANSFORMERS

Most buildings with 480/277-V electrical systems obtain 208/120-V power by means of two-winding (or insulating) transformers. The same function can be obtained from autotransformers. Figure 7-31 shows a comparison of the two-winding hookup versus the autotransformer hookup.

In a large, modern commercial, institutional, or industrial building, the usual practice is to locate small step-down transformers—generally in the 15- to 45-kVA range, with 480-V delta primary and 208/120-V wye secondary windings—at regular intervals to provide power for convenience outlets and miscellaneous small equipment. And, quite often, relatively large amounts of 208/120-V power may be required for larger load concentrations such as computer equipment and kitchen loads. In such cases, transformers rated up to 500 kVA are commonly used.

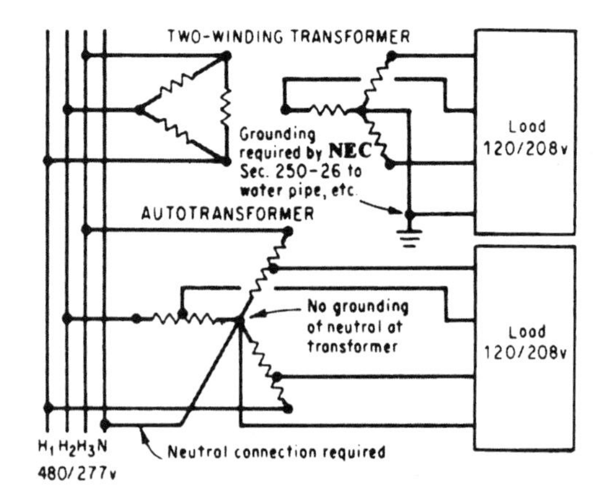

FIG. 7-31 Autotransformer may be used to derive a 120/208-V, 3-phase, 4-wire system from a 480/277-V system.

The following comparison of one manufacturer's 225-kVA dual-winding transformers and autotransformers indicates the relative efficiency, weight, and sound-level advantages of the autotransformer for the common 480/277- to 208/120-V transformation.

	Dual-winding transformer	Autotransformer	
Full-load loss, W	5800	3200	(55%)
Sound level, dB	47	44	(94%)
Weight, lb	1750	1050	(60%)

NEC Sec. 210-9, headed "Circuits Derived from Autotransformers," says "branch circuits shall not be supplied by autotransformer" (transformers "in which a part of the winding is common to both primary and secondary circuits") unless the system supplied has an identified grounded conductor which is solidly connected to a similar identified grounded conductor of the system supplying the autotransformer. In effect, this section requires that a grounded neutral conductor be included in the supply circuit to the primary side of a 480/277- to 208/120-V autotransformer. And the secondary side of the autotransformer must also have a grounded neutral which is connected to the primary grounded neutral.

The addition of this fourth conductor for the primary supply to an autotransformer is not a significant disadvantage in the usual layout of 480-277-V systems. Dry-type transformers for stepping down to 208/120 V are located very close to (commonly within a few feet of) the 480/277-V panel. The neutral feeder conductor must be brought to this panel to provide for the 277-V grounded circuits for fluorescent lighting. Then the only additional requirement for supplying an autotransformer instead of a two-winding transformer is the short run of neutral conductor from the 480/277-V panel to the autotransformer.

Although the 480-V primary of a two-winding transformer (normally a delta-wye connection) requires only connections of the three phase conductors (no neutral is needed), an NEC rule requires a service-type grounding connection to be made at each two-winding transformer (as described previously for two-winding transformers, which are "separately derived systems" as covered by NEC Sec. 250-26). The definite grounding connection required by the NEC for all two-winding units adds labor and material costs when two-winding transformers are used. These added grounding costs for two-winding units are generally greater than the cost of neutral connection for an autotransformer primary.

In specifying autotransformers to derive 4-wire, 208/120-V systems from 480-V feeders, it is very desirable to require three-legged core-type units. This construction leads to reduced third harmonic currents and is more tolerant of unbalanced phase loads. Autotransformers are also lighter and, therefore, easier and less expensive to install. Since they are more efficient, they require less ventilation. They are also significantly quieter in the larger sizes.

Figure 7-32 shows two other applications for autotransformers. The top diagram shows how a 110-V system for lighting may be derived from a 220-V system by means of an autotransformer. The 220-V system may be either a single-phase system or one leg of a 3-phase system. In the case illustrated, the "supplied" system has the required grounded wire solidly connected to a grounded wire of the "supplying" system: a 220-V single-phase system with one conductor grounded.

Exception 2 of NEC Sec. 210-9 permits the use of an autotransformer in existing installations for an individual branch circuit without connection to a similar identified grounded conductor where it is used for transforming from 208 to 240 V or vice versa, as shown at the bottom of Fig. 7-32. Typical applications are concerned with cooking

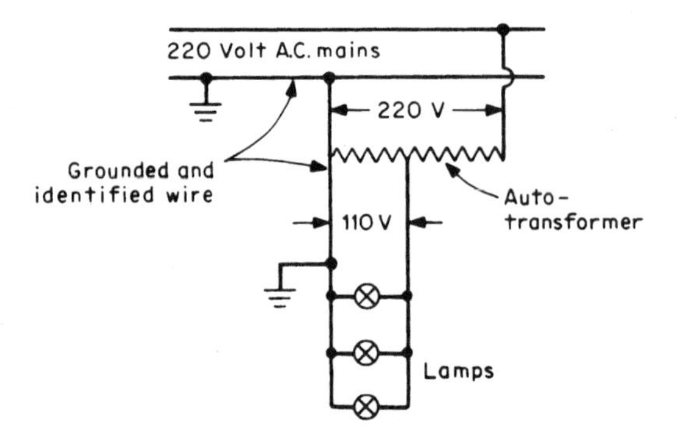

Autotransformer used to derive a two-wire 110-V system for lighting from a 220-V power system.

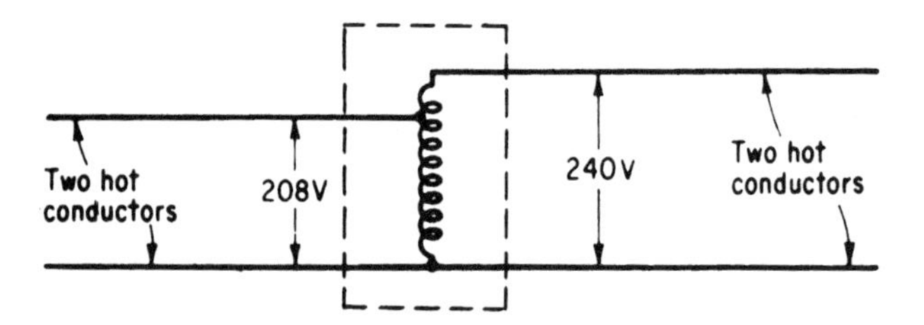

Autotransformers without grounded conductors are recognized.

FIG. 7-32 Two common uses of autotransformers, mainly on branch circuits.

equipment, heaters, motors, and air-conditioning equipment. For such applications, transformers are commonly used. This is a long-established practice for voltage ranges where a hazard is not considered to exist.

Buck or *boost* transformers are designed for use on single- or 3-phase circuits to supply 12/24- or 16/32-V secondaries with a 120/240-V primary. When connected as autotransformers, they will handle kilovoltampere loads that are large in comparison to their physical size and relative cost.

Grounding Autotransformers

In recent years, autotransformers have been used increasingly to convert existing ungrounded 480-V industrial distribution systems to grounded operation. There are two basic ways to convert an ungrounded system to a grounded type:

First, one of three phase legs of the 480-V delta can be intentionally connected to a grounding-electrode conductor that is then run to a suitable grounding electrode. Such grounding gives the two ungrounded phases (*A* and *B*) a voltage of 480 V to ground. The system then operates as a grounded system, so that a ground fault (phase to conduit or other enclosure) on the secondary can cause fault-current flow that opens a circuit protective device to clear the faulted circuit.

But corner grounding of a delta system does not give the lowest possible phase-to-ground voltage. In fact, the voltage to ground of a corner-grounded delta system is the same as it is for an ungrounded delta system, because the voltage to ground for ungrounded circuits is defined as the greatest voltage between the given conductor and any other conductor of the circuit. Thus, the voltage to ground for an ungrounded delta system is the maximum voltage between any two conductors, on the assumption that an accidental ground on any one phase puts the other two phases at full line-to-line voltage above ground.

In recognition of increasing emphasis on the safety of grounded systems over ungrounded systems, NEC Sec. 450-4 covers the use of zigzag grounding autotransformers to convert 3-phase, 3-wire, ungrounded delta systems to grounded wye systems. Such grounding of a 480-V delta system, therefore, lowers the voltage to ground from 480 V (when ungrounded) to 277 V (the phase-to-grounded-neutral voltage) when the system is converted to a wye system as shown in Fig. 7-33.

The *zigzag* grounding autotransformer gets its name from the angular phase differences among the six windings that are divided among the three legs of the transformer's laminated magnetic-core assembly. The actual hookup of the six windings is an interconnection of two wye configurations, with specific polarities and locations for each winding. Just as a wye or delta transformer hookup has a graphic representation that looks like the letter Y or the Greek letter Δ, so a zigzag grounding autotransformer is represented as two wye hookups with pairs of windings in series but phase-displaced, as in Fig. 7-34.

With no ground fault on any leg of the 3-phase system, current flow in the transformer windings is balanced, because equal impedances are connected across each pair of phase legs. The net impedance of the transformer under balanced conditions is very high, so that only a low level of magnetizing current flows through the windings. But when a ground fault develops on one leg of the 3-phase system, the transformer windings assume a very low impedance in the fault path, permitting a large fault current to flow and operate the circuit protective device—just as it would on a conventional grounded-neutral wye system, as shown in Fig. 7-35.

Because the kilovoltampere rating of a grounding autotransformer is based on short-time fault current, the selection of such a transformer is much different from the sizing

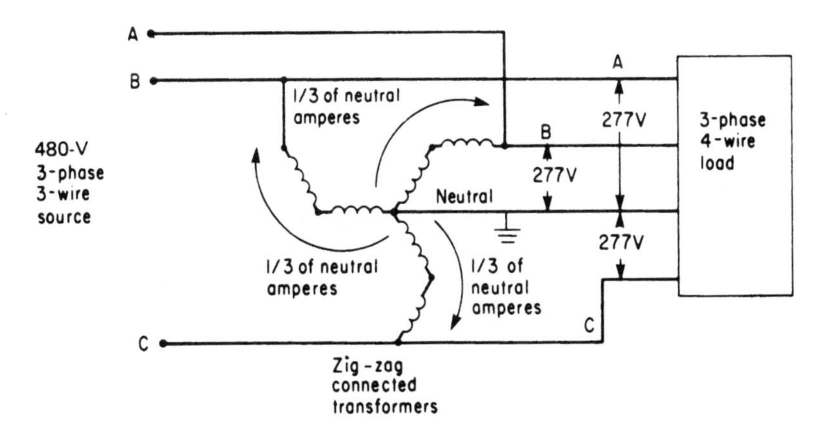

FIG. 7-33 Zigzag grounding autotransformer can be used to change voltage to ground from 480 to 277 V.

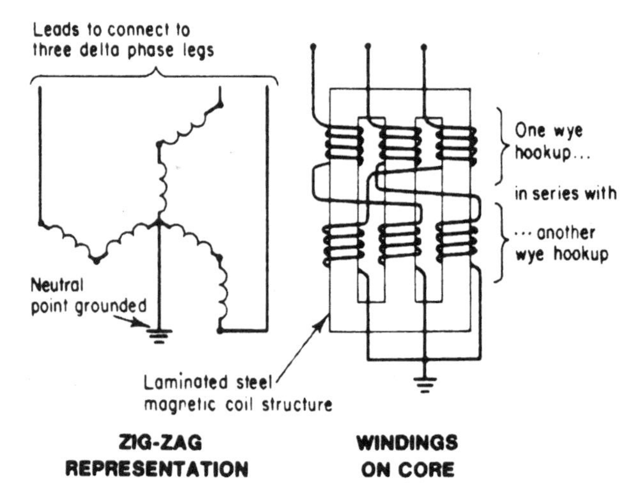

FIG. 7-34 In zigzag connection, a path is provided for the flow of fault current or neutral current.

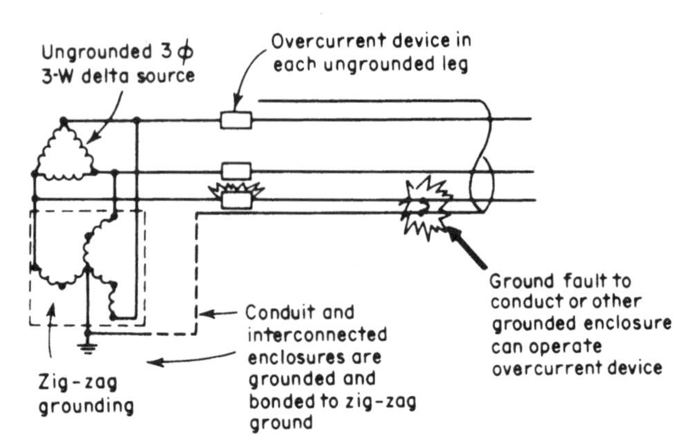

FIG. 7-35 Zigzag autotransformer provides the same automatic fault clearing as obtained on a solidly grounded wye system.

of a conventional two-winding transformer to supply a load. Careful consultation with a manufacturer's sales engineer should precede any decisions about the use of these transformers.

NEC Sec. 450-4 points out that a grounding autotransformer may be used to provide a neutral reference for grounding purposes or for the purpose of converting a 3-phase, 3-wire delta system to a 3-phase, 4-wire, grounded wye system. In the latter case, a neutral conductor can be taken from the transformer to supply loads connected phase to neutral, such as 277-V loads on a 480-V delta system that is converted to a 480Y/277-V system.

Such transformers must have a continuous rating and a continuous neutral-current rating. The phase current in a grounding autotransformer is one-third the neutral current, as shown in Fig. 7-32.

The same NEC rule requires the use of a 3-pole CB rated at 125 percent of the transformer phase current. The requirement for "common-trip" in the overcurrent device excludes the conventional use of fuses in a switch as overcurrent protection. A 3-pole CB prevents single-phase opening of the circuit.

INDEX

ABOUT THE AUTHORS

Joseph F. McPartland is the former publisher of *Electrical Design and Installation* magazine. He is the principal author of the Twenty-First Edition of *McGraw-Hill's National Electrical Code® Handbook* and of McGraw-Hill's *Handbook of Practical Electrical Design.*

Brian J. McPartland is President of the McPartland Electrical Education Institute, where he is the principal consultant/lecturer. He conducts seminars and workshops—primarily on the National Electrical Code (NEC) and the OSHA safety standards that apply to electrical systems—for contractors, design engineers, electrical consulting engineers, plant/facility electrical personnel, as well as inspectors and other interested safety professionals in all parts of the nation. In addition to writing numerous magazine articles, Brian has also authored or co-authored more than 10 books on electrical design and construction, including *McGraw-Hill's National Electrical Code® Handbook, Handbook of Practical Electrical Design,* and *Yearbook Supplements to the National Electrical Code Handbook.*

Steven P. McPartland is an electrical programs instructor for the International Union of Operating Engineers, a former instructor for the New Jersey State Apprenticeship Training Program, and assistant editor of the *Handbook of Practical Electrical Design.*

Jack E. Pullizzi is the Chief Facilities Engineer for AT&T Bell Laboratories, and instructor in electrical systems at Ocean County College, and a contributing editor to *Electrical Contractor* magazine.